Geographical Information Systems and Science

Geographical Information Systems and Science

2nd Edition

Paul A. Longley *University College London, UK*

Michael F. Goodchild *University of California, Santa Barbara, USA*

David J. Maguire *ESRI Inc., Redlands, USA*

David W. Rhind *City University, London, UK*

John Wiley & Sons, Ltd

Copyright © 2005   John Wiley & Sons Ltd, The Atrium, Southern Gate, Chichester,
West Sussex PO19 8SQ, England

Telephone (+44) 1243 779777

Email (for orders and customer service enquiries): cs-books@wiley.co.uk
Visit our Home Page on www.wiley.com

ESRI Press logo is the trademark of ESRI and is used herein under licence.

Main cover image and first box from bottom, courtesy of NASA.
Second box, reproduced from Ordnance Survey.
Third box, reproduced by permission of National Geographic Maps.
Fourth box, reproduced from Ordnance Survey, courtesy @Last.

Other Wiley Editorial Offices

John Wiley & Sons Inc., 111 River Street, Hoboken, NJ 07030, USA

Jossey-Bass, 989 Market Street, San Francisco, CA 94103-1741, USA

Wiley-VCH Verlag GmbH, Boschstr. 12, D-69469 Weinheim, Germany

John Wiley & Sons Australia Ltd, 33 Park Road, Milton, Queensland 4064, Australia

John Wiley & Sons (Asia) Pte Ltd, 2 Clementi Loop #02-01, Jin Xing Distripark, Singapore 129809

John Wiley & Sons Canada Ltd, 22 Worcester Road, Etobicoke, Ontario, Canada M9W 1L1

Wiley also publishes its books in a variety of electronic formats. Some content that appears
in print may not be available in electronic books.

British Library Cataloguing in Publication Data

A catalogue record for this book is available from the British Library

ISBN 0-470-87000-1 (HB)
ISBN 0-470-87001-X (PB)

Typeset in 9/10.5pt Times by Laserwords Private Limited, Chennai, India
Printed and bound in Spain by Grafos S.A., Barcelona, Spain
This book is printed on acid-free paper responsibly manufactured from sustainable forestry
in which at least two trees are planted for each one used for paper production.

Contents

Foreword

At the time of writing, the first edition of *Geographic Information Systems and Science* (GIS&S) has sold well over 25 000 copies – the most, it seems, of any GIS textbook. Its novel structure, content, and 'look and feel' expanded the very idea of what a GIS *is*, what it involves, and its pervasive importance. In so doing, the book introduced thousands of readers to the field in which we have spent much of our working lifetimes. Being human, we take pleasure in that achievement – but it is not enough. Convinced as we are of the benefits of thinking and acting geographically, we are determined to enthuse and involve many more people. This and the high rate of change in GIS&S (Geographic Information Systems *and* Science) demands a new edition that benefits from the feedback we have received on the first one.

Setting aside the (important) updates, the major changes reflect our changing world. The use of GIS was pioneered in the USA, Canada, various countries in Europe, and Australia. But it is expanding rapidly – and in innovative ways – in South East Asia, Latin America and Eastern Europe, for example. We have recognized this by broadening our geography of examples. The world of 2005 is not the same as that prior to 11 September 2001. Almost all countries are now engaged in seeking to protect their citizens against the threat of terrorism. Whilst we do not seek to exaggerate the contribution of GIS, there are many ways in which these systems and our geographic knowledge can help in this, the first duty of a national government. Finally, the sheen has come off much information technology and information systems: they have become consumer goods, ubiquitous in the market place. Increasingly they are recognized as a necessary underpinning of government and commerce – but one where real advantage is conferred by their ease of use and low price, rather than the introduction of exotic new functions. As we demonstrate in this book, GIS&S was never simply hardware and software. It has also always been about people and, in preparing this second edition, we have taken the decision to present an entirely new set of current GIS protagonists. This has inevitably meant that all boxes from the first edition pertaining to living individuals have been removed in order to create space: we hope that the individuals concerned will understand, and we congratulate them on their longevity! This second edition, then, remains about hardware, software, people – and also about geographic information, some real science, a clutch of partnerships, and much judgment. Yet we recognize the progressive 'consumerization' of our basic tool set and welcome it, for it means more can be done for greater numbers of beneficiaries for less money. Our new book reflects the continuing shift from tools to understanding and coping with the fact that, in the real world, 'everything is connected to everything else'!

We asked Joe Lobley, an individual unfamiliar with political correctness and with a healthy scepticism about the utterances of GIS gurus, to write the foreword for the first edition. To our delight, he is now cited in various academic papers and reviews as a stimulating, fresh, and lateral thinker. Sadly, at the time of going to press, Joe had not responded to our invitation to repeat his feat. He was last heard of on location as a GIS consultant in Afghanistan. So this Foreword is somewhat less explosive than last time. We hope the book is no less valuable.

Paul A. Longley
Michael F. Goodchild
David J. Maguire
David W. Rhind
October 2004

Addendum

Hi again! Greetings from Afghanistan, where I am temporarily resident in the sort of hotel that offers direct access to GPS satellite signals through the less continuous parts of its roof structure. Global communications mean I can stay in touch with the GIS world from almost anywhere. Did you know that when Abraham Lincoln was assassinated it took 16 days for the news to reach Britain? But when William McKinley was assassinated only 36 years later the telegraph (the first Internet) ensured it took only seconds for the news to reach Old and New Europe. Now I can pull down maps and images of almost anything I want, almost anywhere. Of course, I get lots of crap as well – the curse of the age – and some of the information is rubbish. What does Kabul's premier location prospector want with botox? But technology makes good (and bad) information available, often without payment (which I like), to all those with telecoms and access to a computer. Sure, I know that's still a small fraction of mankind but boy is that fraction growing daily. It's helped of course by the drop in price of hardware and even software: GIS tools are increasingly becoming like washing machines – manufactured in bulk and sold on price though there is a lot more to getting success than buying the cheapest.

I've spent lots of time in Asia since last we communicated and believe me there are some smart things going on there with IS and GIS. Fuelled by opportunism (and possibly a little beer) the guys writing this book have seen the way the wind is blowing and made a good stab at representing the *whole* world of GIS. So what else is new in this revised edition of what they keep telling me is the world's best-selling GIS textbook? I like the way homeland security issues are built in. All of us have to live with terrorist threats these days and GIS can help as a data and intelligence integrator. I like the revised structure, the continuing emphasis on business benefit and institutions and the new set of role models they have chosen (though 'new' is scarcely the word I would have used for Roger Tomlinson...). I like the same old unstuffy ways these guys write in proper American English, mostly avoiding jargon.

On the down-side, I still think they live in a rose-tinted world where they believe government and academia actually do useful things. If you share their strange views, tell me what the great National Spatial Data Infrastructure movement has *really* achieved worldwide except hype and numerous meetings in nice places? Wise up guys! You don't have to pretend. Now I do like the way that the guys recognize that places are unique (boy, my hotel is...), but don't swallow the line that digital representations of space are any less valid, ethical or usable than digital measures of time or sound. Boast a little more, and, while you're at it, say less about 'the' digital divide and more about digital differentiation. And keep well clear of patronizing, social theory stroking, box-ticking, self-congratulatory claptrap. The future of that is just people with spectacles who write books in garden sheds. Trade up from the caves of the pre-digital era and educate the wannabes that progress can be a good thing. And wise up that the real benefits of GIS do not depend on talking shops or gravy trains. What makes GIS unstoppable is what we can do with the tools, with decent data and with our native wit and training to make the world a better and more efficient place. Business and markets (mostly) will do that for you!

Joe Lobley

Preface

The field of geographic information systems (GIS) is concerned with the description, explanation, and prediction of patterns and processes at geographic scales. GIS is a science, a technology, a discipline, and an applied problem solving methodology. There are perhaps 50 other books on GIS now on the world market. We believe that this one has become one of the fastest selling and most used because we see GIS as providing a gateway to *science* and problem solving (geographic information systems 'and science' in general), and because we relate available software for handling *geographic information* to the scientific principles that should govern its use (geographic information: 'systems and science'). GIS is of enduring importance because of its central co-ordinating *principles*, the specialist *techniques* that have been developed to handle spatial data, the special *analysis* methods that are key to spatial data, and because of the particular *management* issues presented by geographic information (GI) handling. Each section of this book investigates the unique, complex, and difficult problems that are posed by geographic information, and together they build into a holistic understanding of all that is important about GIS.

Our approach

GIS is a proven technology and the basic operations of GIS today provide secure and established foundations for measurement, mapping, and analysis of the real world. GIScience provides us with the ability to devise GIS-based analysis that is robust and defensible. GI technology facilitates analysis, and continues to evolve rapidly, especially in relation to the Internet, and its likely successors and its spin-offs. Better technology, better systems, and better science make better management and exploitation of GI possible.

Fundamentally, GIS is an applications-led technology, yet successful applications need appropriate scientific foundations. Effective use of GIS is impossible if they are simply seen as black boxes producing magic. GIS is applied rarely in controlled, laboratory-like conditions. Our messy, inconvenient, and apparently haphazard real world is the laboratory for GIS, and the science of real-world application is the difficult kind – it can rarely control for, or assume away, things that we would prefer were not there and that get in the way of almost any given application. Scientific understanding of the inherent uncertainties and imperfections in representing the world makes us able to judge whether the conclusions of our analysis are sustainable, and is essential for everything except the most trivial use of GIS. GIScience is also founded on a search for understanding and predictive power in a world where human factors interact with those relating to the physical environment. Good science is also ethical and clearly communicated science, and thus the ways in which we analyze and depict geography also play an important role.

Digital geographic information is central to the practicality of GIS. If it does not exist, it is expensive to collect, edit, or update. If it does exist, it cuts costs and time – assuming it is fit for the purpose, or good enough for the particular task in hand. It underpins the rapid growth of trading in geographic information (g-commerce). It provides possibilities not only for local business but also for entering new markets or for forging new relationships with other organizations. It is a foolish individual who sees it only as a commodity like baked beans or shaving foam. Its value relies upon its coverage, on the strengths of its representation of diversity, on its truth within a constrained definition of that word, and on its availability.

Few of us are hermits. The way in which geographic information is created and exploited through GIS affects us as citizens, as owners of enterprises, and as employees. It has increasingly been argued that GIS is only a part – albeit a part growing in importance and size – of the Information, Communications, and Technology (ICT) industry. This is a limited perception, typical of the ICT supply-side industry which tends to see itself as the sole progenitor of change in the world (wrongly). Actually, it is much more sensible to take a balanced demand- and supply-side perspective: GIS and geographic information can and do underpin many operations of many organizations, but how GIS works in detail differs between different cultures, and can often also partly depend on whether an organization is in the private or public sector. Seen from this perspective, management of GIS facilities is crucial to the success of organizations – businesses as we term them later. The management of the organizations using our tools, information, knowledge, skills, and commitment is therefore what will ensure the ultimate local and global success of GIS. For this reason we devote an entire section of this book to management issues. We go far beyond how to choose, install, and run a GIS; that is only one part of the enterprise. We try to show how to use GIS and geographic information to contribute to the business success of your organization (whatever it is), and have it recognized as doing just that. To achieve that, you need to know what drives organizations and how they operate in the reality of their business environments. You need to know something about assets, risks, and constraints on actions – and how to avoid the last two and nurture the first. And you need to be exposed – for that is reality – to the inter-dependencies in any organization and the tradeoffs in decision making in which GIS can play a major role.

Our audience

Originally, we conceived this book as a 'student companion' to a very different book that we also produced as a team – the second edition of the 'Big Book' of GIS (Longley et al 1999). This reference work on GIS provided a defining statement of GIS at the end of the last millennium: many of the chapters that are of enduring relevance are now available as an advanced reader in GIS (Longley et al 2005). These books, along with the first 'Big Book' of GIS (Maguire et al 1991) were designed for those who were already very familiar with GIS, and desired an advanced understanding of enduring GIS principles, techniques, and management practices. They were not designed as books for those being introduced to the subject.

This book is the companion for everyone who desires a rich understanding of how GIS is used in the real world. GIS today is both an increasingly mature technology and a strategically important interdisciplinary meeting place. It is taught as a component of a huge range of undergraduate courses throughout the world, to students that already have different skills, that seek different disciplinary perspectives on the world, and that assign different priorities to practical problem solving and the intellectual curiosities of science. This companion can be thought of as a textbook, though not in a conventionally linear way. We have not attempted to set down any kind of rigid GIS curriculum beyond the core organizing principles, techniques, analysis methods, and management practices that we believe to be important. We have structured the material in each of the sections of the book in a cumulative way, yet we envisage that very few students will start at Chapter 1 and systematically work through to Chapter 21 – much of learning is not like that any more (if ever it was), and most instructors will navigate a course between sections and chapters of the book that serves their particular disciplinary, curricular, and practical priorities. The ways in which three of us use the book in our own undergraduate and postgraduate settings are posted on the book's website (**www.wiley.com/go/longley**), and we hope that other instructors will share their best practices with us as time goes on (please see the website for instructions on how to upload instructor lists and offer feedback on those that are already there!). Our Instructor Manual (see **www.wiley.com/go/longley**) provides suggestions as to the use of this book in a range of disciplines and educational settings. The linkage of the book to reference material (specifically Longley et al (2005) and Maguire et al (1991) at **www.wiley.com/go/longley**) is a particular strength for GIS postgraduates and professionals. Such users might desire an up-to-date overview of GIS to locate their own particular endeavors, or (particularly if their previous experience lies outside the mainstream geographic sciences) a fast track to get up-to-speed with the range of principles, techniques, and practice issues that govern real-world application.

The format of the book is intended to make learning about GIS fun. GIS is an important transferable skill because people successfully use it to solve real-world problems. We thus convey this success through use of real (not contrived, conventional text-book like) applications, in clearly identifiable boxes throughout the text. But even this does not convey the excitement of learning about GIS that only comes from doing. With this in mind, an on-line series of laboratory classes have been created to accompany the book. These are available, free of charge, to any individual working in an institution that has an ESRI site license (see **www.esri.com**). They are cross-linked in detail to individual chapters and sections in the book, and provide learners with the opportunity to refresh the concepts and techniques that they have acquired through classes and reading, and the opportunity to work through extended examples using ESRI ArcGIS. This is by no means the only available software for learning GIS: we have chosen it for our own lab exercises because it is widely used, because one of us works for ESRI Inc. (Redlands, CA., USA) and because ESRI's cooperation enabled us to tailor the lab exercises to our own material. There are, however, many other options for lab teaching and distance learning from private and publicly funded bodies such as the UNIGIS consortium, the Worldwide Universities Network, and Pennsylvania State University in its World Campus (**www.worldcampus.psu.edu/pub/index.shtml**).

GIS is not just about machines, but also about people. It is very easy to lose touch with what is new in GIS, such is the scale and pace of development. Many of these developments have been, and continue to be, the outcome of work by motivated and committed individuals – many an idea or implementation of GIS would not have taken place without an individual to champion it. In the first edition of this book, we used boxes highlighting the contributions of a number of its champions to convey that GIS is a living, breathing subject. In this second edition, we have removed all of the living champions of GIS and replaced them with a completely new set – not as any intended slight upon the remarkable contributions that these individuals have made, but as a necessary way of freeing up space to present vignettes of an entirely new set of committed, motivated individuals whose contributions have also made a difference to GIS.

As we say elsewhere in this book, human attention is valued increasingly by business, while students are also seemingly required to digest ever-increasing volumes of material. We have tried to summarize some of the most important points in this book using short 'factoids', such as that below, which we think assist students in recalling core points.

Short, pithy, statements can be memorable.

We hope that instructors will be happy to use this book as a core teaching resource. We have tried to provide a number of ways in which they can encourage their students to learn more about GIS through a range of assessments. At the end of each chapter we provide four questions in the following sequence that entail:

- **Student-centred learning** by doing.
- A **review** of material contained in the chapter.
- A **review and research** task – involving integration of issues discussed in the chapter with those discussed in additional external sources.
- A **compare and research** task – similar to the review and research task above, but additionally entailing linkage with material from one or more other chapters in the book.

The on-line lab classes have also been designed to allow learning in a self-paced way, and there are self-test exercises at the end of each section for use by learners working alone or by course evaluators at the conclusion of each lab class.

As the title implies, this is a book about geographic information systems, the practice of science in general, and the principles of geographic information science (GIScience) in particular. We remain convinced of the need for high-level understanding and our book deals with ideas and concepts – as well as with actions. Just as scientists need to be aware of the complexities of interactions between people and the environment, so managers must be well-informed by a wide range of knowledge about issues that might impact upon their actions. Success in GIS often comes from dealing as much with people as with machines.

The new learning paradigm

This is not a traditional textbook because:

- It recognizes that GISystems and GIScience do not lend themselves to traditional classroom teaching alone. Only by a combination of approaches can such crucial matters as principles, technical issues, practice, management, ethics, and accountability be learned. Thus the book is complemented by a website (**www.wiley.com/go/longley**) and by exercises that can be undertaken in laboratory or self-paced settings.
- It brings the principles and techniques of GIScience to those learning about GIS for the first time – and as such represents part of the continuing evolution of GIS.
- The very nature of GIS as an underpinning technology in huge numbers of applications, spanning different fields of human endeavor, ensures that learning has to be tailored to individual or small-group needs. These are addressed in the Instructor Manual to the book (**www.wiley.com/go/longley**).
- We have recognized that GIS is driven by real-world applications and real people, that respond to real-world needs. Hence, information on a range of applications and GIS champions is threaded throughout the text.

- We have linked our book to online learning resources throughout, notably the ESRI Virtual Campus.
- The book that you have in your hands has been completely restructured and revised, while retaining the best features of the (highly successful) first edition published in 2001.

Summary

This is a book that recognizes the growing commonality between the concerns of science, government, and business. The examples of GIS people and problems that are scattered through this book have been chosen deliberately to illuminate this commonality, as well as the interplay between organizations and people from different sectors. To differing extents, the five sections of the book develop common concerns with effectiveness and efficiency, by bringing together information from disparate sources, acting within regulatory and ethical frameworks, adhering to scientific principles, and preserving good reputations. This, then, is a book that combines the basics of GIS with the solving of problems which often have no single, ideal solution – the world of business, government, and interdisciplinary, mission-orientated holistic science.

In short, we have tried to create a book that remains attuned to the way the world works now, that understands the ways in which most of us increasingly operate as knowledge workers, and that grasps the need to face complicated issues that do not have ideal solutions. As with the first edition of the book, this is an unusual enterprise and product. It has been written by a multinational partnership, drawing upon material from around the world. One of the authors is an employee of a leading software vendor and two of the other three have had business dealings with ESRI over many years. Moreover, some of the illustrations and examples come from the customers of that vendor. We wish to point out, however, that neither ESRI (nor Wiley) has ever sought to influence our content or the way in which we made our judgments, and we have included references to other software and vendors throughout the book. Whilst our lab classes are part of ESRI's Virtual Campus, we also make reference to similar sources of information in both paper and digital form. We hope that we have again created something novel but valuable by our lateral thinking in all these respects, and would very much welcome feedback through our website (**www.wiley.com/go/longley**).

Conventions and organization

We use the acronym *GIS* in many ways in the book, partly to emphasize one of our goals, the interplay between geographic information *systems* and geographic information *science*; and at times we use two other possible interpretations of the three-letter acronym: geographic information *studies* and geographic information *services*. We distinguish between the various meanings where appropriate, or where the context fails to make the meaning clear, especially in Section 1.6 and in the Epilog. We also use the acronym in both singular and plural senses, following what is now standard practice in the field, to refer as appropriate to a single geographic information system or to geographic information systems in general. To complicate matters still further, we have noted the increasing use of 'geospatial' rather than 'geographic'. We use 'geospatial' where other people use it as a proper noun/title, but elsewhere use the more elegant and readily intelligible 'geographic'.

We have organized the book in five major but interlocking sections: after two chapters that establish the foundations to GI Systems and Science and the real world of applications, the sections appear as Principles (Chapters 3 through 6), Techniques (Chapters 7 through 11), Analysis (12 through 16) and Management and Policy (Chapters 17 through 20). We cap the book off with an Epilog that summarizes the main topics and looks to the future. The boundaries between these sections are in practice permeable, but remain in large part predicated upon providing a systematic treatment of enduring principles – ideas that will be around long after today's technology has been relegated to the museum – and the knowledge that is necessary for an understanding of today's technology, and likely near-term developments. In a similar way, we illustrate how many of the analytic methods have had reincarnations through different manual and computer technologies in the past, and will doubtless metamorphose further in the future.

We hope you find the book stimulating and helpful. Please tell us – either way!

Acknowledgments

We take complete responsibility for all the material contained herein. But much of it draws upon contributions made by friends and colleagues from across the world, many of them outside the academic GIS community. We thank them all for those contributions and the discussions we have had over the years. We cannot mention all of them but would particularly like to mention the following.

We thanked the following for their direct and indirect inputs to the first edition of this book: Mike Batty, Clint Brown, Nick Chrisman, Keith Clarke, Andy Coote, Martin Dodge, Danny Dorling, Jason Dykes, Max Egenhofer, Pip Forer, Andrew Frank, Rob Garber, Gayle Gaynor, Peter Haggett, Jim Harper, Rich Harris, Les Hepple, Sophie Hobbs, Andy Hudson-Smith, Karen Kemp, Chuck Killpack, Robert Laurini, Vanessa Lawrence, John Leonard, Bob Maher, Nick Mann, David Mark, David Martin, Elanor McBay, Ian McHarg, Scott Morehouse, Lou Page, Peter Paisley, Cath Pyke, Jonathan Raper, Helen Ridgway, Jan Rigby, Christopher Roper, Garry Scanlan, Sarah Sheppard, Karen Siderelis, David Simonett, Roger Tomlinson, Carol Tullo, Dave Unwin, Sally Wilkinson, David Willey, Jo Wood, Mike Worboys.

Many of those listed above also helped us in our work on the second edition. But this time around we additionally acknowledge the support of: Tessa Anderson, David Ashby, Richard Bailey, Brad Baker, Bob Barr, Elena Besussi, Dick Birnie, John Calkins, Christian Castle, David Chapman, Nancy Chin, Greg Cho, Randy Clast, Rita Colwell, Sonja Curtis, Jack Dangermond, Mike de Smith, Steve Evans, Andy Finch, Amy Garcia, Hank Gerie, Muki Haklay, Francis Harvey, Denise Lievesley, Daryl Lloyd, Joe Lobley, Ian Masser, David Miller, Russell Morris, Doug Nebert, Hugh Neffendorf, Justin Norry, Geof Offen, Larry Orman, Henk Ottens, Jonathan Rhind, Doug Richardson, Dawn Robbins, Peter Schaub, Sorin Scortan, Duncan Shiell, Alex Singleton, Aidan Slingsby, Sarah Smith, Kevin Schürer, Josef Strobl, Larry Sugarbaker, Fraser Taylor, Bethan Thomas, Carolina Tobón, Paul Torrens, Nancy Tosta, Tom Veldkamp, Peter Verburg, and Richard Webber. Special thanks are also due to Lyn Roberts and Keily Larkins at John Wiley and Sons for successfully guiding the project to fruition. Paul Longley's contribution to the book was carried out under ESRC AIM Fellowship RES-331-25-0001, and he also acknowledges the guiding contribution of the CETL Center for Spatial Literacy in Teaching (Splint).

Each of us remains indebted in different ways to Stan Openshaw, for his insight, his energy, his commitment to GIS, and his compassion for geography.

Finally, thanks go to our families, especially Amanda, Fiona, Heather, and Christine.

Paul Longley, University College London
Michael Goodchild, University of California
Santa Barbara
David Maguire, ESRI Inc., Redlands CA
David Rhind, City University, London

October 2004

Further reading

Maguire D.J., Goodchild M.F., and Rhind D.W. (eds) 1991 *Geographical Information Systems*. Harlow: Longman.

Longley P.A., Goodchild M.F., Maguire D.W., and Rhind D.W. (eds) 1999 *Geographical Information Systems: Principles, Techniques, Management and Applications (two volumes)*. New York, NJ: Wiley.

Longley P.A., Goodchild M.F., Maguire D.W., and Rhind D.W. (eds) 2005 *Geographical Information Systems: Principles, Techniques, Management and Applications (abridged edition)*. Hoboken, NJ: Wiley.

List of Acronyms and Abbreviations

AA Automobile Association
ABM agent-based model
AGI Association for Geographic Information
AGILE Association of Geographic Information Laboratories in Europe
AHP Analytical Hierarchy Process
AM automated mapping
AML Arc Macro Language
API application programming interface
ARPANET Advanced Research Projects Agency Network
ASCII American Standard Code for Information Interchange
ASP Active Server Pages
AVIRIS Airborne Visible InfraRed Imaging Spectrometer
BBC British Broadcasting Corporation
BLM Bureau of Land Management
BLOB binary large object
CAD Computer-Aided Design
CAMA Computer Assisted Mass Appraisal
CAP Common Agricultural Policy
CASA Centre for Advanced Spatial Analysis
CASE computer-aided software engineering
CBD central business district
CD compact disc
CEN Comité Européen de Normalisation
CERN Conseil Européen pour la Recherche Nucléaire
CGIS Canada Geographic Information System
CGS Czech Geological Survey
CIA Central Intelligence Agency
CLI Canada Land Inventory
CLM collection-level metadata
COGO coordinate geometry
COM component object model
COTS commercial off-the-shelf
CPD continuing professional development
CSDGM Content Standards for Digital Geospatial Metadata
CSDMS Centre for Spatial Database Management and Solutions
CSO color separation overlay
CTA Chicago Transit Authority
DARPA Defense Advanced Research Projects Agency
DBA database administrator
DBMS database management system
DCL data control language
DCM digital cartographic model
DCW Digital Chart of the World
DDL data definition language
DEM digital elevation model
DGPS Differential Global Positioning System
DHS Department of Homeland Security
DIME Dual Independent Map Encoding
DLG digital line graph

DLM digital landscape model
DML data manipulation language
DRG digital raster graphic
DST Department of Science and Technology
DXF drawing exchange format
EBIS ESRI Business Information Solutions
EC European Commission
ECU Experimental Cartography Unit
EDA exploratory data analysis
EOSDIS Earth Observing System Data and Information System
EPA Environmental Protection Agency
EPS encapsulated postscript
ERDAS Earth Resource Data Analysis System
ERP Enterprise Resource Planning
ERTS Earth Resources Technology Satellite
ESDA exploratory spatial data analysis
ESRI Environmental Systems Research Institute
EU European Union
EUROGI European Umbrella Organisation for Geographic Information
FAO Food and Agriculture Organization
FEMA Federal Emergency Management Agency
FGDC Federal Geographic Data Committee
FIPS Federal Information Processing Standard
FM facility management
FOIA Freedom of Information Act
FSA Forward Sortation Area
GAO General Accounting Office
GBF-DIME Geographic Base Files – Dual Independent Map Encoding
GDI GIS data industry
GIO Geographic Information Officer
GIS geographic(al) information system
GIScience geographic(al) information science
GML Geography Markup Language
GNIS Geographic Names Information System
GOS geospatial one-stop
GPS Global Positioning System
GRASS Geographic Resources Analysis Support System
GSDI global spatial data infrastructure
GUI graphical user interface
GWR geographically weighted regression
HLS hue, lightness, and saturation
HTML hypertext markup language
HTTP hypertext transmission protocol
ICMA International City/County Management Association
ICT Information and Communication Technology
ID identifier
IDE Integrated Development Environment
IDW inverse-distance weighting
IGN Institut Géographique National

IMW International Map of the World
INSPIRE Infrastructure for Spatial Information in Europe
IP Internet protocol
IPR intellectual property rights
IS information system
ISCGM International Steering Committee for Global Mapping
ISO International Standards Organization
IT information technology
ITC International Training Centre for Aerial Survey
ITS intelligent transportation systems
JSP Java Server Pages
KE knowledge economy
KRIHS Korea Research Institute for Human Settlements
KSUCTA Kyrgyz State University of Construction, Transportation and Architecture
LAN local area network
LBS location-based services
LiDAR light detection and ranging
LISA local indicators of spatial association
LMIS Land Management Information System
MAT point of minimum aggregate travel
MAUP Modifiable Areal Unit Problem
MBR minimum bounding rectangle
MCDM multicriteria decision making
MGI Masters in Geographic Information
MIT Massachusetts Institute of Technology
MOCT Ministry of Construction and Transportation
MrSID Multiresolution Seamless Image Database
MSC Mapping Science Committee
NASA National Aeronautics and Space Administration
NATO North Atlantic Treaty Organization
NAVTEQ Navigation Technologies
NCGIA National Center for Geographic Information and Analysis
NGA National Geospatial-Intelligence Agency
NGIS National GIS
NILS National Integrated Land System
NIMA National Imagery and Mapping Agency
NIMBY not in my back yard
NMO national mapping organization
NMP National Mapping Program
NOAA National Oceanic and Atmospheric Administration
NPR National Performance Review
NRC National Research Council
NSDI National Spatial Data Infrastructure
NSF National Science Foundation
OCR optical character recognition
ODBMS object database management system
OEM Office of Emergency Management
OGC Open Geospatial Consortium
OLM object-level metadata
OLS ordinary least squares
OMB Office of Management and Budget
ONC Operational Navigation Chart
ORDBMS object-relational database management system
PAF postcode address file

PASS Planning Assistant for Superintendent Scheduling
PCC percent correctly classified
PCGIAP Permanent Committee on GIS Infrastructure for Asia and the Pacific
PDA personal digital assistant
PE photogrammetric engineering
PERT Program, Evaluation, and Review Techniques
PLSS Public Land Survey System
PPGIS public participation in GIS
RDBMS relational database management system
RFI Request for Information
RFP Request for Proposals
RGB red-green-blue
RMSE root mean square error
ROMANSE Road Management System for Europe
RRL Regional Research Laboratory
RS remote sensing
SAP spatially aware professional
SARS severe acute respiratory syndrome
SDE Spatial Database Engine
SDI spatial data infrastructure
SDSS spatial decision support systems
SETI Search for Extraterrestrial Intelligence
SIG Special Interest Group
SOHO small office/home office
SPC State Plane Coordinates
SPOT Système Probatoire d'Observation de la Terre
SQL Structured/Standard Query Language
SWMM Storm Water Management Model
SWOT strengths, weaknesses, opportunities, threats
TC technical committee
TIGER Topologically Integrated Geographic Encoding and Referencing
TIN triangulated irregular network
TINA there is no alternative
TNM The National Map
TOID Topographic Identifier
TSP traveling-salesman problem
TTIC Traffic and Travel Information Centre
UCAS Universities Central Admissions Service
UCGIS University Consortium for Geographic Information Science
UCSB University of California, Santa Barbara
UDDI Universal Description, Discovery, and Integration
UDP Urban Data Processing
UKDA United Kingdom Data Archive
UML Unified Modeling Language
UN United Nations
UNIGIS UNIversity GIS Consortium
UPS Universal Polar Stereographic
URISA Urban and Regional Information Systems Association
USGS United States Geological Survey
USLE Universal Soil Loss Equation
UTC urban traffic control
UTM Universal Transverse Mercator
VBA Visual Basic for Applications
VfM value for money
VGA video graphics array

ViSC visualization in scientific computing
VPF vector product format
WAN wide area network
WIMP windows, icons, menus, and pointers
WIPO World Intellectual Property Organization
WSDL Web Services Definition Language

WTC World Trade Center
WTO World Trade Organization
WWF World Wide Fund for Nature
WWW World Wide Web
WYSIWYG what you see is what you get
XML extensible markup language

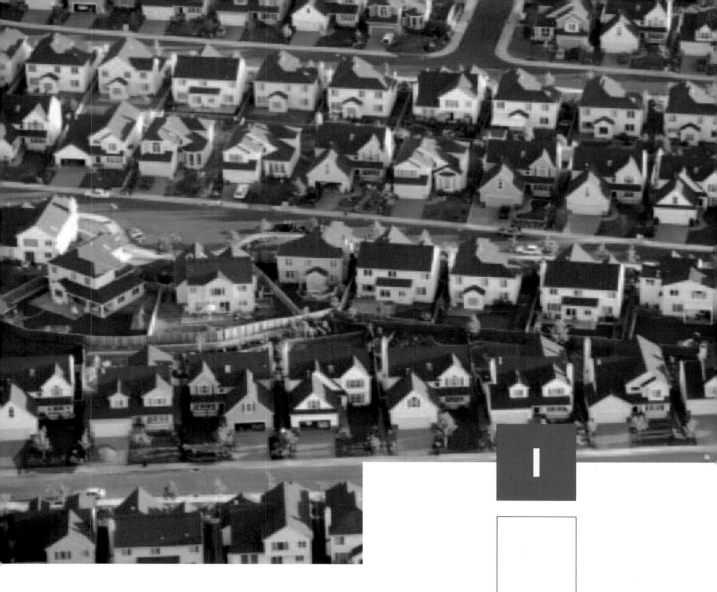

Introduction

I

1 *Systems, science, and study*

This chapter introduces the conceptual framework for the book, by addressing several major questions:

- What exactly is geographic information, and why is it important? What is special about it?

- What is information generally, and how does it relate to data, knowledge, evidence, wisdom, and understanding?

- What kinds of decisions make use of geographic information?

- What is a geographic information system, and how would I know one if I saw one?

- What is geographic information science, and how does it relate to the use of GIS for scientific purposes?

- How do scientists use GIS, and why do they find it helpful?

- How do companies make money from GIS?

Geographic Information Systems and Science, 2nd edition   Paul Longley, Michael Goodchild, David Maguire, and David Rhind.
© 2005 John Wiley & Sons, Ltd.   ISBNs: 0-470-87000-1 (HB); 0-470-87001-X (PB)

Learning Objectives

At the end of this chapter you will:

- Know definitions of the terms used throughout the book, including GIS itself;

- Be familiar with a brief history of GIS;

- Recognize the sometimes invisible roles of GIS in everyday life, and the roles of GIS in business;

- Understand the significance of geographic information science, and how it relates to geographic information systems;

- Understand the many impacts GIS is having on society, and the need to study those impacts.

1.1 Introduction: why does GIS matter?

Almost everything that happens, happens somewhere. Largely, we humans are confined in our activities to the surface and near-surface of the Earth. We travel over it and in the lower levels of the atmosphere, and through tunnels dug just below the surface. We dig ditches and bury pipelines and cables, construct mines to get at mineral deposits, and drill wells to access oil and gas. Keeping track of all of this activity is important, and knowing where it occurs can be the most convenient basis for tracking. Knowing where something happens is of critical importance if we want to go there ourselves or send someone there, to find other information about the same place, or to inform people who live nearby. In addition, most (perhaps all) decisions have geographic consequences, e.g., adopting a particular funding formula creates geographic winners and losers, especially when the process entails zero sum gains. Therefore geographic location is an important attribute of activities, policies, strategies, and plans. Geographic information systems are a special class of information systems that keep track not only of events, activities, and things, but also of *where* these events, activities, and things happen or exist.

Almost everything that happens, happens somewhere. Knowing where something happens can be critically important.

Because location is so important, it is an issue in many of the problems society must solve. Some of these are so routine that we almost fail to notice them – the daily question of which route to take to and from work, for example. Others are quite extraordinary occurrences, and require rapid, concerted, and coordinated responses by a wide range of individuals and organizations – such as the events of September 11 2001 in New York (Box 1.1). Problems that involve an aspect of location, either in the information used to solve them, or in the solutions themselves, are termed *geographic problems*. Here are some more examples:

- Health care managers solve geographic problems (and may create others) when they decide where to locate new clinics and hospitals.
- Delivery companies solve geographic problems when they decide the routes and schedules of their vehicles, often on a daily basis.
- Transportation authorities solve geographic problems when they select routes for new highways.
- Geodemographics consultants solve geographic problems when they assess and recommend where best to site retail outlets.
- Forestry companies solve geographic problems when they determine how best to manage forests, where to cut, where to locate roads, and where to plant new trees.
- National Park authorities solve geographic problems when they schedule recreational path maintenance and improvement (Figure 1.3).
- Governments solve geographic problems when they decide how to allocate funds for building sea defenses.
- Travelers and tourists solve geographic problems when they give and receive driving directions, select hotels in unfamiliar cities, and find their way around theme parks (Figure 1.4).
- Farmers solve geographic problems when they employ new information technology to make better decisions about the amounts of fertilizer and pesticide to apply to different parts of their fields.

If so many problems are geographic, what distinguishes them from each other? Here are three bases for classifying geographic problems. First, there is the question of scale, or level of geographic detail. The architectural design of a building can present geographic problems, as in disaster management (Box 1.1), but only at a very detailed or local scale. The information needed to service the building is also local – the size and shape of the parcel, the vertical and subterranean extent of the building, the slope of the land, and its accessibility using normal and emergency infrastructure. The global diffusion of the 2003 severe acute respiratory syndrome (SARS) epidemic, or of bird flu in 2004 were problems at a much broader and coarser scale, involving information about entire national populations and global transport patterns.

Scale or level of geographic detail is an essential property of any GIS project.

Second, geographic problems can be distinguished on the basis of intent, or purpose. Some problems are strictly practical in nature – they must often be solved as quickly as possible and/or at minimum cost, in order to achieve such practical objectives as saving money, avoiding fines by regulators, or coping with an emergency. Others are better characterized as driven by human curiosity. When geographic data are used to verify the theory of continental drift, or to map distributions of glacial deposits, or to analyze the historic movements of people in anthropological or archaeological research (Box 1.2 and Figure 1.5), there is no sense of an immediate problem that needs to be solved – rather, the intent is the advancement of human understanding of the world, which we often recognize as the intent of science.

Although science and practical problem solving are often seen as distinct human activities, it is often argued that there is no longer any effective distinction between their methods. The tools and methods used by a scientist in a government agency to ensure the protection of an endangered species are essentially the same as the tools used by an academic ecologist to advance our scientific knowledge of biological systems. Both use the most accurate measurement devices, use terms whose meanings have been widely shared and agreed, insist that their results be replicable by others, and in general follow all of the principles of science that have evolved over the past centuries.

The use of GIS for both forms of activity certainly reinforces this idea that science and practical problem solving are no longer distinct in their methods, as does the fact that GIS is used widely in all kinds of organizations, from academic institutions to government agencies and corporations. The use of similar tools and methods right across science and problem solving is part of a shift from the pursuit of curiosity within traditional academic disciplines to solution centered, interdisciplinary team work.

Applications Box **1.1**

September 11 2001

Almost everyone remembers where they were when they learned of the terrorist atrocities in New York on September 11 2001. Location was crucial in the immediate aftermath and the emergency response, and the attacks had locational repercussions at a range of spatial

Figure 1.1 GIS in the Office of Emergency Management (OEM), first set up in the World Trade Center (WTC) complex immediately following the 2001 terrorist attacks on New York (Courtesy ESRI)

(geographic) and temporal (short, medium, and long time periods) *scales*. In the short term, the incidents triggered local emergency evacuation and disaster recovery procedures and global shocks to the financial system through the suspension of the New York Stock Exchange; in the medium term they blocked part of the New York subway system (that ran underneath the Twin Towers), profoundly changed regional work patterns (as affected workers became telecommuters) and had calamitous effects on the local retail economy; and in the

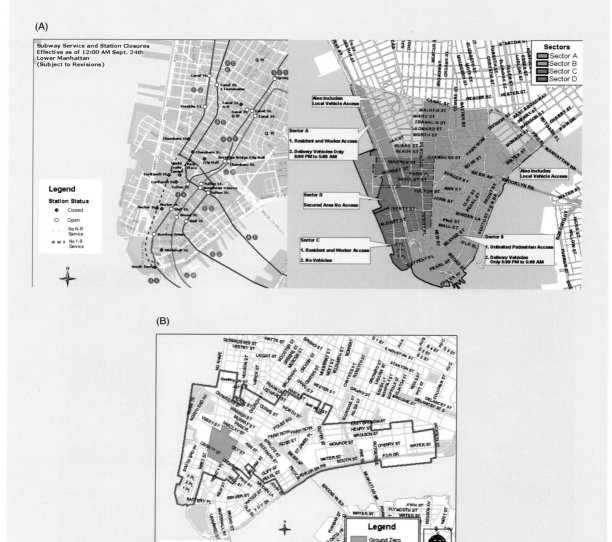

Figure 1.2 GIS usage in emergency management following the 2001 terrorist attacks on New York: (A) subway, pedestrian and vehicular traffic restrictions; (B) telephone outages; and (C) surface dust monitoring three days after the disaster. (Courtesy ESRI)

(C)

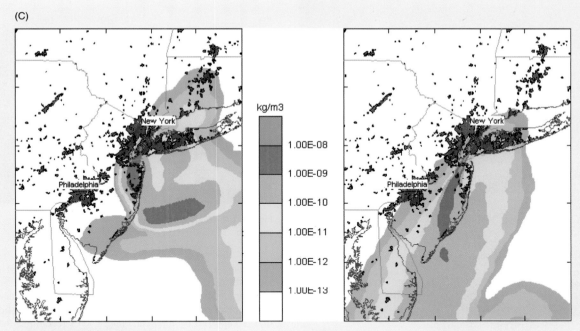

Figure 1.2 (*continued*)

long term, they have profoundly changed the way that we think of emergency response in our heavily networked society. Figures 1.1 and 1.2 depict some of the ways in which GIS was used for emergency management in New York in the immediate aftermath of the attacks. But the events also have much wider implications for the handling and management of geographic information, that we return to in Chapter 20.

At some points in this book it will be useful to distinguish between applications of GIS that focus on design, or so-called *normative* uses, and applications that advance science, or so-called *positive* uses (a rather confusing meaning of that term, unfortunately, but the one commonly used by philosophers of science – its use implies that science confirms theories by finding *positive* evidence in support of them, and rejects theories when negative evidence is found). Finding new locations for retailers is an example of a normative application of GIS, with its focus on design. But in order to predict how consumers will respond to new locations it is necessary for retailers to analyze and model the actual patterns of behavior they exhibit. Therefore, the models they use will be grounded in observations of messy reality that have been tested in a positive manner.

With a single collection of tools, GIS is able to bridge the gap between curiosity-driven science and practical problem-solving.

Third, geographic problems can be distinguished on the basis of their time scale. Some decisions are *operational*, and are required for the smooth functioning of an organization, such as how to control electricity inputs into grids that experience daily surges and troughs in usage (see Section 10.6). Others are *tactical*, and concerned with medium-term decisions, such as where to cut trees in next year's forest harvesting plan. Others are *strategic*, and are required to give an organization long-term direction, as when retailers decide to expand or rationalize their store networks (Figure 1.7). These terms are explored in the context of logistics applications of GIS in Section 2.3.4.6. The real world is somewhat more complex than this, of course, and these distinctions may blur – what is theoretically and statistically the 1000-year flood influences strategic and tactical considerations but may possibly arrive a year after the previous one! Other problems that interest geophysicists, geologists, or evolutionary biologists may occur on time scales that are much longer than a human lifetime, but are still geographic in nature, such as predictions about the future physical environment of Japan, or about the animal populations of Africa. Geographic databases are often *transactional* (see Sections 10.2.1 and 10.9.1), meaning

Figure 1.3 Maintaining and improving footpaths in National Parks is a geographic problem

Figure 1.4 Navigating tourist destinations is a geographic problem

that they are constantly being updated as new information arrives, unlike maps, which stay the same once printed.

Chapter 2 contains a more detailed discussion of the range and remits of GIS applications, and a view of how GIS pervades many aspects of our daily lives. Other applications are discussed to illustrate particular principles, techniques, analytic methods, and management practices as these arise throughout the book.

1.1.1 Spatial is special

The adjective *geographic* refers to the Earth's surface and near-surface, and defines the subject matter of this book, but other terms have similar meaning. *Spatial* refers to any space, not only the space of the Earth's surface, and it is used frequently in the book, almost always

with the same meaning as *geographic*. But many of the methods used in GIS are also applicable to other non-geographic spaces, including the surfaces of other planets, the space of the cosmos, and the space of the human body that is captured by medical images. GIS techniques have even been applied to the analysis of genome sequences on DNA. So the discussion of analysis in this book is of *spatial* analysis (Chapters 14 and 15), not geographic analysis, to emphasize this versatility.

Another term that has been growing in usage in recent years is *geospatial* – implying a subset of spatial applied specifically to the Earth's surface and near-surface. The former National Intelligence and Mapping Agency was renamed as the National Geospatial-Intelligence Agency in late 2003 by the US President and the Web portal for US Federal Government data is called Geospatial One-Stop. In this book we have tended to avoid geospatial, preferring geographic, and spatial where we need to emphasize generality (see Section 21.2.2).

People who encounter GIS for the first time are sometimes driven to ask why geography is so important – why *is* spatial special? After all, there is plenty of information around about geriatrics, for example, and in principle one could create a geriatric information system. So why has geographic information spawned an entire industry, if geriatric information hasn't to anything like the same extent? Why are there no courses in universities specifically in geriatric information systems? Part of the answer should be clear already – almost all human

Where did your ancestors come from?

As individuals, many of us are interested in *where* we came from – socially and geographically. Some of the best clues to our ancestry come from our (family) surnames, and Western surnames have different

types of origins – many of which are explicitly or implicitly geographic in origin (such clues are less important in some Eastern societies where family histories are generally much better documented). Research at

University College London is using GIS and historic censuses and records to investigate the changing local and regional geographies of surnames within the UK since the late 19th century (Figure 1.5).

This tells us quite a lot about migration, changes in local and regional economies, and even about measures of local economic health and vitality. Similar GIS-based analysis can be used to generalize about

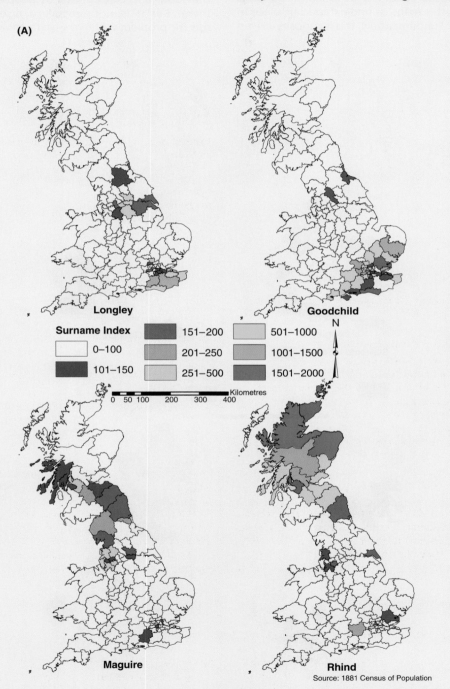

Figure 1.5 The UK geography of the Longleys, the Goodchilds, the Maguires, and the Rhinds in (A) 1881 and (B) 1998 (Reproduced with permission of Daryl Lloyd)

the characteristics of international emigrants (for example to North America, Australia, and New Zealand: Figure 1.6), or the regional naming patterns of immigrants to the US from the Indian sub-continent or China. In all kinds of senses, this helps us understand our place in the world. Fundamentally, this is curiosity-driven research: it is interesting to individuals to understand more about their origins, and it is interesting to everyone with planning or policy concerns with any particular place to understand the social and cultural mix of people that live there. But it is not central to resolving any specific problem within a specific timescale.

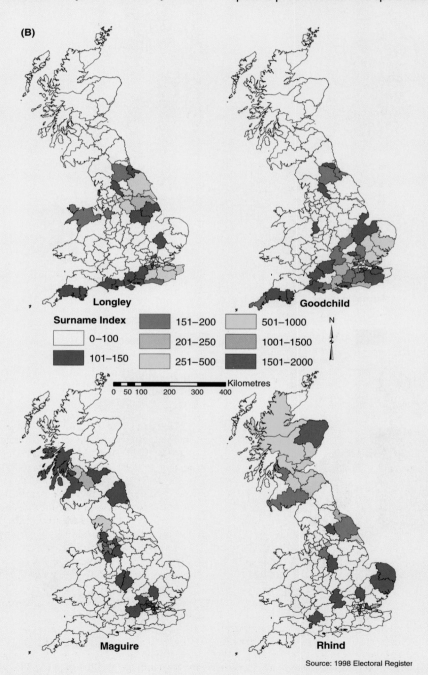

Figure 1.5 (*continued*)

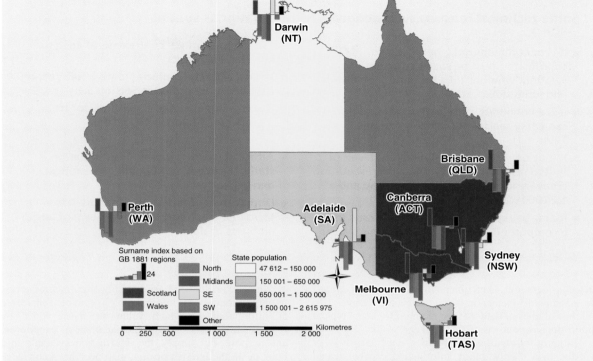

Figure 1.6 The geography of British emigrants to Australia (bars beneath the horizontal line indicate low numbers of migrants to the corresponding destination) (Reproduced with permission of Daryl Lloyd)

Figure 1.7 Store location principles are very important in the developing markets of Europe, as with Tesco's successful investment in Budapest, Hungary

activities and decisions involve a geographic component, and the geographic component is important. Another reason will become apparent in Chapter 3 – working with geographic information involves complex and difficult choices that are also largely unique. Other, more-technical reasons will become clear in later chapters, and are briefly summarized in Box 1.3.

1.2 Data, information, evidence, knowledge, wisdom

Information systems help us to manage *what we know*, by making it easy to organize and store, access and retrieve, manipulate and synthesize, and apply knowledge to the solution of problems. We use a variety of terms to describe what we know, including the five that head this section and that are shown in Table 1.2. There are no universally agreed definitions of these terms, the first two of which are used frequently in the GIS arena. Nevertheless it is worth trying to come to grips with their various meanings, because the differences between them can often be significant, and what follows draws upon many sources, and thus provides the basis for the use of these terms throughout the book. Data clearly refers to the most mundane kind of information, and wisdom to the most substantive.

Data consist of numbers, text, or symbols which are in some sense neutral and almost context-free. Raw geographic facts (see Box 18.7), such as the temperature at a specific time and location, are examples of data. When data are transmitted, they are treated as a stream of bits; a crucial requirement is to preserve the integrity of the dataset. The internal meaning of the data is irrelevant in

Technical Box **1.3**

Some technical reasons why geographic information is special

- It is multidimensional, because *two* coordinates must be specified to define a location, whether they be *x* and *y* or latitude and longitude.
- It is voluminous, since a geographic database can easily reach a terabyte in size (see Table 1.1).
- It may be represented at different levels of spatial resolution, e.g., using a representation equivalent to a 1:1 million scale map and a 1:24 000 scale one (see Box 4.2).
- It may be represented in different ways inside a computer (Chapter 3) and how this is done can strongly influence the ease of analysis and the end results.
- It must often be projected onto a flat surface, for reasons identified in Section 5.7.
- It requires many special methods for its analysis (see Chapters 14 and 15).
- It can be time-consuming to analyze.
- Although much geographic information is static, the process of updating is complex and expensive.
- Display of geographic information in the form of a map requires the retrieval of large amounts of data.

such considerations. Data (the noun is the plural of *datum*) are assembled together in a *database* (see Chapter 10), and the volumes of data that are required for some typical applications are shown in Table 1.1.

The term *information* can be used either narrowly or broadly. In a narrow sense, information can be treated as devoid of meaning, and therefore as essentially synonymous with data, as defined in the previous paragraph. Others define information as *anything* which can be digitized, that is, represented in digital form (Chapter 3), but also argue that information is differentiated from data by implying some degree of selection, organization, and preparation for particular purposes – information is data serving some *purpose*, or data that have been given some degree of *interpretation*. Information is often costly to produce, but once digitized it is cheap to reproduce and distribute. Geographic datasets, for example, may be very expensive to collect and assemble, but very cheap to copy and disseminate. One other characteristic of information is that it is easy to add value to it through processing, and through merger with other information. GIS provides an excellent example of the latter, because of the tools it provides for combining information from different sources (Section 18.3).

GIS does a better job of sharing data and information than knowledge, which is more difficult to detach from the knower.

Knowledge does not arise simply from having access to large amounts of information. It can be considered as information to which value has been added by interpretation based on a particular context, experience, and purpose. Put simply, the information available in a book or on the Internet or on a map becomes knowledge only when it has been read and understood. How the information is interpreted and used will be different for different readers depending on their previous experience, expertise, and needs. It is important to distinguish two types of knowledge: *codified* and *tacit*. Knowledge is codifiable if it can be written down and transferred relatively easily to others. Tacit knowledge is often slow to acquire and much more difficult to transfer. Examples include the knowledge built up during an apprenticeship, understanding of how a particular market works, or familiarity with using a particular technology or language. This difference in transferability means that codified and tacit knowledge need to be managed and rewarded quite differently. Because of its nature, tacit knowledge is often a source of competitive advantage.

Some have argued that knowledge and information are fundamentally different in at least three important respects:

- Knowledge entails a knower. Information exists independently, but knowledge is intimately related to people.

Table 1.1 Potential GIS database volumes for some typical applications (volumes estimated to the nearest order of magnitude). Strictly, bytes are counted in powers of 2 – 1 kilobyte is 1024 bytes, not 1000

1 megabyte	1 000 000	Single dataset in a small project database
1 gigabyte	1 000 000 000	Entire street network of a large city or small country
1 terabyte	1 000 000 000 000	Elevation of entire Earth surface recorded at 30 m intervals
1 petabyte	1 000 000 000 000 000	Satellite image of entire Earth surface at 1 m resolution
1 exabyte	1 000 000 000 000 000 000	A future 3-D representation of entire Earth at 10 m resolution?

Table 1.2 A ranking of the support infrastructure for decision making

Decision-making support infrastructure	Ease of sharing with everyone	GIS example
Wisdom ↑	*Impossible*	Policies developed and accepted by stakeholders
Knowledge ↑	*Difficult, especially tacit knowledge*	Personal knowledge about places and issues
Evidence ↑	*Often not easy*	Results of GIS analysis of many datasets or scenarios
Information ↑	*Easy*	Contents of a database assembled from raw facts
Data	*Easy*	Raw geographic facts

- Knowledge is harder to detach from the knower than information; shipping, receiving, transferring it between people, or quantifying it are all much more difficult than for information.
- Knowledge requires much more assimilation – we digest it rather than hold it. While we may hold conflicting information, we rarely hold conflicting knowledge.

Evidence is considered a half way house between information and knowledge. It seems best to regard it as a multiplicity of information from different sources, related to specific problems and with a consistency that has been validated. Major attempts have been made in medicine to extract evidence from a welter of sometimes contradictory sets of information, drawn from worldwide sources, in what is known as *meta-analysis*, or the comparative analysis of the results of many previous studies.

Wisdom is even more elusive to define than the other terms. Normally, it is used in the context of decisions made or advice given which is disinterested, based on all the evidence and knowledge available, but given with some understanding of the likely consequences. Almost invariably, it is highly individualized rather than being easy to create and share within a group. Wisdom is in a sense the top level of a hierarchy of decision-making infrastructure.

1.3 The science of problem solving

How are problems solved, and are geographic problems solved any differently from other kinds of problems? We humans have accumulated a vast storehouse about the world, including information both on how it *looks*, or its *forms*, and how it *works*, or its dynamic *processes*. Some of those processes are natural and built into the design of the planet, such as the processes of tectonic movement that lead to earthquakes, and the processes of atmospheric circulation that lead to hurricanes. Others are

human in origin, reflecting the increasing influence that we have on our natural environment, through the burning of fossil fuels, the felling of forests, and the cultivation of crops (Figure 1.8). Others are imposed by us, in the form of laws, regulations, and practices. For example, zoning regulations affect the ways in which specific parcels of land can be used.

Knowledge about how the world works is more valuable than knowledge about how it looks, because such knowledge can be used to predict.

These two types of information differ markedly in their degree of generality. Form varies geographically, and the Earth's surface looks dramatically different in different places – compare the settled landscape of northern England with the deserts of the US Southwest (Figure 1.9). But processes can be very general. The ways in which the burning of fossil fuels affects the atmosphere are essentially the same in China as in Europe, although the two landscapes look very different. Science has always valued such general knowledge over knowledge of the specific, and hence has valued process knowledge over knowledge of form. Geographers in particular have witnessed a longstanding debate, lasting

Figure 1.8 Social processes, such as carbon dioxide emissions, modify the Earth's environment

Figure 1.9 The form of the Earth's surface shows enormous variability, for example, between the deserts of the southwest USA and the settled landscape of northern England

centuries, between the competing needs of *idiographic* geography, which focuses on the description of form and emphasizes the unique characteristics of places, and *nomothetic* geography, which seeks to discover general processes. Both are essential, of course, since knowledge of general process is only useful in solving specific problems if it can be combined effectively with knowledge of form. For example, we can only assess the impact of soil erosion on agriculture in New South Wales if we know both how soil erosion is generally impacted by such factors as slope and specifically how much of New South Wales has steep slopes, and where they are located (Figure 1.10).

One of the most important merits of GIS as a tool for problem solving lies in its ability to combine the general with the specific, as in this example from New South Wales. A GIS designed to solve this problem would contain knowledge of New South Wales's slopes, in the form of computerized maps, and the programs executed by the GIS would reflect general knowledge of how slopes affect soil erosion. The *software* of a GIS captures and implements general knowledge, while the *database* of a GIS represents specific information. In that sense a GIS resolves the old debate between nomothetic and idiographic camps, by accommodating both.

GIS solves the ancient problem of combining general scientific knowledge with specific information, and gives practical value to both.

General knowledge comes in many forms. Classification is perhaps the simplest and most rudimentary, and is widely used in geographic problem solving. In many parts of the USA and other countries efforts have been made to limit development of wetlands, in the interests of preserving them as natural habitats and avoiding excessive impact on water resources. To support these efforts, resources have been invested in mapping wetlands, largely from aerial photography and satellite imagery. These maps simply classify land, using established rules that define what is and what is not a wetland (Figure 1.11).

Figure 1.10 Predicting landslides requires general knowledge of processes and specific knowledge of the area – both are available in a GIS (Reproduced with permission of PhotoDisc, Inc.)

More sophisticated forms of knowledge include *rule sets* – for example, rules that determine what use can be made of wetlands, or what areas in a forest can be legally logged. Rules are used by the US Forest Service

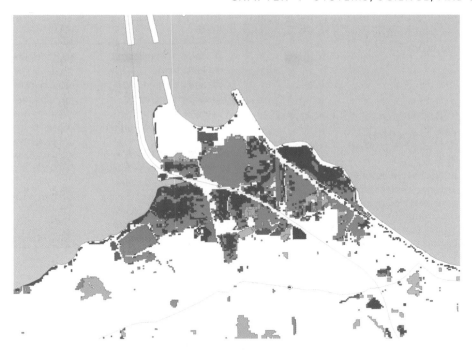

Figure 1.11 A wetland map of part of Erie County, Ohio, USA. The map has been made by classifying Landsat imagery at 30 m resolution. Brown = woods on hydric soil, dark blue = open water (excludes Lake Erie), green = shallow marsh, light blue = shrub/scrub wetland, blue-green = wet meadow, pink = farmed wetland. *Source*: Ohio Department of Natural Resources, **www.dnr.state.oh.us**

to define wilderness, and to impose associated regulations regarding the use of wilderness, including prohibition on logging and road construction.

Much of the knowledge gathered by the activities of scientists suggests the term *law*. The work of Sir Isaac Newton established the Laws of Motion, according to which all matter behaves in ways that can be perfectly predicted. From Newton's laws we are able to predict the motions of the planets almost perfectly, although Einstein later showed that certain observed deviations from the predictions of the laws could be explained with his Theory of Relativity. Laws of this level of predictive quality are few and far between in the geographic world of the Earth's surface. The real world is the only geographic-scale 'laboratory' that is available for most GIS applications, and considerable uncertainty is generated when we are unable to control for all conditions. These problems are compounded in the socioeconomic realm, where the role of human agency makes it almost inevitable that any attempt to develop rigid laws will be frustrated by isolated exceptions. Thus, while market researchers use spatial interaction models, in conjunction with GIS, to predict how many people will shop at each shopping center in a city, substantial errors will occur in the predictions. Nevertheless the results are of great value in developing location strategies for retailing. The Universal Soil Loss Equation, used by soil scientists in conjunction with GIS to predict soil erosion, is similar in its relatively low predictive power, but again the results are sufficiently accurate to be very useful in the right circumstances.

Solving problems involves several distinct components and stages. First, there must be an *objective*, or a goal that the problem solver wishes to achieve. Often this is a desire to maximize or minimize – find the solution of least cost, or shortest distance, or least time, or greatest profit; or to make the most accurate prediction possible. These objectives are all expressed in *tangible* form, that is, they can be measured on some well-defined scale. Others are said to be *intangible*, and involve objectives that are much harder, if not impossible to measure. They include maximizing *quality of life* and *satisfaction*, and minimizing *environmental impact*. Sometimes the only way to work with such intangible objectives is to involve human subjects, through surveys or focus groups, by asking them to express a preference among alternatives. A large body of knowledge has been acquired about such human-subjects research, and much of it has been employed in connection with GIS. For an example of the use of such mixed objectives see Section 16.4.

Often a problem will have *multiple objectives*. For example, a company providing a mobile snack service to construction sites will want to maximize the number of sites that can be visited during a daily operating schedule, and will also want to maximize the expected returns by visiting the most lucrative sites. An agency charged with locating a corridor for a new power transmission line may decide to minimize cost, while at the same time seeking to minimize environmental impact. Such problems employ methods known as *multicriteria decision making* (MCDM).

Many geographic problems involve multiple goals and objectives, which often cannot be expressed in commensurate terms.

1.4 The technology of problem solving

The previous sections have presented GIS as a technology to support both science and problem solving, using both specific and general knowledge about geographic reality. GIS has now been around for so long that it is, in many senses, a background technology, like word processing. This may well be so, but what exactly is this technology called GIS, and how does it achieve its objectives? In what ways is GIS more than a technology, and why does it continue to attract such attention as a topic for scientific journals and conferences?

Many definitions of GIS have been suggested over the years, and none of them is entirely satisfactory, though many suggest much more than a technology. Today, the label GIS is attached to many things: amongst them, a software product that one can buy from a vendor to carry out certain well-defined functions (*GIS software*); digital representations of various aspects of the geographic world, in the form of datasets (*GIS data*); a community of people who use and perhaps advocate the use of these tools for various purposes (the *GIS community*); and the activity of using a GIS to solve problems or advance science (*doing GIS*). The basic label works in all of these ways, and its meaning surely depends on the context in which it is used.

Nevertheless, certain definitions are particularly helpful (Table 1.3). As we describe in Chapter 3, GIS is much more than *a container of maps in digital form*. This can be a misleading description, but it is a helpful definition to give to someone looking for a simple explanation – a guest at a cocktail party, a relative, or a seat neighbor on an airline flight. We all know and appreciate the value of maps, and the notion that maps could be processed by a computer is clearly analogous to the use of word processing or spreadsheets to handle other types of information. A GIS is also *a computerized tool for solving geographic problems*, a definition that speaks to the purposes of GIS, rather than to its functions or physical form – an idea that is expressed in another definition, *a spatial decision support system*. A GIS is *a mechanized inventory of geographically distributed features and facilities*, the definition that explains the value of GIS to the utility industry, where it is used to keep track of such entities as underground pipes, transformers, transmission lines, poles, and customer accounts. A GIS is *a tool for revealing what is otherwise invisible in geographic information* (see Section 2.3.4.4), an interesting definition that emphasizes the power of a GIS as an analysis engine, to examine data and reveal its patterns, relationships, and anomalies – things that might not be apparent to someone looking at a map. A GIS is *a tool for performing*

Table 1.3 Definitions of a GIS, and the groups who find them useful

A container of maps in digital form	The general public
A computerized tool for solving geographic problems	Decision makers, community groups, planners
A spatial decision support system	Management scientists, operations researchers
A mechanized inventory of geographically distributed features and facilities	Utility managers, transportation officials, resource managers
A tool for revealing what is otherwise invisible in geographic information	Scientists, investigators
A tool for performing operations on geographic data that are too tedious or expensive or inaccurate if performed by hand	Resource managers, planners

operations on geographic data that are too tedious or expensive or inaccurate if performed by hand, a definition that speaks to the problems associated with manual analysis of maps, particularly the extraction of simple measures, of area for example.

Everyone has their own favorite definition of a GIS, and there are many to choose from.

1.4.1 A brief history of GIS

As might be expected, there is some controversy about the history of GIS since parallel developments occurred in North America, Europe, and Australia (at least). Much of the published history focuses on the US contributions. We therefore do not yet have a well-rounded history of our subject. What is clear, though, is that the extraction of simple measures largely drove the development of the first real GIS, the Canada Geographic Information System or CGIS, in the mid-1960s (see Box 17.1). The Canada Land Inventory was a massive effort by the federal and provincial governments to identify the nation's land resources and their existing and potential uses. The most useful results of such an inventory are measures of area, yet area is notoriously difficult to measure accurately from a map (Section 14.3). CGIS was planned and developed as a measuring tool, a producer of tabular information, rather than as a mapping tool.

The first GIS was the Canada Geographic Information System, designed in the mid-1960s as a computerized map measuring system.

A second burst of innovation occurred in the late 1960s in the US Bureau of the Census, in planning the tools needed to conduct the 1970 Census of Population. The DIME program (Dual Independent Map Encoding) created digital records of all US streets, to support automatic referencing and aggregation of census records. The similarity of this technology to that of CGIS was recognized immediately, and led to a major program at Harvard University's Laboratory for Computer Graphics and Spatial Analysis to develop a general-purpose GIS that could handle the needs of both applications – a project that led eventually to the ODYSSEY GIS of the late 1970s.

Early GIS developers recognized that the same basic needs were present in many different application areas, from resource management to the census.

In a largely separate development during the latter half of the 1960s, cartographers and mapping agencies had begun to ask whether computers might be adapted to their needs, and possibly to reducing the costs and shortening the time of map creation. The UK Experimental Cartography Unit (ECU) pioneered high-quality computer mapping in 1968; it published the world's first computer-made map in a regular series in 1973 with the British Geological Survey (Figure 1.12); the ECU also pioneered GIS work in education, post and zip codes as geographic references, visual perception of maps, and much else. National mapping agencies, such as Britain's Ordnance Survey, France's Institut Géographique National, and the US Geological Survey

and the Defense Mapping Agency (now the National Geospatial-Intelligence Agency) began to investigate the use of computers to support the editing of maps, to avoid the expensive and slow process of hand correction and redrafting. The first automated cartography developments occurred in the 1960s, and by the late 1970s most major cartographic agencies were already computerized to some degree. But the magnitude of the task ensured that it was not until 1995 that the first country (Great Britain) achieved complete digital map coverage in a database.

Remote sensing also played a part in the development of GIS, as a source of technology as well as a source of data. The first military satellites of the 1950s were developed and deployed in great secrecy to gather intelligence, but the declassification of much of this material in recent years has provided interesting insights into the role played by the military and intelligence communities in the development of GIS. Although the early spy satellites used conventional film cameras to record images, digital remote sensing began to replace them in the 1960s, and by the early 1970s civilian remote sensing systems such as Landsat were beginning to provide vast new data resources on the appearance of the planet's surface from space, and to exploit the technologies of image classification and pattern recognition that had been developed earlier for military applications. The military was also responsible for the development in the 1950s of the world's first uniform system of measuring location, driven by the need for accurate targeting of intercontinental ballistic missiles, and this development led directly to the methods of

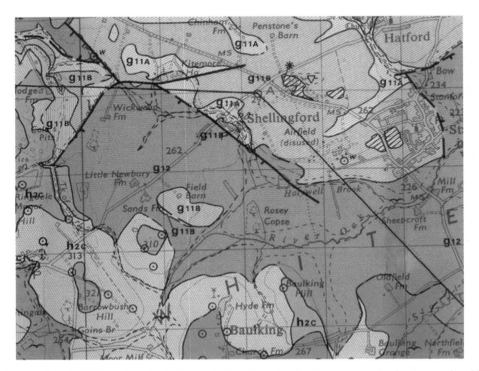

Figure 1.12 Section of the 1:63 360 scale geological map of Abingdon – the first known example of a map produced by automated means and published in a standard map series to established cartographic standards. (Reproduced by permission of the British Geological Survey and Ordnance Survey © NERC. All right reserved. IPR/59-13C)

positional control in use today (Section 5.6). Military needs were also responsible for the initial development of the Global Positioning System (GPS; Section 5.8).

Many technical developments in GIS originated in the Cold War.

GIS really began to take off in the early 1980s, when the price of computing hardware had fallen to a level that could sustain a significant software industry and cost-effective applications. Among the first customers were forestry companies and natural-resource agencies, driven by the need to keep track of vast timber resources, and to regulate their use effectively. At the time a modest computing system – far less powerful than today's personal computer – could be obtained for about $250 000, and the associated software for about $100 000. Even at these prices the benefits of consistent management using GIS, and the decisions that could be made with these new tools, substantially exceeded the costs. The market for GIS software continued to grow, computers continued to fall in price and increase in power, and the GIS software industry has been growing ever since.

The modern history of GIS dates from the early 1980s, when the price of sufficiently powerful computers fell below a critical threshold.

As indicated earlier, the history of GIS is a complex story, much more complex than can be described in this brief history, but Table 1.4 summarizes the major events of the past three decades.

1.4.2 Views of GIS

It should be clear from the previous discussion that GIS is a complex beast, with many distinct appearances. To some it is a way to automate the production of maps, while to others this application seems far too mundane compared to the complexities associated with solving geographic problems and supporting spatial decisions, and with the power of a GIS as an engine for analyzing data and revealing new insights. Others see a GIS as a tool for maintaining complex inventories, one that adds geographic perspectives to existing information systems, and allows the geographically distributed resources of a forestry or utility company to be tracked and managed. The sum of all of these perspectives is clearly too much for any one software package to handle, and GIS has grown from its initial commercial beginnings as a simple off-the-shelf package to a complex of software, hardware, people, institutions, networks, and activities that can be very confusing to the novice. A major software vendor such as ESRI today sells many distinct products, designed to serve very different needs: a major GIS workhorse (ArcInfo), a simpler system designed for viewing, analyzing, and mapping data (ArcView), an engine for supporting GIS-oriented websites (ArcIMS), an information system with spatial extensions (ArcSDE), and several others. Other vendors specialize in certain niche markets, such as the utility industry, or military

and intelligence applications. GIS is a dynamic and evolving field, and its future is sure to be exciting, but speculations on where it might be headed are reserved for the final chapter.

Today a single GIS vendor offers many different products for distinct applications.

1.4.3 Anatomy of a GIS

1.4.3.1 The network

Despite the complexity noted in the previous section, a GIS does have its well-defined component parts. Today, the most fundamental of these is probably the *network*, without which no rapid communication or sharing of digital information could occur, except between a small group of people crowded around a computer monitor. GIS today relies heavily on the Internet, and on its limited-access cousins, the *intranets* of corporations, agencies, and the military. The Internet was originally designed as a network for connecting computers, but today it is rapidly becoming society's mechanism of information exchange, handling everything from personal messages to massive shipments of data, and increasing numbers of business transactions.

It is no secret that the Internet in its many forms has had a profound effect on technology, science, and society in the last few years. Who could have foreseen in 1990 the impact that the Web, e-commerce, digital government, mobile systems, and information and communication technologies would have on our everyday lives (see Section 18.4.4)? These technologies have radically changed forever the way we conduct business, how we communicate with our colleagues and friends, the nature of education, and the value and transitory nature of information.

The Internet began life as a US Department of Defense communications project called ARPANET (Advanced Research Projects Agency Network) in 1972. In 1980 Tim Berners-Lee, a researcher at CERN, the European organization for nuclear research, developed the hypertext capability that underlies today's World Wide Web – a key application that has brought the Internet into the realm of everyday use. Uptake and use of Web technologies have been remarkably quick, diffusion being considerably faster than almost all comparable innovations (for example, the radio, the telephone, and the television: see Figure 18.5). By 2004, 720 million people worldwide used the Internet (see Section 18.4.4 and Figure 18.8), and the fastest growth rates were to be found in the Middle East, Latin America, and Africa (**www.internetworldstats.com**). However, the global penetration of the medium remained very uneven – for example 62% of North Americans used the medium, but only 1% of Africans (Figure 1.13). Other Internet maps are available at the Atlas of Cyber-geography maintained by Martin Dodge (**www.geog.ucl. ac.uk/casa/martin/atlas/atlas.html**).

Geographers were quick to see the value of the Internet. Users connected to the Internet could zoom in to parts of the map, or pan to other parts, using simple

Table 1.4 Major events that shaped GIS

Date	Type	Event	Notes
		The Era of Innovation	
1957	Application	First known automated mapping produced	Swedish meteorologists and British biologists
1963	Technology	CGIS development initiated	Canada Geographic Information System is developed by Roger Tomlinson and colleagues for Canadian Land Inventory. This project pioneers much technology and introduces the term GIS.
1963	General	URISA established	The Urban and Regional Information Systems Association founded in the US. Soon becomes point of interchange for GIS innovators.
1964	Academic	Harvard Lab established	The Harvard Laboratory for Computer Graphics and Spatial Analysis is established under the direction of Howard Fisher at Harvard University. In 1966 SYMAP, the first raster GIS, is created by Harvard researchers.
1967	Technology	DIME developed	The US Bureau of Census develops DIME-GBF (Dual Independent Map Encoding – Geographic Database Files), a data structure and street-address database for 1970 census.
1967	Academic and general	UK Experimental Cartography Unit (ECU) formed	Pioneered in a range of computer cartography and GIS areas.
1969	Commercial	ESRI Inc. formed	Jack Dangermond, a student from the Harvard Lab, and his wife Laura form ESRI to undertake projects in GIS.
1969	Commercial	Intergraph Corp. formed	Jim Meadlock and four others that worked on guidance systems for Saturn rockets form M&S Computing, later renamed Intergraph.
1969	Academic	'Design With Nature' published	Ian McHarg's book was the first to describe many of the concepts in modern GIS analysis, including the map overlay process (see Chapter 14).
1969	Academic	First technical GIS textbook	Nordbeck and Rystedt's book detailed algorithms and software they developed for spatial analysis.
1972	Technology	Landsat 1 launched	Originally named ERTS (Earth Resources Technology Satellite), this was the first of many major Earth remote sensing satellites to be launched.
1973	General	First digitizing production line	Set up by Ordnance Survey, Britain's national mapping agency.
1974	Academic	AutoCarto 1 Conference	Held in Reston, Virginia, this was the first in an important series of conferences that set the GIS research agenda.
1976	Academic	GIMMS now in worldwide use	Written by Tom Waugh (a Scottish academic), this vector-based mapping and analysis system was run at 300 sites worldwide.
1977	Academic	Topological Data Structures conference	Harvard Lab organizes a major conference and develops the ODYSSEY GIS.
		The Era of Commercialization	
1981	Commercial	ArcInfo launched	ArcInfo was the first major commercial GIS software system. Designed for minicomputers and based on the vector and relational database data model, it set a new standard for the industry.

(continued overleaf)

Table 1.4 (*continued*)

Date	Type	Event	Notes
1984	Academic	'Basic Readings in Geographic Information Systems' published	This collection of papers published in book form by Duane Marble, Hugh Calkins, and Donna Peuquet was the first accessible source of information about GIS.
1985	Technology	GPS operational	The Global Positioning System gradually becomes a major source of data for navigation, surveying, and mapping.
1986	Academic	'Principles of Geographical Information Systems for Land Resources Assessment' published	Peter Burrough's book was the first specifically on GIS principles. It quickly became a worldwide reference text for GIS students.
1986	Commercial	MapInfo Corp. formed	MapInfo software develops into first major desktop GIS product. It defined a new standard for GIS products, complementing earlier software systems.
1987	Academic	*International Journal of Geographical Information Systems*, now *IJGI Science*, introduced	Terry Coppock and others published the first journal on GIS. The first issue contained papers from the USA, Canada, Germany, and UK.
1987	General	Chorley Report	'Handling Geographical Information' was an influential report from the UK government that highlighted the value of GIS.
1988	General	*GISWorld* begins	*GISWorld*, now *GeoWorld*, the first worldwide magazine devoted to GIS, was published in the USA.
1988	Technology	TIGER announced	TIGER (Topologically Integrated Geographic Encoding and Referencing), a follow-on from DIME, is described by the US Census Bureau. Low-cost TIGER data stimulate rapid growth in US business GIS.
1988	Academic	US and UK Research Centers announced	Two separate initiatives, the US NCGIA (National Center for Geographic Information and Analysis) and the UK RRL (Regional Research Laboratory) Initiative show the rapidly growing interest in GIS in academia.
1991	Academic	*Big Book 1* published	Substantial two-volume compendium *Geographical Information Systems*; *principles and applications*, edited by David Maguire, Mike Goodchild, and David Rhind documents progress to date.
1992	Technical	DCW released	The 1.7 GB Digital Chart of the World, sponsored by the US Defense Mapping Agency, (now NGA), is the first integrated 1:1 million scale database offering global coverage.
1994	General	Executive Order signed by President Clinton	Executive Order 12906 leads to creation of US National Spatial Data Infrastructure (NSDI), clearinghouses, and Federal Geographic Data Committee (FGDC).
1994	General	OpenGIS® Consortium born	The OpenGIS® Consortium of GIS vendors, government agencies, and users is formed to improve interoperability.
1995	General	First complete national mapping coverage	Great Britain's Ordnance Survey completes creation of its initial database – all 230 000 maps covering country at largest scale (1:1250, 1:2500 and 1:10 000) encoded.
1996	Technology	Internet GIS products introduced	Several companies, notably Autodesk, ESRI, Intergraph, and MapInfo, release new generation of Internet-based products at about the same time.

Table 1.4 (*continued*)

Date	Type	Event	Notes
1996	Commercial	MapQuest	Internet mapping service launched, producing over 130 million maps in 1999. Subsequently purchased by AOL for $1.1 billion.
1999	General	GIS Day	First GIS Day attracts over 1.2 million global participants who share an interest in GIS.

The Era of Exploitation

Date	Type	Event	Notes
1999	Commercial	IKONOS	Launch of new generation of satellite sensors: IKONOS claims 90 centimeter ground resolution; Quickbird (launched 2001) claims 62 cm resolution.
2000	Commercial	GIS passes $7 bn	Industry analyst Daratech reports GIS hardware, software, and services industry at $6.9 bn, growing at more than 10% per annum.
2000	General	GIS has 1 million users	GIS has more than 1 million core users, and there are perhaps 5 million casual users of GI.
2002	General	Launch of online National Atlas of the United States	Online summary of US national-scale geographic information with facilities for map making (**www.nationalatlas.gov**)
2003	General	Launch of online national statistics for the UK	Exemplar of new government websites describing economy, population, and society at local and regional scales (**www.statistics.gov.uk**)
2003	General	Launch of Geospatial One-Stop	A US Federal E-government initiative providing access to geospatial data and information (**www.geodata.gov/gos**)
2004	General	National Geospatial-Intelligence Agency (NGA) formed	Biggest GIS user in the world, National Imagery and Mapping Agency (NIMA), renamed NGA to signify emphasis on geo-intelligence

mouse clicks in their desktop WWW browser, without ever needing to install specialized software or download large amounts of data. This research project soon gave way to industrial-strength Internet GIS software products from mainstream software vendors (see Chapter 7).

The use of the WWW to give access to maps dates from 1993.

The recent histories of GIS and the Internet have been heavily intertwined; GIS has turned out to be a compelling application that has prompted many people to take advantage of the Web. At the same time, GIS has benefited greatly from adopting the Internet paradigm and the momentum that the Web has generated. Today there are many successful applications of GIS on the Internet, and we have used some of them as examples and illustrations at many points in this book. They range from using GIS on the Internet to disseminate information – a type of electronic yellow pages – (e.g., **www.yell.com**), to selling goods and services (e.g., **www.landseer.com.sg**, Figure 1.14), to direct revenue generation through subscription services (e.g., **www.mapquest.com/solutions/main.adp**), to helping members of the public to participate in important local, regional, and national debates.

The Internet has proven very popular as a vehicle for delivering GIS applications for several reasons. It is an established, widely used platform and accepted standard for interacting with information of many types. It also offers a relatively cost-effective way of linking together distributed users (for example, telecommuters and office workers, customers and suppliers, students and teachers). The interactive and exploratory nature of navigating linked information has also been a great hit with users. The availability of geographically enabled multi-content site gateways (geoportals) with powerful search engines has been a stimulus to further success.

Internet technology is also increasingly portable – this means not only that portable GIS-enabled devices can be used in conjunction with the wireless networks available in public places such as airports and railway stations, but also that such devices may be connected through broadband in order to deliver GIS-based representations on the move. This technology is being exploited in the burgeoning GIService (yet another use of the three-letter acronym GIS) sector, which offers distributed users access to centralized GIS capabilities. Later (Chapter 18 and onwards) we use the term *g-business* to cover all the myriad applications carried out in enterprises in different sectors that have a strong geographical component. The

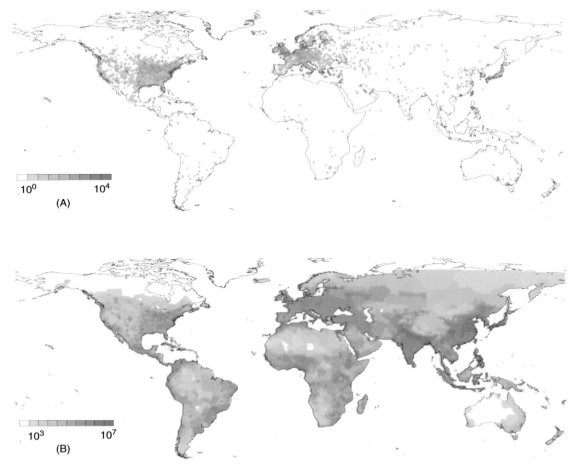

Figure 1.13 (A) The density of Internet hosts (routers) in 2002, a useful surrogate for Internet activity. The bar next to the map gives the range of values encoded by the color code per box (pixel) in the map. (B) This can be compared with the density of population, showing a strong correlation with Internet access in economically developed countries: elsewhere Internet access is sparse and is limited to urban areas. Both maps have a resolution of $1° \times 1°$. (Courtesy Yook S.-H., Jeong H. and Barabsi A.-L. 2002. 'Modeling the Internet's large-scale topology,' *Proceedings of the National Academy of Sciences* **99**, 13382–13386. See **www.nd.edu/~networks/PDF/Modeling%202002.pdf**) (Reproduced with permission of National Academy of Sciences, USA)

more restrictive term *g-commerce* is also used to describe types of electronic commerce (e-commerce) that include location as an essential element. Many GIServices are made available for personal use through mobile and handheld applications as *location-based services* (see Chapter 11). Personal devices, from pagers to mobile phones to Personal Digital Assistants, are now filling the briefcases and adorning the clothing of people in many walks of life (Figure 1.15). These devices are able to provide real-time geographic services such as mapping, routing, and geographic yellow pages. These services are often funded through advertisers, or can be purchased on a pay-as-you go or subscription basis, and are beginning to change the business GIS model for many types of applications.

A further interesting twist is the development of themed geographic networks, such as the US Geospatial One-Stop (**www.geo-one-stop.gov/**: see Box 11.4), which is one of 24 federal e-government initiatives to improve the coordination of government at local, state,

and Federal levels. Its geoportal (**www.geodata.gov/gos**) identifies an integrated collection of geographic information providers and users that interact via the medium of the Internet. On-line content can be located using the interactive search capability of the portal and then content can be directly used over the Internet. This form of Internet application is explored further in Chapter 11.

> **The Internet is increasingly integrated into many aspects of GIS use, and the days of standalone GIS are mostly over.**

1.4.3.2 The other five components of the GIS anatomy

The second piece of the GIS anatomy (Figure 1.16) is the user's hardware, the device that the user interacts with

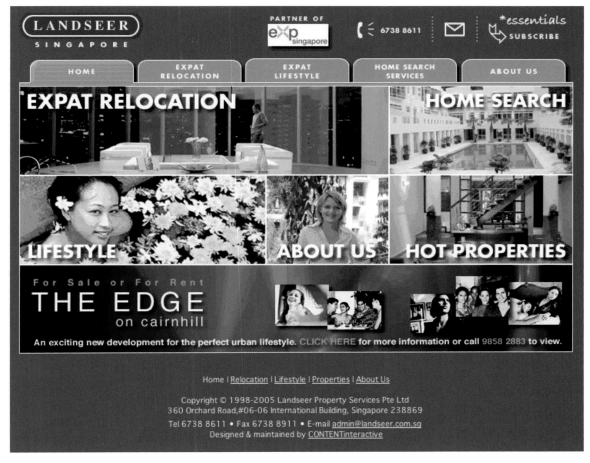

Figure 1.14 Niche marketing of residential property in Singapore (**www.landseer.com.sg**) (Reproduced with permission of Landseer Property Services Pte Ltd.)

Figure 1.15 Wearable computing and personal data assistants are key to the diffusion and use of location-based services

directly in carrying out GIS operations, by typing, pointing, clicking, or speaking, and which returns information by displaying it on the device's screen or generating meaningful sounds. Traditionally this device sat on an office desktop, but today's user has much more freedom, because GIS functions can be delivered through laptops, personal data assistants (PDAs), in-vehicle devices, and even cellular telephones. Section 11.3 discusses the currently available technologies in greater detail. In the language of the network, the user's device is the *client*, connected through the network to a *server* that is probably handling many other user clients simultaneously. The client may be *thick*, if it performs a large part of the work locally, or *thin* if it does little more than link the user to the server. A PC or Macintosh is an instance of a thick client, with powerful local capabilities, while devices attached to TVs that offer little more than Web browser capabilities are instances of thin clients.

The third piece of the GIS anatomy is the software that runs locally in the user's machine. This can be as simple as a standard Web browser (Microsoft Explorer or Netscape) if all work is done remotely using assorted digital services offered on large servers. More likely it is a package bought from one of the GIS vendors, such as Intergraph Corp. (Huntsville, Alabama, USA; **www.ingr.com**), Environmental Systems Research Institute (ESRI; Redlands, California,

Six parts of a GIS

People

Software

Network

Data

Hardware

Procedures

Figure 1.16 The six component parts of a GIS

USA; **www.esri.com**), Autodesk Inc. (San Rafael, California, USA; **www.autodesk.com**), or MapInfo Corp. (Troy, New York, USA; **www.mapinfo.com**). Each vendor offers a range of products, designed for different levels of sophistication, different volumes of data, and different application niches. IDRISI (Clark University, Worcester, Massachusetts, USA, **www.clarklabs.org**) is an example of a GIS produced and marketed by an academic institution rather than by a commercial vendor.

Many GIS tasks must be performed repeatedly, and GIS designers have created tools for capturing such repeated sequences into easily executed *scripts* or *macros* (Section 16.3.1) For example, the agency that needs to predict erosion of New South Wales's soils (Section 1.3) would likely establish a standard script written in the scripting language of its favorite GIS. The instructions in the script would tell the GIS how to model erosion given required data inputs and parameters, and how to output the results in suitable form. Scripts can be used repeatedly, for different areas or for the same area at different times. Support for scripts is an important aspect of GIS software.

GIS software can range from a simple package designed for a PC and costing a few hundred dollars, to a major industrial-strength workhorse designed to serve an entire enterprise of networked computers, and costing tens of thousands of dollars. New products are constantly emerging, and it is beyond the scope of this book to provide a complete inventory.

The fourth piece of the anatomy is the database, which consists of a digital representation of selected aspects of some specific area of the Earth's surface or near-surface, built to serve some problem solving or scientific purpose. A database might be built for one major project, such as the location of a new high-voltage power transmission corridor, or it might be continuously maintained, fed by the daily transactions that occur in a major utility company (installation of new underground pipes, creation of new customer accounts, daily service crew activities). It might be as small as a few megabytes (a few million bytes,

easily stored on a few diskettes), or as large as a terabyte (roughly a trillion bytes, occupying a storage unit as big as a small office). Table 1.1 gives some sense of potential GIS database volumes.

GIS databases can range in size from a megabyte to a petabyte.

In addition to these four components – network, hardware, software, and database – a GIS also requires management. An organization must establish procedures, lines of reporting, control points, and other mechanisms for ensuring that its GIS activities stay within budgets, maintain high quality, and generally meet the needs of the organization. These issues are explored in Chapters 18, 19, and 20.

Finally, a GIS is useless without the people who design, program, and maintain it, supply it with data, and interpret its results. The people of GIS will have various skills, depending on the roles they perform. Almost all will have the basic knowledge needed to work with geographic data – knowledge of such topics as data sources, scale and accuracy, and software products – and will also have a network of acquaintances in the GIS community. We refer to such people in this book as *spatially aware professionals*, or SAPs, and the humor in this term is not intended in any way to diminish their importance, or our respect for what they know – after all, we would like to be recognized as SAPs ourselves! The next section outlines some of the roles played by the people of GIS, and the industries in which they work.

1.5 The business of GIS

Very many people play many roles in GIS, from software development to software sales, and from teaching about GIS to using its power in everyday activities. GIS is big business, and this section looks at the diverse roles that people play in the business of GIS, and is organized by the major areas of human activity associated with GIS.

1.5.1 The software industry

Perhaps the most conspicuous sector, although by no means the largest either in economic or human terms, is the GIS software industry. Some GIS vendors have their roots in other, larger computer applications: thus Intergraph and Autodesk, have roots in computer-assisted design software developed for engineering and architectural applications; and Leica Geosystems (ERDAS IMAGINE: **gis.leica-geosystems.com**) and PCI (**www.pcigeomatics.com**) have roots in remote sensing and image processing. Others began as specialists in GIS. Measured in economic terms, the GIS software industry currently accounts for over $1.8 billion in annual sales, although estimates vary, in part because of the difficulty of defining GIS precisely. The software industry employs several thousand programmers, software designers, systems analysts, application specialists, and sales staff, with backgrounds that include computer science, geography, and many other disciplines.

The GIS software industry accounts for about $1.8 billion in annual sales.

1.5.2 The data industry

The acquisition, creation, maintenance, dissemination, and sale of GIS data also account for a large volume of economic activity. Traditionally, a large proportion of GIS data have been produced centrally, by national mapping agencies such as Great Britain's Ordnance Survey. In most countries the funds needed to support national mapping come from sales of products to customers, and sales now account for almost all of the Ordnance Survey's annual turnover of approximately $200 million. But federal government policy in the US requires that prices be set at no more than the cost of reproduction of data, and sales are therefore only a small part of the income of the US Geological Survey, the nation's premier civilian mapping agency.

In value of annual sales, the GIS data industry is much more significant than the software industry.

In recent years improvements in GIS and related technologies, and reductions in prices, along with various kinds of government stimulus, have led to the rapid growth of a private GIS data industry, and to increasing interest in data sales to customers in the public sector. In the socioeconomic realm, there is continuing investment in the creation and updating of general-purpose geodemographic indicators (Section 2.3.3), created using private sector datasets alongside traditional socioeconomic sources such as the Census. For example, UK data warehouse Experian's (Nottingham, UK) 2003 Mosaic product comprises 54% census data, with the balance of 46% coming from private sector sources and spatial indicators created using GIS. Data may also be packaged with software in order to offer integrated solutions, as with ESRI's Business Analyst product. Private companies are now also licensed to collect high-resolution data using satellites, and to sell it to customers – Space Imaging (**www.spaceimaging.com**) and its IKONOS satellite are a prominent instance (see Table 1.4). Other companies collect similar data from aircraft. Still other companies specialize in the production of high-quality data on street networks, a basic requirement of many delivery companies. Tele Atlas (**www.teleatlas.com** and its North American subsidiary, Geographic Data Technology **www.geographic.com**) is an example of this industry, employing some 1850 staff in producing, maintaining, and marketing high-quality street network data in Europe and North America.

As developments in the information economy gather pace, many organizations are becoming focused upon delivering integrated business solutions rather than raw or value-added data. The Internet makes it possible for GIS users to access routinely collected data from sites that may be remote from locations where more specialized analysis and interpretation functions are performed. In these circumstances, it is no longer incumbent upon an organization to manage either its own data, or those that it buys in from value-added resellers. For example, ESRI offers a data management service, in which client data are managed and maintained for a range of clients that are at liberty to analyze them in quite separate locations. This may lead to greater vertical integration of the software and data industry – ESRI has developed an e-bis division and acquired its own geodemographic system (called Tapestry) to service a range of business needs. As GIS-based data handling becomes increasingly commonplace, so GIS is finding increasing application in new areas of public sector service provision, particularly where large amounts of public money are disbursed at the local level – as in policing, education provision, and public health. Many data warehouses and start-up organizations are beginning to develop public sector data infrastructures particularly where greater investment in public services is taking place.

1.5.3 The GIService industry

The Internet also allows GIS users to access specific functions that are provided by remote sites. For example, the US MapQuest site (**www.mapquest.com**) or the UK Yellow Pages site (**www.yell.com**) provide routing services that are used by millions of people every day to find the best driving route between two points. By typing a pair of street addresses, the user can execute a routing analysis (see Section 15.3.2) and receive the results in the form of a map and a set of written driving or walking directions (see Figure 1.17B). This has several advantages over performing the same analysis on one's own PC – there is no need to buy software to perform the analysis, there is no need to buy the necessary data, and the data are routinely updated by the GIService provider. There are clear synergies of interest between GIService providers and organizations providing location-based services (Section 1.4.3.1 and Chapter 11), and both activities are part of what we will describe as g-business in Chapter 19. Many sites that provide access to raw GIS data also provide GIServices.

GIServices are a rapidly growing form of electronic commerce.

GIServices continue to develop rapidly. In today's world one of the most important commodities is attention – the fraction of a second of attention given to a billboard, or the audience attention that a TV station sells to its advertisers. The value of attention also depends on the degree of fit between the message and the recipient – an advertiser will pay more for the attention of a small number of people if it knows that they include a large proportion of its typical customers. Advertising directed at the individual, based on an individual

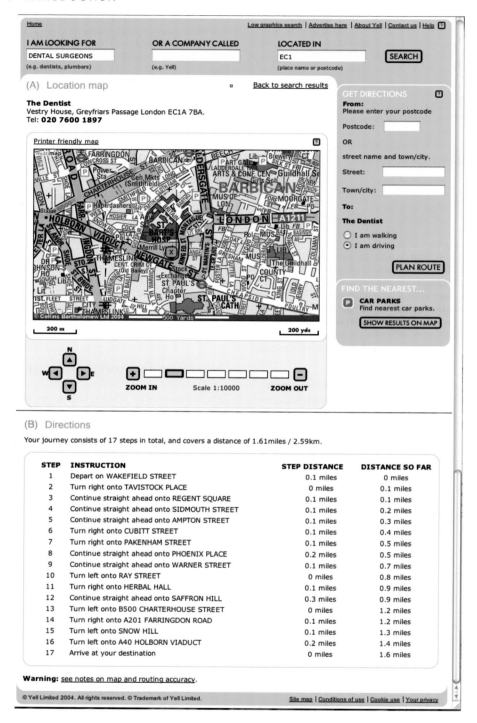

Figure 1.17 A GIS-enabled London electronic yellow pages: (A) location map of a dentist near St. Paul's Cathedral; and (B) written directions of how to get there from University College London Department of Geography

profile, is even more attractive to the advertiser. Direct-mail companies have exploited the power of geographic location to target specific audiences for many years, basing their strategies on neighborhood profiles constructed from census records. But new technologies offer to take this much further. For example, the technology already exists to identify the buying habits of a customer who stops at a gas pump and uses a credit card, and to direct targeted advertising through a TV screen at the pump.

1.5.4 The publishing industry

Much smaller, but nevertheless highly influential in the world of GIS, is the publishing industry, with its magazines, books, and journals. Several magazines are directed at the GIS community, as well as some increasingly significant news-oriented websites (see Box 1.4).

Several journals have appeared to serve the GIS community, by publishing new advances in GIS research. The oldest journal specifically targeted at the community is the *International Journal of Geographical Information Science*, established in 1987. Other older journals in areas such as cartography now regularly accept GIS articles, and several have changed their names and shifted focus significantly. Box 1.5 gives a list of the journals that emphasize GIS research.

1.5.5 GIS education

The first courses in GIS were offered in universities in the early 1970s, often as an outgrowth of courses

---

Technical Box **1.4**

Magazines and websites offering GIS news and related services

ArcNews and ArcUser Magazine (published by ESRI), see **www.esri.com**

Directions Magazine (Internet-centered and weekly newsletter publication by **directionsmag.com**), available online at **www.directionsmag.com**

GEO:connexion UK Magazine published quarterly by GEO:connexion Ltd., with website at **www.geoconnexion.com**

GEOInformatics published eight times a year by Cmedia Productions BV, with website at **www.geoinformatics.com**

GeoSpatial Solutions (published monthly by Advanstar Communications), and see their website **www.geospatial-online.com**. The company also publishes *GPSWorld GeoWorld* (published monthly by GEOTEC Media), available online at **www.geoplace.com**

GIS@development (published monthly for an Asian readership by GIS Development, India), with website at **www.GISDevelopment.net**

Spatial Business Online (published fortnightly in hard and electronic copy form by South Pacific Science Press), available online at **www.gisuser.com.au**

Some websites offering online resources for the GIS community:

www.gisdevelopment.net

www.geoconnexion.com

www.gis.com

www.giscafe.com

gis.about.com

www.geocomm.com and **www.spatialnews.com**

www.directionsmag.com

www.opengis.org/press/

Technical Box **1.5**

Some scholarly journals emphasizing GIS research

Annals of the Association of American Geographers

Cartography and Geographic Information Science

Cartography – The Journal

Computers and Geosciences

Computers, Environment and Urban Systems

Geographical Analysis

GeoInformatica

International Journal of Geographical Information Science (formerly

International Journal of Geographical Information Systems)

ISPRS Journal of Photogrammetry and Remote Sensing

Journal of Geographical Systems

Photogrammetric Engineering and Remote Sensing (PE&RS)

Terra Forum

The Photogrammetric Record

Transactions in GIS

URISA Journal

Technical Box **1.6**

Sites offering Web-based education and training programs in GIS

Birkbeck College (University of London) GIScOnline M.Sc. in Geographic Information Science at **www.bbk.ac.uk**

City University (London) MGI – Masters in Geographic Information – a course with face-to-face or distance learning options at **www.city.ac.uk**

Curtin University's distance learning programs in geographic information science at **www.cage.curtin.edu.au**

ESRI's Virtual Campus at **campus.esri.com**

Kingston Centre for GIS, Distance Learning Programme at **www.kingston.ac.uk**

Pennsylvania State University Certificate Program in Geographic Information Systems at **www.worldcampus.psu.edu**

UNIGIS International, Postgraduate Courses in GIS at **www.unigis.org**

University of Southern California GIS distance learning certificate program at **www.usc.edu**

in cartography or remote sensing. Today, thousands of courses can be found in universities and colleges all over the world. Training courses are offered by the vendors of GIS software, and increasing use is made of the Web in various forms of remote GIS education and training (Box 1.6).

Often, a distinction is made between education and training in GIS – training in the use of a particular software product is contrasted with education in the fundamental principles of GIS. In many university courses, lectures are used to emphasize fundamental principles while computer-based laboratory exercises emphasize training. In our view, an education should be for life, and the material learned during an education should be applicable for as far into the future as possible. Fundamental principles tend to persist long after software has been replaced with new versions, and the skills learned in running one software package may be of very little value when a new technology arrives. On the other hand much of the fun and excitement of GIS comes from actually working with it, and fundamental principles can be very dry and dull without hands-on experience.

1.6 GISystems, GIScience, and GIStudies

Geographic information systems are useful tools, helping everyone from scientists to citizens to solve geographic problems. But like many other kinds of tools, such as computers themselves, their use raises questions that are sometimes frustrating, and sometimes profound. For example, how does a GIS user know that the results obtained are accurate? What principles might help a GIS user to design better maps? How can location-based services be used to help users to navigate and understand human and natural environments? Some of these are questions of GIS design, and others are about

GIS data and methods. Taken together, we can think of them as questions that arise from the use of GIS – that are stimulated by exposure to GIS or to its products. Many of them are addressed in detail at many points in this book, and the book's title emphasizes the importance of both systems and science.

The term *geographic information science* was coined in a paper by Michael Goodchild published in 1992. In it, the author argued that these questions and others like them were important, and that their systematic study constituted a science in its own right. Information science studies the fundamental issues arising from the creation, handling, storage, and use of information – similarly, GIScience should study the fundamental issues arising from geographic information, as a well-defined class of information in general. Other terms have much the same meaning: *geomatics* and *geoinformatics*, *spatial information science*, *geoinformation engineering*. All suggest a scientific approach to the fundamental issues raised by the use of GIS and related technologies, though they all have different roots and emphasize different ways of thinking about problems (specifically geographic or more generally spatial, emphasizing engineering or science, etc.).

GIScience has evolved significantly in recent years. It is now part of the title of several renamed research journals (see Box 1.5), and the focus of the US University Consortium for Geographic Information Science (**www.ucgis.org**), an organization of roughly 60 research universities that engages in research agenda setting (Box 1.7), lobbying for research funding, and related activities. An international conference series on GIScience has been held in the USA biannually since 2000 (see **www.giscience.org**). The Varenius Project (**www.ncgia.org**) provides one disarmingly simple way to view developments in GIScience (Figure 1.18). Here, GIScience is viewed as anchored by three concepts – the individual, the computer, and society. These form the vertices of a triangle, and GIScience lies at its core. The various terms that are used to describe GIScience activity can be used to populate this triangle. Thus research about the individual is dominated by cognitive science, with its concern for understanding of spatial concepts,

The 2002 research agenda of the US University Consortium for Geographic Information Science (www.ucgis.org), and related chapters in this book

1. **Long-term research challenges**

   a. Spatial ontologies (Chapters 3 and 6)
   b. Geographic representation (Chapter 3)
   c. Spatial data acquisition and integration (Chapters 9 and 10)
   d. Scale (Chapter 4)
   e. Spatial cognition (Chapter 3)
   f. Space and space/time analysis and modeling (Chapters 4, 14, 15, and 16)
   g. Uncertainty in geographic information (Chapter 6)
   h. Visualization (Chapters 12 and 13)
   i. GIS and society (Chapters 1 and 17)
   j. Geographic information engineering (Chapters 11 and 20)

2. **Short-term research priorities**

   a. GIS and decision making (Chapters 2, 17, and 18)
   b. Location-based services (Chapters 7 and 11)
   c. Social implications of LBS (Chapter 11)
   d. Identification of spatial clusters (Chapters 13 and 14)
   e. Geospatial semantic Web (Chapters 1 and 11)
   f. Incorporating remotely sensed data and information in GIS (Chapters 3 and 9)

   g. Geographic information resource management (Chapters 17 and 18)
   h. Emergency data acquisition and analysis (Chapter 9)
   i. Gradation and indeterminate boundaries (Chapter 6)
   j. Geographic information security (Chapter 17)
   k. Geospatial data fusion (Chapters 2 and 11)
   l. Institutional aspects of SDIs (Chapters 19 and 20)
   m. Geographic information partnering (Chapter 20)
   n. Geocomputation (Chapter 16)
   o. Global representation and modeling (Chapter 3)
   p. Spatialization (Chapters 3 and 13)
   q. Pervasive computing (Chapter 11)
   r. Geographic data mining and knowledge discovery (Chapter 14)
   s. Dynamic modeling (Chapter 16)

   More detail on all of these topics, and additional topics added at more recent UCGIS assemblies, can be found at **www.ucgis.org/priorities/research/2002researchagenda.htm**

learning and reasoning about geographic data, and interaction with the computer. Research about the computer is dominated by issues of representation, the adaptation of new technologies, computation, and visualization. And finally, research about society addresses issues of impacts and societal context. Others have developed taxonomies of challenges facing the nascent discipline of GIScience, such as the US University Consortium for Geographic Information Science (Box 1.7). It is possible to imagine how the themes presented in Box 1.7 could be used to populate Figure 1.18 in relation to the three vertices of this triangle.

There are important respects in which GIScience is about using the software environment of GIS to redefine, reshape, and resolve pre-existing problems. Many of the research topics in GIScience are actually much older than GIS. The need for methods of spatial analysis, for example, dates from the first maps, and many methods were developed long before the first GIS appeared on the scene in the mid-1960s. Another way to look at GIScience is to see it as the body of knowledge

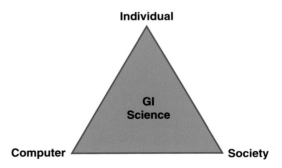

Figure 1.18 The remit of GIScience, according to Project Varenius (**www.ncgia.org**)

that GISystems implement and exploit. Map projections (Chapter 5), for example, are part of GIScience, and are used and transformed in GISystems. Another area of great importance to GIS is cognitive science, and

Reg Golledge, Behavioral Geographer

Reg Golledge was born in Australia but has worked in the US since completing his Ph.D. at the University of Iowa in 1966. He has worked at The Ohio State University (1967–1977), and since 1977 at the University of California, Santa Barbara (UCSB).

GIScience revisits many of the classic problems of spatial analysis, most of which assumed that people were rational and were optimizers in a very narrow sense. Over the last four decades, Reg's work has contributed much to our understanding of individual spatial behavior by relaxing these restrictive assumptions yet retaining the power of scientific generalization. Golledge's *analytical behavioral geography* has examined individual behavior using statistical and computational process models, particularly within the domain of transportation GIS (GIS-T: see Section 2.3.4), and has done much to make sense of the complexities and constraints that govern movement within urban systems. Related to this, analytical behavioral geography has also developed our understanding of individual cognitive awareness of urban networks and landmarks.

Figure 1.19 Reg Golledge, behavioral geographer

Reg's work is avowedly interdisciplinary. He has undertaken extensive work with cognitive psychologists at UCSB to develop personal guidance systems for use by visually-impaired travelers. This innovative work has linked GPS (for location and tracking) and GIS (for performing operations such as shortest path calculation, buffering, and orientation: see Chapters 14 and 15) with a novel auditory virtual system that presents users with the spatial relations between nearby environmental features. The device also allows users to personalize their representations of the environment.

Reg's enduring contribution to GIScience has been in modeling, explaining, and predicting disaggregate behaviors of individuals. This has been achieved through researching spatial cognition and cognitive science through GIS applications. He has established the importance of cognitive mapping to reasoning through GIScience, developed our understanding of the ways in which spatial concepts are embedded in GIS technology, and made vital contributions to the development of multimodal interfaces to GIS. These efforts have helped to develop new links to information science, information technology, and multimedia, and suggest ways of bridging the digital divides that threatens to further disadvantage disabled and elderly people. As a visually-impaired individual himself, Reg firmly believes that GIS technology and GIScience research are the most significant contributions that geography can make to truly integrated human and physical sciences, and sees a focus upon cognition as the natural bridge between these approaches to scientific inquiry.

particularly the scientific understanding of how people think about their geographic surroundings. If GISystems are to be easy to use they must fit with human ideas about such topics as driving directions, or how to construct useful and understandable maps. Box 1.8 introduces Reg Golledge, a quantitative and behavioral geographer who has brought diverse threads of cognitive science, transportation modeling, and analysis of geography and disability together under the umbrella of GIScience.

Many of the roots to GIS can be traced to the spatial analysis tradition in the discipline of geography.

In the 1970s it was easy to define or delimit a geographic information system – it was a single piece of software residing on a single computer. With time, and particularly with the development of the Internet and new approaches to software engineering, the old monolithic nature of GIS has been replaced by something much more fluid, and GIS is no longer an activity confined to the desktop (Chapter 11). The emphasis throughout this book is on this new vision of GIS, as the set of coordinated parts discussed earlier in Section 1.4. Perhaps the *system* part of GIS is no longer necessary – certainly the phrase *GIS data* suggests some redundancy, and many people have suggested that we could drop the 'S' altogether in favor of GI, for geographic information. GI*Systems* are only one part of the GI whole, which also includes the fundamental issues of GI*Science*. Much of this book is really about GI*Studies*, which can be defined as the systematic study of society's use of geographic information, including its institutions, standards, and procedures, and many of these topics are addressed in the later chapters. Several of the UCGIS research topics suggest this kind of focus, including *GIS and society* and *geographic information partnering*. In recent years the role of GIS in society – its impacts and its deeper significance – has become the focus of extensive writing in the academic literature, particularly in the discipline of geography, and much of it has been critical of GIS. We explore these critiques in detail in the next section.

The importance of social context is nicely expressed by Nick Chrisman's definition of GIS which might also serve as an appropriate final comment on the earlier discussion of definitions:

> **The organized activity by which people:**
> **1) measure aspects of geographic phenomena and processes; 2) represent these measurements, usually in the form of a computer database, to emphasize spatial themes, entities, and relationships; 3) operate upon these representations to produce more measurements and to discover new relationships by integrating disparate sources; and 4) transform these representations to conform to other frameworks of entities and relationships. These activities reflect the larger context (institutions and cultures) in which these people carry out their work. In turn, the GIS may influence these structures.**
>
> **(Chrisman 2003, p. 13)**

Chrisman's social structures are clearly part of the GIS whole, and as students of GIS we should be aware of the ethical issues raised by the technology we study. This is the arena of GIStudies.

1.7 GIS and geography

GIS has always had a special relationship to the academic discipline of geography, as it has to other disciplines that deal with the Earth's surface, including planning and landscape architecture. This section explores that special relationship and its sometimes tense characteristics. Non-geographers can conveniently skip this section, though much of its material might still be of interest.

Chapter 2 presents a gallery of successful GIS applications. This paints a picture of a field built around low-order concepts that actually stands in rather stark contrast to the scientific tradition in the academic discipline of geography. Here, the spatial analysis tradition has developed during the past 40 years around a range of more-sophisticated operations and techniques, which have a much more elaborate conceptual structure (see Chapters 14 through 16). One of the foremost proponents of the spatial analysis approach is Stewart Fotheringham, whose contribution is discussed in Box 1.9. As we will begin to see in Chapter 14, spatial analysis is the process by which we turn raw spatial data into useful spatial information. For the first half of its history, the principal focus of spatial analysis in most universities was upon development of theory, rather than working applications. Actual data were scarce, as were the means to process and analyze them.

In the 1980s GIS technology began to offer a solution to the problems of inadequate computation and limited data handling. However, the quite sensible priorities of vendors at the time might be described as solving the problems of 80% of their customers 80% of the time, and the integration of techniques based upon higher-order concepts was a low priority. Today's GIS vendors can probably be credited with solving the problems of at least 90% of their customers 90% of the time, and much of the remit of GIScience is to diffuse improved, curiosity-driven scientific understanding into the knowledge base of existing successful applications. But the drive towards improved applications has also been propelled to a significant extent by the advent of GPS and other digital data infrastructure initiatives by the late 1990s. New data handling technologies and new rich sources of digital data open up prospects for refocusing and reinvigorating academic interest in applied scientific problem solving.

Although repeat purchases of GIS technology leave the field with a buoyant future in the IT mainstream, there is enduring unease in some academic quarters about GIS applications and their social implications. Much of this unease has been expressed in the form of critiques, notably from geographers. John Pickles has probably contributed more to the debate than almost anyone else, notably through his 1993 edited volume *Ground Truth: The Social Implications of Geographic Information Systems*. Several types of arguments have surfaced:

- The ways in which GIS represents the Earth's surface, and particularly human society, favor certain phenomena and perspectives, at the expense of others. For example, GIS databases tend to emphasize homogeneity, partly because of the limited space available and partly because of the costs of more accurate data collection (see Chapters 3, 4, and 8). Minority views, and the views of individuals, can be submerged in this process, as can information that differs from the official or consensus view. For example, a soil map represents the geographic variation in soils by depicting areas of constant class, separated by sharp boundaries. This is clearly an approximation, and in Chapter 6 we explore the role of uncertainty in GIS. GIS often forces knowledge into forms more likely to reflect the view of the majority, or the official view of government, and as a result *marginalizes* the opinions of minorities or the less powerful.

- Although in principle it is possible to use GIS for any purpose, in practice it is often used for purposes that may be ethically questionable or may invade individual privacy, such as surveillance and the gathering of military and industrial intelligence. The technology may appear neutral, but it is always used in a social context. As with the debates over the

Stewart Fotheringham, geocomputation specialist

There are many close synonyms for geographic information science (GIScience), one of which is *geocomputation* – a term first coined by the geographer Stan Openshaw to describe the scientific application of computationally-intensive techniques to problems with a spatial dimension. A. Stewart Fotheringham is Science Foundation Ireland Research Professor and Director of the National Centre for Geocomputation at the National University of Ireland in Maynooth. He is a spatial scientist who has considerable previous experience of the Anglo-American university systems – he has worked and studied at the Universities of Newcastle and Aberdeen in the UK, the State University of New York at Buffalo, the University of Florida, and Indiana University in the US, and McMaster University in Canada (Figure 1.20).

Figure 1.20 Stewart Fotheringham, quantitative geographer

Like GIScience, geocomputation is fundamentally about satisfying human curiosity through systematic, scientific problem solving. Many of the roots to the scientific use of GIS in scientific problem solving can be traced to the 'Quantitative Revolution' in geography of the 1960s, which had the effect of popularizing systematic techniques of spatial analysis throughout the discipline – an approach that had its detractors then as well as now (Section 1.7). The Quantitative Revolution has not only bequeathed GIS a rich legacy of methods and techniques, but has also developed into a sustained concern for understanding the nature of spatial variations in relationships. The range of these methods and techniques is described in Stewart's 2000 book *Quantitative Geography: Perspectives on Spatial Data Analysis* (Sage, London: with co-researchers Chris Brunsdon and Martin Charlton), while spatial variations in relationships are considered in detail in his 2002 book *Geographically Weighted Regression: The Analysis of Spatially Varying Relationships* (Wiley, Chichester: also with the same co-authors).

The methods and techniques that Stewart has developed and applied permeate the world of GIS applications that we consider in Chapter 2. Stewart remains evangelical about the importance of space and our need to use GIS to make spatial analysis sensitive to context. He says: 'We know that many spatially aggregated statements, such as the average temperature of the entire US on any given day, actually tell us very little. Yet when we seek to establish relationships between data, we all too often hypothesize that relationships are the same everywhere – that relationships are *spatially invariant*'. Stewart's work on geographically weighted regression (GWR) is part of a growing realization that relationships, or our measurements of them, can vary over space and that we need to investigate this potential non-stationarity further (see Chapter 4). Stewart's geocomputational approach is closely linked to GIS because it uses locational information as inputs and produces geocoded results as outputs that can be mapped and further analyzed. GWR exploits the property of spatial location to the full, and has led to geocomputational analysis of relationships by researchers working in many disciplines.

atomic bomb in the 1940s and 1950s, the scientists who develop and promote the use of GIS surely bear some responsibility for how it is eventually used. The idea that a tool can be inherently neutral, and its developers therefore immune from any ethical debates, is strongly questioned in this literature.

■ The very success of GIS is a cause of concern. There are qualms about a field that appears to be led by technology and the marketplace, rather than by human need. There are fears that GIS has become *too* successful in modeling socioeconomic distributions, and that as a consequence GIS has become a tool of the 'surveillance society'.

■ There are concerns that GIS remains a tool in the hands of the already powerful – notwithstanding the diffusion of technology that has accompanied the plummeting cost of computing and wide adoption of the Internet. As such, it is seen as maintaining the *status quo* in terms of power structures. By implication, any vision of GIS for all of society is seen as unattainable.

■ There appears to be an absence of applications of GIS in *critical* research. This academic perspective is centrally concerned with the connections between human agency and particular social structures and contexts. Some of its protagonists are of the view that such connections are not amenable to digital representation in whole or in part.

■ Some view the association of GIS with the scientific and technical project as fundamentally flawed. More narrowly, there is a view that GIS applications are (like spatial analysis before it) inextricably bound to the philosophy and assumptions of the approach to science known as *logical positivism* (see also the reference to 'positive' in Section 1.1). As such, the argument goes, GIS can never be more than a logical positivist tool and a normative instrument, and cannot enrich other more critical perspectives in geography.

Many geographers remain suspicious of the use of GIS in geography.

We wonder where all this discussion will lead. For our own part, we have chosen a title that includes both systems and science, and certainly much more of this book is about the broader concept of geographic information than about isolated, monolithic software systems *per se*. We believe strongly that effective users of GIS require some awareness of *all* aspects of geographic information, from the basic principles and techniques to concepts of management and familiarity with applications. We hope this book provides that kind of awareness. On the other hand, we have chosen not to include GIStudies in the title. Although the later chapters of the book address many aspects of the social context of GIS, including issues of privacy, the context to GIStudies is rooted in social theory. GIStudies need the kind of focused attention that we cannot give, and we recommend that students interested in more depth in this area explore the specialized texts listed in the guide to further reading.

Questions for further study

1. Examine the geographic data available for the area within 50 miles (80 km) of either where you live or where you study. Use it to produce a short (2500 word) illustrated profile of either the socioeconomic or the physical environment. (See for example **www.geodata.gov/gos**; **www.geographynetwork.com**; **eu-geoportal.jrc.it**; or **www.magic.gov.uk**).

2. What are the distinguishing characteristics of the scientific method? Discuss the relevance of each to GIS.

3. We argued in Section 1.4.3.1 that the Internet has dramatically changed GIS. What are the arguments for and against this view?

4. Locate each of the issues identified in Box 1.7 in two triangular 'GIScience' diagrams like that shown in Figure 1.18 – one for long-term research challenges and one for short-term research priorities. Give short written reasons for your assignments. Compare the distribution of issues within each of your triangles in order to assess the relative importance of the individual, the computer, and society in the development of GIScience over both the short- and long-term.

Further reading

Chrisman N.R. 2003 *Exploring Geographical Information Systems* (2nd edn). Hoboken, NJ: Wiley.

Curry M.R. 1998 *Digital Places: Living with Geographic Information Technologies*. London: Routledge.

Foresman T.W. (ed) 1998 *The History of Geographic Information Systems: Perspectives from the Pioneers*. Upper Saddle River, NJ: Prentice Hall.

Goodchild M.F. 1992 'Geographical information science'. *International Journal of Geographical Information Systems* **6**: 31–45.

Longley P.A. and Batty M. (eds) 2003 *Advanced Spatial Analysis: The CASA Book of GIS*. Redlands, CA: ESRI Press.

Pickles J. 1993 *Ground Truth: The Social Implications of Geographic Information Systems*. New York: Guilford Press.

University Consortium for Geographic Information Science 1996 'Research priorities for geographic information science'. *Cartography and Geographic Information Systems* **23**(3): 115–127.

2 *A gallery of applications*

Fundamentally, GIS is about workable applications. This chapter gives a flavor of the breadth and depth of real-world GIS implementations. It considers:

- How GIS affects our everyday lives;

- How GIS applications have developed, and how the field compares with scientific practice;

- The goals of applied problem solving;

- How GIS can be used to study and solve problems in transportation, the environment, local government, and business.

Geographic Information Systems and Science, 2nd edition   Paul Longley, Michael Goodchild, David Maguire, and David Rhind.
© 2005 John Wiley & Sons, Ltd.   ISBNs: 0-470-87000-1 (HB); 0-470-87001-X (PB)

Learning Objectives

After studying this chapter you will:

- Grasp the many ways in which we interact with GIS in everyday life;

- Appreciate the range and diversity of GIS applications in environmental and social science;

- Be able to identify many of the scientific assumptions that underpin real-world applications;

- Understand how GIS is applied in the representative application areas of transportation, the environment, local government, and business.

2.1 Introduction

2.1.1 One day of life with GIS

7:00 My alarm goes off... The energy to power the alarm comes from the local energy company, which uses a GIS to manage all its assets (e.g., electrical conductors, devices, and structures) so that it can deliver electricity continuously to domestic and commercial customers (Figure 2.1).

7:05 I jump in the shower... The water for the shower is provided by the local water company, which uses a hydraulic model linked to its GIS to predict water usage and ensure that water is always available to its valuable customers (Figure 2.2).

7:35 I open the mail... A property tax bill comes from a local government department that uses a GIS to store property data and automatically produce annual tax bills. This has helped the department to peg increases in property taxes to levels below retail price inflation.

There are also a small number of circulars addressed to me, sometimes called 'junk mail'. We spent our

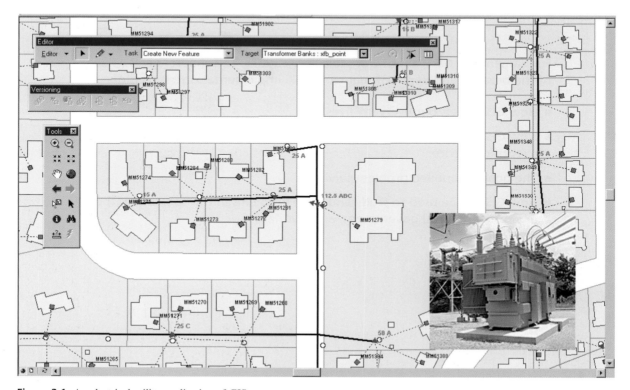

Figure 2.1 An electrical utility application of GIS

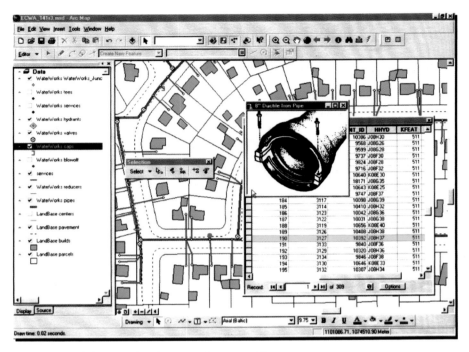

Figure 2.2 Application of a GIS for managing the assets of a water utility

vacation in Southlands and Santatol last year, and the holiday company uses its GIS to market similar destinations to its customer base – there are good deals for the Gower and Northampton this season. A second item is a special offer for property insurance, from a firm that uses its GIS to target neighborhoods with low past-claims histories. We get less junk mail than we used to (and we don't want to opt out of all programs), because geodemographic and lifestyles GIS is used to target mailings more precisely, thus reducing waste and saving time.

8:00 The other half leaves for work. . . He teaches GIS at one of the city community colleges. As a lecturer on one of the college's most popular classes he has a full workload and likes to get to work early.

8:15 I walk the kids to the bus stop. . . Our children attend the local middle school that is three miles away. The school district administrators use a GIS to optimize the routing of school buses (Figure 2.3). Introduction of this service enabled the district to cut their annual school busing costs by 16% and the time it takes the kids to get to school has also been reduced.

8:45 I catch a train to work. . . At the station the current location of trains is displayed on electronic maps on the platforms using a real-time feed from global positioning (GPS) receivers mounted on the trains. The same information is also available on the Internet so I was able to check the status of trains before I left the house.

9:15 I read the newspaper on the train. . . The paper for the newspaper comes from sustainable forests managed by a GIS. The forestry information system used by the forest products company indicates which areas are available for logging, the best access routes, and the likely yield (Figure 2.4).

9:30 I arrive at work. . . I am GIS Manager for the local City government. Today I have meetings to review annual budgets, plan for the next round of hardware and software acquisition, and deal with a nasty copyright infringement claim.

12:00 I grab a sandwich for lunch. . . The price of bread fell in real terms for much of the past decade. In some small part this is because of the increasing use of GIS in precision agriculture. This has allowed real-time mapping of soil nutrients and yield, and means that farmers can apply just the right amount of fertilizer in the right location and at the right time.

6:30 Shop till you drop. . . After work we go shopping and use some of the discount coupons that were in the morning mail. The promotion is to entice customers back to the renovated downtown Tesbury Center. We usually go to MorriMart on the far side of town, but thought we'd participate in the promotion. We actually bump into a few of our neighbors at Tesbury – I suspect the promotion was targeted by linking a marketing GIS to Tesbury's own store loyalty card data.

10:30 The kids are in bed. . . I'm on the Internet to try and find a new house. . . We live in a good neighborhood with many similarly articulate, well-educated folk, but it has become noisier since the new distributor road was routed close by. Our resident association mounted a vociferous campaign of protest, and its members filed numerous complaints to the website where the draft proposals were posted. But

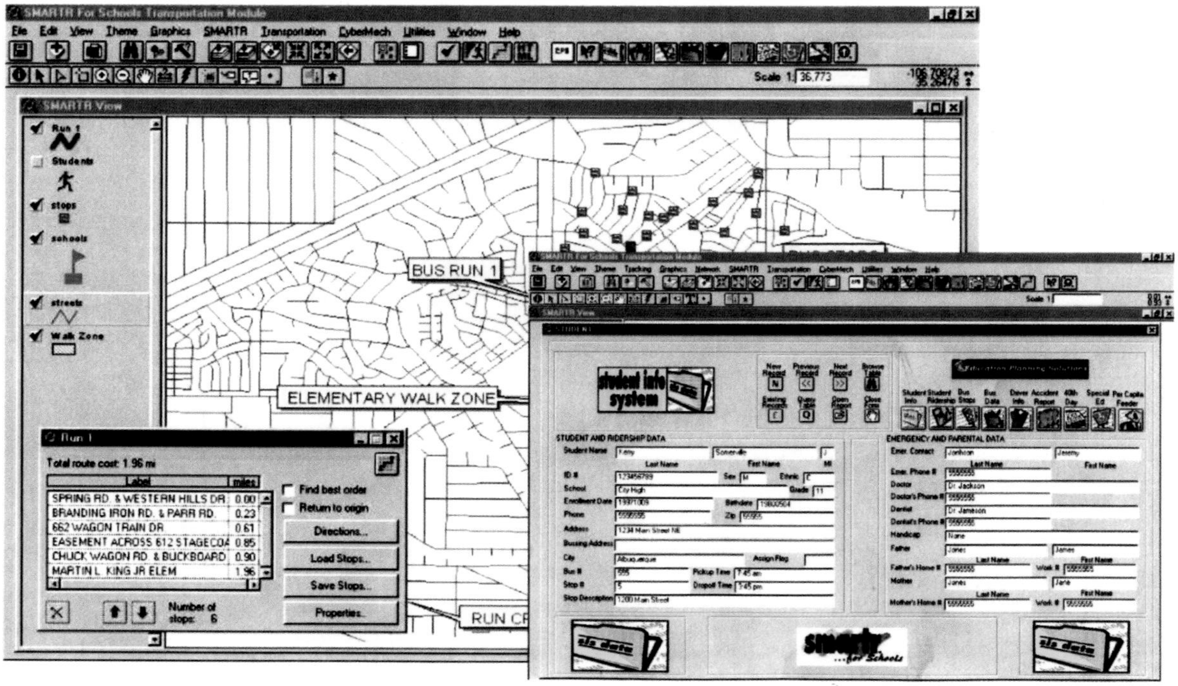

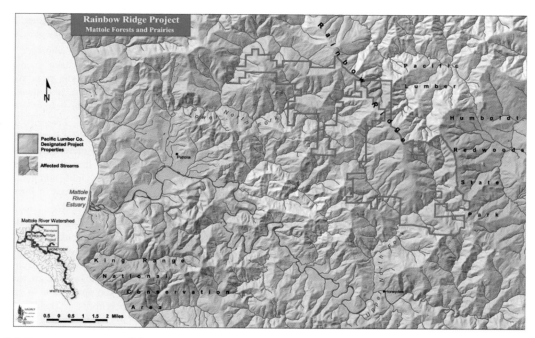

Figure 2.3 A GIS used for school bus routing

Figure 2.4 Forestry management GIS

the benefit-cost analysis carried out using the local authority's GIS clearly demonstrated that it was either a bit more noise for us, or the physical dissection of a vast swathe of housing elsewhere, and that we would have to grin and bear it. Post GIS, I guess that narrow interest NIMBY (Not In My Back Yard) protests don't

get such a free run as they once did. So here I am using one of the on-line GIS-powered websites to find properties that match our criteria (similar to that in Figure 1.14). Once we have found a property, other mapping sites provide us with details about the local and regional facilities.

GIS is used to improve many of our day-to-day working and living arrangements.

This diary is fictitious of course, but most of the things described in it are everyday occurrences repeated hundreds and thousands of times around the world. It highlights a number of key things about GIS. GIS

- affects each of us, every day;
- can be used to foster effective short- and long-term decision making;
- has great practical importance;
- can be applied to many socio-economic and environmental problems;
- supports mapping, measurement, management, monitoring, and modeling operations;
- generates measurable economic benefits;
- requires key management skills for effective implementation;
- provides a challenging and stimulating educational experience for students;
- can be used as a source of direct income;
- can be combined with other technologies; and
- is a dynamic and stimulating area in which to work.

At the same time, the examples suggest some of the elements of the critique that has been leveled at GIS in recent years (see Section 1.7). Only a very small fraction of the world's population has access to information technologies of any kind, let alone high-speed access to the Internet. At the global scale, information technology can exacerbate the differences between developed and less-developed nations, across what has been called the *digital divide*, and there is also *digital differentiation* between rich and poor communities within nations. Uses of GIS for marketing often involve practices that border on invasion of privacy, since they allow massive databases to be constructed from what many would regard as personal information. It is important that we understand and reflect on issues like these while exploring GIS.

2.1.2 Why GIS?

Our day of life with GIS illustrates the unprecedented frequency with which, directly or indirectly, we interact with digital machines. Today, more and more individuals and organizations find themselves using GIS to answer the fundamental question, *where*? This is because of:

- Wider availability of GIS through the Internet, as well as through organization-wide local area networks.
- Reductions in the price of GIS hardware and software, because economies of scale are realized by a fast-growing market.
- Greater awareness that decision making has a geographic dimension.
- Greater ease of user interaction, using standard windowing environments.

- Better technology to support applications, specifically in terms of visualization, data management and analysis, and linkage to other software.
- The proliferation of geographically referenced digital data, such as those generated using Global Positioning System (GPS) technology or supplied by value-added resellers (VARs) of data.
- Availability of packaged applications, which are available commercially off-the-shelf (COTS) or 'ready to run out of the box'.
- The accumulated experience of applications that *work*.

2.2 Science, geography, and applications

2.2.1 Scientific questions and GIS operations

As we saw in Section 1.3, one objective of science is to solve problems that are of real-world concern. The range and complexity of scientific principles and techniques that are brought to bear upon problem solving will clearly vary between applications. Within the spatial domain, the goals of applied problem solving include, but are not restricted to:

- Rational, effective, and efficient allocation of resources, in accordance with clearly stated criteria – whether, for example, this entail physical construction of infrastructure in utilities applications, or scattering fertilizer in precision agriculture.
- Monitoring and understanding observed spatial distributions of attributes – such as variation in soil nutrient concentrations, or the geography of environmental health.
- Understanding the difference that *place* makes – identifying which characteristics are inherently similar between places, and what is distinctive and possibly unique about them. For example, there are regional and local differences in people's surnames (see Box 1.2), and regional variations in voting patterns are the norm in most democracies.
- Understanding of processes in the natural and human environments, such as processes of coastal erosion or river delta deposition in the natural environment, and understanding of changes in residential preferences or store patronage in the social.
- Prescription of strategies for environmental maintenance and conservation, as in national park management.

Understanding and resolving these diverse problems entails a number of general data handling operations – such as inventory compilation and analysis, mapping, and spatial database management – that may be successfully undertaken using GIS.

> **GIS is fundamentally about solving real-world problems.**

GIS has always been fundamentally an applications-led area of activity. The accumulated experience of applications has led to borrowing and creation of particular conventions for representing, visualizing, and to some extent analyzing data for particular classes of applications. Over time, some of these conventions have become useful in application areas quite different from those for which they were originally intended, and software vendors have developed general-purpose routines that may be customized in application-specific ways, as in the way that spatial data are visualized. The way that accumulated experience and borrowed practice becomes formalized into standard conventions makes GIS essentially an inductive field.

In terms of the definition and remit of GIScience (Section 1.6) the conventions used in applications are based on very straightforward concepts. Most data-handling operations are routine and are available as adjuncts to popular word-processing packages (e.g., Microsoft MapPoint: **www.microsoft.com/mappoint**). They work and are very widely used (e.g., see Figure 2.5), yet may not always be readily adaptable to scientific problem solving in the sense developed in Section 1.3.

2.2.2 GIScience applications

Early GIS was successful in depicting how the world looks, but shied away from most of the bigger questions concerning how the world works. Today GIScience is developing this extensive experience of applications into a bigger agenda – and is embracing a full range of conceptual underpinnings to successful problem solving.

GIS nevertheless remains fundamentally an applications-led technology, and many applications remain modest in both the technology that they utilize and the scientific tasks that they set out to accomplish. There is nothing fundamentally wrong with this, of course, as the most important test of geographic science and technology is whether or not it is useful for exploring and understanding the world around us. Indeed the broader relevance of geography as a discipline can only be sustained in relation to this simple goal, and no amount of scientific and technological ingenuity can salvage geographic representations of the world that are too inaccurate, expensive, cumbersome, or opaque to reveal anything new. In practice, this means that GIS applications must be grounded in sound concepts and theory if they are to resolve any but the most trivial of questions.

> **GIS applications need to be grounded in sound concepts and theory.**

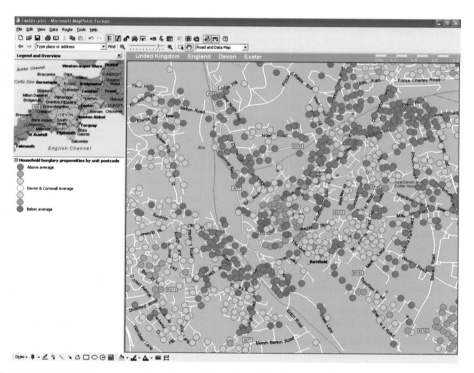

Figure 2.5 Microsoft MapPoint Europe mapping of spreadsheet data of burglary rates in Exeter, England using an adjunct to a standard office software package (courtesy D. Ashby. © 1988–2001 Microsoft Corp. and/or its suppliers. All rights reserved. © 2000 Navigation Technologies B.V. and its suppliers. All rights reserved. Selected Road Maps © 2000 by AND International Publishers N.V. All rights reserved. © Crown Copyright 2000. All rights reserved. License number 100025500. Additional demographic data courtesy of Experian Limited. © 2004 Experian Limited. All rights reserved.)

2.3 Representative application areas and their foundations

2.3.1 Introduction and overview

There is, quite simply, a huge range of applications of GIS, and indeed several pages of this book could be filled with a list of application areas. They include topographic base mapping, socio-economic and environmental modeling, global (and interplanetary!) modeling, and education. Applications generally set out to fulfill the five *Ms* of GIS: mapping, measurement, monitoring, modeling, management.

The five Ms of GIS application are mapping, measurement, monitoring, modeling, and management.

In very general terms, GIS applications may be classified as traditional, developing, and new. Traditional GIS application fields include military, government, education, and utilities. The mid-1990s saw the wide development of business uses, such as banking and financial services, transportation logistics, real estate, and market analysis. The early years of the 21st century are seeing new forward-looking application areas in small office/home office (SOHO) and personal or consumer applications, as well as applications concerned with security, intelligence, and counter-terrorism measures. This is a somewhat rough-and-ready classification, however, because the applications of some agencies (such as utilities) fall into more than one class.

A further way to examine trends in GIS applications is to examine the diffusion of GIS use. Figure 2.6 shows the classic model of GIS diffusion originally developed by Everett Rogers. Rogers' model divides the adopters of an innovation into five categories:

- Venturesome Innovators – willing to accept risks and sometimes regarded as oddballs.

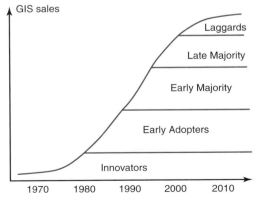

- Respectable Early Adopters – regarded as opinion formers or 'role models'.
- Deliberate Early Majority – willing to consider adoption only after peers have adopted.
- Skeptical Late Majority – overwhelming pressure from peers is needed before adoption occurs.
- Traditional Laggards – people oriented to the past.

GIS is moving into the Late Majority stage, although some areas of application are more comprehensively developed than others. The Innovators who dominated the field in the 1970s were typically based in universities and research organizations. The Early Adopters were the users of the 1980s, many of whom were in government and military establishments. The Early Majority, typically in private businesses, came to the fore in the mid-1990s. The current question for potential users appears to be: do you want to gain competitive advantage by being part of the Majority user base or wait until the technology is completely accepted and contemplate joining the GIS community as a Laggard?

A wide range of motivations underpins the use of GIS, although it is possible to identify a number of common themes. Applications dealing with day-to-day issues typically focus on very practical concerns such as cost effectiveness, service provision, system performance, competitive advantage, and database creation, access, and use. Other, more strategic applications are more concerned with creating and evaluating scenarios under a range of circumstances.

Many applications involve use of GIS by large numbers of people. It is not uncommon for a large government agency, university, or utility to have more than 100 GIS seats, and a significant number have more than 1000. Once GIS applications become established within an organization, usage often spreads widely. Integration of GIS with corporate information system (IS) policy and with forward planning policy is an essential prerequisite for success in many organizations.

The scope of these applications is best illustrated with respect to representative application areas, and in the remainder of this chapter we consider:

1. Government and public service (Section 2.3.2)
2. Business and service planning (Section 2.3.3)
3. Logistics and transportation (Section 2.3.4)
4. Environment (Section 2.3.5)

We begin by identifying the range of applications within each of the four domains. Next, we go on to focus upon one application within each domain. Each application is chosen, first, for simplicity of exposition but also, second, for the scientific questions that it raises. In this book, we try to relate science and application in two ways. First, we flag the sections elsewhere in the book where the scientific issues raised by the applications are discussed. Second, the applications discussed here, and others like them, provide the illustrative material for our discussion of principles, techniques, analysis, and practices in the other chapters of the book.

A recurrent theme in each of the application classes is the importance of geographic location, and hence

Figure 2.6 The classic Rogers model of innovation diffusion applied to GIS (After Rogers E.M. 2003 *Diffusion of Innovations* (5th edn). New York: Simon and Schuster.)

what is *special* about the handling of georeferenced data (Section 1.1.1). The gallery of applications that we set out here intends to show how geographic data can provide crucial context to decision making.

2.3.2 Government and public service

2.3.2.1 Applications overview

Government users were among the first to discover the value of GIS. Indeed the first recognized GIS – the Canadian Geographic Information System (CGIS) – was developed for natural resource inventory and management by the Canadian government (see Section 1.4.1). CGIS was a national system and, unlike now, in the early days of GIS it was only national or federal organizations that could afford the technology. Today GIS is used at all levels of government from the national to the neighborhood, and government users still comprise the biggest single group of GIS professionals. It is helping to supplement traditional 'top down' government decision making with 'bottom up' representation of real communities in government decision making at all levels (Figure 2.7). We will see in later chapters how this deployment of GIS applications is consistent with greater supplementation of 'top down' deductivism with 'bottom up' inductivism in science. The importance of spatial variation to government and public service should not be underestimated – 70–80% of local government work should involve GIS in some way.

> **As GIS has become cheaper, so it has come to be used in government decision making at all levels from the nation to the neighborhood.**

Today, local government organizations are acutely aware of the need to improve the quality of their products, processes, and services through ever-increasing efficiency of resource usage (see also Section 15.3). Thus GIS is used to inventory resources and infrastructure, plan transportation routing, improve public service delivery, manage land development, and generate revenue by increasing economic activity.

Local governments also use GIS in unique ways. Because governments are responsible for the long-term health, safety, and welfare of citizens, wider issues need to be considered, including incorporating public values in decision making, delivering services in a fair and equitable manner, and representing the views of citizens by working with elected officials. Typical GIS applications thus include monitoring public health risk, managing public housing stock, allocating welfare assistance funds, and tracking crime. Allied to analysis using geodemographics (see Section 2.3.3) they are also used for operational, tactical, and strategic decision making in law enforcement, health care planning, and managing education systems.

It is convenient to group local government GIS applications on the basis of their contribution to asset inventory, policy analysis, and strategic modeling/planning. Table 2.1 summarizes GIS applications in this way.

These applications can be implemented as centralized GIS or distributed desktop applications. Some will be designed for use by highly trained GIS professionals, while citizens will access others as 'front counter' or Internet systems. Chapter 8 discusses the different implementation models for GIS.

2.3.2.2 Case study application: GIS in tax assessment

Tax mapping and assessment is a classic example of the value of GIS in local government. In many countries local

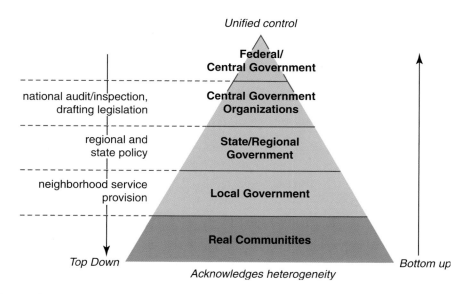

Figure 2.7 The use of GIS at different levels of government decision making

Table 2.1 GIS applications in local government (simplified from O'Looney 2000)

	Inventory Applications (locating property information such as ownership and tax assessments by clicking on a map)	Policy Analysis Applications (e.g., number of features per area, proximity to a feature or land use, correlation of demographic features with geological features)	Management/Policy-Making Applications (e.g., more efficient routing, modeling alternatives, forecasting future needs, work scheduling)
Economic Development	Location of major businesses and their primary resource demands	Analysis of resource demand by potential local supplier	Informing businesses of availability of local suppliers
Transportation and Services Routing	Identification of sanitation truck routes, capacities and staffing by area; identification of landfill and recycling sites	Analysis of potential capacity strain given development in certain areas; analysis of accident patterns by type of site	Identification of ideal high-density development areas based on criteria such as established transportation capacity
Housing	Inventory of housing stock age, condition, status (public, private, rental, etc.), durability, and demographics	Analysis of public support for housing by geographic area, drive time from low-income areas to needed service facilities, etc.	Analysis of funding for housing rehabilitation, location of related public facilities; planning for capital investment in housing based on population growth projections
Infrastructure	Inventory of roads, sidewalks, bridges, utilities (locations, names, conditions, foundations, and most recent maintenance)	Analysis of infrastructure conditions by demographic variables such as income and population change	Analysis to schedule maintenance and expansion
Health	Locations of persons with particular health problems	Spatial, time-series analysis of the spread of disease; effects of environmental conditions on disease	Analysis to pinpoint possible sources of disease
Tax Maps	Identification of ownership data by land plot	Analysis of tax revenues by land use within various distances from the city center	Projecting tax revenue change due to land-use changes
Human Services	Inventory of neighborhoods with multiple social risk indicators; location of existing facilities and services designated to address these risks	Analysis of match between service facilities and human services needs and capacities of nearby residents	Facility siting, public transportation routing, program planning, and place-based social intervention
Law Enforcement	Inventory of location of police stations, crimes, arrests, convicted perpetrators, and victims; plotting police beats and patrol car routing; alarm and security system locations	Analysis of police visibility and presence; officers in relation to density of criminal activity; victim profiles in relation to residential populations; police experience and beat duties	Reallocation of police resources and facilities to areas where they are likely to be most efficient and effective; creation of random routing maps to decrease predictability of police beats
Land-use Planning	Parcel inventory of zoning areas, floodplains, industrial parks, land uses, trees, green space, etc.	Analysis of percentage of land used in each category, density levels by neighborhoods, threats to residential amenities, proximity to locally unwanted land uses	Evaluation of land-use plan based on demographic characteristics of nearby population (e.g., will a smokestack industry be sited upwind of a respiratory disease hospital?)

(continued overleaf)

Table 2.1 (*continued*)

	Inventory Applications	Policy Analysis Applications	Management/Policy-Making Applications
Parks and Recreation	Inventory of park holdings/playscapes, trails by type, etc.	Analysis of neighborhood access to parks and recreation opportunities, age-related proximity to relevant playscapes	Modeling population growth projections and potential future recreational needs/playscape uses
Environmental Monitoring	Inventory of environmental hazards in relation to vital resources such as groundwater; layering of nonpoint pollution sources	Analysis of spread rates and cumulative pollution levels; analysis of potential years of life lost in a particular area due to environmental hazards	Modeling potential environmental harm to specific local areas; analysis of place-specific multilayered pollution abatement plans
Emergency Management	Location of key emergency exit routes, their traffic flow capacity and critical danger points (e.g., bridges likely to be destroyed by an earthquake)	Analysis of potential effects of emergencies of various magnitudes on exit routes, traffic flow, etc.	Modeling effect of placing emergency facilities and response capacities in particular locations
Citizen Information/ Geodemographics	Location of persons with specific demographic characteristics such as voting patterns, service usage and preferences, commuting routes, occupations	Analysis of voting characteristics of particular areas	Modeling effect of placing information kiosks at particular locations

government agencies have a mandate to raise revenue from property taxes. The amount of tax payable is partly or wholly determined by the value of taxable land and property. A key part of this process is evaluating the value of land and property fairly to ensure equitable distribution of a community's tax burden. In the United States the task of determining the taxable value of land and property is performed by the Tax Assessor's Office, which is usually a separate local government department. The Valuation Office Agency fulfills a similar role in the UK. The tax department can quickly get overwhelmed with requests for valuation of new properties and protests about existing valuations.

The Tax Assessor's Office is often the first home of GIS in local government.

Essentially, a Tax Assessor's role is to assign a value to properties using three basic methods: cost, income, and market. The cost method is based on the replacement cost of the property and the value of the land. The Tax Assessor must examine data on construction costs and vacant land values. The income method takes into consideration how much income a property would generate if it were rented. This requires details on current market rents, vacancy rates, operating expenses, taxes, insurance, maintenance, and other costs. The market method is the most popular. It compares the property to other recent sales that have a similar location, size, condition, and quality.

Collecting, storing, managing, analyzing, and displaying all this information is a very time-consuming activity and not surprisingly GIS has had a major impact on the way Tax Assessors go about their business.

2.3.2.3 Method

Tax Assessors, working in a Tax Assessor's Office, are responsible for accurately, uniformly, and fairly judging the value of all taxable properties in their jurisdiction. Details about properties are maintained on a tax assessment roll that includes information such as ownership, address, land and building value, and tax exemptions. The Assessor's Office is also responsible for processing applications for tax abatement, in cases of overvaluation, and exemptions for surviving spouses, veterans, and the elderly. Figure 2.8 shows some aspects of a tax assessment GIS in Ohio, USA.

A GIS is used to collect and manage the geographic boundaries and associated information about properties. Typically, data associated with properties is held in a Computer Assisted Mass Appraisal (CAMA) system that is responsible for sale analysis, evaluation, data management, and administration, and for generating notices to owners. CAMA systems are usually implemented on top of a database management system (DBMS) and can be linked to the parcel database using a common key (see Section 10.2 for further discussion of how this works).

The basic tax assessment task involves a geographic database query to locate all sales of similar properties within a predetermined distance of a given property. The property to be valued is first identified in the property database. Next, a geographic query is used to ascertain the values of all comparable properties within a predetermined search radius (typically one mile) of the property. These properties are then displayed on the assessor's screen. The assessor can then compare the

(A)

(B)

Figure 2.8 Lucas County, Ohio, USA tax assessment GIS: (A) tax map; (B) property attributes and photograph

characteristics of these properties (lot size, sales price and date of sale, neighborhood status, property improvements, etc.) and value the property.

2.3.2.4 Scientific foundations: principles, techniques, and analysis

Scientific foundations

Critical to the success of the tax assessment process is a high-quality, up-to-date geographic database that can be linked to a CAMA system. Considerable effort must expended to design, implement, and maintain the geographic database. Even for a small community of 50 000 properties it can take several months to assemble the geographic descriptions of property parcels with their associated attributes. Chapters 9 and 10 explain the processes involved in managing geographic databases such as this. Linking GIS and CAMA systems can be quite straightforward providing that both systems are based on DBMS technology and use a common identifier to effect linkage between a map feature and a property record. Typically, a unique parcel number (in the US) or unique property reference number (in the UK) is used.

A high-quality geographic database is essential to tax assessment.

Clearly, the system is dependent on an unambiguous definition of parcels, and common standards about how different characteristics (such as size, age, and value of improvements) are represented. The GIS can help enforce coding standards and can be used to derive some characteristics automatically in an objective fashion. For example, GIS makes it straightforward to calculate the area of properties using boundary coordinates.

Fundamentally, this application, like many others in GIS, depends upon an unambiguous and accurate inventory of geographic extent. To be effective it must link this with clear, intelligible, and stable attribute descriptors. These are all core characteristics of scientific investigation, and although the application is driven by results rather than scientific curiosity, it nevertheless follows scientific procedures of controlled comparison.

Principles

Tax assessment makes the assumption that, other things being equal, properties close together in space will have similar values. This is an application of Tobler's First Law of Geography, introduced in Section 3.1. However, it is left to the Assessor to identify comparator properties and to weight their relative importance. This seems rather straightforward, but in practice can prove very difficult – particularly where the exact extent of the effects of good and bad neighborhood attributes cannot be precisely delineated. In practice the value of location in a given neighborhood is often assumed to be uniform (see Section 4.7), and properties of a given construction type are also assumed to be identical. This assumption may be valid in areas where houses were constructed at the same time according to common standards; however, in older areas where infill has been common, properties of a given type vary radically in quality over short distances.

Techniques

Tax assessment requires a good database, a plan for system management and administration, and a workflow design. These procedures are set out in Chapters 9, 10 and 17. The alternative of manually sorting paper records, or even tabular data in a CAMA system, is very laborious and time-consuming, and thus the automated approach of GIS is very cost-effective.

Analysis

Tax assessment actually uses standard GIS techniques such as proximity analysis, and geographic and attribute query, mapping, and reporting. These must be robust and defensible when challenged by individuals seeking reductions in assessments. Chapter 14 sets out appropriate procedures, while Chapter 12 describes appropriate conventions for representing properties and neighborhoods cartographically.

2.3.2.5 Generic scientific questions arising from the application

This is not perhaps the most glamorous application of GIS, but its operational value in tax assessment cannot be overestimated. It requires an up-to-date inventory of properties and information from several sources about sales and sale prices, improvements, and building programs.

To help tax assessors understand geographic variations in property characteristics it is also possible to use GIS for more strategic modeling activities. The many tools in GIS for charting, reporting, mapping, and exploratory data analysis help assessors to understand the variability of property value within their jurisdictions. Some assessors have also built models of property valuations and have clustered properties based on multivariate criteria (see Section 4.7). These help assessors to gain knowledge of the structure of communities and highlight unusually high or low valuations. Once a property database has been created, it becomes a very valuable asset, not just for the tax assessor's department, but also for many other departments in a local government agency. Public works departments may seek to use it to label access points for repairs and meter reading, housing departments may use it to maintain data on property condition, and many other departments may like shared access to a common address list for record keeping and mailings.

A property database is useful for many purposes besides tax assessment.

2.3.2.6 Management and policy

Tax assessment is a key local government application because it is a direct revenue generator. It is easy to develop a cost-benefit case for this application (Chapter 17) and it can pay for a complete department or corporate GIS implementation quickly (Chapter 18). Tax assessment is a service offered directly to members of the public. As such, the service must be reliable and achieve a quick turnaround (usually with one week). It is quite common for citizens to question the assessed value for their property, since this is the principal determinant of the amount of tax they will have to pay. A tax assessor must, therefore, be able to justify the method and data used to determine property values. A GIS is a great help and often convinces people of the objectivity involved (sometimes over-impressing people that it is totally scientific). As such, GIS is an important tool for efficiency and equitable local government.

2.3.3 Business and service planning

2.3.3.1 Applications overview

Business and service planning (sometimes called retailing) applications focus upon the use of geographic data to provide operational, tactical, and strategic context to decisions that involve the fundamental question, *where*? *Geodemographics* is a shorthand term for composite indicators of consumer behavior that are available at the small-area level (e.g., census output area, or postal zone). Figure 2.9 illustrates the profile of one geodemographic type from a UK classification called Mosaic, developed by market researcher and academic Richard Webber. The current version of Mosaic divides the UK population into 11

Figure 2.9 A geodemographic profile: Town Gown Transition (a Type within the Urban Intelligence Group of the 2001 MOSAIC classification). (Courtesy of Experian Limited. © 2004 Experian Limited. All rights reserved)

Groups (such as 'Happy Families', 'Urban Intelligence', and 'Blue Collar Enterprise'), which in turn are subdivided into a total of 61 Types. Geodemographic data are frequently used in business applications to identify geographic variations in the incidences of customer types. They are often supplemented by *lifestyles* data on the consumption choices and shopping habits of individuals who fill out questionnaires or participate in store loyalty programs. The term *market area analysis* describes the activity of assessing the distribution of retail outlets relative to the greatest concentrations of potential customers. The approach is increasingly being adapted to improving public service planning, in areas such as health, education, and law enforcement (see Box 2.1 and Section 2.3.2).

Geodemographic data are the basis for much market area analysis.

The tools of business applications typically range from simple desktop mapping to sophisticated decision support systems. Tools are used to analyze and inform the range of *operational*, *tactical*, and *strategic* functions of an organization. These tools may be part of standard GIS software, or they may be developed in-house by the organization, or they may be purchased (with or without accompanying data) as a 'business solution' product. We noted in Section 1.1 that operational functions concern the day-to-day processing of routine transactions and inventory analysis in an organization, such as stock management. Tactical functions require the allocation of resources to address specific (usually short term) problems, such as store sales promotions. Strategic functions contribute to the organization's longer-term goals and mission, and entail problems such as opening new stores or rationalizing existing store networks. Early business applications were simply concerned with mapping spatially referenced data, as a general descriptive indicator of the retail environment. This remains the first stage in most business applications, and in itself adds an important dimension to analysis of organizational function. More

Marc Farr, geodemographer

'*City Adventurers* are young, well educated and open to new ideas and influences. They are cosmopolitan in their tastes and liberal in their social attitudes. Few have children. Many are in further education while others are moving into full-time employment. Most do not feel ready to make permanent commitments, whether to partners, professions or to specific employers. As higher education has become internationalized, the *City Adventurers* group has acquired many foreign-born residents, which further encourages ethnic and cultural variety.'

This is the geodemographic profile of the neighborhood in Hove, UK, where Marc Farr lives. Marc read economics and marketing at Lancaster University before going to work as a market researcher in London, first at the TMS Partnership and then at Experian. His work involved use of geodemographic data to analyze retail catchments, measure insurance risk, and analyze household expenditure patterns.

Over time, Marc gained increasing consultancy responsibilities for public sector clients in education, health, and law enforcement. As a consequence of his developing interests in the problems that they face, five years after

Figure 2.10 Marc Farr, geodemographer

graduating, he began to work on a Ph.D. that used geodemographics to analyze the ways in which prospective students in the UK choose the universities at which they want to study. He did this work in association with the UK Universities Central Admissions Service (UCAS). Speaking about his Ph.D., which was completed after five years' part time study, Marc says: 'I question the assumptions that the massive increases in numbers of people entering UK higher education during the late 1990s will reduce inequality between different socio-economic groups, or that they will necessarily improve economic and social mobility. Geodemographic analysis also suggests that we need to better understand the relationship between the geography of demand for higher education and its physical supply.'

Marc now works for the Dr. Foster consultancy firm, where he has responsibilities for the calculation of hospital and health authority performance statistics.

---

recently, decision support tools used by Spatially Aware Professionals (SAPs, Section 1.4.3.2) have created mainstream research and development roles for business GIS applications.

Some of the operational roles of GIS in business are discussed under the heading of logistics applications in Section 2.3.4. These include stock flow management systems and distribution network management, the specifics of which vary from industry sector to sector. Geodemographic analysis is an important operational tool in market area analysis, where it is used to plan marketing campaigns. Each of these applications can be described as assessing the circumstances of an organization.

The most obvious strategic application concerns the spatial *expansion* of a new entrant across a retail market. Expansion in a market poses fundamental spatial problems – such as whether to expand through *contagious diffusion* across space, or *hierarchical diffusion* down a settlement structure, or to pursue some combination of the two (Figure 2.11). Many organizations periodically experience spatial *consolidation* and branch *rationalization*. Consolidation and rationalization may occur: (a) when two organizations with overlapping networks

merge; (b) in response to competitive threat; or (c) in response to changes in the retail environment. Changes in the retail environment may be short term and cyclic, as in the response to the recession phase of business cycles, or structural, as with the rationalization of clearing bank branches following the development of personal, telephone, and Internet-based banking (see Section 18.4.4). Still other organizations undergo spatial *restructuring*, as in the market repositioning of bank branches to supply a wider range of more profitable financial services. Spatial restructuring is often the consequence of technological change. For example, a 'clicks and mortar' strategy might be developed by a chain of conventional bookstores, whereby their retail outlets might be reconfigured to offer reliable pick-up points for Internet and telephone orders – perhaps in association with location-based services (Sections 1.4.3.1 and 11.3.2). This may confer advantage over new, purely Internet-based entrants. A final type of strategic operation involves *distribution* of goods and services, as in the case of so-called 'e-tailers', who use the Internet for merchandizing, but must create or buy into viable distribution networks. These various strategic operations require a range of spatial analytic

(A)

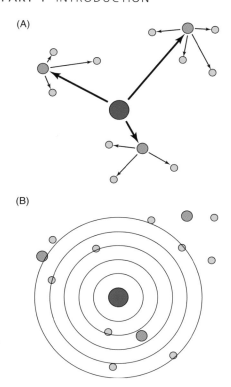

(B)

Figure 2.11 (A) Hierarchical and (B) contagious spatial diffusion

tools and data types, and entail a move from 'what-is' visualization to 'what-if' forecasts and predictions.

2.3.3.2 Case study application: hierarchical diffusion and convenience shopping

Tesco is, by some margin, the most successful grocery (food) retailer in the UK, and has used its knowledge of the home market to launch successful initiatives in Asia and the developing markets of Eastern Europe (see Figure 1.7). Achieving real sales growth in its core business of groceries is difficult, particularly in view of a strict national planning regime that prevents widespread development of new stores, and legislation to prevent the emergence, through acquisitions, of local spatial monopolies of supply. One way in which Tesco has succeeded in sustaining market growth in the domestic market in these circumstances is through strategic diversification into consumer durables and clothing in its largest (high order, colored dark red in Figure 2.11) stores. A second driver to growth has been the successful development of a store loyalty card program, which rewards members with money-off coupons or leisure experiences according to their weekly spend. This program generates *lifestyles* data as a very useful by-product, which enables Tesco to identify consumption profiles of its customers, not unlike the Mosaic geodemographic system. This enables the company to identify, for example, whether customers are 'value driven' and should be directed to budget food offerings, or whether they are principally motivated by

quality and might be encouraged to purchase goods from the company's 'Finest' range. This is a very powerful marketing tool, although unlike geodemographic discriminators these data tell the company rather little about those households that are not their customers, or the products that their own customers buy elsewhere.

A third driver to growth entails the creation or acquisition of much smaller neighborhood stores (low order, in the lightest red in Figure 2.11). These provide a local community service and are not very intrusive on the retail landscape, and are thus much easier to create within the constraints of the planning system. The 'Express' format store shown in Figure 2.12A opened in 2003 and was planned by Tesco's in-house store location team using GE Smallworld GIS. Figure 2.12B shows its location (labeled T3) in Bournemouth, UK, in relation to the edge of the town and the locations of five competitor chains.

GIS can be used to predict the success of a retailer in penetrating a local market area.

2.3.3.3 Method

The location is in a suburban residential neighborhood, and as such it was anticipated that its customer base would be mainly local – that is, resident within a 1 km radius of the store. A budget was allocated for promoting the new establishment, in order to encourage repeat patronage. An established means of promoting new stores is through leaflet drops or enclosures with free local newspapers. However, such tactical interventions are limited by the coarseness of distribution networks – most organizations that deliver circulars will only undertake deliveries for complete postal sectors (typically 20 000 population size) and so this represents a rather crude and wasteful medium.

A second strategy would be to use the GIS to identify all of the households resident within a 1 km radius of the store. Each UK unit postcode (roughly equivalent to a US Zip+4 code, and typically comprising 18–22 addresses) is assigned the grid reference of the first mail delivery point on the 'mailman's walk'. Thus one of the quickest ways of identifying the relevant addresses entails plotting

(A)

Figure 2.12 (A) The site and (B) the location of a new Tesco Express store

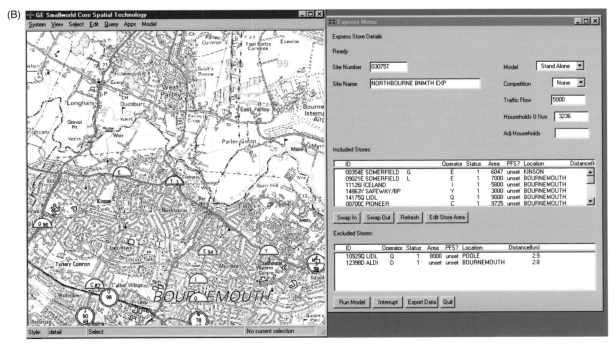

Figure 2.12 (*continued*)

the unit postcode addresses and selecting those that lie within the search radius. Matching the unit postcodes with the full postcode address file (PAF) then suggests that there are approximately 3236 households resident in the search area. Each of these addresses might then be mailed a circular, thus eliminating the largely wasteful activity of contacting around 16 800 households that were unlikely to use the store.

Yet even this tactic can be refined. Sending the same packet of money-off coupons to all 3236 households assumes that each has identical disposable incomes and consumption habits. There may be little point, for example, incentivizing a domestic-beer drinker to buy premium champagne, or vice-versa. Thus it makes sense to overlay the pattern of geodemographic profiles onto the target area, in order to tailor the coupon offerings to the differing consumption patterns of 'blue collar enterprise' neighborhoods versus those classified as belonging to 'suburban comfort', for example.

There is a final stage of refinement that can be developed for this analysis. Using its lifestyles (storecard) data, Tesco can identify those households who already prefer to use the chain, despite the previous nonavailability of a local convenience store. Some of these customers will use Tesco for their main weekly shop, but may 'top up' with convenience or perishable goods (such as bread, milk, or cut flowers) from a competitor. Such households might be offered stronger incentives to purchase particular ranges of goods from the new store, without the wasteful 'cannibalizing' activity of offering coupons towards purchases that are already made from Tesco in the weekly shop.

2.3.3.4 Scientific foundations: geographic principles, techniques, and analysis

The following assumptions and organizing principles are inherent to this case study.

Scientific foundations

Fundamental to the application is the assumption that the closer a customer lives to the store, the more likely he or she is to patronize it. This is formalized as Tobler's First Law of Geography in Section 3.1 and is accommodated into our *representations* as a distance decay effect. The *nature* (Chapter 4) of such distance decay effects does not have to be linear – in Section 4.5 we will introduce a range of *non-linear* effects.

The science of geodemographic profiling can be stated succinctly as 'birds of a feather flock together' – that is, the differences in the observed social and economic characteristics of residents *between* neighborhoods are greater than differences observed *within* them. The use of small-area geodemographic profiles to mix the coupon incentives that might be sent to prospective customers assumes that each potential customer is equally and utterly typical of the post code in which he or she resides. The individual resident *in* an area is thus assigned the characteristics *of* the area. In practice, of course, individuals within households have different characteristics, as do households within streets, zones, and any other aggregation. The practice of confounding characteristics of areas with individuals resident in them is known as committing the *ecological fallacy* – the term does not refer to any branch of biology, but shares with ecology a primary concern with describing the *linkage* of living organisms (individuals) to their geographical surroundings. This is inevitable in most socio-economic GIS applications because data that enable sensitive characteristics of individuals must be kept confidential (see the point about GIS and the surveillance society in Section 1.7). Whilst few could take offence at error in mis-targeting money-off coupons as in this example, ecological analysis has the potential to cause distinctly unethical outcomes if individuals are penalized because of where they live – for example if individuals find it difficult to gain credit because of the credit histories of their neighborhoods. Such discriminatory activity is usually prevented by industry codes of conduct or even legislation. The use of lifestyles data culled from store loyalty card records enables individuals to be targeted precisely, but such individuals might well be geographically or socially *unrepresentative* of the population at large or the market as a whole.

More generally, geography is a science that has very few natural units of analysis – what, for example, is the natural unit for measuring a soil profile? In socio-economic applications, even if we have disaggregate data we might remain uncertain as to whether we should consider the individual or the household as the basic unit of analysis – sometimes one individual in a household always makes the important decisions, while in others this is a shared responsibility. We return to this issue in our discussion of *uncertainty* (Chapter 6).

The use of lifestyle data from store loyalty programs allows the retailer to enrich geodemographic profiling with information about its own customers. This is a cutting-edge marketing activity, but one where there is plenty of scope for relevant research that is able to take 'what is' information about existing customer characteristics and use it to conduct 'what if' analysis of behavior given a different constellation of retail outlets. We return to the issue of defining appropriate *predictor variables* and *measurement error* in our discussions of spatial dependence (Section 4.7) and uncertainty (Section 6.3). More fundamentally still, is it acceptable (predictively and ethically) to represent the behavior of consumers using any measurable socio-economic variables?

Principles

The definition of the primary market area that is to receive incentives assumes that a *linear* radial distance measure is intrinsically meaningful in terms of defining market area. In practice, there are a number of severe shortcomings in this. The simplest is that *spatial structure* will distort the radial measure of market area – the market is likely to extend further along the more important travel arteries, for example, and will be restricted by physical obstacles such as blocked-off streets and rivers, and by traffic management devices such as stop lights. Impediments to access may be perceived as well as real – it may be that residents of West Parley (Figure 2.12B) would never think of going into north Bournemouth to shop, for example, and that the store's customers will remain overwhelmingly drawn from the area south of the store. Such perceptions of *psychological distance* can be very important yet difficult to accommodate in representations. We return to the issue of appropriate distance metrics in Sections 4.6 and 14.3.1.

Techniques

The assignment of unit postcode coordinates to the catchment zone is performed through a procedure known as *point in polygon* analysis, which is considered in our discussion of *transformation* (Section 14.4.2).

The analysis as described here assumes that the principles that underpin consumer behavior in Bournemouth, UK, are essentially the same as those operating anywhere else on Planet Earth. There is no attempt to accommodate regional and local factors. These might include: adjusting the attenuating effect of distance (see above) to accommodate the different distances people are prepared to travel to find a convenience store (e.g., as between an urban and a rural area); and adjusting the likely attractiveness of the outlet to take account of ease of access, forecourt size, or a range of qualitative factors such as layout or branding. A range of spatial techniques is now available for making the general properties of spatial analysis more *sensitive to context*.

Analysis

Our stripped down account of the store location problem has not considered the *competition* from the other stores shown in Figure 2.12B – despite that fact that all residents almost certainly already purchase convenience goods somewhere! Our description does, however, address the phenomenon of *cannibalizing* – whereby new outlets of a chain poach customers from its existing sites. In practice, both of these issues may be addressed through analysis of the *spatial interactions* between stores. Although this is beyond the scope of this book, Mark Birkin and colleagues have described how the tradition of *spatial interaction modeling* is ideally suited to the problems of defining realistic catchment areas and estimating store revenues. A range of analytic solutions can be devised in order to accommodate the fact that store catchments often overlap.

2.3.3.5 Generic scientific questions arising from the application

A dynamic retail sector is fundamental to the functioning of all advanced economies, and many investments in location are so huge that they cannot possibly be left to chance. Doing nothing is simply not an option. Intuition tells us that the effects of distance to outlet, and the organization of existing outlets in the retail hierarchy *must* have some kind of impact upon patterns of store patronage. But, in intensely competitive consumer-led markets, the important question is *how much* impact?

Human decision making is complex, but predicting even a small part of it can be very important to a retailer.

Consumers are sophisticated beings and their shopping behavior is often complex. Understanding local patterns of convenience shopping is perhaps quite straightforward, when compared with other retail decisions that involve stores that have a wider range of attributes, in terms of floor space, range and quality of goods and services, price, and customer services offered. Different consumer groups find different retailer attributes attractive, and hence it is the mix of *individuals* with particular characteristics that largely determines the likely store turnover of a particular location. Our example illustrates the kinds of simplifying assumptions that we may choose to make using the best available data in order to represent consumer characteristics and store attributes. However, it is important to remember that even blunt-edged tools can increase the effectiveness of operational and strategic R&D (research and development) activities many-fold. An untargeted leafleting campaign might typically achieve a 1% hit rate, while one informed by even quite rudimentary market area analysis might conceivably achieve a rate that is five times higher. The pessimist might dwell on the 95% failure rate that a supposedly scientific approach entails, yet the optimist should be more than happy with the fivefold increase in the efficiency of use of the marketing budget!

2.3.3.6 Management and policy

The geographic development of retail and business organizations has sometimes taken place in a haphazard way. However, the competitive pressures of today's markets require an understanding of branch location networks, as well as their abilities to anticipate and respond to threats from new entrants. The role of Internet technologies in the development of 'e-tailing' is important too, and these introduce further spatial problems to retailing – for example, in developing an understanding of the geographies of engagement with new information and communications technologies and in working out the logistics of delivering goods and services ordered in cyberspace to the geographic locations of customers (see Section 2.3.4).

Thus the role of the Spatially Aware Professional is increasingly as a mainstream manager alongside accountants, lawyers, and general business managers. SAPs complement understanding of corporate performance in national and international markets with performance at the regional and local levels. They have key roles in such areas of organizational activity as marketing, store revenue predictions, new product launch, improving retail networks, and the assimilation of pre-existing components into combined store networks following mergers and acquisitions.

Spatially Aware Professionals do much more than simple mapping of data.

Simple mapping packages alone provide insufficient scientific grounding to resolve retail location problems. Thus a range of *GIServices* have been developed – some in house, by large retail corporations (such as Tesco, above), some by software vendors that provide analytical and data services to retailers, and some by specialist consultancy services. There is ongoing debate as to which of these solutions is most appropriate to retail applications. The resolution of this debate lies in understanding the nature of particular organizations, their range of goods and services, and the priority that organizations assign to operational, tactical, and strategic concerns.

2.3.4 Logistics and transportation

2.3.4.1 Applications overview

Knowing where things are can be of enormous importance for the fields of logistics and transportation, which deal with the movement of goods and people from one place to another, and the infrastructure (highways, railroads, canals) that moves them. Highway authorities need to decide what new routes are needed and where to build them, and later need to keep track of highway condition. Logistics companies (e.g., parcel delivery companies, shipping companies) need to organize their operations, deciding where to place their central sorting warehouses and the facilities that transfer goods from one mode to another (e.g., from truck to ship), how to route parcels from origins to destinations, and how to route delivery trucks. Transit authorities need to plan routes and schedules, to keep track of vehicles and to deal with incidents that delay them, and to provide information on the system to the traveling public. All of these fields employ GIS, in a mixture of operational, tactical, and strategic applications.

The field of logistics addresses the shipping and transportation of goods.

Each of these applications has two parts: the static part that deals with the fixed infrastructure, and the dynamic part that deals with the vehicles, goods, and people that move on the static part. Of course, not even a highway network is truly static, since highways are often rebuilt, new highways are added, and highways are even sometimes moved. But the minute-to-minute timescale of vehicle movement is sharply different from the year-to-year changes in the infrastructure. Historically, GIS has been easier to apply to the static part, but recent developments in the technology are making it much more powerful as a tool to address the dynamic part as well. Today, it is possible to use GPS (Section 5.8) to track vehicles as they move around, and transit authorities increasingly use such systems to inform their users of the locations of buses and trains (Section 11.3.2 and Box 13.4).

GPS is also finding applications in dealing with emergency incidents that occur on the transportation network (Figure 2.13). The OnStar system (**www.onstar.com**) is one of several products that make use of the ability of GPS to determine location accurately virtually anywhere. When installed in a vehicle, the system is programmed to transmit location automatically to a central office whenever the vehicle is involved in an accident and its airbags deploy. This can be life-saving if the occupants of the vehicle do not know where they are, or are otherwise unable to call for help.

Many applications in transportation and logistics involve optimization, or the design of solutions to meet specified objectives. Section 15.3 discusses this type of analysis in detail, and includes several examples dealing with transportation and logistics. For example, a delivery company may need to deliver parcels to 200 locations in a given shift, dividing the work between 10 trucks. Different ways of dividing the work, and routing the

Figure 2.14 Hurricane Frances approaching the coast of Florida, USA, September 3, 2004 (Courtesy US National Oceanic and Atmospheric Administration, NOAA)

Figure 2.13 Systems such as OnStar allow information on the location of an accident, determined by a GPS unit in the vehicle, to be sent to a central office and compared to a GIS database of highways and streets, to determine the incident location so that emergency teams can respond

vehicles, can result in substantial differences in time and cost, so it is important for the company to use the most efficient solution (see Box 15.4 for an example of the daily workload of an elevator repair company). Logistics and related applications of GIS have been known to save substantially over traditional, manual ways of determining routes.

> **GIS has helped many service and delivery companies to substantially reduce their operating costs in the field.**

2.3.4.2 Case study application: planning for emergency evacuation

Modern society is at risk from numerous types of disasters, including terrorist attacks, extreme weather events such as hurricanes, accidental spills of toxic chemicals resulting from truck collisions or train derailments, and earthquakes. In recent years several major events have required massive evacuation of civilian populations – for example, 800 000 people evacuated in Florida in advance of Hurricane Frances in 2004 (Figure 2.14).

In response to the threat of such events, most communities attempt to plan. But planning is made particularly difficult because the magnitude and location of the event can rarely be anticipated. Suppose, for example, that we attempt to develop a plan for dealing with a spill of a volatile toxic chemical resulting from a train derailment. It might make sense to plan for the worst case, for example the spillage of the entire contents of several cars loaded with chlorine gas. But the derailment might occur anywhere on the rail network, and the impact will depend on the strength and direction of the wind. Possible scenarios might involve people living within tens of kilometers of any point on the track network (see Section 14.4.1 for details of the buffer operation, which

would be used in such cases to determine areas lying within a specified distance). Locations can be anticipated for some disasters, such as those resulting from fire in buildings known to be storing toxic chemicals, but hurricanes and earthquakes can impact almost anywhere within large areas.

> **The magnitude and location of a disaster can rarely be anticipated.**

To illustrate the value of GIS in evacuation planning, we have chosen the work of Tom Cova, an academic expert on GIS in emergency management. Tom's early work was strongly motivated by the problems that occurred in the Oakland Hills fire of October 1991 in Northern California, USA, which destroyed approximately 1580 acres and over 2700 structures in the East Bay Hills. This became the most expensive fire disaster in Californian history (Figure 2.15), taking 25 lives and causing over \$1.68 billion in damages.

Cova has developed a planning tool that allows neighborhoods to rate the potential for problems associated with evacuation, and to develop plans accordingly. The tool

Figure 2.15 The Oakland Hills fire of October, 1991, which took 25 lives, in part because of the difficulty of evacuation

uses a GIS database containing information on the distribution of population in the neighborhood, and the street pattern. The result is an evacuation vulnerability map. Because the magnitude of a disaster cannot be known in advance, the method works by identifying the worst-case scenario that could affect a given location.

Suppose a specific household is threatened by an event that requires evacuation, such as a wildfire, and assume for the moment that one vehicle is needed to evacuate each household. If the house is in a cul-de-sac, the number of vehicles needing to exit the cul-de-sac will be equal to the number of households on the street. If the entire neighborhood of streets has only one exit, all vehicles carrying people from the neighborhood will need to use that one exit. Cova's method works by looking further and further from the household location, to find the most important bottleneck – the one that has to handle the largest amount of traffic. In an area with a dense network of streets traffic will disperse among several exits, reducing the bottleneck effect. But a densely packed neighborhood with only a single exit can be the source of massive evacuation problems, if a disaster requires the rapid evacuation of the entire neighborhood. In the Oakland Hills fire there were several critical bottlenecks – one-lane streets that normally carry traffic in both directions, but became hopelessly clogged in the emergency.

Figure 2.16 shows a map of Santa Barbara, California, USA, with streets colored according to Cova's measure of evacuation vulnerability. The color assigned to any location indicates the number of vehicles that would have to pass through the critical bottleneck in the worst-case evacuation, with red indicating that over 500 vehicles per lane would have to pass through the bottleneck. The red area near the shore in the lower left is a densely packed area of student housing, with very few routes out of the neighborhood. An evacuation of the entire neighborhood would produce a very heavy flow of vehicles on these exit routes. The red area in the upper left has a much lower population density, but has only one narrow exit.

2.3.4.3 Method

Two types of data are required for the analysis. Census data are used to determine population and household counts, and to estimate the number of vehicles involved in an evacuation. Census data are available as aggregate counts for areas of a few city blocks, but not for individual houses, so there will be some uncertainty regarding the exact numbers of vehicles needing to leave a specific street, though estimates for entire neighborhoods will be much more accurate. The locations of streets are obtained from so-called *street centerline* files, which give the geographic locations, names, and other details of individual streets (see Sections 9.4 and 10.8 for overviews of geographic data sources). The TIGER (Topologically Integrated Geographic Encoding and Referencing) files, produced by the US Bureau of the Census and the US Geological Survey and readily available from many sites on the Internet, are one free source of such data for the USA, and many private companies also offer such data, many adding new information such as traffic flow volumes or directions (for US sources, see, for example, GDT Inc., Lebanon, New Hampshire, now part of Tele Atlas, **www.geographic.com**; and NAVTEQ, formerly Navigation Technologies, Chicago, Illinois, **www.navteq.com**).

> Street centerline files are essential for many applications in transportation and logistics.

The analysis proceeds by beginning at every street intersection, and working outwards following the street connections to reach new intersections. Every connection is tested to see if it presents a bottleneck, by dividing the total number of vehicles that would have to move out

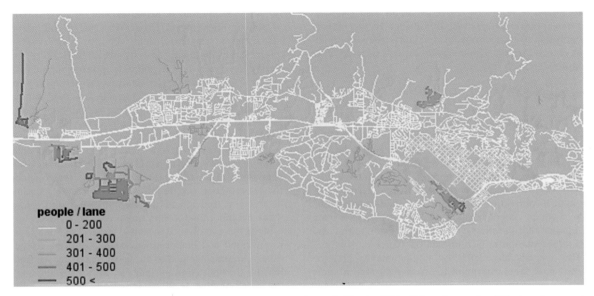

people / lane
— 0 - 200
— 201 - 300
— 301 - 400
— 401 - 500
— 500 <

Figure 2.16 Evacuation vulnerability map of the area of Santa Barbara, California, USA. Colors denote the difficulty of evacuating an area based on the area's worst-case scenario (Reproduced by permission of Tom Cova)

of the neighborhood by the number of exit lanes. After all streets have been searched out to a specified distance from the start, the worst-case value (vehicles per lane) is assigned to the starting intersection. Finally, the entire network is colored by the worst-case value.

2.3.4.4 Scientific foundations: geographic principles, techniques, and analysis

Scientific foundations

Cova's example is one of many applications that have been found for GIS in the general areas of logistics and transportation. As a planning tool, it provides a way of rating areas against a highly uncertain form of risk, a major evacuation. Although the worst-case scenario that might affect an area may never occur, the tool nevertheless provides very useful information to planners who design neighborhoods, giving them graphic evidence of the problems that can be caused by lack of foresight in street layout. Ironically, the approach points to a major problem with the modern style of street layout in subdivisions, which limits the number of entrances to subdivisions from major streets in the interests of creating a sense of community, and of limiting high-speed through traffic. Cova's analysis shows that such limited entrances can also be bottlenecks in major evacuations.

The analysis demonstrates the value of readily available sources of geographic data, since both major inputs – demographics and street layout – are available in digital form. At the same time we should note the limitations of using such sources. Census data are aggregated to areas that, while small, nevertheless provide only aggregated counts of population. The street layouts of TIGER and other sources can be out of date and inaccurate, particularly in new developments, although users willing to pay higher prices can often obtain current data from the private sector. And the essentially geometric approach cannot deal with many social issues: evacuation of the disabled and elderly, and issues of culture and language that may impede evacuation. In Chapter 16 we look at this problem using the tools of dynamic simulation modeling, which are much more powerful and provide ways of addressing such issues.

Principles

Central to Cova's analysis is the concept of *connectivity*. Very little would change in the analysis if the input maps were stretched or distorted, because what matters is how the network of streets is connected to the rest of the world. Connectivity is an instance of a *topological* property, a property that remains constant when the spatial framework is stretched or distorted. Other examples of topological properties are *adjacency* and *intersection*, both of which cannot be destroyed by stretching a map. We discuss the importance of topological properties and their representation in GIS in Section 10.7.1.

The analysis also relies on being able to find the *shortest path* from one point to another through a street network, and it assumes that people will follow such paths when they evacuate. Many forms of GIS analysis rely on being able to find shortest paths, and we discuss some

of them in Section 15.3.3. Many WWW sites will find shortest paths between two street addresses (Figure 1.17). In practice, people will often not use the shortest path, preferring routes that may be quicker but longer, or routes that are more scenic.

Techniques

The techniques used in this example are widely available in GIS. They include *spatial interpolation* techniques, which are needed to assign worst-case values to the streets, since the analysis only produces values for the intersections. Spatial interpolation is widely applied in GIS to use information obtained at a limited number of sample points to guess values at other points, and is discussed in general in Box 4.3, and in detail in Section 14.4.4.

The shortest path methods used to route traffic are also widely available in GIS, along with other functions needed to create, manage, and visualize information about networks.

Analysis

Cova's technique is an excellent example of the use of GIS analysis to *make visible what is otherwise invisible*. By processing the data and mapping the results in ways that would be impossible by hand, he succeeds in exposing areas that are difficult to evacuate and draws attention to potential problems. This idea is so central to GIS that it has sometimes been claimed as the primary purpose of the technology, though that seems a little strong, and ignores many of the other applications discussed in this chapter.

2.3.4.5 Generic scientific questions arising from the application

Logistic and transportation applications of GIS rely heavily on representations of networks, and often must ignore off-network movement. Drivers who cut through parking lots, children who cross fields on their way to school, houses in developments that are not aligned along linear streets, and pedestrians in underground shopping malls all confound the network-based analysis that GIS makes possible. Humans are endlessly adaptable, and their behavior will often confound the simplifying assumptions that are inherent to a GIS model. For example, suppose a system is developed to warn drivers of congestion on freeways, and to recommend alternative routes on neighborhood streets. While many drivers might follow such recommendations, others will reason that the result could be severe congestion on neighborhood streets, and reduced congestion on the freeway, and ignore the recommendation. Residents of the neighborhood streets might also be tempted to try to block the use of such systems, arguing that they result in unwanted and inappropriate traffic, and risk to themselves. Arguments such as these are based on the notion that the transportation system can only be addressed as a whole, and that local modifications based on limited perspectives, such as the addition of a new freeway or bypass, may create more problems than they solve.

2.3.4.6 Management and policy

GIS is used in all three modes – operational, tactical, and strategic – in logistics and transportation. This section concludes with some examples in all three categories.

In *operational* systems, GIS is used:

- To monitor the movement of mass transit vehicles, in order to improve performance and to provide improved information to system users.
- To route and schedule delivery and service vehicles on a daily basis to improve efficiency and reduce costs.

In *tactical* systems:

- To design and evaluate routes and schedules for public bus systems, school bus systems, garbage collection, and mail collection and delivery.
- To monitor and inventory the condition of highway pavement, railroad track, and highway signage, and to analyze traffic accidents.

In *strategic* systems:

- To plan locations for new highways and pipelines, and associated facilities.
- To select locations for warehouses, intermodal transfer points, and airline hubs.

2.3.5 Environment

2.3.5.1 Applications overview

Although it is the last area to be discussed here, the environment drove some of the earliest applications of GIS, and was a strong motivating force in the development of the very first GIS in the mid-1960s (Section 1.4.1). Environmental applications are the subject of several GIS texts, so only a brief overview will be given here for the purposes of illustration.

The development of the Canada Geographic Information System in the 1960s was driven by the need for policies over the use of land. Every country's land base is strictly limited (although the Dutch have managed to expand theirs very substantially by damming and draining), and alternative uses must compete for space. Measures of area are critical to effective strategy – for example, how much land is being lost to agriculture through urban development and sprawl, and how will this impact upon the ability of future generations to feed themselves? Today, we have very effective ways of monitoring land use change through remote sensing from space, and are able to get frequent updates on the loss of tropical forest in the Amazon basin (see Figure 15.14). GIS is also allowing us to devise measures of urban sprawl in historically separate national settlement systems in Europe (Figure 2.17).

GIS allows us to compare the environmental conditions prevailing in different nations.

Generally, it is understood that the 21st century will see increasing proportions of the world's population resident in cities and towns, and so understanding of the environmental impacts of urban settlements is an increasingly important focus of attention in science and policy. Researchers have used GIS to investigate and understand how urban sprawl occurs, in order to understand the environmental consequences of sprawl and to predict its future consequences. Such predictions can be based on historic patterns of growth, together with information on the locations of roads, steeply sloping land unsuitable for development, land that is otherwise protected from urban use, and other factors that encourage or restrict urban development. Each of these factors may be represented in map form, as a layer in the GIS, while specialist software can be designed to simulate the processes that drive growth. These urban growth models are examples of *dynamic simulation models*, or computer programs designed to simulate the operation of some part of the human or environmental system. Figure 2.18, taken from the work of geographer Paul Torrens, presents a simple simulation of urban growth in the American Mid-West under four rather different growth scenarios: (A) uncontrolled suburban sprawl; (B) growth restricted to existing travel arteries; (C) 'leap-frog' development, occurring because of local zoning controls; and (D) development that is constrained to some extent.

Other applications are concerned with the simulation of processes principally in the natural environment. Many models have been coupled with GIS in the past decade, to simulate such processes as soil erosion, forest growth, groundwater movement, and runoff. Dynamic simulation modeling is discussed in detail in Chapter 16.

2.3.5.2 Case study application: deforestation on Sibuyan Island, the Philippines

If the increasing extent of urban areas, described above, is one side of the development coin, then the reduction in the extent of natural land cover is frequently the other. Deforestation is one important manifestation of land use change, and poses a threat to the habitat of many species in tropical and temperate forest areas alike. Ecologists, environmentalists, and urban geographers are therefore using GIS in interdisciplinary investigations to understand the local conditions that lead to deforestation, and to understand its consequences. Important evidence of the rate and patterning of deforestation has been provided through analysis of remote sensing images (again, see Figure 15.14 for the case of the Amazon), and these analyses of *pattern* need to be complemented by analysis at detailed levels of the causes and underlying driving factors of the *processes* that lead to deforestation. The negative environmental impacts of deforestation can be ameliorated by adequate spatial planning of natural parks and land development schemes. But the more strategic objective of sustainable development can only be achieved if a holistic approach is taken to ecological, social, and economic needs. GIS provides the medium of choice for integrating knowledge of natural and social processes in the interests of integrated environmental planning.

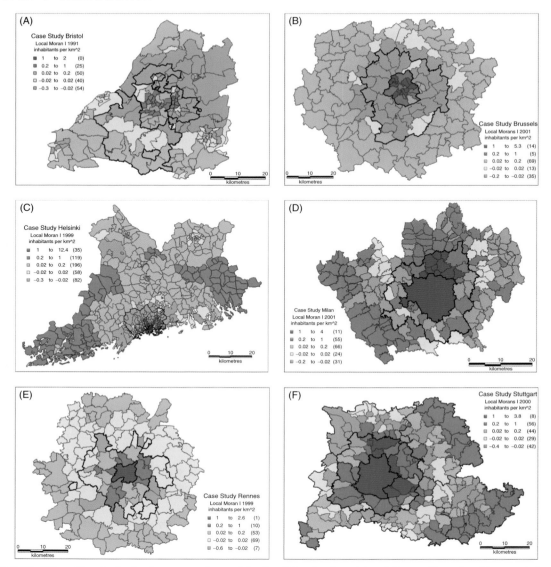

Figure 2.17 GIS enables standardized measures of sprawl for the different nation states of Europe (Reproduced by permission of Guenther Haag & Elena Besussi)

Working at the University of Wageningen in the Netherlands, Peter Verburg and Tom Veldkamp coordinate a research program that is using GIS to understand the sometimes complex interactions that exist between socio-economic and environmental systems, and to gauge their impact upon land use change in a range of different regions of the world (see **www.cluemodel.nl**). One of their case study areas is Sibuyan Island in the Philippines (Figure 2.19A), where deforestation poses a major threat to biodiversity. Sibuyan is a small island (area 456 km^2) of steep forested mountain slopes (Figure 2.19B) and gently sloping coastal land that is used mainly for agriculture, mining, and human settlement. The island has remarkable biodiversity – there are an estimated 700 plant species, of which 54 occur only on Sibuyan Island, and a unique local fauna. The objective of this case study was to identify a range of different development *scenarios* that

make it possible to anticipate future land use and habitat change, and hence also anticipate changes in biodiversity.

2.3.5.3 Method

The initial stage of Verburg and Veldkamp's research was a *qualitative* investigation, involving interviews with different stakeholders on the island to identify a list of factors that are likely to influence land use patterns. Table 2.2 lists the data that provided direct or indirect indicators of pressure for land use change. For example, the suitability of the soil for agriculture or the accessibility of a location to local markets can increase the likelihood that a location will be stripped of forest and used for agriculture. They then used these data in a *quantitative* GIS-based analysis to calculate the probabilities of land use transition under three different

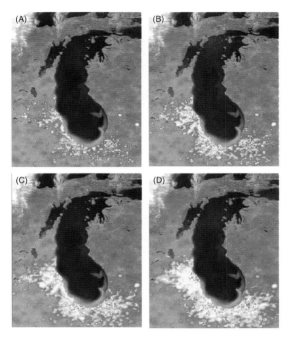

Figure 2.18 Growth in the American Mid-West under four different urban growth scenarios. Horizontal extent of image is 400 km (*Source*: P. Torrens 2005 'Simulating sprawl with geographic automata models', reproduced with permission of Paul Torrens)

scenarios of land use change – each of which was based on different spatial planning policies. Scenario 1 assumes no effective protection of the forests on the island (and a consequent piecemeal pattern of illegal logging), Scenario 2 assumes protection of the designated natural park area alone, and Scenario 3 assumes protection not only of the natural park but also a GIS-defined buffer zone. Figure 2.20 illustrates the forecasted remaining forest area under each of the scenarios at the end of a twenty-year simulation period (1999–2019). The three different scenarios not only resulted in different forest *areas* by 2019 but also different *spatial patterning* of the remaining forest. For example, gaps in the forest area under Scenario 1 were mainly caused by shifting cultivation and illegal logging within the area of primary rainforest, while most deforestation under Scenario 2 occurred in the lowland areas. Qualitative interpretations of the outcomes and aggregate statistics are supplemented by numerical spatial indices such as *fractal dimensions* in order to anticipate the effects of changes upon ecological processes – particularly the effects of disturbance at the edges of the remaining forest area. Such statistics make it possible to define the relative sizes of core and fragmented forest areas (for example, Scenario 1 in Figure 2.20 leads to the greatest fragmentation of the forest area), and this in turn makes it possible to measure the effects of development on biodiversity. Fragmentation statistics are discussed in Section 15.2.5.

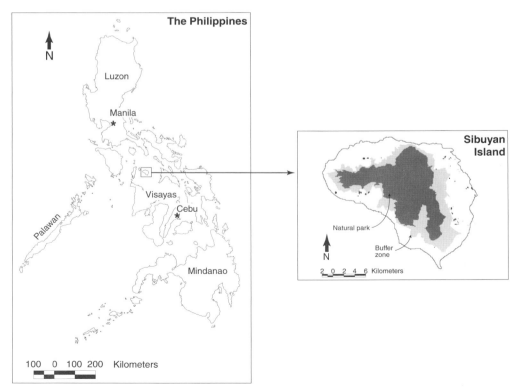

Figure 2.19 (A) Location of Sibuyan Island in the Philippines, showing location of the park and buffer zone, and (B) typical forested mountain landscape of Sibuyan Island

Figure 2.19 (*continued*)

Table 2.2 Data sources used in Verburg and Veldkamp's ecological analysis

Land use
Mangroves
Coconut plantations
Wetland rice cultivation
Grassland
Secondary forest
Swidden agriculture
Primary rainforest and mossy forest
Location factors
Accessibility of roads, rivers and populated places
Altitude
Slope
Aspect
Geology
Geomorphology
Population density
Population pressure
Land tenure
Spatial policies

2.3.5.4 Scientific foundations: geographic principles, techniques, and analysis

Scientific foundations

The goal of the research was to predict changes in the biodiversity of the island. Existing knowledge of a set of ecological *processes* led the researchers to the view that biodiversity would be compromised by changes in overall size of natural forest, and by changes in the *patterning* of the forest that remained. It was hypothesized that changes in patterning would be *caused* by the combined effects of a further set of physical, biological, and human *processes*. Thus the researchers take the observed existing land use pattern, and use understanding of the physical, biological, and human processes to predict future land use changes. The different forecasts of land use (based on different

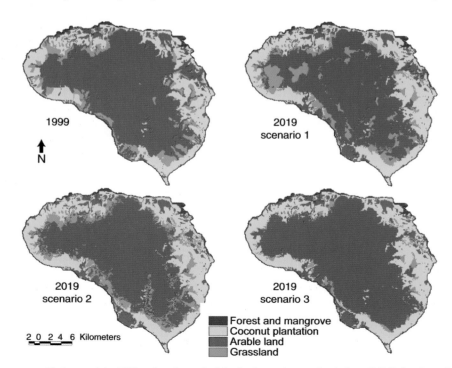

Figure 2.20 Forest area (dark green) in 1999 and at the end of the land use change simulations (2019) for three different scenarios

scenarios) can then be used see whether the functioning of ecological processes on the island will be changed in the future. This theme of inferring *process* from *pattern*, or *function* from *form*, is a common characteristic of GIScience applications.

Of course, it is not possible to identify a uniform set of physical, biological, and human processes that is valid in all regions of the world (the pure *nomothetic* approach of Section 1.3). But conversely it does not make sense to treat every location as unique (the *idiographic* approach of Section 1.3) in terms of the processes extant upon it. The art and science of ecological modeling requires us to make a good call not just on the range of relevant determining factors, but also their importance in the specific case study, with due consideration to the appropriate scale range at which each is relevant.

Geographic principles

GIS makes it possible to incorporate diverse physical, biological, and human elements, and to forecast the size, shape, scale, and dimension of land use parcels. Therefore, it is possible in this case study to predict habitat fragmentation and changes in biodiversity. Fundamental to the end use of this analysis is the assumption that the ecological consequences of future deforestation can be reliably predicted using a forecast land use map. The forecasting procedure also assumes that the various indicators of land development pressure are *robust, accurate*, and *reliable*. There are inevitable uncertainties in the ways in which these indicators are conceived and measured. Further uncertainties are generated by the scale of analysis that is carried out (Section 6.4), and, like our retailing example above, the qualitative importance of local context may be important.

Land use change is deemed to be a measurable response to a wide range of locationally variable factors. These factors have traditionally been the remit of different disciplines that have different intellectual traditions of measurement and analysis. As Peter Verburg says: 'The research assumes that GIS can provide a sort of "Geographic Esperanto" – that is, a common language to integrate diverse, geographically variable factors. It makes use of the core GIS idea that the world can be understood as a series of layers of different types of information, that can be added together meaningfully through overlay analysis to arrive at conclusions.'

Techniques

The *multicriteria* techniques used to harmonize the different location factors into a composite spatial indicator of development pressure are widely available in GIS, and are discussed in detail in Section 16.4. The individual component indicators are acquired using techniques such as *on-screen digitizing* and *classification of imagery* to obtain a land use map. All data are converted to a *raster* data structure with common resolution and extent. Relations between the location factors and land use are quantified using *correlation* and *regression* analysis based on the spatial dataset (Section 4.7).

Analysis

The GIS is used to simulate scenarios of future land use change based on different spatial policies. The application is predicated on the premise that changes in ecological *process* can be reliably inferred from predicted changes in land use *pattern*. Process is inferred not just through size measures, but also through spatial measures of connectivity and fragmentation – since these latter aspects affect the ability of a species to mix and breed without disturbance. The analysis of the extent and ways in which different land uses fill space is performed using specialized software (see Section 15.2.5). Although such spatial indices are useful tools, they may be more relevant to some aspects of biodiversity than others, since different species are vulnerable to different aspects of habitat change.

2.3.5.5 Generic scientific questions arising from the application

GIS applications need to be based on sound science. In environmental applications, this knowledge base is unlikely to be the preserve of any single academic discipline. Many environmental applications require recourse to use of GIS in the field, and field researchers often require multidisciplinary understanding of the full range of processes leading to land use change.

Irrespective of the quality of the measurement process, uncertainty will always creep into any prediction, for a number of reasons. Data are never perfect, being subject to measurement error (Chapter 6), and uncertainty arising out of the need selectively to generalize, abstract, and approximate (Chapter 3). Furthermore, simulations of land use change are subject to changes in exogenous forces such as the world economy. Any forecast can only be a selectively simplified representation of the real world and the processes operating within it. GIS users need to be aware of this, because the forecasts produced by a GIS will always appear to be precise in numerical terms, and spatial representations will usually be displayed using crisp lines and clear mapped colors.

GIS users should not think of systems as *black boxes*, and should be aware that explicit spatial forecasts may have been generated by invoking assumptions about process and data that are not as explicit. User awareness of these important issues can be improved through appropriate *metadata* and documentation of research procedures – particularly when interdisciplinary teams may be unaware of the disciplinary conventions that govern data creation and analysis in parts of the research. Interdisciplinary science and the cumulative development of algorithms and statistical procedures can lead GIS applications to conflict with an older principle of scientific reporting, that the results of analysis *should always be reported in sufficient detail to allow someone else to replicate them*. Today's science is complex, and all of us from time to time may find ourselves using tools developed by others that we do not fully understand. It is up to all of us to demand to know as many of the details of GIS analysis as is reasonably possible.

Users of GIS should always know exactly what the system is doing to their data.

2.3.5.6 Management and policy

GIS is now widely used in all areas of environmental science, from ecology to geology, and from oceanography to alpine geomorphology. GIS is also helping to reinvent environmental science as a discipline grounded in field observation, as data can be captured using battery-powered personal data assistants (PDAs) and notebooks, before being analyzed on a battery-powered laptop in a field tent, and then uploaded via a satellite link to a home institution. The art of scientific forecasting (by no means a contradiction in terms) is developing in a cumulative way, as interdisciplinary teams collaborate in the development and sharing of applications that range in sophistication from simple composite mapping projects to intensive numerical and statistical simulation experiments.

2.4 Concluding comments

This chapter has presented a selection of GIS application areas and specific instances within each of the selected areas. Throughout, the emphasis has been on the range of contexts, from day-to-day problem solving to curiosity-driven science. The principles of the scientific method have been stressed throughout – the need to maintain an enquiring mind, constantly asking questions about what is going on, and what it means; the need to use terms that are well-defined and understood by others, so that knowledge can be communicated; the need to describe procedures in sufficient detail so that they can be replicated by others; and the need for accuracy, in observations, measurements, and predictions. These principles are valid whether the context is a simple inventory of the assets of a utility company, or the simulation of complex biological systems.

Questions for further study

1. Devise a diary for your own activity patterns for a typical (or a special) day, like that described in Section 2.1.1, and speculate how GIS might affect your own daily activities. What activities are not influenced by GIS, and how might its use in some of these contexts improve your daily quality of life?

2. Compare and contrast the operational, tactical, and strategic priorities of the GIS specialists responsible for the specific applications described in Sections 2.3.2, 2.3.3, 2.3.4 and 2.3.5.

3. Look at one of the applications chapters in the CD in the Longley et al (2005) volume in the references below. To what extent do you believe that the author of your chapter has demonstrated that GIS has been 'successful' in application? Suggest some of the implicit and explicit assumptions that are made in order to achieve a 'successful' outcome.

4. Look at one of the applications areas in the CD in the Longley et al (2005) volume in the references below. Then re-examine the list of critiques of GIS at the end of Section 1.7. To what extent do you think that the critiques are relevant to the applications that you have studied?

Further reading

Birkin M., Clarke G.P., and Clarke M. 2002 *Retail Geography and Intelligent Network Planning*. Chichester, UK: Wiley.

Chainey S., and Ratcliffe J. 2005 *GIS and Crime Mapping*. Chichester, UK: Wiley.

Greene R.W. 2000 *GIS in Public Policy*. Redlands, CA: ESRI Press.

Harris R., Sleight P., and Webber R. 2005 *Geodemographics, GIS and Neighbourhood Targeting*. Chichester, UK: Wiley.

Johnston C.A. 1998 *Geographic Information Systems in Ecology*. Oxford: Blackwell.

Longley P.A., Goodchild M.F., Maguire D.J. and Rhind D.W. (eds) 2005 *Geographical Information Systems: Principles; Techniques; Management and Applications (abridged edition)*. Hoboken, N.J.: Wiley.

O'Looney J. 2000 *Beyond Maps: GIS and Decision Making in Local Government*. Redlands, CA: ESRI Press.

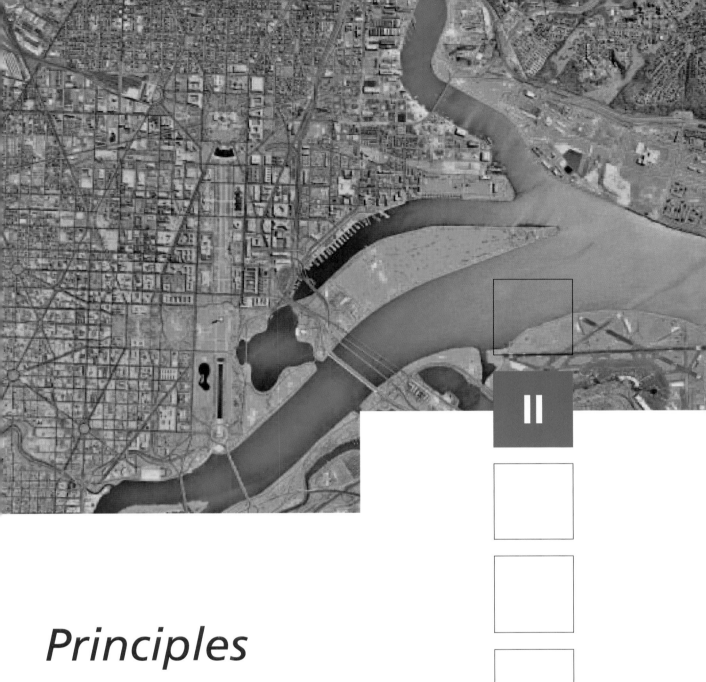

Principles

3 *Representing geography*

This chapter introduces the concept of representation, or the construction of a digital model of some aspect of the Earth's surface. Representations have many uses, because they allow us to learn, think, and reason about places and times that are outside our immediate experience. This is the basis of scientific research, planning, and many forms of day-to-day problem solving.

The geographic world is extremely complex, revealing more detail the closer one looks, almost *ad infinitum*. So in order to build a representation of any part of it, it is necessary to make choices, about what to represent, at what level of detail, and over what time period. The large number of possible choices creates many opportunities for designers of GIS software. Generalization methods are used to remove detail that is unnecessary for an application, in order to reduce data volume and speed up operations.

Geographic Information Systems and Science, 2nd edition   Paul Longley, Michael Goodchild, David Maguire, and David Rhind.
© 2005 John Wiley & Sons, Ltd.   ISBNs: 0-470-87000-1 (HB); 0-470-87001-X (PB)

Learning Objectives

After reading this chapter you will know:

■ The importance of understanding representation in GIS;

■ The concepts of fields and objects and their fundamental significance;

■ Raster and vector representation and how they affect many GIS principles, techniques, and applications;

■ The paper map and its role as a GIS product and data source;

■ The importance of generalization methods and the concept of representational scale;

■ The art and science of representing real-world phenomena in GIS.

3.1 Introduction

We live on the surface of the Earth, and spend most of our lives in a relatively small fraction of that space. Of the approximately 500 million square kilometers of surface, only one third is land, and only a fraction of that is occupied by the cities and towns in which most of us live. The rest of the Earth, including the parts we never visit, the atmosphere, and the solid ground under our feet, remains unknown to us except through the information that is communicated to us through books, newspapers, television, the Web, or the spoken word. We live lives that are almost infinitesimal in comparison with the 4.5 billion years of Earth history, or the over 10 billion years since the universe began, and know about the Earth before we were born only through the evidence compiled by geologists, archaeologists, historians, etc. Similarly, we know nothing about the world that is to come, where we have only predictions to guide us.

Because we can observe so little of the Earth directly, we rely on a host of methods for learning about its other parts, for deciding where to go as tourists or shoppers, choosing where to live, running the operations of corporations, agencies, and governments, and many other activities. Almost all human activities at some time require knowledge (Section 1.2) about parts of the Earth that are outside our direct experience, because they occur either elsewhere in space, or elsewhere in time.

Sometimes this knowledge is used as a *substitute* for directly sensed information, creating a *virtual* reality (see Section 11.3.1). Increasingly it is used to *augment* what we can see, touch, hear, feel, and smell, through the use of mobile information systems that can be carried around.

Our knowledge of the Earth is not created entirely freely, but must fit with the mental concepts we began to develop as young children – concepts such as containment (Paris is *in* France) or proximity (Dallas and Fort Worth are *close*). In digital representations, we formalize these concepts through *data models* (Chapter 8), the structures and rules that are programmed into a GIS to accommodate data. These concepts and data models together constitute our *ontologies*, the frameworks that we use for acquiring knowledge of the world.

Almost all human activities require knowledge about the Earth – past, present, or future.

One such ontology, a way to structure knowledge of movement through time, is a three-dimensional diagram, in which the two horizontal axes denote location on the Earth's surface, and the vertical axis denotes time. In Figure 3.1, the daily lives of a sample of residents of Lexington, Kentucky, USA are shown as they move by car through space and time, from one location to another, while going about their daily business of shopping, traveling to work, or dropping children at school. The diagram is crude, because each journey is represented by a series of straight lines between locations measured with GPS, and if we were able to examine each track or trajectory in more detail we would see the effects of having to follow streets, stopping at traffic lights, or slowing for congestion. If we looked even closer we might see details of each person's walk to and from the car. Each closer perspective would display more information, and a vast storehouse would be required to capture the precise trajectories of all humans throughout even a single day.

The real trajectories of the individuals shown in Figure 3.1 are complex, and the figure is only a representation of them – a model on a piece of paper, generated by a computer from a database. We use the terms *representation* and *model* because they imply a simplified relationship between the contents of the figure and the database, and the real-world trajectories of the individuals. Such representations or models serve many useful purposes, and occur in many different forms. For example, representations occur:

■ in the human mind, when our senses capture information about our surroundings, such as the images captured by the eye, or the sounds captured by the ear, and memory preserves such representations for future use;

■ in photographs, which are two-dimensional models of the light emitted or reflected by objects in the world into the lens of a camera;

■ in spoken descriptions and written text, in which people describe some aspect of the world in language, in the form of travel accounts or diaries; or

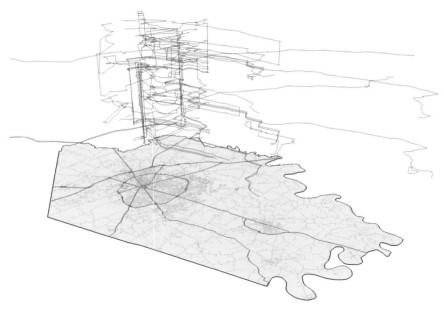

Figure 3.1 Schematic representation of the daily journeys of a sample of residents of Lexington, Kentucky, USA. The horizontal dimensions represent geographic space and the vertical dimension represents time of day. Each person's track plots as a three-dimensional line, beginning at the base in the morning and ending at the top in the evening. (Reproduced with permission of Mei-Po Kwan)

■ in the numbers that result when aspects of the world are measured, using such devices as thermometers, rulers, or speedometers.

By building representations, we humans can assemble far more knowledge about our planet than we ever could as individuals. We can build representations that serve such purposes as planning, resource management and conservation, travel, or the day-to-day operations of a parcel delivery service.

> **Representations help us assemble far more knowledge about the Earth than is possible on our own.**

Representations are reinforced by the rules and laws that we humans have learned to apply to the unobserved world around us. When we encounter a fallen log in a forest we are willing to assert that it once stood upright, and once grew from a small shoot, even though no one actually observed or reported either of these stages. We predict the future occurrence of eclipses based on the laws we have discovered about the motions of the Solar System. In GIS applications, we often rely on methods of spatial interpolation to guess the conditions that exist in places where no observations were made, based on the rule (often elevated to the status of a First Law of Geography and attributed to Waldo Tobler) that all places are similar, but nearby places are more similar than distant places.

> **Tobler's First Law of Geography: Everything is related to everything else, but near things are more related than distant things.**

3.2 Digital representation

This book is about one particular form of representation that is becoming increasingly important in our society – representation in digital form. Today, almost all communication between people through such media as the telephone, FAX, music, television, newspapers and magazines, or email is at some time in its life in digital form. Information technology based on digital representation is moving into all aspects of our lives, from science to commerce to daily existence. Almost half of all households in some industrial societies now own at least one powerful digital information processing device (a computer); a large proportion of all work in offices now occurs using digital computing technology; and digital technology has invaded many devices that we use every day, from the microwave oven to the automobile.

One interesting characteristic of digital technology is that the representation itself is rarely if ever seen by the user, because only a few technical experts ever see the individual elements of a digital representation. What we see instead are *views*, designed to present the contents of the representation in a form that is meaningful to us.

The term *digital* derives from *digits*, or the fingers, and our system of counting based on the ten digits of the human hand. But while the counting system has ten symbols (0 through 9), the representation system in digital computers uses only two (0 and 1). In a sense, then, the term *digital* is a misnomer for a system that represents all information using some combination of the two symbols 0 and 1, and the more exact term *binary* is more appropriate. In this book we follow the convention of using *digital* to refer to electronic technology based on binary representations.

Computers represent phenomena as binary digits. Every item of useful information about the Earth's surface is ultimately reduced by a GIS to some combination of 0s and 1s.

Over the years many standards have been developed for converting information into digital form. Box 3.1 shows the standards that are commonly used in GIS to store data, whether they consist of whole or decimal numbers or text. There are many competing coding standards for images and photographs (GIF, JPEG, TIFF, etc.) and for movies (e.g., MPEG) and sound (e.g., MIDI, MP3). Much of this book is about the coding systems used to represent geographic data, especially Chapter 8, and as you might guess that turns out to be comparatively complicated.

Digital technology is successful for many reasons, not the least of which is that all kinds of information share a common basic format (0s and 1s), and can be handled in ways that are largely independent of their actual meaning.

The Internet, for example, operates on the basis of packets of information, consisting of strings of 0s and 1s, which are sent through the network based on the information contained in the packet's header. The network needs to know only what the header means, and how to read the instructions it contains regarding the packet's destination. The rest of the contents are no more than a collection of bits, representing anything from an email message to a short burst of music or highly secret information on its way from one military installation to another, and are almost never examined or interpreted during transmission. This allows one digital communications network to serve every need, from electronic commerce to chatrooms, and it allows manufacturers to build processing and storage technology for vast numbers of users who have very different applications in mind. Compare this to earlier ways of communicating, which required printing presses and delivery trucks for one application (newspapers) and networks of copper wires for another (telephone).

Digital representations of geography hold enormous advantages over previous types – paper maps, written reports from explorers, or spoken accounts. We can use

Technical Box **3.1**

The binary counting system

The binary counting system uses only two symbols, 0 and 1, to represent numerical information. A group of eight binary digits is known as a *byte*, and volume of storage is normally measured in bytes rather than bits (Table 1.1). There are only two options for a single digit, but there are four possible combinations for two digits (00, 01, 10, and 11), eight possible combinations for three digits (000, 001, 010, 011, 100, 101, 110, 111), and 256 combinations for a full byte. Digits in the binary system (known as binary digits, or *bits*) behave like digits in the decimal system but using powers of two. The rightmost digit denotes units, the next digit to the left denotes twos, the next to the left denotes fours, etc. For example, the binary number 11001 denotes one unit, no twos, no fours, one eight, and one sixteen, and is equivalent to 25 in the normal (decimal) counting system. We call this the *integer* digital representation of 25, because it represents 25 as a whole number, and is readily amenable to arithmetic operations. Whole numbers are commonly stored in GIS using either *short* (2-byte or 16-bit) or *long* (4-byte or 32-bit) options. Short integers can range from −65535 to +65535, and long integers from −4294967295 to +4294967295.

The 8-bit ASCII (American Standard Code for Information Interchange) system assigns codes to each symbol of text, including letters, numbers, and common symbols. The number

2 is assigned ASCII code 48 (00110000 in binary), and the number 5 is 53 (00110101), so if 25 were coded as two characters using 8-bit ASCII its digital representation would be 16 bits long (0011000000110101). The characters 2 = 2 would be coded as 48, 61, 48 (001100000011110100110000). ASCII is used for coding text, which consists of mixtures of letters, numbers, and punctuation symbols.

Numbers with decimal places are coded using *real* or *floating-point* representations. A number such as 123.456 (three decimal places and six significant digits) is first transformed by powers of ten so that the decimal point is in a standard position, such as the beginning (e.g., 0.123456×10^3). The fractional part (0.123456) and the power of 10 (3) are then stored in separate sections of a block of either 4 bytes (32 bits, *single* precision) or 8 bytes (64 bits, *double* precision). This gives enough precision to store roughly 7 significant digits in single precision, or 14 in double precision.

Integer, ASCII, and real conventions are adequate for most data, but in some cases it is desirable to associate images or sounds with places in GIS, rather than text or numbers. To allow for this GIS designers have included a BLOB option (standing for binary large object), which simply allocates a sufficient number of bits to store the image or sound, without specifying what those bits might mean.

the same cheap digital devices – the components of PCs, the Internet, or mass storage devices – to handle every type of information, independent of its meaning. Digital data are easy to copy, they can be transmitted at close to the speed of light, they can be stored at high density in very small spaces, and they are less subject to the physical deterioration that affects paper and other physical media. Perhaps more importantly, data in digital form are easy to transform, process, and analyze. Geographic information systems allow us to do things with digital representations that we were never able to do with paper maps: to measure accurately and quickly, to overlay and combine, and to change scale, zoom, and pan without respect to map sheet boundaries. The vast array of possibilities for processing that digital representation opens up is reviewed in Chapters 14 through 16, and is also covered in the applications that are distributed throughout the book.

Digital representation has many uses because of its simplicity and low cost.

3.3 Representation for what and for whom?

Thus far we have seen how humans are able to build representations of the world around them, but we have not yet discussed why representations are useful, and why humans have become so ingenious at creating and sharing them. The emphasis here and throughout the book is on one type of representation, termed *geographic*, and defined as a representation of some part of the Earth's surface or near-surface, at scales ranging from the architectural to the global.

Geographic representation is concerned with the Earth's surface or near-surface.

Geographic representations are among the most ancient, having their roots in the needs of very early societies. The tasks of hunting and gathering can be much more efficient if hunters are able to communicate the details of their successes to other members of their group – the locations of edible roots or game, for example. Maps must have originated in the sketches early people made in the dirt of campgrounds or on cave walls, long before language became sufficiently sophisticated to convey equivalent information through speech. We know that the peoples of the Pacific built representations of the locations of islands, winds, and currents out of simple materials to guide each other, and that very simple forms of representation are used by social insects such as bees to communicate the locations of food resources.

Hand-drawn maps and speech are effective media for communication between members of a small group, but much wider communication became possible with the invention of the printing press in the 15th century. Now large numbers of copies of a representation could be made

and distributed, and for the first time it became possible to imagine that something could be known by every human being – that knowledge could be the common property of humanity. Only one major restriction affected what could be distributed using this new mechanism: the representation had to be flat. If one were willing to accept that constraint, however, paper proved to be enormously effective; it was cheap, light and thus easily transported, and durable. Only fire and water proved to be disastrous for paper, and human history is replete with instances of the loss of vital information through fire or flood, from the burning of the Alexandria Library in the 7th century that destroyed much of the accumulated knowledge of classical times to the major conflagrations of London in 1666, San Francisco in 1906, or Tokyo in 1945, and the flooding of the Arno that devastated Florence in 1966.

One of the most important periods for geographic representation began in the early 15th century in Portugal. Henry the Navigator (Box 3.2) is often credited with originating the Age of Discovery, the period of European history that led to the accumulation of large amounts of information about other parts of the world through sea voyages and land explorations. Maps became the medium for sharing information about new discoveries, and for administering vast colonial empires, and their value was quickly recognized. Although detailed representations now exist of all parts of the world, including Antarctica, in a sense the spirit of the Age of Discovery continues in the explorations of the oceans, caves, and outer space, and in the process of re-mapping that is needed to keep up with constant changes in the human and natural worlds.

It was the creation, dissemination, and sharing of accurate representations that distinguished the Age of Discovery from all previous periods in human history (and it would be unfair to ignore its distinctive negative consequences, notably the spread of European diseases and the growth of the slave trade). Information about other parts of the world was assembled in the form of maps and journals, reproduced in large numbers using the recently invented printing press, and distributed on paper. Even the modest costs associated with buying copies were eventually addressed through the development of free public lending libraries in the 19th century, which gave access to virtually everyone. Today, we benefit from what is now a longstanding tradition of free and open access to much of humanity's accumulated store of knowledge about the geographic world, in the form of paper-based representations, through the institution of libraries and the copyright doctrine that gives people rights to material for personal use (see Chapter 18 for a discussion of laws affecting ownership and access). The Internet has already become the delivery mechanism for providing distributed access to geographic information.

In the Age of Discovery maps became extremely valuable representations of the state of geographic knowledge.

It is not by accident that the list of important applications for geographic representations closely follows the list of applications of GIS (see Section 1.1 and Chapter 2),

Prince Henry the Navigator

Prince Henry of Portugal, who died in 1460, was known as Henry the Navigator because of his keen interest in exploration. In 1433 Prince Henry sent a ship from Portugal to explore the west coast of Africa in an attempt to find a sea route to the Spice Islands. This ship was the first to travel south of Cape Bojador (latitude 26 degrees 20 minutes N). To make this and other voyages Prince Henry assembled a team of map-makers, sea captains, geographers, ship builders, and many other skilled craftsmen. Prince Henry showed the way for Vasco da Gama and other famous 15th century explorers. His management skills could be applied in much the same way in today's GIS projects.

Figure 3.2 Prince Henry the Navigator, originator of the Age of Discovery in the 15th century, and promoter of a systematic approach to the acquisition, compilation, and dissemination of geographic knowledge

since representation is at the heart of our ability to solve problems using digital tools. Any application of GIS requires clear attention to questions of *what* should be represented, and *how*. There is a multitude of possible ways of representing the geographic world in digital form, none of which is perfect, and none of which is ideal for all applications.

> **The key GIS representation issues are what to represent and how to represent it.**

One of the most important criteria for the usefulness of a representation is its *accuracy*. Because the geographic world is seemingly of infinite complexity, there are always choices to be made in building any representation – what to include, and what to leave out. When US President Thomas Jefferson dispatched Meriwether Lewis to explore and report on the nature of the lands from the upper Missouri to the Pacific, he said Lewis possessed 'a fidelity to the truth so scrupulous that whatever he should report would be as certain as if seen by ourselves'. But he clearly didn't expect Lewis to report everything he saw in complete detail: Lewis exercised a large amount of judgment about what to report, and what to omit. The question of accuracy is taken up at length in Chapter 6.

One more vital interest drives our need for representations of the geographic world, and also the need for representations in many other human activities. When a pilot must train to fly a new type of aircraft, it is much cheaper and less risky for him or her to work with a flight simulator than with the real aircraft. Flight simulators can represent a much wider range of conditions than a pilot will normally experience in flying. Similarly,

when decisions have to be made about the geographic world, it is effective to experiment first on models or representations, exploring different scenarios. Of course this works only if the representation behaves as the real aircraft or world does, and a great deal of knowledge must be acquired about the world before an accurate representation can be built that permits such simulations. But the use of representations for training, exploring future scenarios, and recreating the past is now common in many fields, including surgery, chemistry, and engineering, and with technologies like GIS is becoming increasingly common in dealing with the geographic world.

> **Many plans for the real world can be tried out first on models or representations.**

3.4 The fundamental problem

Geographic data are built up from atomic elements, or facts about the geographic world. At its most primitive, an atom of geographic data (strictly, a datum) links a place, often a time, and some descriptive property. The first of these, place, is specified in one of several ways that are discussed at length in Chapter 5, and there are also many ways of specifying the second, time. We often use the term *attribute* to refer to the last of these three. For example, consider the statement 'The temperature at local noon on December 2nd 2004 at latitude 34 degrees

45 minutes north, longitude 120 degrees 0 minutes west, was 18 degrees Celsius'. It ties location and time to the property or attribute of atmospheric temperature.

Geographic data link place, time, and attributes.

Other facts can be broken down into their primitive atoms. For example, the statement 'Mount Everest is 8848 m high' can be derived from two atomic geographic facts, one giving the location of Mt Everest in latitude and longitude, and the other giving the elevation at that latitude and longitude. Note, however, that the statement would not be a geographic fact to a community that had no way of knowing where Mt Everest is located.

Many aspects of the Earth's surface are comparatively static and slow to change. Height above sea level changes slowly because of erosion and movements of the Earth's crust, but these processes operate on scales of hundreds or thousands of years, and for most applications except geophysics we can safely omit time from the representation of elevation. On the other hand atmospheric temperature changes daily, and dramatic changes sometimes occur in minutes with the passage of a cold front or thunderstorm, so time is distinctly important, though such climatic variables as mean annual temperature can be represented as static.

The range of attributes in geographic information is vast. We have already seen that some vary slowly and some rapidly. Some attributes are physical or environmental in nature, while others are social or economic. Some attributes simply *identify* a place or an entity, distinguishing it from all other places or entities – examples include street addresses, social security numbers, or the parcel numbers used for recording land ownership. Other attributes measure something at a location and perhaps at a time (e.g., atmospheric temperature or elevation), while others classify into categories (e.g., the class of land use, differentiating between agriculture, industry, or residential land). Because attributes are important outside the domain of GIS there are standard terms for the different types (see Box 3.3).

Geographic attributes are classified as nominal, ordinal, interval, ratio, and cyclic.

But this idea of recording atoms of geographic information, combining location, time, and attribute, misses a fundamental problem, which is that the world is in effect infinitely complex, and the number of atoms required for a complete representation is similarly infinite. The closer we look at the world, the more detail it reveals – and it seems that this process extends *ad infinitum*. The shoreline of Maine appears complex on a map, but even more complex when examined in greater detail, and as more detail is revealed the shoreline appears to get longer and longer,

---

Technical Box 3.3

Types of attributes

The simplest type of attribute, termed *nominal*, is one that serves only to identify or distinguish one entity from another. Placenames are a good example, as are names of houses, or the numbers on a driver's license – each serves only to identify the particular instance of a class of entities and to distinguish it from other members of the same class. Nominal attributes include numbers, letters, and even colors. Even though a nominal attribute can be numeric it makes no sense to apply arithmetic operations to it: adding two nominal attributes, such as two drivers' license numbers, creates nonsense.

Attributes are *ordinal* if their values have a natural order. For example, Canada rates its agricultural land by classes of soil quality, with Class 1 being the best, Class 2 not so good, etc. Adding or taking ratios of such numbers makes little sense, since 2 is not twice as much of anything as 1, but at least ordinal attributes have inherent order. Averaging makes no sense either, but the *median*, or the value such that half of the attributes are higher-ranked and half are lower-ranked, is an effective substitute for the average for ordinal data as it gives a useful central value.

Attributes are *interval* if the differences between values make sense. The scale of Celsius temperature is interval, because it makes sense to say that 30 and 20 are as different as 20 and 10. Attributes are *ratio* if the ratios between values make sense. Weight is ratio, because it makes sense to say that a person of 100 kg is twice as heavy as a person of 50 kg; but Celsius temperature is only interval, because 20 is not twice as hot as 10 (and this argument applies to all scales that are based on similarly arbitrary zero points, including longitude).

In GIS it is sometimes necessary to deal with data that fall into categories beyond these four. For example, data can be directional or *cyclic*, including flow direction on a map, or compass direction, or longitude, or month of the year. The special problem here is that the number following 359 degrees is 0. Averaging two directions such as 359 and 1 yields 180, so the average of two directions close to North can appear to be South. Because cyclic data occur sometimes in GIS, and few designers of GIS software have made special arrangements for them, it is important to be alert to the problems that may arise.

and more and more convoluted (see Figure 4.18). To characterize the world completely we would have to specify the location of every person, every blade of grass, and every grain of sand – in fact, every subatomic particle, clearly an impossible task, since the Heisenberg uncertainty principle places limits on the ability to measure precise positions of subatomic particles. So in practice any representation must be partial – it must limit the level of detail provided, or ignore change through time, or ignore certain attributes, or simplify in some other way.

The world is infinitely complex, but computer systems are finite. Representations must somehow limit the amount of detail captured.

One very common way of limiting detail is by throwing away or ignoring information that applies only to small areas, in other words not looking too closely. The image you see on a computer screen is composed of a million or so basic elements or *pixels*, and if the whole Earth were displayed at once each pixel would cover an area roughly 10 km on a side, or about 100 sq km. At this level of detail the island of Manhattan occupies roughly 10 pixels, and virtually everything on it is a blur. We would say that such an image has a *spatial resolution* of about 10 km, and know that anything much less than 10 km across is virtually invisible. Figure 3.3 shows Manhattan at a spatial resolution of 250 m, detailed enough to pick out the shape of the island and Central Park.

It is easy to see how this helps with the problem of too much information. The Earth's surface covers about 500 million sq km, so if this level of detail is sufficient for an application, a property of the surface such as elevation can be described with only 5 million pieces of information, instead of the 500 million it would take to describe elevation with a resolution of 1 km, and the 500 trillion (500 000 000 000 000) it would take to describe elevation with 1 m resolution.

Another strategy for limiting detail is to observe that many properties remain constant over large areas. For

example, in describing the elevation of the Earth's surface we could take advantage of the fact that roughly two-thirds of the surface is covered by water, with its surface at sea level. Of the 5 million pieces of information needed to describe elevation at 10 km resolution, approximately 3.4 million will be recorded as zero, a colossal waste. If we could find an efficient way of identifying the area covered by water, then we would need only 1.6 million real pieces of information.

Humans have found many ingenious ways of describing the Earth's surface efficiently, because the problem we are addressing is as old as representation itself, and as important for paper-based representations as it is for binary representations in computers. But this ingenuity is itself the source of a substantial problem for GIS: there are many ways of representing the Earth's surface, and users of GIS thus face difficult and at times confusing choices. This chapter discusses some of those choices, and the issues are pursued further in subsequent chapters on uncertainty (Chapter 6) and data modeling (Chapter 8). Representation remains a major concern of GIScience, and researchers are constantly looking for ways to extend GIS representations to accommodate new types of information (Box 3.5).

3.5 Discrete objects and continuous fields

3.5.1 Discrete objects

Mention has already been made of the level of detail as a fundamental choice in representation. Another, perhaps even more fundamental choice, is between two conceptual schemes. There is good evidence that we as humans like to simplify the world around us by naming things, and seeing individual things as instances of broader categories. We prefer a world of black and white, of good guys and bad guys, to the real world of shades of gray.

The two fundamental ways of representing geography are discrete objects and continuous fields.

This preference is reflected in one way of viewing the geographic world, known as the *discrete object* view. In this view, the world is empty, except where it is occupied by objects with well-defined boundaries that are instances of generally recognized categories. Just as the desktop is littered with books, pencils, or computers, the geographic world is littered with cars, houses, lampposts, and other discrete objects. Thus the landscape of Minnesota is littered with lakes, and the landscape of Scotland is littered with mountains. One characteristic of the discrete object view is that objects can be counted, so license plates issued by the State of Minnesota carry

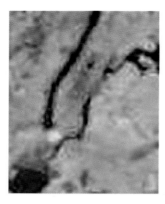

Figure 3.3 An image of Manhattan taken by the MODIS instrument on board the TERRA satellite on September 12, 2001. MODIS has a spatial resolution of about 250 m, detailed enough to reveal the coarse shape of Manhattan and to identify the Hudson and East Rivers, the burning World Trade Center (white spot), and Central Park (the gray blur with the Jacqueline Kennedy Onassis Reservoir visible as a black dot)

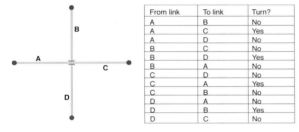

From link	To link	Turn?
A	B	No
A	C	Yes
A	D	No
B	C	No
B	D	Yes
B	A	No
C	D	No
C	A	Yes
C	B	No
D	A	No
D	B	Yes
D	C	No

Figure 3.4 The problems of representing a three-dimensional world using a two-dimensional technology. The intersection of links A, B, C, and D is an overpass, so no turns are possible between such pairs as A and B

the legend '10 000 lakes', and climbers know that there are exactly 284 mountains in Scotland over 3000 ft (the so-called Munros, from Sir Hugh Munro who originally listed 277 of them in 1891 – the count was expanded to 284 in 1997).

The discrete object view represents the geographic world as objects with well-defined boundaries in otherwise empty space.

Biological organisms fit this model well, and this allows us to count the number of residents in an area of a city, or to describe the behavior of individual bears. Manufactured objects also fit the model, and we have little difficulty counting the number of cars produced in a year, or the number of airplanes owned by an airline. But other phenomena are messier. It is not at all clear what constitutes a mountain, for example, or exactly how a mountain differs from a hill, or when a mountain with two peaks should be counted as two mountains.

Geographic objects are identified by their dimensionality. Objects that occupy area are termed two-dimensional, and generally referred to as areas. The term *polygon* is also common for technical reasons explained later. Other objects are more like one-dimensional lines, including roads, railways, or rivers, and are often represented as one-dimensional objects and generally referred to as lines. Other objects are more like zero-dimensional points, such as individual animals or buildings, and are referred to as points.

Of course, in reality, all objects that are perceptible to humans are three dimensional, and their representation in fewer dimensions can be at best an approximation. But the ability of GIS to handle truly three-dimensional objects as volumes with associated surfaces is very limited. Some GIS allow for a third (vertical) coordinate to be specified for all point locations. Buildings are sometimes represented by assigning height as an attribute, though if this option is used it is impossible to distinguish flat roofs from any other kind. Various strategies have been used for representing overpasses and underpasses in transportation networks, because this information is vital for navigation but not normally represented in strictly two-dimensional network representations. One common strategy is to represent turning options at every intersection – so an overpass appears in the database as an intersection with no turns (Figure 3.4).

Figure 3.5 Bears are easily conceived as discrete objects, maintaining their identity as objects through time and surrounded by empty space

The discrete object view leads to a powerful way of representing geographic information about objects. Think of a class of objects of the same dimensionality – for example, all of the Brown bears (Figure 3.5) in the Kenai Peninsula of Alaska. We would naturally think of these objects as points. We might want to know the sex of each bear, and its date of birth, if our interests were in monitoring the bear population. We might also have a collar on each bear that transmitted the bear's location at regular intervals. All of this information could be expressed in a table, such as the one shown in Table 3.1, with each row corresponding to a different discrete object, and each column to an attribute of the object. To reinforce a point made earlier, this is a very efficient way of capturing raw geographic information on Brown bears.

But it is not perfect as a representation for all geographic phenomena. Imagine visiting the Earth from another planet, and asking the humans what they chose as a representation for the infinitely complex and beautiful environment around them. The visitor would hardly be impressed to learn that they chose tables, especially when the phenomena represented were natural phenomena such as rivers, landscapes, or oceans. Nothing on the natural Earth looks remotely like a table. It is not at all clear how the properties of a river should be represented as a table, or the properties of an ocean. So while the discrete object

Table 3.1 Example of representation of geographic information as a table: the locations and attributes of each of four Brown bears in the Kenai Peninsula of Alaska. Locations have been obtained from radio collars. Only one location is shown for each bear, at noon on July 31 2003 (imaginary data)

Bear ID	Sex	Estimated year of birth	Date of collar installation	Location, noon on 31 July 2003
001	M	1999	02242003	−150.6432, 60.0567
002	F	1997	03312003	−149.9979, 59.9665
003	F	1994	04212003	−150.4639, 60.1245
004	F	1995	04212003	−150.4692, 60.1152

view works well for some kinds of phenomena, it misses the mark badly for others.

3.5.2 Continuous fields

While we might think of terrain as composed of discrete mountain peaks, valleys, ridges, slopes, etc., and think of listing them in tables and counting them, there are unresolvable problems of definition for all of these objects. Instead, it is much more useful to think of terrain as a continuous surface, in which elevation can be defined rigorously at every point (see Box 3.4). Such continuous surfaces form the basis of the other common view of geographic phenomena, known as the *continuous field* view (and not to be confused with other meanings of the word field). In this view the geographic world can be described by a number of *variables*, each measurable at any point on the Earth's surface, and changing in value across the surface.

The continuous field view represents the real world as a finite number of variables, each one defined at every possible position.

Objects are distinguished by their dimensions, and naturally fall into categories of points, lines, or areas. Continuous fields, on the other hand, can be distinguished by what varies, and how smoothly. A continuous field of elevation, for example, varies much more smoothly in a landscape that has been worn down by glaciation or flattened by blowing sand than one recently created by cooling lava. Cliffs are places in continuous fields where elevation changes suddenly, rather than smoothly. Population density is a kind of continuous field, defined everywhere as the number of people per unit area, though the definition breaks down if the field is examined so closely that the individual people become visible. Continuous fields can also be created from classifications of land, into categories of land use, or soil type. Such fields change suddenly at the boundaries between different classes. Other types of fields can be defined by continuous variation along lines, rather than across space. Traffic density, for example, can be defined everywhere on a road network, and flow volume can be defined everywhere on a river. Figure 3.6 shows some examples of field-like phenomena.

Continuous fields can be distinguished by *what* is being measured at each point. Like the attribute types discussed in Box 3.3, the variable may be nominal, ordinal, interval, ratio, or cyclic. A *vector* field assigns two variables, magnitude and direction, at every point in space, and is used to represent flow phenomena such as winds or currents; fields of only one variable are termed *scalar* fields.

Here is a simple example illustrating the difference between the discrete object and field conceptualizations. Suppose you were hired for the summer to count the number of lakes in Minnesota, and promised that your answer would appear on every license plate issued by the state. The task sounds simple, and you were happy to get the job. But on the first day you started to run into difficulty (Figure 3.7). What about small ponds, do they count as lakes? What about wide stretches of rivers? What about swamps that dry up in the summer? What about a lake with a narrow section connecting two wider parts, is it one lake or two? Your biggest dilemma concerns the scale of mapping, since the number of lakes shown on a map clearly depends on the map's level of detail – a more detailed map almost certainly will show more lakes.

Your task clearly reflects a discrete object view of the phenomenon. The action of counting implies that lakes are discrete, two-dimensional objects littering an otherwise empty geographic landscape. In a continuous field view, on the other hand, all points are either lake or non-lake. Moreover, we could refine the scale a little to take account

---

Technical Box **3.4**

2.5 dimensions

Areas are two-dimensional objects, and volumes are three dimensional, but GIS users sometimes talk about '2.5-D'. Almost without exception the elevation of the Earth's surface has a single value at any location (exceptions include overhanging cliffs). So elevation is conveniently thought of as a continuous field, a variable with a value everywhere in two dimensions, and a full 3-D representation is only necessary in areas with an abundance of overhanging cliffs or caves, if these are important features. The idea of dealing with a three-dimensional phenomenon by treating it as a single-valued function of two horizontal variables gives rise to the term '2.5-D'. Figure 3.6B shows an example, in this case an elevation surface.

(A)

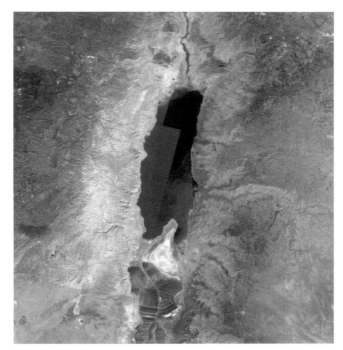

(B)

Figure 3.6 Examples of field-like phenomena. (A) Image of part of the Dead Sea in the Middle East. The lightness of the image at any point measures the amount of radiation captured by the satellite's imaging system. (B) A simulated image derived from the Shuttle Radar Topography Mission, a new source of high-quality elevation data. The image shows the Carrizo Plain area of Southern California, USA, with a simulated sky and with land cover obtained from other satellite sources (Courtesy NASA/JPL–Caltech)

of marginal cases; for example, we might define the scale shown in Table 3.2, which has five degrees of lakeness. The complexity of the view would depend on how closely we looked, of course, and so the scale of mapping would still be important. But all of the problems of defining a lake as a discrete object would disappear (though there would still be problems in defining the levels of the scale). Instead of counting, our strategy would be to lay a grid over the map, and assign each grid cell a score on the lakeness scale. The size of the grid cell would determine how accurately the result approximated the value we could theoretically obtain by visiting every one of the infinite

Figure 3.7 Lakes are difficult to conceptualize as discrete objects because it is often difficult to tell where a lake begins and ends, or to distinguish a wide river from a lake

Table 3.2 A scale of lakeness suitable for defining lakes as a continuous field

Lakeness	Definition
1	Location is always dry under all circumstances
2	Location is sometimes flooded in Spring
3	Location supports marshy vegetation
4	Water is always present to a depth of less than 1 m
5	Water is always present to a depth of more than 1 m

number of points in the state. At the end, we would tabulate the resulting scores, counting the number of cells having each value of lakeness, or averaging the lakeness score. We could even design a new and scientifically more reasonable license plate – 'Minnesota, 12% lake' or 'Minnesota, average lakeness 2.02'.

The difference between objects and fields is also illustrated well by photographs (e.g., Figure 3.6A). The image in a photograph is created by variation in the chemical state of the material in the photographic film – in early photography, minute particles of silver

were released from molecules of silver nitrate when the unstable molecules were exposed to light, thus darkening the image in proportion to the amount of incident light. We think of the image as a field of continuous variation in color or darkness. But when we look at the image, the eye and brain begin to infer the presence of discrete objects, such as people, rivers, fields, cars, or houses, as they interpret the content of the image.

3.6 Rasters and vectors

Continuous fields and discrete objects define two conceptual views of geographic phenomena, but they do not solve the problem of digital representation. A continuous field view still potentially contains an infinite amount of information if it defines the value of the variable at every point, since there is an infinite number of points in any defined geographic area. Discrete objects can also require an infinite amount of information for full description – for example, a coastline contains an infinite amount of information if it is mapped in infinite detail. Thus continuous fields and discrete objects are no more than conceptualizations, or ways in which we think about geographic phenomena; they are not designed to deal with the limitations of computers.

Two methods are used to reduce geographic phenomena to forms that can be coded in computer databases, and we call these raster and vector. In principle, both can be used to code both fields and discrete objects, but in practice there is a strong association between raster and fields, and between vector and discrete objects.

Raster and vector are two methods of representing geographic data in digital computers.

3.6.1 Raster data

In a raster representation space is divided into an array of rectangular (usually square) cells (Figure 3.8). All geographic variation is then expressed by assigning properties or attributes to these cells. The cells are sometimes called pixels (short for *picture elements*).

Raster representations divide the world into arrays of cells and assign attributes to the cells.

One of the commonest forms of raster data comes from remote-sensing satellites, which capture information in this form and send it to ground to be distributed and analyzed. Data from the Landsat Thematic Mapper, for example, which are commonly used in GIS applications, come in cells that are 30 m a side on the ground, or approximately 0.1 hectare in area. Other similar data can be obtained from sensors mounted on aircraft. Imagery varies according to the spatial resolution (expressed as the length of a cell side as measured on the ground), and also according to the timetable of image capture by the

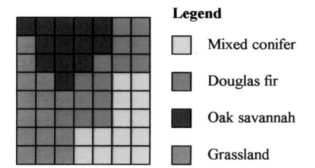

Legend

☐ Mixed conifer

▨ Douglas fir

■ Oak savannah

▨ Grassland

Figure 3.8 Raster representation. Each color represents a different value of a nominal-scale variable denoting land cover class

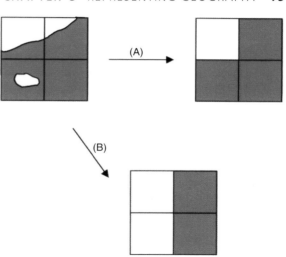

Figure 3.9 Effect of a raster representation using (A) the largest share rule and (B) the central point rule

sensor. Some satellites are in *geostationary* orbit over a fixed point on the Earth, and capture images constantly. Others pass over a fixed point at regular intervals (e.g., every 12 days). Finally, sensors vary according to the part or parts of the spectrum that they sense. The visible parts of the spectrum are most important for remote sensing, but some invisible parts of the spectrum are particularly useful in detecting heat, and the phenomena that produce heat, such as volcanic activities. Many sensors capture images in several areas of the spectrum, or *bands*, simultaneously, because the relative amounts of radiation in different parts of the spectrum are often useful indicators of certain phenomena, such as green leaves, or water, on the Earth's surface. The AVIRIS (Airborne Visible InfraRed Imaging Spectrometer) captures no fewer than 224 different parts of the spectrum, and is being used to detect particular minerals in the soil, among other applications. Remote sensing is a complex topic, and further details are available in Chapter 9.

Square cells fit together nicely on a flat table or a sheet of paper, but they will not fit together neatly on the curved surface of the Earth. So just as representations on paper require that the Earth be flattened, or projected, so too do rasters (because of the distortions associated with flattening, the cells in a raster can never be perfectly equal in shape or area on the Earth's surface). Projections, or ways of flattening the Earth, are described in Section 5.7. Many of the terms that describe rasters suggest the laying of a tile floor on a flat surface – we talk of raster cells *tiling* an area, and a raster is said to be an instance of a *tesselation*, derived from the word for a mosaic. The mirrored ball hanging above a dance floor recalls the impossibility of covering a spherical object like the Earth perfectly with flat, square pieces.

When information is represented in raster form all detail about variation within cells is lost, and instead the cell is given a single value. Suppose we wanted to represent the map of the counties of Texas as a raster. Each cell would be given a single value to identify a county, and we would have to decide the rule to apply when a cell falls in more than one county. Often the rule is that the county with the *largest share* of the cell's area gets the cell. Sometimes the rule is based on the *central point* of the cell, and the county at that point is

assigned to the whole cell. Figure 3.9 shows these two rules in operation. The largest share rule is almost always preferred, but the central point rule is sometimes used in the interests of faster computing, and is often used in creating raster datasets of elevation.

3.6.2 Vector data

In a vector representation, all lines are captured as points connected by precisely straight lines (some GIS software allows points to be connected by curves rather than straight lines, but in most cases curves have to be approximated by increasing the density of points). An area is captured as a series of points or *vertices* connected by straight lines as shown in Figure 3.10. The straight edges between vertices explain why areas in vector representation are often called *polygons*, and in GIS-speak the terms polygon and area are often used

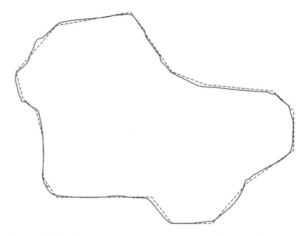

Figure 3.10 An area (red line) and its approximation by a polygon (blue line)

interchangeably. Lines are captured in the same way, and the term *polyline* has been coined to describe a curved line represented by a series of straight segments connecting vertices.

To capture an area object in vector form, we need only specify the locations of the points that form the vertices of a polygon. This seems simple, and also much more efficient than a raster representation, which would require us to list all of the cells that form the area. These ideas are captured succinctly in the comment 'Raster is vaster, and vector is correcter'. To create a precise approximation to an area in raster, it would be necessary to resort to using very small cells, and the number of cells would rise proportionately (in fact, every halving of the width and height of each cell would result in a quadrupling of the number of cells). But things are not quite as simple as they seem. The apparent precision of vector is often unreasonable, since many geographic phenomena simply cannot be located with high accuracy. So although raster data may look less attractive, they may be more honest to the inherent quality of the data. Also, various methods exist for compressing raster data that can greatly reduce the capacity needed to store a given dataset (see Chapter 8). So the choice between raster and vector is often complex, as summarized in Table 3.3.

3.6.3 Representing continuous fields

While discrete objects lend themselves naturally to representation as points, lines, or areas using vector methods, it is less obvious how the continuous variation of a field can be expressed in a digital representation. In GIS six alternatives are commonly implemented (Figure 3.11):

A. capturing the value of the variable at each of a grid of regularly spaced sample points (for example, elevations at 30 m spacing in a DEM);

B. capturing the value of the field variable at each of a set of irregularly spaced sample points (for example, variation in surface temperature captured at weather stations);

C. capturing a single value of the variable for a regularly shaped cell (for example, values of reflected radiation in a remotely sensed scene);

D. capturing a single value of the variable over an irregularly shaped area (for example, vegetation cover class or the name of a parcel's owner);

E. capturing the linear variation of the field variable over an irregularly shaped triangle (for example, elevation captured in a triangulated irregular network or TIN, Section 9.2.3.4);

F. capturing the isolines of a surface, as digitized lines (for example, digitized contour lines representing surface elevation).

Each of these methods succeeds in compressing the potentially infinite amount of data in a continuous field to a finite amount, using one of the six options, two of which (A and C) are raster, and four (B, D, E, and F) are vector. Of the vector methods one (B) uses points, two (D and E) use polygons, and one (F) uses lines to express the continuous spatial variation of the field in terms of a finite set of vector objects. But unlike the discrete object conceptualization, the objects used to represent a field are not real, but simply artifacts of the representation of something that is actually conceived as spatially continuous. The triangles of a TIN representation (E), for example, exist only in the digital representation, and cannot be found on the ground, and neither can the lines of a contour representation (F).

Table 3.3 Relative advantages of raster and vector representation

Issue	Raster	Vector
Volume of data	Depends on cell size	Depends on density of vertices
Sources of data	Remote sensing, imagery	Social and environmental data
Applications	Resources, environmental	Social, economic, administrative
Software	Raster GIS, image processing	Vector GIS, automated cartography
Resolution	Fixed	Variable

3.7 The paper map

The paper map has long been a powerful and effective means of communicating geographic information. In contrast to digital data, which use coding schemes such as ASCII, it is an instance of an *analog* representation, or a physical model in which the real world is scaled – in the case of the paper map, part of the world is scaled to fit the size of the paper. A key property of a paper map is its *scale* or *representative fraction*, defined as the ratio of distance on the map to distance on the Earth's surface. For example, a map with a scale of 1:24 000 reduces everything on the Earth to one 24 000th of its real size. This is a bit misleading, because the Earth's surface is curved and a paper map is flat, so scale cannot be exactly constant.

A paper map is: a source of data for geographic databases; an analog product from a GIS; and an effective communication tool.

Maps have been so important, particularly prior to the development of digital technology, that many of the ideas associated with GIS are actually inherited directly from paper maps. For example, scale is often cited as a property of a digital database, even though the definition of scale makes no sense for digital data – ratio of distance *in the*

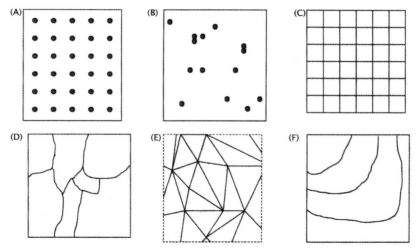

Figure 3.11 The six approximate representations of a field used in GIS. (A) Regularly spaced sample points. (B) Irregularly spaced sample points. (C) Rectangular cells. (D) Irregularly shaped polygons. (E) Irregular network of triangles, with linear variation over each triangle (the Triangulated Irregular Network or TIN model; the bounding box is shown dashed in this case because the unshown portions of complete triangles extend outside it). (F) Polylines representing contours (see the discussion of isopleth maps in Box 4.3) (Courtesy US Geological Survey)

Biographical Box **3.5**

May Yuan and new forms of representation

May Yuan received her Bachelor of Science degree in Geography from the National Taiwan University, where she was attracted to the fields of geomorphology and climatology. Continuing her fundamental interest in evolution of processes, she studied geographic representation and temporal GIS and earned both her Masters and PhD degrees in Geography from the State University of New York at Buffalo.

Currently, May is an Associate Professor of Geography at the University of Oklahoma. Severe weather in the Southern Plains of the United States (Figure 3.12) has inspired her to re-evaluate GIS representation of geographic dynamics, the complexity of events and processes at spatial and temporal scales, and GIS applications in meteorology (i.e., weather and climate). She investigates meteorological cases (e.g., convective storms and flash floods) to develop new ideas of using events and processes as the basis to integrate spatial and temporal data in GIS. Her publications address theoretical issues on representation of geographic dynamics and

Figure 3.13 May Yuan, developer of new forms of representation

offer conceptual models and a prototype GIS to support spatiotemporal queries and analysis of dynamic geographic phenomena. Her temporal GIS research goes beyond merely considering time as an attribute or annotation of spatial objects to incorporate much richer spatiotemporal meaning. In her case study on convective storms, she has demonstrated that, by modeling storms as data objects, GIS is able to support information query about storm evolution, storm behaviors, and interactions with environments.

May developed a strong interest in physics in early childhood. Newton's theory of universal gravitation sparked her appreciation for simple principles that can explain how things work and for the use of graphical and symbolic representation to conceptualize complex processes. Planck's quantum theory and Heisenberg's uncertainty principle further stimulated her thinking on the nature of matter and its behavior at different scales of observations. Shaped by Einstein's theory of relativity, May developed her world view as a four-dimensional space-time continuum populated with events and phenomena. Before she pursued a career in GIScience, May studied fluvial processes and developed a model to classify waterfalls and explain

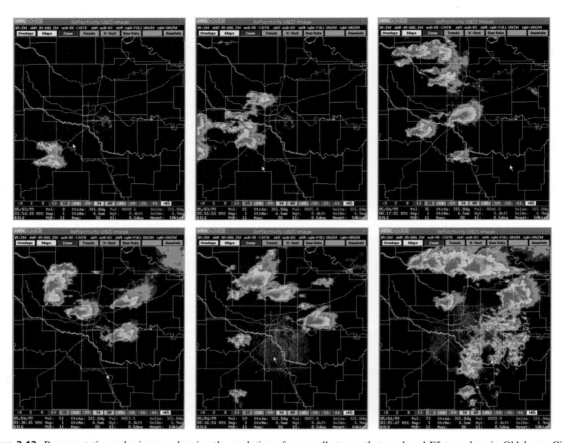

Figure 3.12 Representative radar images showing the evolution of supercell storms that produced F5 tornadoes in Oklahoma City, May 3, 1999. WSR-88D radar TKLX scanned the supercells every five minutes, but the images shown here were selected approximately every two hours

their formation. She went on to study paleoclimatology by analyzing soil and speleothem sediments. Both studies, as well as her dissertation research on wildfire representation, reinforced her interest in developing conceptual models of processes and examining the relationships between space and time. Since she moved to the University of Oklahoma, a suite of world-class meteorological research initiatives has offered her unique opportunities to extend her interest in physics to fundamental research in GIScience through meteorological applications. Weather and climate offer rich cases that emphasize movement, processes, and evolution and pose grand challenges to GIScience research regarding representation, object–field duality, and uncertainty. May enjoys the challenges that ultimately connect to her fundamental interest in how things work.

computer to distance on the ground; how can there be distances in a computer? What is meant is a little more complicated: when a scale is quoted for a digital database it is usually the scale of the map that formed the source of the data. So if a database is said to be at a scale of 1:24 000 one can safely assume that it was created from a paper map at that scale, and includes representations of the features that are found on maps at that scale. Further discussion of scale can be found in Box 4.2 and

in Chapter 6, where it is important to the concept of uncertainty.

There is a close relationship between the contents of a map and the raster and vector representations discussed in the previous section. The US Geological Survey, for example, distributes two digital versions of its topographic maps, one in raster form and one in vector form, and both attempt to capture the contents of the map as closely as possible. In the raster form, or

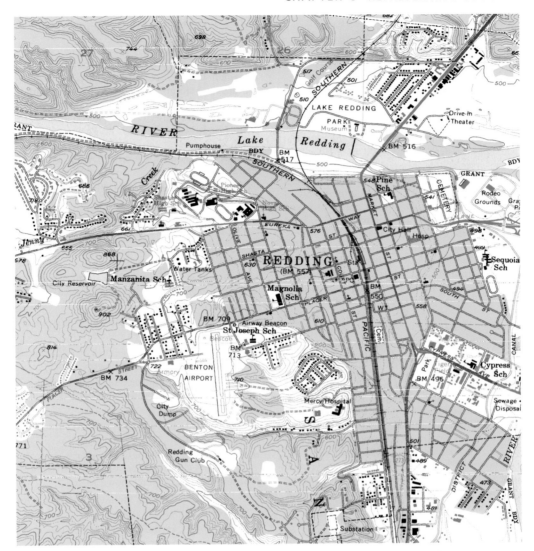

Figure 3.14 Part of a Digital Raster Graphic, a scan a US Geological Survey 1:24 000 topographic map

digital raster graphic (DRG), the map is scanned at a very high density, using very small pixels, so that the raster looks very much like the original (Figure 3.14). The coding of each pixel simply records the color of the map picked up by the scanner, and the dataset includes all of the textual information surrounding the actual map.

In the vector form, or *digital line graph* (DLG), every geographic feature shown on the map is represented as a point, polyline, or polygon. The symbols used to represent point features on the map, such as the symbol for a windmill, are replaced in the digital data by points with associated attributes, and must be regenerated when the data are displayed. Contours, which are shown on the map as lines of definite width, are replaced by polylines of no width, and given attributes that record their elevations.

In both cases, and especially in the vector case, there is a significant difference between the analog representation of the map and its digital equivalent. So it is quite misleading to think of the contents of a digital representation as a map, and to think of a GIS as a container of digital maps. Digital representations can include information that would be very difficult to show on maps. For example, they can represent the curved surface of the Earth, without the need for the distortions associated with flattening. They can represent changes, whereas maps must be static because it is very difficult to change their contents once they have been printed or drawn. Digital databases can represent all three spatial dimensions, including the vertical, whereas maps must always show two-dimensional views. So while the paper map is a useful metaphor for the contents of a geographic database, we must be careful not to let it limit our thinking about what is possible in the way of representation. This issue is pursued at greater length in Chapter 8, and map production is discussed in detail in Chapter 12.

3.8 Generalization

In Section 3.4 we saw how thinking about geographic information as a collection of atomic links – between a place, a time (not always, because many geographic facts are stated as if they were permanently true), and a property – led to an immediate problem, because the potential number of such atomic facts is infinite. If seen in enough detail, the Earth's surface is unimaginably complex, and its effective description impossible. So instead, humans have devised numerous ways of simplifying their view of the world. Instead of making statements about each and every point, we describe entire areas, attributing uniform characteristics to them, even when areas are not strictly uniform; we identify features on the ground and describe their characteristics, again assuming them to be uniform; or we limit our descriptions to what exists at a finite number of sample points, hoping that these samples will be adequately representative of the whole (Section 4.4).

A geographic database cannot contain a perfect description – instead, its contents must be carefully selected to fit within the limited capacity of computer storage devices.

From this perspective some degree of generalization is almost inevitable in all geographic data. But cartographers often take a somewhat different approach, for which this observation is not necessarily true. Suppose we are tasked to prepare a map at a specific scale, say 1:25 000, using the standards laid down by a national mapping agency, such as the Institut Géographique National (IGN) of France. Every scale used by IGN has its associated rules of representation. For example, at a scale of 1:25 000 the rules lay down that individual buildings will be shown only in specific circumstances, and similar rules apply to the 1:24 000 series of the US Geological Survey. These rules are known by various names, including *terrain nominal* in the case of IGN, which translates roughly but not very helpfully to 'nominal ground', and is perhaps better translated as 'specification'. From this perspective a map that represents the world by following the rules of a specification precisely can be perfectly accurate *with respect to the specification*, even though it is not a perfect representation of the full detail on the ground.

A map's specification defines how real features on the ground are selected for inclusion on the map.

Consider the representation of vegetation cover using the rules of a specification. For example, the rules might state that at a scale of 1:100 000, a vegetation cover map should not show areas of vegetation that cover less than 1 hectare. But small areas of vegetation almost certainly exist, so deleting them inevitably results in information loss. But under the principle discussed above, a map that adheres to this rule must be accurate, *even though it differs substantively from the truth as observed on the ground.*

3.8.1 Methods of generalization

A GIS dataset's level of detail is one of its most important properties, as it determines both the degree to which the dataset approximates the real world, and the dataset's complexity. It is often necessary to remove detail, in the interests of compressing data, fitting them into a storage device of limited capacity, processing them faster, or creating less confusing visualizations that emphasize general trends. Consequently many methods have been devised for generalization, and several of the more important are discussed in this section.

McMaster and Shea (1992) identify the following types of generalization rules:

- *simplification*, for example by weeding out points in the outline of a polygon to create a simpler shape;
- *smoothing*, or the replacement of sharp and complex forms by smoother ones;
- *aggregation*, or the replacement of a large number of distinct symbolized objects by a smaller number of new symbols;
- *amalgamation*, or the replacement of several area objects by a single area object;
- *merging*, or the replacement of several line objects by a smaller number of line objects;
- *collapse*, or the replacement of an area object by a combination of point and line objects;
- *refinement*, or the replacement of a complex pattern of objects by a selection that preserves the pattern's general form;
- *exaggeration*, or the relative enlargement of an object to preserve its characteristics when these would be lost if the object were shown to scale;
- *enhancement*, through the alteration of the physical sizes and shapes of symbols; and
- *displacement*, or the moving of objects from their true positions to preserve their visibility and distinctiveness.

The differences between these types of rules are much easier to understand visually and Figure 3.15 reproduces McMaster's and Shea's original example drawings. In addition, they describe two forms of generalization of attributes, as distinct from geometric forms of generalization. *Classification* generalization reclassifies the attributes of objects into a smaller number of classes, while *symbolization* generalization changes the assignment of symbols to objects. For example, it might replace an elaborate symbol including the words 'Mixed Forest' with a color identifying that class.

3.8.2 Weeding

One of the commonest forms of generalization in GIS is the process known as weeding, or the simplification of the representation of a line represented as a polyline. The process is an instance of McMaster and Shea's simplification. Standard methods exist in GIS for doing

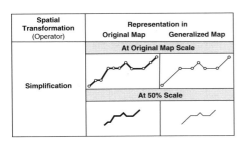

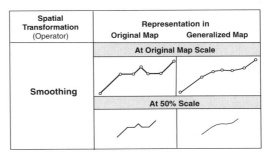

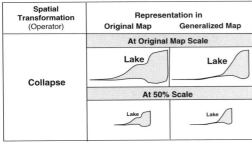

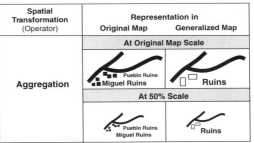

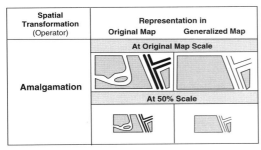

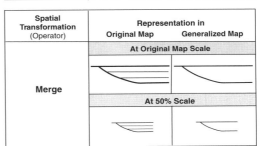

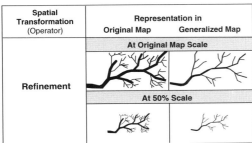

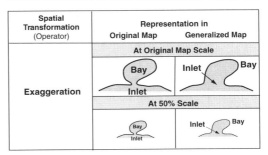

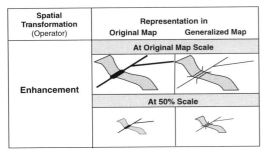

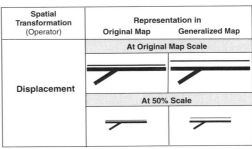

Figure 3.15 Illustrations from McMaster and Shea (1992) of their ten forms of generalization. The original feature is shown at its original level of detail, and below it at 50% coarser scale. Each generalization technique resolves a specific problem of display at coarser scale and results in the acceptable version shown in the lower right

Figure 3.16 The Douglas–Poiker algorithm is designed to simplify complex objects like this shoreline by reducing the number of points in its polyline representation

this, and the commonest by far is the method known as the Douglas–Poiker algorithm (Figure 3.16) after its inventors, David Douglas and Tom Poiker. The operation of the Douglas–Poiker weeding algorithm is shown in Figure 3.17.

> **Weeding is the process of simplifying a line or area by reducing the number of points in its representation.**

Note that the algorithm relies entirely on the assumption that the line is represented as a polyline, in other words as a series of straight line segments. GIS increasingly support other representations, including arcs of circles, arcs of ellipses, and Bézier curves, but there is little consensus to date on appropriate methods for weeding or generalizing them, or on methods of analysis that can be applied to them.

3.9 Conclusion

Representation, or more broadly ontology, is a fundamental issue in GIS, since it underlies all of our efforts to express useful information about the surface of the Earth in a digital computer. The fact that there are so many ways of doing this makes GIS at once complex and interesting,

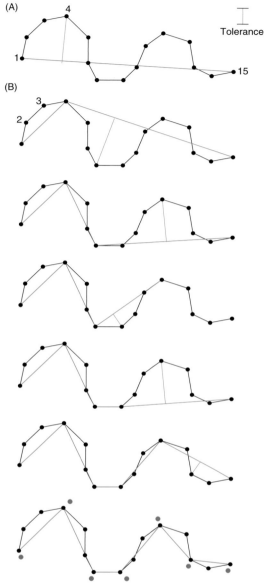

Figure 3.17 The Douglas–Poiker line simplification algorithm in action. The original polyline has 15 points. In (A) Points 1 and 15 are connected (red), and the furthest distance of any point from this connection is identified (blue). This distance to Point 4 exceeds the user-defined tolerance. In (B) Points 1 and 4 are connected (green). Points 2 and 3 are within the tolerance of this line. Points 4 and 15 are connected, and the process is repeated. In the final step 7 points remain (identified with green disks), including 1 and 15. No points are beyond the user-defined tolerance distance from the line

a point that will become much clearer on reading the technical chapter on data modeling, Chapter 8. But the broader issues of representation, including the distinction between field and object conceptualizations, underlie not only that chapter but many other issues as well, including uncertainty (Chapter 6), and Chapters 14 through 16 on analysis and modeling.

Questions for further study

1. What fraction of the Earth's surface have you experienced in your lifetime? Make diagrams like that shown in Figure 3.1, at appropriate levels of detail, to show a) where you have lived in your lifetime, b) how you spent last weekend. How would you describe what is missing from each of these diagrams?

2. Table 3.3 summarized some of the arguments between raster and vector representations. Expand on these arguments, providing examples, and add any others that would be relevant in a GIS application.

3. The early explorers had limited ways of communicating what they saw, but many were very effective at it. Examine the published diaries, notebooks, or dispatches of one or two early explorers and look at the methods they used to communicate with others. What words did they use to describe unfamiliar landscapes and how did they mix words with sketches?

4. Identify the limits of your own neighborhood, and start making a list of the discrete objects you are familiar with in the area. What features are hard to think of as discrete objects? For example, how will you divide up the various roadways in the neighborhood into discrete objects – where do they begin and end?

Further reading

Chrisman N.R. 2002 *Exploring Geographic Information Systems* (2nd edn). New York: Wiley.

McMaster R.B. and Shea K.S. 1992 *Generalization in Digital Cartography*. Washington, DC: Association of American Geographers.

National Research Council 1999 *Distributed Geolibraries: Spatial Information Resources*. Washington, DC: National Academy Press. Available: **www.nap.edu**.

4 *The nature of geographic data*

This chapter elaborates on the *spatial is special* theme by examining the nature of geographic data. It sets out the distinguishing characteristics of geographic data, and suggests a range of guiding principles for working with them. Many geographic data are correctly thought of as sample observations, selected from the larger universe of possible observations that could be made. This chapter describes the main principles that govern scientific sampling, and the principles that are invoked in order to infer information about the gaps between samples. When devising spatial sample designs, it is important to be aware of the nature of spatial variation, and here we learn how this is formalized and measured as spatial autocorrelation. Another key property of geographic information is the level of detail that is apparent at particular scales of analysis. The concept of fractals provides a solid theoretical foundation for understanding scale when building geographic representations.

Geographic Information Systems and Science, 2nd edition   Paul Longley, Michael Goodchild, David Maguire, and David Rhind.
© 2005 John Wiley & Sons, Ltd.   ISBNs: 0-470-87000-1 (HB); 0-470-87001-X (PB)

Learning Objectives

After reading this chapter you will understand:

■ How Tobler's First Law of Geography is formalized through the concept of spatial autocorrelation;

■ The relationship between scale and the level of geographic detail in a representation;

■ The principles of building representations around geographic samples;

■ How the properties of smoothness and continuous variation can be used to characterize geographic variation;

■ How fractals can be used to measure and simulate surface roughness.

4.1 Introduction

In Chapter 1 we identified the central motivation for scientific applications of GIS as the development of representations, not only of how the world *looks*, but also how it *works*. Chapter 3 established three governing principles that help us towards this goal, namely that:

1. the representations we build in GIS are of *unique places*;

2. our representations of them are necessarily *selective* of reality, and hence incomplete;

3. in building representations, it is useful to think of the world as either comprising continuously varying *fields* or as an empty space littered with *objects* that are crisp and well-defined.

In this chapter we build on these principles to develop a fuller understanding of the ways in which the *nature of spatial variation* is represented in GIS. We do this by asserting three further principles:

4. that proximity effects are key to understanding spatial variation, and to joining up incomplete representations of unique places;

5. that issues of geographic scale and level of detail are key to building appropriate representations of the world;

6. that different measures of the world *co-vary*, and understanding the nature of co-variation can help us to predict.

Implicit in all of this is one further principle, that we will develop in Chapter 6:

7. because almost all representations of the world are necessarily incomplete, they are *uncertain*.

GIS is about representing spatial and temporal phenomena in the real world and, because the real world is complicated, this task is difficult and error prone. The real world provides an intriguing laboratory in which to examine phenomena, but is one in which it can be impossible to control for variation in all characteristics – be they relevant to landscape evolution, consumer behavior, urban growth, or whatever. In the terminology of Section 1.3, generalized *laws* governing spatial distributions and temporal dynamics are therefore most unlikely to work perfectly. We choose to describe the seven points above as 'principles', rather than 'laws' (see Section 1.3) because, like our discussion in Chapter 2, this chapter is grounded in empirical generalization about the real world. A more elevated discussion of the way that these principles build into 'fundamental laws of GIScience' has been published by Goodchild.

4.2 The fundamental problem revisited

Consider for a moment a GIS-based representation of your own life history to date. It is infinitesimally small compared with the geographic extent and history of the world but, as we move to finer spatial and temporal scales than those shown in Figure 3.1, nevertheless very intricate in detail. Viewed in aggregate, human behavior where you live exhibits structure in geographic space, as the aggregated outcomes of day-to-day (often repetitive) decisions about where to go, what to do, how much time to spend doing it, and longer-term (one-off) decisions about where to live, how to achieve career objectives, and how to balance work, leisure, and family pursuits. It is helpful to distinguish between *controlled* and *uncontrolled* variation – the former oscillates around a steady state (daily, weekly) pattern, while the latter (career changes, residential moves) does not.

When relating our own daily regimes and life histories, or indeed any short or long term *time series* of events, we are usually mindful of the contexts in which our decisions (to go to work, to change jobs, to marry) are made – 'the past is the key to the present' aptly summarizes the effect of temporal context upon our actions. The day-to-day operational context to our activities is very much determined by where we live and work. The longer-term strategic context may well be provided by where we were born, grew up, and went to college.

Our behavior in geographic space often reflects past patterns of behavior.

The relationship between consecutive events in *time* can be formalized in the concept of *temporal autocorrelation*. The analysis of time series data is in some senses straightforward, since the direction of causality is only one way – past events are sequentially related to the present and to the future. This chapter (and book) is principally concerned with spatial, rather than temporal, autocorrelation. Spatial autocorrelation shares some similarities with its temporal counterpart. Yet time moves in one direction only (forward), making temporal autocorrelation one-dimensional, while spatial events can potentially have consequences anywhere in two-dimensional or even three-dimensional space.

Explanation in time need only look to the past, but explanation in space must look in all directions simultaneously.

Assessment of spatial autocorrelation can be informed by knowledge of the degree and nature of *spatial heterogeneity* – the tendency of geographic places and regions to be different from each other. Everyone would recognize the extreme difference of landscapes between such regions as the Antarctic, the Nile delta, the Sahara desert, or the Amazon basin, and many would recognize the more subtle differences between the Central Valley of California, the Northern Plain of China, and the valley of the Ganges in India. Heterogeneity occurs both in the way the landscape looks, and in the way processes act on the landscape (the form/process distinction of Section 1.3). While the spatial variation in some processes simply oscillates about an average (controlled variation), other processes vary ever more the longer they are observed (uncontrolled variation). For example, controlled variation characterizes the operational environment of GIS applications in utility management (Section 2.1.1), or the tactical environment of retail promotions (Section 2.3.3), while longer-term processes such as global warming or deforestation may exhibit uncontrolled variation. As a general rule, spatial data exhibit an increasing range of values, hence increased heterogeneity, with increased distance. In this chapter we focus on the ways in which phenomena vary across space, and the general nature of geographic variation: later, in Chapter 14, we return to the techniques for measuring spatial heterogeneity.

Also, this requires us to move beyond thinking of GIS data as abstracted only from the continuous spatial distributions implied by the Tobler Law (Section 3.1) and from sequences of events over continuous time. Some events, such as the daily rhythm of the journey to work, are clearly incremental extensions of past practice, while others, such as residential relocation, constitute sudden breaks with the past. Similarly, landscapes of gently undulating terrain are best thought of as smooth and continuous, while others (such as the landscapes developed about fault systems, or mountain ranges) are best conceived as discretely bounded, jagged, and irregular. Smoothness and irregularity turn out to be among the most important distinguishing characteristics of geographic data.

Some geographic phenomena vary smoothly across space, while others can exhibit extreme irregularity, in violation of Tobler's Law.

Finally, it is highly likely that a representation of the real world that is suitable for predicting future change will need to incorporate information on how two or more factors *co-vary*. For example, planners seeking to justify improvements to a city's public transit system might wish to point out how house prices increase with proximity to existing rail stops. It is highly likely that patterns of spatial autocorrelation in one variable will, to a greater or lesser extent, be mirrored in another. Whilst this is helpful in building representations of the real world, the property of spatial autocorrelation can frustrate our attempts to build inferential statistical models of the co-variation of geographic phenomena.

Spatial autocorrelation helps us to build representations, but frustrates our efforts to predict.

The nature of geographic variation, the scale at which uncontrolled variation occurs, and the way in which different geographic phenomena co-vary are all key to building effective representations of the real world. These principles are of practical importance and guide us to answering questions such as: What is an appropriate scale or level of detail at which to build a representation for a particular application? How do I design my spatial sample? How do I generalize from my sample measurements? And what formal methods and techniques can I use to relate key spatial events and outcomes to one another?

Each of these questions is a facet of the fundamental problem of GIS, that is of selecting what to leave in and what to take out of our digital representations of the real world (Section 3.2). The Tobler Law (Section 3.1), that everything is related to everything else, but near things are more related than distant things, amounts to a succinct definition of spatial autocorrelation. An understanding of the nature of the spatial autocorrelation that characterizes a GIS application helps us to *deduce* how best to collect and assemble data for a representation, and also how best to develop inferences between events and occurrences.

The concept of geographic *scale* or level of detail will be fundamental to observed measures of the likely strength and nature of autocorrelation in any given application. Together, the scale and spatial structure of a particular application suggest ways in which we should *sample* geographic reality, and the ways in which we should *weight* sample observations in order to build our representation. We will return to the key concepts of scale, sampling, and weighting throughout much of this book.

4.3 Spatial autocorrelation and scale

In Chapter 3 (Box 3.3) we classified attribute data into the nominal, ordinal, interval, ratio, and cyclic scales of measurement. Objects existing in space are described by locational (spatial) descriptors, and are conventionally classified using the taxonomy shown in Box 4.1.

Types of spatial objects

We saw in Section 3.4 that geographic objects are classified according to their *topological dimension*, which provides a measure of the way they fill space. For present purposes we assume that dimensions are restricted to *integer* (whole number) values, though in later sections (Sections 4.8 and 15.2.5) we relax this constraint and consider geographic objects of non-integer (fractional, or *fractal*) dimension. All geometric objects can be used to represent occurrences at absolute locations (*natural* objects), or they may be used to summarize spatial distributions (*artificial* objects).

A *point* has neither length nor breadth nor depth, and hence is said to be of dimension 0. Points may be used to indicate spatial occurrences or events, and their spatial patterning. *Point pattern analysis* is used to identify whether occurrences or events are inter-related – as in the analysis of the incidence of crime, or in identifying whether patterns of disease infection might be related to environmental or social factors (Section 15.2.3). The *centroid* of an area object is an artificial point reference, which is located so as to provide a summary measure of the location of the object (Section 15.2.1).

Lines have length, but not breadth or depth, and hence are of dimension 1. They are used to represent linear entities such as roads, pipelines, and cables, which frequently build together into networks. They can also be used to measure distances between spatial objects, as in the measurement of inter-centroid distance. In order to reduce the burden of data capture and storage, lines are often held in GIS in *generalized* form (see Section 3.8).

Area objects have the two dimensions of length and breadth, but not depth. They may be used to represent natural objects, such as agricultural fields, but are also commonly used to represent artificial aggregations, such as census tracts (see below). Areas may bound linear features and enclose points, and GIS functions can be used to identify

whether a given area encloses a given point (Section 14.4.2).

Volume objects have length, breadth, and depth, and hence are of dimension 3. They are used to represent natural objects such as river basins, or artificial phenomena such as the population potential of shopping centers or the density of resident populations (Section 14.4.5).

Time is often considered to be the fourth dimension of spatial objects, although GIS remains poorly adapted to the modeling of temporal change.

The relationship between higher- and lower-dimension spatial objects is analogous to that between higher- and lower-order attribute data, in that lower-dimension objects can be derived from those of higher dimension but not vice versa. Certain phenomena, such as population, may be held as natural or artificially imposed spatial object types. The chosen way of representing phenomena in GIS not only defines the apparent nature of geographic variation, but also the way in which geographic variation may be analyzed. Some objects, such as agricultural fields or digital terrain models, are represented in their natural state. Others are transformed from one spatial object class to another, as in the transformation of population data from individual points to census tract areas, for reasons of confidentiality or convention. Some high-order representations are created by interpolation between lower-order objects, as in the creation of digital terrain models (DTMs) from spot height data (Chapter 8).

The classification of spatial phenomena into object types is dependent fundamentally upon scale. For example, on a less-detailed map of the world, New York is represented as a one-dimensional point. On a more-detailed map such as a road atlas it will be represented as a two-dimensional area. Yet if we visit the city, it is very much experienced as a three-dimensional entity, and virtual reality systems seek to represent it as such (see Section 13.4.2).

Spatial autocorrelation measures attempt to deal simultaneously with similarities in the location of spatial objects (Box 4.1) and their attributes (Box 3.3). If features that are similar in location are also similar in attributes, then the pattern as a whole is said to exhibit *positive spatial autocorrelation*. Conversely, *negative spatial autocorrelation* is said to exist when features which are

close together in space tend to be more dissimilar in attributes than features which are further apart (in opposition to Tobler's Law). Zero autocorrelation occurs when attributes are independent of location. Figure 4.1 presents some simple field representations of a geographic variable in 64 cells that can each take one of two values, coded blue and white. Each of the five illustrations contains the same set of attributes, 32 white cells and 32 blue cells, yet the spatial arrangements are very different. Figure 4.1A presents the familiar chess board, and illustrates extreme negative spatial autocorrelation between neighboring cells. Figure 4.1E presents the opposite extreme of positive autocorrelation, when blue and white cells cluster together in homogeneous regions. The other illustrations show arrangements which exhibit intermediate levels of autocorrelation. Figure 4.1C corresponds to spatial independence, or no autocorrelation, Figure 4.1B shows a relatively dispersed arrangement, and Figure 4.1D a relatively clustered one.

Spatial autocorrelation is determined both by similarities in position, and by similarities in attributes.

The patterns shown in Figure 4.1 are examples of a particular case of spatial autocorrelation. In terms of the classification developed in Chapter 3 (Box 3.3) the attribute data are *nominal* (blue and white simply identify two different possibilities, with no implied order and no possibility of difference, or ratio) and their spatial distribution is conceived as a field, with a single value everywhere. The figure gives no clue as to the true dimensions of the area being represented. Usually, similarities in attribute values may be more precisely measured on higher-order measurement scales, enabling continuous measures of spatial variation (See Section 6.3.2.2 for a discussion of precision). As we see below, the way in which we define what we mean by *neighboring* in

investigating spatial arrangements may be more or less sophisticated. In considering the various arrangements shown in Figure 4.1, we have only considered the relationship between the attributes of a cell and those of its four *immediate* neighbors. But we could include a cell's four diagonal neighbors in the comparison, and more generally there is no reason why we should not interpret Tobler's Law in terms of a gradual incremental attenuating effect of distance as we traverse successive cells.

We began this chapter by considering a time series analysis of events that are highly, even perfectly, repetitive in the short term. Activity patterns often exhibit strong positive temporal autocorrelation (where you were at this time last week, or this time yesterday is likely to affect where you are now), but only if measures are made at the same time every day – that is, at the temporal scale of the daily interval. If, say, sample measurements were taken every 17 hours, measures of the temporal autocorrelation of your activity patterns would likely be much lower. Similarly, if the measures of the blue/white property were made at intervals that did not coincide with the dimensions of the squares of the chess boards in Figure 4.1, then the spatial autocorrelation measures would be different. Thus the issue of *sampling interval* is of direct importance in the measurement of spatial autocorrelation, because spatial events and occurrences may or may not accommodate spatial structure. In general, measures of spatial and temporal autocorrelation are *scale dependent* (see Box 4.2). Scale is often integral to the trade off between the level of spatial resolution and the

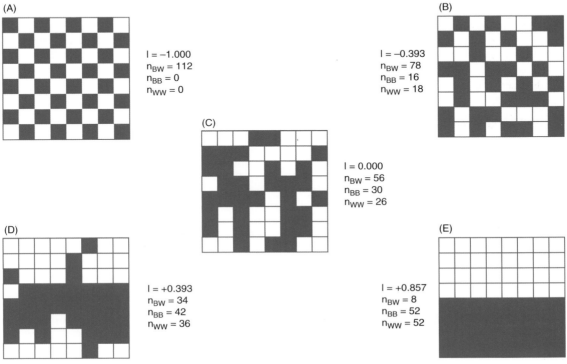

Figure 4.1 Field arrangements of blue and white cells exhibiting: (A) extreme negative spatial autocorrelation; (B) a dispersed arrangement; (C) spatial independence; (D) spatial clustering; and (E) extreme positive spatial autocorrelation. The values of the *I* statistic are calculated using the equation in Section 4.6 (*Source*: Goodchild 1986 CATMOG, GeoBooks, Norwich)

(A) 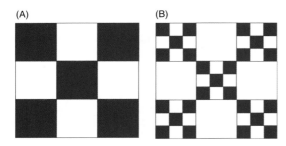 (B)

Figure 4.2 A Sierpinski carpet at two levels of resolution: (A) coarse scale and (B) finer scale

Figure 4.3 Individual rocks may resemble larger-scale structures, such as the mountains from which they are broken, in form

degree of attribute detail that can be stored in a given application – as in the trade off between spatial and spectral resolution in remote sensing.

Quattrochi and Goodchild have undertaken an extensive discussion of these and other meanings of scale (e.g., the degree of spectral or temporal coarseness), and their implications.

A further important property is that of *self-similarity*. This is illustrated using a mosaic of squares in Figure 4.2. Figure 4.2A presents a coarse-scale representation of attributes in nine squares, and a pattern of negative spatial autocorrelation. However, the pattern is self-replicating at finer scales, and in Figure 4.4B, a finer-scale representation reveals that the smallest blue cells replicate the pattern of the whole area in a recursive manner. The pattern of spatial autocorrelation at the coarser scale is replicated at the finer scale, and the overall pattern is said to exhibit the property of *self-similarity*. Self-similar structure is characteristic of natural as well as social systems: for example, a rock may resemble the physical form of the mountain from which it was broken (Figure 4.3), small coastal features may resemble larger

bays and inlets in structure and form, and neighborhoods may be of similar population size and each offer similar ranges of retail facilities right across a metropolitan area. Self-similarity is a core concept of fractals, a topic introduced in Section 4.8.

4.4 Spatial sampling

The quest to represent the myriad complexity of the real world requires us to abstract, or sample, events and occurrences from a *sample frame*, defined as the universe of eligible elements of interest. Thus the process of sampling elements from a sample frame can very much determine the apparent nature of geographic data. A spatial sampling frame might be bounded by the extent

---

Technical Box **4.2**

The many meanings of *scale*

Unfortunately the word *scale* has acquired too many meanings in the course of time. Because they are to some extent contradictory, it is best to use other terms that have clearer meaning where appropriate.

Scale is in the details. Many scientists use scale in the sense of spatial resolution, or the level of spatial detail in data. Data are fine-scaled if they include records of small objects, and coarse-scaled if they do not.

Scale is about extent. Scale is also used by scientists to talk about the geographic extent or scope of a project: a large-scale project covers a large area, and a small-scale project covers a small area. Scale can also refer to other aspects of the project's scope,

including the cost, or the number of people involved.

The scale of a map. Geographic data are often obtained from maps, and often displayed in map form. Cartographers use the term scale to refer to a map's *representative fraction* (the ratio of distance on the map to distance on the ground – see Section 3.7). Unfortunately this leads to confusion (and often bemusement) over the meaning of *large* and *small* with respect to scale. To a cartographer a large scale corresponds to a large representative fraction, in other words to plenty of geographic detail. This is exactly the opposite of what an average scientist understands by a large-scale study. In this book we have tried to avoid this problem by using *coarse* and *fine* instead.

of a field of interest, or by the combined extent of a set of areal objects. We can think of sampling as the process of selecting points from a continuous field or, if the field has been represented as a mosaic of areal objects, of selecting some of these objects while discarding others. Scientific sampling requires that each element in the sample frame has a known and prespecified chance of selection.

In some important senses, we can think of any geographic representation as a kind of sample, in that the elements of reality that are retained are abstracted from the real world in accordance with some overall design. This is the case in remote sensing, for example (see Section 3.6.1), in which each pixel value is a spatially averaged reflectance value calculated at the spatial resolution characteristic of the sensor. In many situations, we will need consciously to select some observations, and not others, in order to create a generalizable abstraction. This is because, as a general rule, the resources available to any given project do not stretch to measuring every single one of the elements (soil profiles, migrating animals, shoppers) that we know to make up our population of interest. And even if resources were available, science tells us that this would be wasteful, since procedures of *statistical inference* allow us to infer from samples to the populations from which they were drawn. We will return to the process of statistical inference in Sections 4.7 and 15.4. Here, we will confine ourselves to the question, how do we ensure a good sample?

Geographic data are only as good as the sampling scheme used to create them.

Classical statistics often emphasizes the importance of randomness in sound sample design. The purest form, simple random sampling, is well known: each element in the sample frame is assigned a unique number, and a prespecified number of elements are selected using a random number generator. In the case of a spatial sample from continuous space, x, y coordinate pairs might be randomly sampled within the range of x and y values (see Section 5.7 for information on coordinate systems). Since each randomly selected element has a known and prespecified probability of selection, it is possible to make robust and defensible generalizations to the population from which the sample was drawn. A spatially random sample is shown in Figure 4.4A. Random sampling is integral to probability theory, and this enables us to use the distribution of values in our sample to tell us something about the likely distribution of values in the parent population from which the sample was drawn.

However, sheer bad luck can mean that randomly drawn elements are disproportionately concentrated amongst some parts of the population at the expense of others, particularly when the size of our sample is small relative to the population from which it was drawn. For example, a survey of household incomes might happen to select households with unusually low incomes. Spatially systematic sampling aims to circumvent this problem and ensure greater evenness of coverage across the sample frame. This is achieved by identifying a regular sampling interval k (equal to the reciprocal of

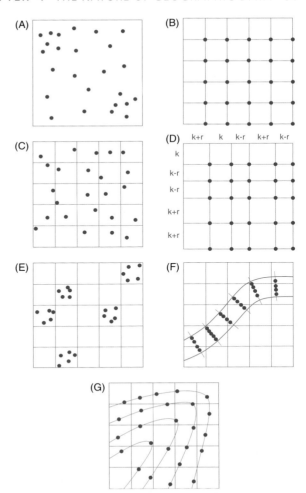

Figure 4.4 Spatial sample designs: (A) simple random sampling; (B) stratified sampling; (C) stratified random sampling; (D) stratified sampling with random variation in grid spacing; (E) clustered sampling; (F) transect sampling; and (G) contour sampling

the sampling fraction N/n, where n is the required sample size and N is the size of the population) and proceeding to select every kth element. In spatial terms, the sampling interval of spatially systematic samples maps into a regularly spaced grid, as shown in Figure 4.4B. This advantage over simple random sampling may be two-edged, however, if the sampling interval and the spatial structure of the study area coincide, that is, the sample frame exhibits *periodicity*. A sample survey of urban land use along streets originally surveyed under the US Public Land Survey System (PLSS: Section 5.5) would be ill-advised to take a sampling interval of one mile, for example, for this was the interval at which blocks within townships were originally laid out, and urban structure is still likely to be repetitive about this original design. In such instances, there may be a consequent failure to detect the true extent of heterogeneity of population attributes (Figure 4.4B) – for example, it is extremely unlikely that the attributes of street intersection locations would be representative of land uses elsewhere in the block structure. A number of systematic and quasi-systematic sample designs have been devised to get around the vulnerability of spatially systematic sample designs to periodicity, and the danger that simple random sampling may generate freak samples. These include stratified random sampling to ensure evenness of coverage (Figure 4.4C) and periodic random changes in the grid width of a spatially systematic sample (Figure 4.4D), perhaps subject to minimum spacing intervals.

In certain circumstances, it may be more efficient to restrict measurement to a specified range of sites – because of the prohibitive costs of transport over large areas, for example. Clustered sample designs, such as that shown in Figure 4.4E, may be used to generalize about attributes if the cluster presents a microcosm of surrounding conditions. In fact this provides a legitimate use of a comprehensive study of one area to say something about conditions beyond it – so long as the study area is known to be representative of the broader study region. For example, political opinion polls are often taken in shopping centers where shoppers can be deemed broadly representative of the population at large. However, instances where they provide a comprehensive detailed picture of spatial structure are likely to be the exception rather than the rule.

Use of either simple random or spatially systematic sampling presumes that each observation is of equal importance, and hence of equal weight, in building a representation. As such, these sample designs are suitable for circumstances in which spatial structure is weak or non-existent, or where (as in circumstances fully described by Tobler's Law) the attenuating effect of distance is constant in all directions. They are also suitable in circumstances where spatial structure is unknown. Yet in most practical applications, spatial structure is (to some extent at least) known, even if it cannot be wholly explained by Tobler's Law. These conditions make it both more efficient and necessary to devise application-specific sample designs. This makes for improved quality of representation, with minimum resource costs of collecting data. Relevant sample designs include sampling along a transect, such as a soil profile (Figure 4.4F), or along a contour line (Figure 4.4G).

Consider the area of Leicestershire, UK, illustrated in Figure 4.5. It depicts a landscape in which the hilly relief of an upland area falls away sharply towards a river's flood plain. In identifying the sample spot heights that we might measure and hold in a GIS to create a representation of this area, we would be advised to sample a disproportionate number of observations in the upland area of the study area where the local variability of heights is greatest.

In a socio-economic context, imagine that you are required to identify the total repair cost of bringing all housing in a city up to a specified standard. (Such applications are common, for example, in forming bids for Federal or Central Government funding.) A GIS that showed the time period in which different neighborhoods were developed (such as the Mid-West settlements simulated in Figure 2.18) would provide a useful guide to effective use of sampling resources. Newer houses

are all likely to be in more or less the same condition, while the repair costs of the older houses are likely to be much more variable and dependent upon the attention that the occupants have lavished upon them. As a general rule, the older neighborhoods warrant a greater sampling frequency than the newer ones, but other considerations may also be accommodated into the sampling design as well – such as construction type (duplex versus apartment, etc.) and local geology (as an indicator of risk of subsidence).

In any application, where the events or phenomena that we are studying are spatially heterogeneous, we will require a large sample to capture the full variability of attribute values at all possible locations. Other parts of the study area may be much more homogeneous in attributes, and a sparser sampling interval may thus be more appropriate. Both simple random and systematic sample designs (and their variants) may be adapted in order to allow a differential sampling interval over a given study area (see Section 6.3.2 for more on this issue with respect to sampling vegetation cover). Thus it may be sensible to partition the sample frame into sub-areas, based on our knowledge of spatial structure – specifically our knowledge of the likely variability of the attributes that we are measuring.

Other application-specific special circumstances include:

- whether source data are ubiquitous or must be specially collected;
- the resources available for any survey undertaking; and
- the accessibility of all parts of the study area to field observation (still difficult even in the era of ubiquitous availability of Global Positioning System receivers: Section 5.8).

> **Stratified sampling designs attempt to allow for the unequal abundance of different phenomena on the Earth's surface.**

It is very important to be aware that this discussion of sampling is appropriate to problems where there is a large hypothetical population of evenly distributed locations (elements, in the terminology of sampling theory, or atoms of information in the terminology of Section 3.4), that each have a known and prespecified probability of selection. Random selection of elements plays a part in each of the sample designs illustrated in Figure 4.4, albeit that the probability of selecting an element may be greater for clearly defined sub-populations – that lie along a contour line or across a soil transect, for example. In circumstances where spatial structure is either weak or is explicitly incorporated through clear definition of sub-populations, standard statistical theory provides a robust framework for inferring the attributes of the population from those of the sample. But the reality is somewhat messier. In most GIS applications, the population of elements (animals, glacial features, voters) may not be large, and its distribution across space may be far from random and independent. In these circumstances, conventional wisdom suggests a number of 'rules of

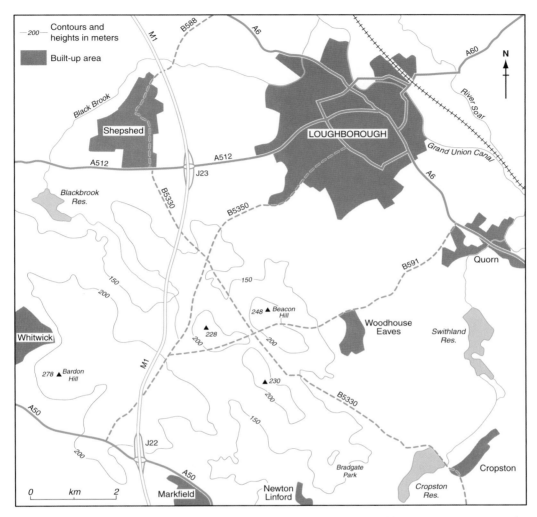

Figure 4.5 An example of physical terrain in which differential sampling would be advisable to construct a representation of elevation (Reproduced by permission of M. Langford, University of Glamorgan)

thumb' to compensate for the likely increase in error in estimating the true population value – as in clustered sampling, where slightly more than doubling the sample size is usually taken to accommodate the effects of spatial autocorrelation within a spatial cluster. However, it may be considered that the existence of spatial autocorrelation fundamentally undermines the inferential framework and invalidates the process of generalizing from samples to populations. We return to discuss this in more detail in our discussion of inference and hypothesis testing in Section 15.4.1.

Finally, it is also worth noting that this discussion assumes that we have the luxury of collecting our own data for our own particular purpose. The reality of analysis in our data-rich world is that more and more of the data that we use are collected by other parties for other purposes: in such cases the metadata of the dataset are crucially important in establishing their provenance for the particular investigation that we may wish to undertake (see Section 11.2.1).

4.5 Distance decay

In selectively abstracting, or sampling, part of reality to hold within a representation, judgment is required to fill in the gaps between the observations that make up a representation. This requires understanding of the likely attenuating effect of distance between the sample observations, and thus of the nature of geographic data (Figure 4.6). That is to say, we need to make an informed judgment about an appropriate *interpolation* function and how to *weight* adjacent observations. A literal interpretation of Tobler's Law implies a continuous, smooth, attenuating effect of distance upon the attribute values of adjacent or contiguous spatial objects, or incremental variation in attribute values as we traverse a field. The polluting effect of a chemical or oil spillage decreases in a predictable (and in still waters, uniform) fashion with distance from the point source; aircraft noise

Figure 4.6 We require different ways of interpolating between points, as well as different sample designs, for representing mountains and forested hillsides

decreases in a linear fashion with distance from the flight path; and the number of visits to a National Park decreases at a regular rate as we traverse the counties that adjoin it. This section focuses on principles, and introduces some of the functions that are used to describe effects over distance, or the nature of geographic variation, while Section 14.4.4 discusses ways in which the principles of distance decay are embodied in techniques of spatial interpolation.

The precise nature of the function used to represent the effects of distance is likely to vary between applications, and Figure 4.7 illustrates several hypothetical types. In mathematical terms, we take b as a parameter that affects the rate at which the weight w_{ij} declines with distance: a small b produces a slow decrease, and a large b a more rapid one. In most applications, the choice of distance attenuation function is the outcome of past experience, the fit of a particular application dataset, and convention. Figure 4.7A presents the simple case of linear distance decay, given by the expression:

$$w_{ij} = a - bd_{ij},$$

for $d_{ij} < a/b$, as might reflect the noise levels experienced across a transect perpendicular to an aircraft flight path. Figure 4.7B presents a negative power distance decay function, given by the expression:

$$w_{ij} = d_{ij}^{-b},$$

which has been used by some researchers to describe the decline in the density of resident population with distance from historic central business district (CBD) areas. Figure 4.7C illustrates a negative exponential statistical fit, given by the expression:

$$w_{ij} = e^{-bd_{ij}},$$

conventionally used in human geography to represent the decrease in retail store patronage with distance from it.

Each of the attenuation functions illustrated in Figure 4.7 is idealized, in that the effects of distance are presumed to be regular, continuous, and *isotropic* (uniform in every direction). This may be appropriate for many applications. The notion of smooth and continuous variation underpins many of the representational traditions in cartography, as in the creation of *isopleth* (or isoline) maps. This is described in Box 4.3. To some extent at least, high school math also conditions us to think of spatial variation as continuous, and as best represented by interpolating smooth curves between everything. Yet our understanding of spatial structure tells us that variation is often far from smooth and continuous. The Earth's surface and geology, for example, are discontinuous at cliffs and fault lines, while the socio-economic geography of cities can be similarly characterized by abrupt changes. Some illustrative physical and social issues pertaining to the catchment of a grocery store are presented in Figure 4.8. A naïve GIS analysis might assume that the maximum extent of the catchment of the store is bounded by an approximately circular area, and that within this area the likelihood (or probability) of shoppers using the store decreases the further away from it that they live. On this basis we might assume a negative exponential distance decay function (Figure 4.7C) and, for practical purposes, an absolute cut-off in patronage beyond a ten

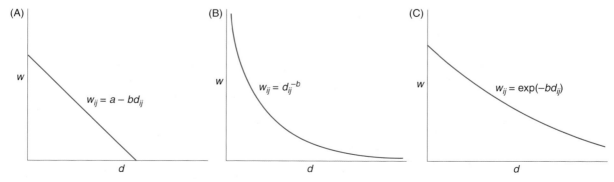

Figure 4.7 The attenuating effect of distance: (A) linear distance decay, $w_{ij} = a - bd_{ij}$; (B) negative power distance decay, $w_{ij} = d_{ij}^{-b}$; and (C) negative exponential distance decay, $w_{ij} = \exp(-bd_{ij})$

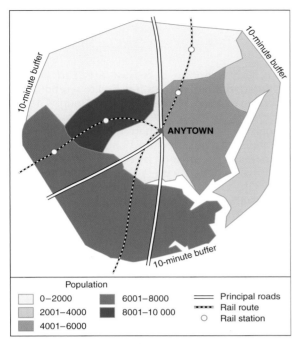

Figure 4.8 Discontinuities in a retail catchment (*Source*: Adapted from Birkin M., Clarke G. P., Clarke M. and Wilson A. G. (1996) *Intelligent GIS*. Cambridge, UK: GeoInformation International). Reproduced by permission of John Wiley & Sons Inc.

minute drive time at average speed. Yet in practice, the catchment also depends upon:

- physical factors, such as rivers and relief;
- road and rail infrastructure and associated capacity, congestion, and access (e.g., rail stations and road access ramps);
- socio-economic factors, that are manifest in differences in customer store preferences;
- administrative geographies, that modify the shape of the circle because census counts of population are only available for administrative zones;
- overlapping trade areas of competing stores, that are likely to truncate the trade area from particular directions;
- a demand constraint, that requires probabilities of patronizing all available stores at any point to sum to 1 (unless people opt out of shopping).

Additionally, the population base to Figure 4.8 raises an important issue of the representation of spatial structure. Remember that we said that the circular retail catchment had to be adapted to fit the administrative geography of population enumeration. The distribution of population is shown using choropleth mapping (Box 4.3), which implicitly assumes that the mapped property is uniformly distributed within zones and that the only important changes in distribution take place at zone boundaries. Such representations can obscure continuous variations and mask the true pattern of distance attenuation.

4.6 Measuring distance effects as spatial autocorrelation

An understanding of spatial structure helps us to deduce a good sampling strategy, to use an appropriate means of interpolating between sampled points, and hence to build a spatial representation that is fit for purpose. Knowledge of the actual or likely nature of spatial autocorrelation can thus be used *deductively* in order to help build a spatial representation of the world. However, in many applications we do not understand enough about geographic variability, distance effects, and spatial structure to invoke deductive reasoning. A further branch of spatial analysis thus emphasizes the *measurement* of spatial autocorrelation as an end in itself. This amounts to a more *inductive* approach to developing an understanding of the nature of a geographic dataset.

Induction reasons from data to build up understanding, while deduction begins with theory and principle as a basis for looking at data.

In Section 4.3 we saw that spatial autocorrelation measures the extent to which similarities in position match similarities in attributes. Methods of measuring spatial autocorrelation depend on the types of objects used as the basis of a representation, and as we saw in Section 4.2, the scale of attribute measurement is important too. Interpretation depends on how the objects relate to our conceptualization of the phenomena they represent. If the phenomenon of interest is conceived as a field, then spatial autocorrelation measures the smoothness of the field using data from the sample points, lines, or areas that represent the field. If the phenomena of interest are conceived as discrete objects, then spatial autocorrelation measures how the attribute values are distributed among the objects, distinguishing between arrangements that are clustered, random, and locally contrasting. Figure 4.11 shows examples of each of the four object types, with associated attributes, chosen to represent situations in which a scientist might wish to measure spatial autocorrelation. The point data in Figure 4.11A comprise data on well bores over an area of 30 km^2, and together provide information on the depth of an aquifer beneath the surface (the blue shading identifies those within a given threshold). We would expect values to exhibit strong spatial autocorrelation, with departures from this indicative of changes in bedrock structure or form. The line data in Figure 4.11B present numbers of accidents for links of road over a lengthy survey period in the Southwestern Ontario, Canada, provincial highway network. Low spatial autocorrelation in these statistics implies that local causative factors (such as badly laid out junctions) account for most accidents, whereas strong spatial autocorrelation would imply a more regional scale of variation, implying a link between accident rates and lifestyles, climate, or population density. The area data in Figure 4.11C illustrate the socio-economic patterning of the south east of England, and beg the question of

Isopleth and choropleth maps

Isopleth maps are used to visualize phenomena that are conceptualized as fields, and measured on interval or ratio scales. An *isoline* connects points with equal attribute values, such as contour lines (equal height above sea level), *isohyets* (points of equal precipitation), *isochrones* (points of equal travel time), or *isodapanes* (points of equal transport cost). Figure 4.9 illustrates the procedures that are used to create a surface about a set of point measurements (Figure 4.9A), such as might be collected from rain gauges across a study region (and see Section 14.4.4 for more technical detail on the process of spatial interpolation). A parsimonious number of user-defined values is identified to define the contour intervals (Figure 4.9B). The GIS then interpolates a contour between point observations of greater and lesser value (Figure 4.9C) using standard procedures of inference, and the other contours are then interpolated using the same procedure (Figure 4.9D). Hue or shading can be added to improve user interpretability (Figure 4.9E).

Choropleth maps are constructed from values describing the properties of non-overlapping areas, such as counties or census tracts. Each area is colored, shaded, or cross-hatched to symbolize the value of a specific variable, as in Figure 4.10. Geographic rules define what happens to the properties of objects when they are split or merged (e.g., see Section 13.3.3). Figure 4.10 compares a map of total population (a spatially extensive variable) with a map of population density (a spatially intensive variable). Spatially extensive variables take values that are true only of entire areas, such as total population, or total number of children under 5 years of age. They are highly misleading – the same color is applied uniformly to each part of an area, yet we know that the mapped property cannot be true of each part of the area. The values taken by spatially intensive variables could potentially be true of every part of an area, if the area were homogeneous – examples include densities, rates, or proportions. Conceptually, a spatially intensive variable is a field, averaged over each area, whereas a spatially extensive variable is a field of density whose values are summed or integrated to obtain each area's value.

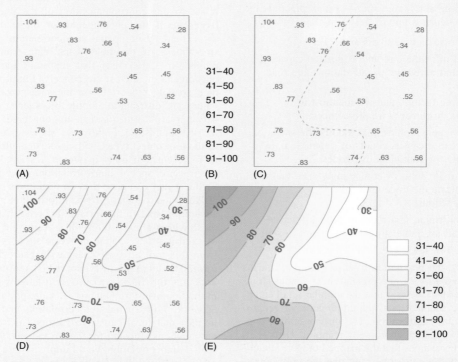

Figure 4.9 The creation of isopleth maps: (A) point attribute values; (B) user-defined classes; (C) interpolation of class boundary between points; (D) addition and labeling of other class boundaries; and (E) use of hue to enhance perception of trends (after Kraak and Ormeling 1996: 161)

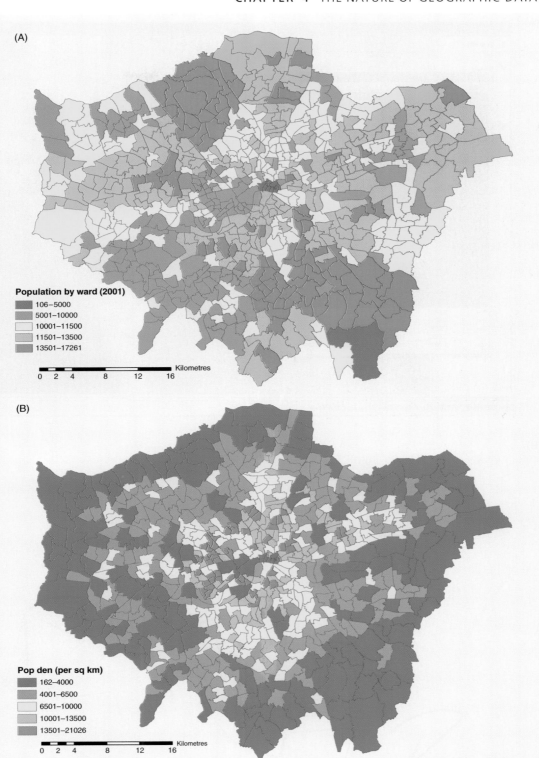

Figure 4.10 Choropleth maps of (A) a spatially extensive variable, total population, and (B) a related but spatially intensive variable, population density. Many cartographers would argue that (A) is misleading, and that spatially extensive variables should always be converted to spatially intensive form (as densities, ratios, or proportions) before being displayed as choropleth maps. (Reproduced with permission of Daryl Lloyd)

(A)

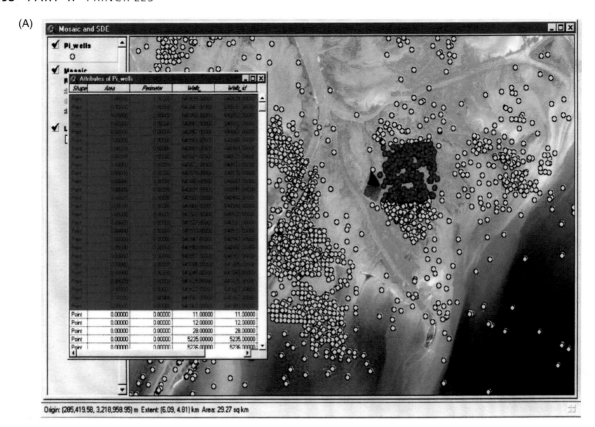

(B)

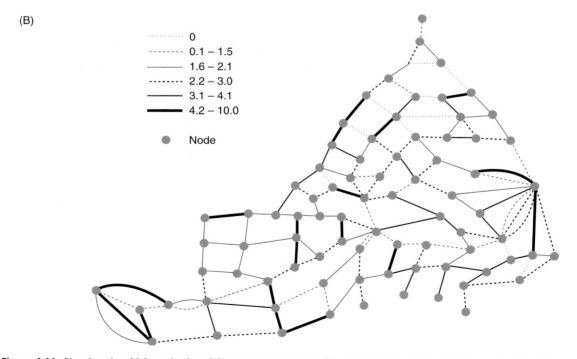

Figure 4.11 Situations in which a scientist might want to measure spatial autocorrelation: (A) point data (wells with attributes stored in a spreadsheet); (B) line data (accident rates in the Southwestern Ontario provincial highway network); (C) area data (percentage of population that are old age pensioners (OAPs) in South East England); and (D) volume data (elevation and volume of buildings in Seattle). (A) and (D) courtesy ESRI; (C) reproduced with permission of Daryl Lloyd

(C)

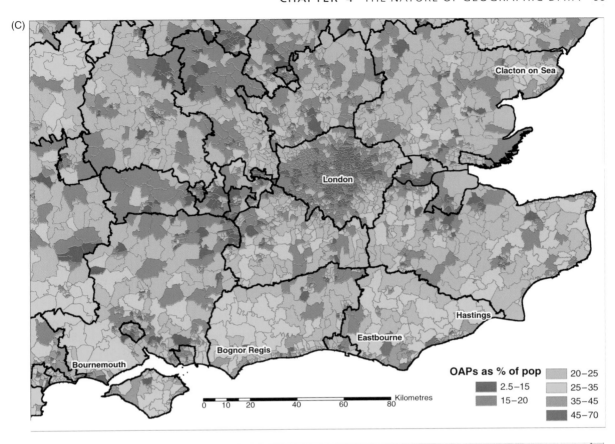

OAPs as % of pop
- 2.5–15
- 15–20
- 20–25
- 25–35
- 35–45
- 45–70

Kilometres
0 10 20 40 60 80

(D)

Figure 4.11 (*continued*)

Technical Box **4.4**

Measuring similarity between neighbors

In the simple example shown in Figure 4.12, we compare neighboring values of spatial attributes by defining a weights matrix **W** in which each element w_{ij} measures the locational similarity of i and j (i identifies the row and j the column of the matrix). We use a simple measure of contiguity, coding $w_{ij} = 1$ if regions i and j are contiguous and $w_{ij} = 0$ otherwise. w_{ii} is set equal to 0 for all i. This is shown in Table 4.1.

Figure 4.12 A simple mosaic of zones

Table 4.1 The weights matrix **W** derived from the zoning system shown in Figure 4.12

	1	2	3	4	5	6	7	8
1	0	1	1	1	0	0	0	0
2	1	0	1	0	0	1	1	0
3	1	1	0	1	1	1	0	0
4	1	0	1	0	1	0	0	0
5	0	0	1	1	0	1	0	1
6	0	1	1	0	1	0	1	1
7	0	1	0	0	0	1	0	1
8	0	0	0	0	1	1	1	0

The weights matrix provides a simple way of representing similarities between location and attribute values, in a region of contiguous areal objects. Autocorrelation is identified by the presence of neighboring cells or zones that take the same (binary) attribute value. More sophisticated measures of w_{ij} include a decreasing function (such as one of those shown in Figure 4.7) of the straight line distance between points at the centers of zones, or the lengths of common boundaries. A range of different spatial metrics may also be used, such as existence of linkage by air, or a decreasing function of travel time by air, road, or rail, or the strength of linkages between individuals or firms on some (non-spatial) network.

The weights matrix makes it possible to develop measures of spatial autocorrelation

using four of the attribute types (nominal, ordinal, interval, and ratio, but not, in practice, cyclic) in Box 3.3 and the dimensioned classes of spatial objects in Box 4.1. Any measure of spatial autocorrelation seeks to compare a set of locational similarities w_{ij} (contained in a weights matrix) with a corresponding set of attribute similarities c_{ij}, combining them into a single index in the form of a cross-product:

$$\sum_i \sum_j c_{ij} w_{ij},$$

This expression is the total obtained by multiplying every cell in the **W** matrix with its corresponding entry in the **C** matrix, and summing.

There are different ways of measuring the attribute similarities, c_{ij}, depending upon whether they are measured on the nominal, ordinal, interval, or ratio scale. For nominal data, the usual approach is to set c_{ij} to 1 if i and j take the same attribute value, and zero otherwise. For ordinal data, similarity is usually based on comparing the ranks of i and j. For interval and ratio data, the attribute of interest is denoted z_i, and the product $(z_i - \bar{z})(z_j - \bar{z})$ is calculated, where $\bar{z}$ denotes the average of the zs.

One of the most widely used spatial autocorrelation statistics for the case of area objects and interval-scale attributes is the Moran Index. This is positive when nearby areas tend to be similar in attributes, negative when they tend to be more dissimilar than one might expect, and approximately zero when attribute values are arranged randomly and independently in space. It is given by the expression:

$$I = \frac{n \sum_i \sum_j w_{ij}(z_i - \bar{z})(z_j - \bar{z})}{\sum_i \sum_j w_{ij} \sum_i (z_i - \bar{z})^2}$$

where n is the number of areal objects in the set. This brief exposition is provided at this point to emphasize the way in which spatial autocorrelation measures are able to accommodate attributes scaled as nominal, ordinal, interval, and ratio data, and to illustrate that there is flexibility in the nature of contiguity (or adjacency) relations that may be specified. Further techniques for measuring spatial autocorrelation are reviewed in connection with spatial interpolation in Section 14.4.4.

whether, at a regional scale, there are commonalties in household structure. The volume data in Figure 4.11D allow some measure of the spatial autocorrelation of high-rise structures to be made, perhaps as part of a study of the way that the urban core of Seattle functions. The way that spatial autocorrelation might actually be calculated for the data used to construct Figure 4.11C is described in Box 4.4.

4.7 Establishing dependence in space

Spatial autocorrelation measures tell us about the inter-relatedness of phenomena across space, one attribute at a time. Another important facet to the nature of geographic data is the tendency for relationships to exist between different phenomena at the same place – between the values of two different fields, between two attributes of a set of discrete objects, or between the attributes of overlapping discrete objects. This section introduces one of the ways of describing such relationships (see also Box 4.5).

How the various properties of a location are related is an important aspect of the nature of geographic data.

In a formal statistical sense, regression analysis allows us to identify the *dependence* of one variable upon one or more *independent* variables. For example, we might hypothesize that the value of individual properties in a city is dependent upon a number of variables such as floor area, distance to local facilities such as parks and schools, standard of repair, local pollution levels, and so forth. Formally this may be written:

$$Y = f(X_1, X_2, X_3, \ldots, X_K)$$

where Y is the dependent variable and X_1 through X_K are all of the possible independent variables that might impact upon property value. It is important to note that it is the independent variables that together affect the dependent variable, and that the hypothesized causal relationship is one way – that is, that property value is *responsive* to floor area, distance to local facilities, standard of repair, and pollution, and not vice versa. For this reason the dependent variable is termed the *response* variable and the independent variables are termed *predictor* variables in some statistics textbooks.

In practice, of course, we will never successfully predict the exact values of any sample of properties. We can identify two broad classes of reasons why this might be the case. First, a property price outcome is the response to a huge range of factors, and it is likely that we will have evidence of and be able to measure only a small

---

Biographical Box **4.5**

Dawn Wright, marine geographer

Dawn Wright (a.k.a. 'Deepsea Dawn' by colleagues and friends: Figure 4.13) is a professor of Geography and Oceanography at Oregon State University (OrSt) in Corvallis, Oregon, USA, where she also directs Davey Jones Locker, a seafloor mapping and marine GIS research laboratory.

Shortly after the deepsea vehicle *Argo I* was used to discover the *HMS Titanic* in 1986, Dawn used some of the first GIS datasets that it collected to develop her Ph.D. at the University of California, Santa Barbara. It was then that she became acutely aware of the challenges of applying GIS to deep ocean environments. When we discuss the nature of the Earth's surface (this chapter) and the way in which it is georeferenced (Chapter 5), we implicitly assume that it is above sea level. Dawn has written widely on the nature of geographic data with regard to the entirety of the Earth's surface – especially the 70% covered by water. Research issues endemic to oceanographic applications of GIS include the handling of spatial data structures that can vary their relative positions and values over time, geostatistical interpolation (Box 4.3 and Section 14.4.4) of data that are sparser in one dimension as compared to the others, volumetric analysis, and the input and management of very large spatial databases. Dawn's research has described the range of these issues and applications, as well as recent advances in marine map-making, charting, and scientific visualization.

Figure 4.13 Dawn Wright, marine geographer, and friend Lydia

Dawn remains a strong advocate of the potential of these issues to not only advance the body of knowledge in GIS design and architecture, but also to inform many of the long-standing research challenges of geographic information science. She says, 'The ocean forces us to think about the nature of geographic data in different ways and to consider radically different ways of representing space and time – we have to go "super-dimensional" to get our minds and our maps around the natural processes at work. We cannot fully rely on the absolute coordinate systems that are so familiar to us in a GIS, or ignore the dissimilarity between the horizontal and the vertical dimension when measuring geographic features and objects. How deep is the ocean at any precise moment in time? How do we represent all of the relevant attributes of the habitat of marine mammals? How can we enforce marine protected area boundaries at depth? Much has been written about the importance of error and uncertainty in geographic analysis. The challenge of gathering data in dynamic marine environments using platforms that are constantly in motion in all directions (roll, pitch, yaw, heave), or of tracking fish, mammals, and birds at sea, creates critical challenges in managing uncertainty in marine position.' These issues of uncertainty (see Chapter 6) also have implications for the establishment of dependence in space (Section 4.7). Dawn and her students continue to develop methods, techniques, and tools for handling data in GIS, but with a unique oceanographic take on data modeling, geocomputation, and the incorporation of spatio-temporal data standards and protocols. Take a dive into Davey Jones Locker to learn more (**dusk.geo.orst.edu/djl**).

subset of these. Second, even if we were able to identify and measure every single relevant independent variable, we would in practice only be able to do so to a given level of measurement precision (for a more detailed discussion of what we mean by precision see Box 6.3 and Section 6.3.2.2). Such caveats do not undermine the wider rationale for trying to generalize, since any assessment of the effects of variables we know about is better than no assessment at all. But our conceptual solution to the problems of unknown and imprecisely measured variables is to subsume them all within a statistical error term, and to revise our regression model so that it looks like this:

$$Y = f(X_1, X_2, X_3, \ldots, X_K) + \varepsilon$$

where ε denotes the error term.

We assume that this relationship holds for each case (which we denote using the subscript i) in our population of interest, and thus:

$$Y_i = f(X_{i1}, X_{i2}, X_{i3}, \ldots, X_{iK}) + \varepsilon_i$$

The essential task of regression analysis is to identify the *direction* and *strength* of the association implied by this equation. This becomes apparent if we rewrite it as:

$$Y_i = b_0 + b_1 X_{i1} + b_2 X_{i2} + b_3 X_{i3} + \cdots + b_K X_{iK} + \varepsilon_i$$

where b_1 through b_K are termed regression *parameters*, which measure the direction and strength of the influence of the independent variables X_1 through X_K on Y. b_0 is termed the *constant* or *intercept* term. This is illustrated in simplified form as a scatterplot in Figure 4.14. Here, for reasons of clarity, the values of the dependent (Y) variable (property value) are regressed and plotted against just one independent (X) variable (floorspace; for more on scatterplots see Section 14.2). The scatter of points

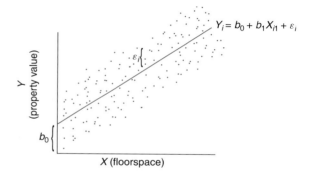

Figure 4.14 The fit of a regression line to a scatter of points, showing intercept, slope and error terms

exhibits an upward trend, suggesting that the response to increased floorspace is a higher property price. A *best fit* line has been drawn through this scatter of points. The gradient of this line is calculated as the b parameter of the regression, and the upward trend of the regression line means that the gradient is positive. The greater the magnitude of the b parameter, the stronger the (in this case positive) effect of marginal increases in the X variable. The value where the regression line intersects the Y axis identifies the property value when floorspace is zero (which can be thought of as the value of the land parcel when no property is built upon it), and gives us the intercept value b_0. The more general multiple regression case works by extension of this principle, and each of the b parameters gauges the marginal effects of its respective X variable.

This kind of effect of floorspace area upon property value is intuitively plausible, and a survey of any sample of individual properties is likely to yield the kind of well-behaved plot illustrated in Figure 4.14. In other cases the overall trend may not be as unambiguous. Figure 4.15A presents a hypothetical plot of the effect of

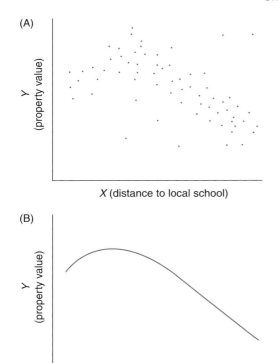

Figure 4.15 (A) A scatterplot and (B) hypothetical relationship between distance to local school and domestic property value

distance to a local school (measured perhaps as straight line distance; see Section 14.3.1 for more on measuring distance in a GIS) upon property value. Here the plot is less well behaved: the overall fit of the regression line is not as good as it might be, and a number of poorly fitting observations (termed high-residual and high-leverage points) present exceptions to a weak general trend. A number of formal statistical measures (notably t statistics and the R^2 measure) as well as less formal diagnostic procedures exist to gauge the statistical fit of the regression model to the data. Details of these can be found in any introductory statistics text, and will not be examined here.

It is easiest to assume that a relationship between two variables can be described by a straight line or linear equation, and that assumption has been followed in this discussion. But although a straight line may be a good first approximation to other functional forms (curves, for example), there is no reason to suppose that linear relationships represent the *truth*, in other words, how the world's social and physical variables are actually related. For example, it might be that very close proximity to the school has a negative effect upon property value (because of noise, car parking, and other localized nuisance) and it is properties at intermediate distances that gain the greatest positive neighborhood effect from this amenity. This is shown in Figure 4.15B: these and

other effects might be accommodated by changing the intrinsic *functional form* of the model.

A straight line or linear distance relationship is the easiest assumption to make and analyze, but it may not be the correct one.

Figure 4.16 identifies the discrepancy between one observed property value and the value that is predicted by the regression line. This difference can be thought of as the error term for individual property i (strictly speaking, it is termed a *residual* when the scatterplot depicts a sample and not a population). The precise slope and intercept of the best fit line is usually identified using the principle of *ordinary least squares* (OLS). OLS regression fits the line through the scatter of points such that the sum of squared residuals across the entire sample is minimized. This procedure is robust and statistically efficient, and yields estimates of the b parameters. But in many situations it is common to try to go further, by *generalizing* results. Suppose the data being analyzed can be considered a representative sample of some larger group. In the field case, sample points might be representative of all of the infinite number of sample points one might select to represent the continuous variation of the field variable (for example, weather stations measuring temperature in new locations, or more soil pits dug to measure soil properties). In the discrete object case, the data analyzed might be only a selection of all of the objects. If this is the case, then statistics provides methods for making accurate and unbiased statements about these larger populations.

Generalization is the process of reasoning from the nature of a sample to the nature of a larger group.

For this to work, several conditions have to hold. First, the sample must be representative, which means for example that every case (element) in the larger group or *population* has a prespecified and independent chance of being selected. The sampling designs discussed in Section 4.4 are one way of ensuring this. But all too

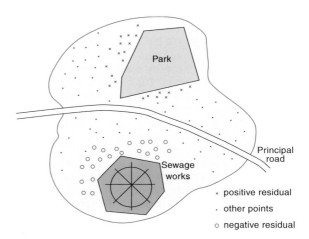

Figure 4.16 A hypothetical spatial pattern of residuals from a regression analysis

often in the analysis of geographic data it turns out to be difficult or impossible to imagine such a population. It is inappropriate, for example, to try to generalize from one study to statements about all of the Earth's surface, if the study was conducted in one area. Generalizations based on samples taken in Antarctica are clearly not representative of all of the Earth's surface. Often GIS provide complete coverage of an area, allowing us to analyze all of the census tracts in a city, or all of the provinces of China. In such cases the apparatus of generalization from samples to populations is unnecessary, and indeed, becomes meaningless.

In addition, the statistical apparatus that allows us to make inferences assumes that there is no *autocorrelation* between errors across space or time. This assumption clearly does not accord with Tobler's Law, where the greater relatedness of near things to one another than distant things is manifest in positive spatial autocorrelation. If strong (positive or negative) spatial autocorrelation is present, the inference apparatus of the ordinary least squares regression procedure rapidly breaks down. The consequence of this is that estimates of the population b parameters become imprecise and the statistical validity of the tests used to confirm the strength and direction of apparent relationships is seriously weakened.

The assumption of zero spatial autocorrelation that is made by many methods of statistical inference is in direct contradiction to Tobler's Law.

The spatial patterning of residuals can provide clues as to whether the structure of space has been correctly specified in the regression equation. Figure 4.16 illustrates the hypothetical spatial distribution of residuals in our property value example – the high clustering of negative residuals around the school suggests that some distance threshold should be added to the specification, or some function that negatively weights property values that are very close to the school. The spatial clustering of residuals can also help to suggest omitted variables that should have been included in the regression specification. Such variables might include the distance to a neighborhood facility that might have strong positive (e.g., a park) or negative (e.g., a sewage works) effect upon values in our property example.

A second assumption of the multiple regression model is that there is no intercorrelation between the independent variables, that is, that no two or more variables essentially measure the same construct. The statistical term for such intercorrelation is *multicollinearity*, and this is a particular problem in GIS applications. GIS is a powerful technology for combining information about a place, and for examining relationships between attributes, whether they be conceptualized as fields, or as attributes of discrete objects. The implication is that each attribute makes a distinct contribution to the total picture of geographic variability. In practice, however, geographic layers are almost always highly correlated. It is very difficult to imagine that two fields representing different variables over the same geographic area would not somehow reveal their common geographic location through similar patterns. For example, a map of rainfall and a map of population density would

clearly have similarities, whether population was dependent on agricultural production and thus rainfall or tended to avoid steep slopes and high elevations where rainfall was also highest.

It is almost impossible to imagine that two maps of different phenomena over the same area would not reveal some similarities.

From this brief overview, it should be clear that there are many important questions about the applicability of such procedures to establishing statistical relationships using spatial data. We return to discuss these in more detail in Section 15.4.

4.8 Taming geographic monsters

Thus far in our discussion of the nature of geographic data we have assumed that spatial variation is smooth and continuous, apart from when we encounter abrupt truncations and discrete shifts at boundaries. However, much spatial variation is not smooth and continuous, but rather is jagged and apparently irregular. The processes which give rise to the form of a mountain range produce features that are spatially autocorrelated (for example, the highest peaks tend to be clustered), yet it would be wholly inappropriate to represent a mountainscape using smooth interpolation between peaks and valley troughs.

Jagged irregularity is a property which is also often observed across a range of scales, and detailed irregularity may resemble coarse irregularity in shape, structure, and form. We commented on this in Section 4.3 when we suggested that a rock broken off a mountain may, for reasons of lithology, represent the mountain in form, and this property is often termed *self-similarity*. Urban geographers also recognize that cities and city systems are also self-similar in organization across a range of scales, and the ways in which this echoes many of the earlier ideas of Christaller's Central Place Theory have been discussed in the academic literature. It is unlikely that idealized smooth curves and conventional mathematical functions will provide useful representations for self-similar, irregular spatial structures: at what scale, if any, does it become appropriate to approximate the San Andreas Fault system by a continuous curve? Urban geographers, for example, have long sought to represent the apparent decline in population density with distance from historic central business districts (CBDs), yet the three-dimensional profiles of cities are characterized by urban canyons between irregularly spaced high-rise buildings (Figure 4.11D). Each of these phenomena is characterized by spatial trends (the largest faults, the largest mountains, and the largest skyscrapers tend to be close to one another), but they are not contiguous and smoothly joined, and the kinds of surface functions shown in Figure 4.7 present inappropriate generalizations of their structure.

For many years, such features were considered geometrical monsters that defied intuition. More recently, however, a more general geometry of the irregular, termed *fractal geometry* by Benoît Mandelbrot, has come to provide a more appropriate and general means of summarizing the structure and character of spatial objects. Fractals can be thought of as geometric objects that are, literally, between Euclidean dimensions, as described in Box 4.6.

In a self-similar object, each part has the same nature as the whole.

Fractal ideas are important, and for many phenomena a measurement of fractal dimension is as important as measures of spatial autocorrelation, or of medians and modes in standard statistics. An important application of

Technical Box 4.6

The strange story of the lengths of geographic objects

How long is the coastline of Maine (Figure 4.17)? (Benoît Mandelbrot, a French mathematician, originally posed this question in 1967 with regard to the coastline of Great Britain.)

Figure 4.17 Part of the Maine coastline

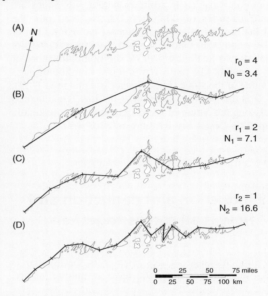

Figure 4.18 The coastline of Maine, at three levels of recursion: (A) the base curve of the coastline; (B) approximation using 100 km steps; (C) 50 km step approximation; and (D) 25 km step approximation

We might begin to measure the stretch of coastline shown in Figure 4.18A. With dividers set to measure 100 km intervals, we would take approximately 3.4 swings and record a length of 340 km (Figure 4.18B).

If we then halved the divider span so as to measure 50 km swings, we would take approximately 7.1 swings and the measured length would increase to 355 km (Figure 4.18C).

If we halved the divider span once again to measure 25 km swings, we would take approximately 16.6 swings and the measured length would increase still further to 415 km (Figure 4.18D).

And so on until the divider span was so small that it picked up all of the detail on this particular representation of the coastline. But even that would not be the end of the story.

If we were to resort instead to field measurement, using a tape measure or the Distance Measuring Instruments (DMIs) used by highway departments, the length would increase still further, as we picked up detail that even the most detailed maps do not seek to represent.

If we were to use dividers, or even microscopic measuring devices, to measure every last grain of sand or earth particle, our recorded length measurement would stretch towards infinity, seemingly without limit.

In short, the answer to our question is that the length of the Maine coastline is indeterminate. More helpfully, perhaps, any approximation is *scale-dependent* – and thus any measurement must also specify scale. The line representation of the coastline also possesses two other properties. First, where small deviations about the overall trend of the coastline resemble larger deviations in form, the coast is said to be *self-similar*. Second, as the path of the coast traverses space, its intricate structure comes to fill up more space than a one-dimensional straight line but less space than a two-dimensional area. As such, it is said to be of *fractional dimension* (and is termed a *fractal*) between 1 (a line) and 2 (an area).

fractal concepts is discussed in Section 15.2.5, and we return again to the issue of length estimation in GIS in Section 14.3.1. Ascertaining the fractal dimension of an object involves identifying the scaling relation between its length or extent and the yardstick (or level of detail) that is used to measure it. Regression analysis, as described in the previous section, provides one (of many) means of establishing this relationship. If we return to the Maine coastline example in Figures 4.17 and 4.18, we might obtain scale dependent coast length estimates (L) of 13.6 (4×3.4), 14.1 (2×7.1) and 16.6 (1×16.6) units for the step lengths (r) used in Figures 4.18B, 4.18C and 4.18D respectively. (It is arbitrary whether the steps are measured in miles or kilometers.) If we then plot the natural log of L (on the y-axis) against the natural log or r for these and other values, we will build up a scatterplot like that shown in Figure 4.19. If the points lie more or less on a straight line and we fit a regression line through it, the value of the slope (b) parameter is equal to ($1 - D$), where D is the fractal dimension of the line. This method for analyzing the nature of geographic lines was originally developed by Lewis Fry Richardson (Box 4.7).

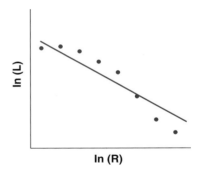

Figure 4.19 The relationship between recorded length (L) and step length (R)

Tobler's Law presents an elementary general rule about spatial structure, and a starting point for the measurement and simulation of spatially autocorrelated structures. This in turn assists us in devising appropriate spatial sampling schemes and creating improved representations, which tell us still more about the real world and how we might represent it. A goal of GIS is often to establish causality between different geographically referenced data, and the multiple regression model potentially provides one means of relating spatial variables to one another, and of inferring from samples to the properties of the populations from which they were drawn. Yet statistical techniques often need to be recast in order to accommodate the special properties of spatial data, and regression analysis is no exception in this regard.

Spatial data provide the foundations to operational and strategic applications of GIS, foundations that must

4.9 Induction and deduction and how it all comes together

The abiding message of this chapter is that spatial is special – that geographic data have a unique nature.

Biographical Box 4.7

Lewis Fry Richardson

Lewis Fry Richardson (1881–1953: Figure 4.20) was one of the founding fathers of the ideas of scaling and fractals. He was brought up a Quaker, and after earning a degree at Cambridge University went to work for the Meteorological Office, but his pacifist beliefs forced him to leave in 1920 when the Meteorological Office was militarized under the Air Ministry. His early work on how atmospheric turbulence is related at different scales established his scientific reputation. Later he became interested in the causes of war and human conflict, and in order to pursue one of his investigations found that he needed a rigorous way of defining the length of a boundary between two states. Unfortunately published lengths tended to vary dramatically, a specific instance being the difference between the lengths of the Spanish–Portuguese border as stated by Spain and by Portugal. He developed a method of walking a pair of dividers along a mapped line, and analyzed the relationship between the length estimate and the setting of the dividers, finding remarkable predictability. In the 1960s Benoît Mandelbrot's concept of fractals finally provided the theoretical framework needed to understand this result.

Figure 4.20 Lewis Fry Richardson: the formalization of scale effects

be used creatively yet rigorously if they are to support the spatial analysis superstructure that we wish to erect. This entails much more than technical competence with software. An understanding of the nature of spatial data allows us to use induction (reasoning from observations) and deduction (reasoning from principles and theory) alongside each other to develop effective spatial representations that are safe to use.

Questions for further study

1. Many jurisdictions tout the number of miles of shoreline in their community – for example, Ottawa County, Ohio, USA claims 107 miles of Lake Erie shoreline. What does this mean, and how could you make it more meaningful?

2. The apparatus of inference was developed by statisticians because they wanted to be able to reason from the results of experiments involving small samples to make conclusions about the results of much larger, hypothetical experiments – for example, in using samples to test the effects of drugs. Summarize the problems inherent in using this apparatus for geographic data in your own words.

3. How many definitions and uses of the word *scale* can you identify?

4. What important aspects of the nature of geographic data have not been covered in this chapter?

Further reading

Batty M. and Longley P.A. 1994 *Fractal Cities: A Geometry of Form and Function*. London: Academic Press.

Mandelbrot B.B. 1983 *The Fractal Geometry of Nature*. San Francisco: Freeman.

Quattrochi D.A. and Goodchild M.F. (eds) 1996 *Scale in Remote Sensing and GIS*. Boca Raton, Florida: Lewis Publishers.

Tate N.J. and Atkinson P.M. (eds) 2001 *Modelling Scale in Geographical Information Science*. Chichester: Wiley.

Wright D. and Bartlett D. (eds) 2000 *Marine and Coastal Geographical Information Systems*. London: Taylor and Francis.

Wright D. 2002 *Undersea with GIS*. Redlands, CA: ESRI Press.

5 *Georeferencing*

Geographic location is the element that distinguishes geographic information from all other types, so methods for specifying location on the Earth's surface are essential to the creation of useful geographic information. Humanity has developed many such techniques over the centuries, and this chapter provides a basic guide for GIS students – what you need to know about georeferencing to succeed in GIS. The first section lays out the principles of georeferencing, including the requirements that any effective system must satisfy. Subsequent sections discuss commonly used systems, starting with the ones closest to everyday human experience, including placenames and street addresses, and moving to the more accurate scientific methods that form the basis of geodesy and surveying. The final sections deal with issues that arise over conversions between georeferencing systems, with the Global Positioning System (GPS), with georeferencing of computers and cellphones, and with the concept of a gazetteer.

Geographic Information Systems and Science, 2nd edition   Paul Longley, Michael Goodchild, David Maguire, and David Rhind.
© 2005 John Wiley & Sons, Ltd.   ISBNs: 0-470-87000-1 (HB); 0-470-87001-X (PB)

Learning Objectives

By the end of this chapter you will:

- Know the requirements for an effective system of georeferencing;

- Be familiar with the problems associated with placenames, street addresses, and other systems used every day by humans;

- Know how the Earth is measured and modeled for the purposes of positioning;

- Know the basic principles of map projections, and the details of some commonly used projections;

- Understand the principles behind GPS, and some of its applications.

5.1 Introduction

Chapter 3 introduced the idea of an atomic element of geographic information: a triple of location, optionally time, and attribute. To make GIS work there must be techniques for assigning values to all three of these, in ways that are understood commonly by people who wish to communicate. Almost all the world agrees on a common calendar and time system, so there are only minor problems associated with communicating that element of the atom when it is needed (although different time zones, different names of the months in different languages, the annual switch to Summer or Daylight Saving Time, and systems such as the classical Japanese convention of dating by the year of the Emperor's reign all sometimes manage to confuse us).

Time is optional in a GIS, but location is not, so this chapter focuses on techniques for specifying location, and the problems and issues that arise. Locations are the basis for many of the benefits of GIS: the ability to map, to tie different kinds of information together because they refer to the same place, or to measure distances and areas. Without locations, data are said to be *non-spatial* or *aspatial* and would have no value at all within a geographic information system.

Time is an optional element in geographic information, but location is essential.

Commonly, several terms are used to describe the act of assigning locations to atoms of information. We use the verbs *to georeference*, *to geolocate*, and *to geocode*, and

say that facts have been *georeferenced* or *geocoded*. We talk about *tagging* records with geographic locations, or about *locating* them. The term *georeference* will be used throughout this chapter.

The primary requirements of a georeference are that it must be *unique*, so that there is only one location associated with a given georeference, and therefore no confusion about the location that is referenced; and that its meaning be *shared* among all of the people who wish to work with the information, including their geographic information systems. For example, the georeference 909 West Campus Lane, Goleta, California, USA points to a single house – there is no other house anywhere on Earth with that address – and its meaning is shared sufficiently widely to allow mail to be delivered to the address from virtually anywhere on the planet. The address may not be meaningful to everyone living in China, but it will be meaningful to a sufficient number of people within China's postal service, so a letter mailed from China to that address will likely be delivered successfully. Uniqueness and shared meaning are sufficient also to allow people to link different kinds of information based on common location: for example, a driving record that is georeferenced by street address can be linked to a record of purchasing. The negative implications of this kind of record linking for human privacy are discussed further by Mark Monmonier (see Box 5.2).

To be as useful as possible a georeference must be *persistent through time*, because it would be very confusing if georeferences changed frequently, and very expensive to update all of the records that depend on them. This can be problematic when a georeferencing system serves more than one purpose, or is used by more than one agency with different priorities. For example, a municipality may expand by incorporating more land, creating problems for mapping agencies, and for researchers who wish to study the municipality through time. Street names sometimes change, and postal agencies sometimes revise postal codes. Changes even occur in the names of cities (Saigon to Ho Chi Minh City), or in their conventional transcriptions into the Roman alphabet (Peking to Beijing).

To be most useful, georeferences should stay constant through time.

Every georeference has an associated spatial resolution (Section 3.4), equal to the size of the area that is assigned that georeference. A mailing address could be said to have a spatial resolution equal to the size of the mailbox, or perhaps to the area of the parcel of land or structure assigned that address. A US state has a spatial resolution that varies from the size of Rhode Island to that of Alaska, and many other systems of georeferencing have similarly wide-ranging spatial resolutions.

Many systems of georeferencing are unique only within an area or *domain* of the Earth's surface. For example, there are many cities with the name Springfield in the USA (18 according to a recent edition of the Rand McNally Road Atlas; similarly there are nine places called Whitchurch in the 2003 AA Road Atlas of the United Kingdom). City name is unique within the domain of a US state, however, a property that was engineered with the advent of the postal system in the 19th century. Today there is no danger of there being two Springfields in Massachusetts, and a driver can confidently ask for directions to 'Springfield, Massachusetts' in the knowledge that there is no danger of being sent to the wrong Springfield. But people living in London, Ontario, Canada are well aware of the dangers of talking about 'London' without specifying the appropriate domain. Even in Toronto, Ontario a reference to 'London' may be misinterpreted as a reference to the older (UK) London on a different continent, rather than to the one 200 km away in the same province (Figure 5.1). Street name is unique in the USA within municipal domains, but not within larger domains such as county or state. The six digits of a UK National Grid reference repeat every 100 km, so additional letters are needed to achieve uniqueness within the national domain (see Box 5.1). Similarly there are 120 places on the Earth's surface with the same Universal Transverse Mercator coordinates (see Section 5.7.2), and a zone number and hemisphere must be added to make a reference unique in the global domain.

While some georeferences are based on simple names, others are based on various kinds of *measurements*, and are called *metric* georeferences. They include latitude and longitude and various kinds of coordinate systems, all of which are discussed in more detail below, and are essential to the making of maps and the display of mapped information in GIS. One enormous advantage of such systems is that they provide the potential for infinitely fine spatial resolution: provided we have sufficiently accurate measuring devices, and use enough decimal places, it is possible with such systems to locate information to any level of accuracy. Another advantage is that from measurements of two or more locations it is possible

Figure 5.1 Placenames are not necessarily unique at the global level – there are many Londons, for example, besides the largest and most prominent one in the UK. People living in other Londons must often add additional information (e.g., London, *Ontario, Canada*) to resolve ambiguity

to compute distances, a very important requirement of georeferencing in GIS.

> **Metric georeferences are much more useful, because they allow maps to be made and distances to be calculated.**

Other systems simply *order* locations. In most countries mailing addresses are ordered along streets, often using the odd integers for addresses on one side and the even integers for addresses on the other. This means that it is possible to say that 3000 State Street and 100 State Street are further apart than 200 State Street and 100 State Street, and allows postal services to sort mail for easy

A national system of georeferencing: the National Grid of Great Britain

The National Grid is administered by the Ordnance Survey of Great Britain, and provides a unique georeference for every point in England, Scotland, and Wales. The first designating letter defines a 500 km square, and the second defines a 100 km square (see Figure 5.2). Within each square, two measurements, called easting and northing, define a location with respect to the lower left corner of the square. The number of digits defines the precision – three digits for easting and three for northing (a total of six) define location to the nearest 100 m.

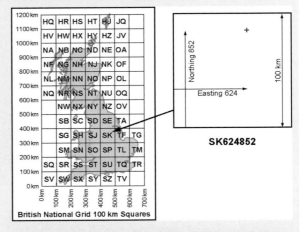

Figure 5.2 The National Grid of Great Britain, illustrating how a point is assigned a grid reference that locates it uniquely to the nearest 100 m (Reproduced by permission of Peter H. Dana)

Table 5.1 Some commonly used systems of georeferencing

System	Domain of uniqueness	Metric?	Example	Spatial resolution
Placename	varies	no	London, Ontario, Canada	varies by feature type
Postal address	global	no, but ordered along streets in most countries	909 West Campus Lane, Goleta, California, USA	size of one mailbox
Postal code	country	no	93117 (US ZIP code); WC1E 6BT (UK unit postcode)	area occupied by a defined number of mailboxes
Telephone calling area	country	no	805	varies
Cadastral system	local authority	no	Parcel 01452954, City of Springfield, Mass, USA	area occupied by a single parcel of land
Public Land Survey System	Western USA only, unique to Prime Meridian	yes	Sec 5, Township 4N, Range 6E	defined by level of subdivision
Latitude/longitude	global	yes	119 degrees 45 minutes West, 34 degrees 40 minutes North	infinitely fine
Universal Transverse Mercator	zones six degrees of longitude wide, and N or S hemisphere	yes	563146E, 4356732N	infinitely fine
State Plane Coordinates	USA only, unique to state and to zone within state	yes	55086.34E, 75210.76N	infinitely fine

delivery. In the western United States, it is often possible to infer estimates of the distance between two addresses on the same street by knowing that 100 addresses are assigned to each city block, and that blocks are typically between 120 m and 160 m long.

This section has reviewed some of the general properties of georeferencing systems, and Table 5.1 shows some commonly used systems. The following sections discuss the specific properties of the systems that are most important in GIS applications.

5.2 Placenames

Giving names to places is the simplest form of georeferencing, and was most likely the one first developed by early hunter-gatherer societies. Any distinctive feature on the landscape, such as a particularly old tree, can serve as a point of reference for two people who wish to share information, such as the existence of good game in the tree's vicinity. Human landscapes rapidly became littered with names, as people sought distinguishing labels to use in describing aspects of their surroundings, and other people adopted them. Today, of course, we have a complex system of naming oceans, continents, cities, mountains, rivers, and other prominent features. Each country maintains a system of authorized naming, often through national or state committees assigned with the task of standardizing geographic names. Nevertheless multiple names are often attached to the same feature, for example when cultures try to preserve the names given to features by

the original or local inhabitants (for example, Mt Everest to many, but Chomolungma to many Tibetans), or when city names are different in different languages (Florence in English, Firenze in Italian)

Many commonly used placenames have meanings that vary between people, and with the context in which they are used.

Language extends the power of placenames through words such as 'between', which serve to refine references to location, or 'near', which serve to broaden them. 'Where State Street crosses Mission Creek' is an instance of combining two placenames to achieve greater refinement of location than either name could achieve individually. Even more powerful extensions come from combining placenames with directions and distances, as in '200 m north of the old tree' or '50 km west of Springfield'.

But placenames are of limited use as georeferences. First, they often have very coarse spatial resolution. 'Asia' covers over 43 million sq km, so the information that something is located 'in Asia' is not very helpful in pinning down its location. Even Rhode Island, the smallest state of the USA, has a land area of over 2700 sq km. Second, only certain placenames are officially authorized by national or subnational agencies. Many more are recognized only locally, so their use is limited to communication between people in the local community. Placenames may even be lost through time: although there are many contenders, we do not know with certainty where the 'Camelot' described in the English legends of King Arthur was located, if indeed it ever existed.

The meaning of certain placenames can become lost through time.

5.3 Postal addresses and postal codes

Postal addresses were introduced after the development of mail delivery in the 19th century. They rely on several assumptions:

- Every dwelling and office is a potential destination for mail;

- Dwellings and offices are arrayed along paths, roads, or streets, and numbered accordingly;

- Paths, roads, and streets have names that are unique within local areas;

- Local areas have names that are unique within larger regions; and

- Regions have names that are unique within countries.

If the assumptions are true, then mail address provides a unique identification for every dwelling on Earth.

Today, postal addresses are an almost universal means of locating many kinds of human activity: delivery of mail, place of residence, or place of business. They fail, of course, in locating anything that is not a potential destination for mail, including almost all kinds of natural features (Mt Everest does not have a postal address, and neither does Manzana Creek in Los Padres National Forest in California, USA). They are not as useful when dwellings are not numbered consecutively along streets, as happens in some cultures (notably in Japan, where street numbering can reflect date of construction, not sequence along the street – it is temporal, rather than spatial) and in large building complexes like condominiums. Many GIS applications rely on the ability to locate activities by postal address, and to convert addresses to some more universal system of georeferencing, such as latitude and longitude, for mapping and analysis.

Postal addresses work well to georeference dwellings and offices, but not natural features.

Postal codes were introduced in many countries in the late 20th century in order to simplify the sorting of mail. In the Canadian system, for example, the first three characters of the six-character code identify a Forward Sortation Area, and mail is initially sorted so that all mail directed to a single FSA is together. Each FSA's incoming mail is accumulated in a local sorting station, and sorted a second time by the last three characters of the code, to allow it to be delivered easily. Figure 5.3 shows a map of the FSAs for an area of the Toronto metropolitan region. The full six characters are unique to roughly ten houses, a single large business, or a single building. Much effort went into ensuring widespread adoption of the coding system by the general public and businesses, and computer programs were developed to assign codes automatically to addresses for large-volume mailers.

Postal codes have proven very useful for many purposes besides the sorting and delivery of mail.

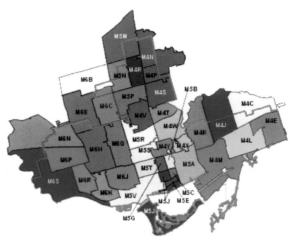

Figure 5.3 Forward Sortation Areas (FSAs) of the central part of the Toronto metropolitan region. FSAs form the first three characters of the six-character Canadian postal code

Although the area covered by a Canadian FSA or a US ZIP code varies, and can be changed whenever the postal authorities want, it is sufficiently constant to be useful for mapping purposes, and many businesses routinely make maps of their customers by counting the numbers present in each postal code area, and dividing by total population to get a picture of market penetration. Figure 5.4 shows an example of summarizing data by ZIP code. Most people know the postal code of their home, and in some instances postal codes have developed popular images (the ZIP code for Beverly Hills, California, 90210, became the title of a successful television series).

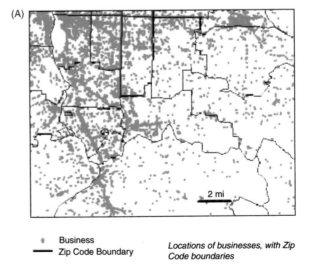

(A)

• Business	*Locations of businesses, with Zip*
—— Zip Code Boundary	*Code boundaries*

Figure 5.4 The use of ZIP codes boundaries as a convenient basis for summarizing data. (A) In this instance each business has been allocated to its ZIP code, and (B) the ZIP code areas have been shaded according to the density of businesses per square mile

(B)

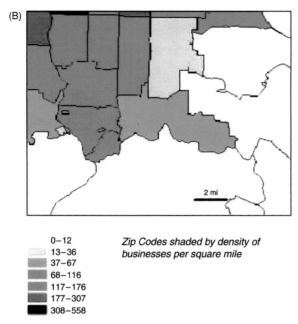

0–12
13–36
37–67
68–116
117–176
177–307
308–558

Zip Codes shaded by density of businesses per square mile

Figure 5.4 (*continued*)

5.4 Linear referencing systems

A linear referencing system identifies location on a network by measuring distance from a defined point of reference along a defined path in the network. Figure 5.5 shows an example, an accident whose location is reported as being a measured distance from a street intersection, along a named street. Linear referencing is closely related to street address, but uses an explicit measurement of distance rather than the much less reliable surrogate of street address number.

Linear referencing is widely used in applications that depend on a linear network. This includes highways (e.g.,

Mile 1240 of the Alaska Highway), railroads (e.g., 25.9 miles from Paddington Station in London on the main line to Bristol, England), electrical transmission, pipelines, and canals. Linear references are used by highway agencies to define the locations of bridges, signs, potholes, and accidents, and to record pavement condition.

Linear referencing systems are widely used in managing transportation infrastructure and in dealing with emergencies.

Linear referencing provides a sufficient basis for georeferencing for some applications. Highway departments often base their records of accident locations on linear references, as well as their inventories of signs and bridges (GIS has many applications in transportation that are known collectively as GIS-T, and in the developing field of intelligent transportation systems or ITS). But for other applications it is important to be able to convert between linear references and other forms, such as latitude and longitude. For example, the Onstar system that is installed in many Cadillacs sold in the USA is designed to radio the position of a vehicle automatically as soon as it is involved in an accident. When the airbags deploy, a GPS receiver determines position, which is then relayed to a central dispatch office. Emergency response centers often use street addresses and linear referencing to define the locations of accidents, so the latitude and longitude received from the vehicle must be converted before an emergency team can be sent to the accident.

Linear referencing systems are often difficult to implement in practice in ways that are robust in all situations. In an urban area with frequent intersections it is relatively easy to measure distance from the nearest one (e.g., on Birch St 87 m west of the intersection with Main St). But in rural areas it may be a long way from the nearest intersection. Even in urban areas it is not uncommon for two streets to intersect more than once (e.g., Birch may have two intersections with Columbia Crescent). There may also be difficulties in defining distance accurately, especially if roads include steep sections where the distance driven is significantly longer than the distance evaluated on a two-dimensional digital representation (Section 14.3.1).

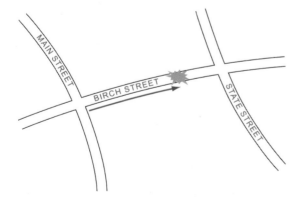

Figure 5.5 Linear referencing – an incident's position is determined by measuring its distance (87 m) along one road (Birch St) from a well-defined point (its intersection with Main St)

5.5 Cadasters and the US Public Land Survey System

The *cadaster* is defined as the map of land ownership in an area, maintained for the purposes of taxing land, or of creating a public record of ownership. The process of *subdivision* creates new parcels by legally subdividing existing ones.

Parcels of land in a cadaster are often uniquely identified, by number or by code, and are also reasonably persistent through time, and thus satisfy the requirements of a georeferencing system. But very few people know the identification code of their home parcel, and use of the cadaster as a georeferencing system is thus limited largely to local officials, with one major exception.

The US Public Land Survey System (PLSS) evolved out of the need to survey and distribute the vast land resource of the Western USA, starting in the early 19th century, and expanded to become the dominant system of cadaster for all of the USA west of Ohio, and all of Western Canada. Its essential simplicity and regularity make it useful for many purposes, and understandable by the general public. Its geometric regularity also allows it to satisfy the requirement of a metric system of georeferencing, because each georeference is defined by measured distances.

The Public Land Survey System defines land ownership over much of western North America, and is a useful system of georeferencing.

To implement the PLSS in an area, a surveyor first lays out an accurate north–south line or *prime meridian*. Rows are then laid out six miles apart and perpendicular to this line, to become the *townships* of the system. Then blocks or *ranges* are laid out in six mile by six mile squares on either side of the prime meridian (see Figure 5.6). Each square is referenced by township number, range number, whether it is to the east or to the west, and the name of the prime meridian. Thirty-six *sections* of one mile by one mile are laid out inside each township, and numbered using a standard system (note how the numbers reverse in every other row). Each section is divided into four quarter-sections of a quarter of a square mile, or 160 acres, the size of the nominal family farm or homestead in the original conception of the PLSS. The process can be continued by subdividing into four to obtain any level of spatial resolution.

The PLSS would be a wonderful system if the Earth were flat. To account for its curvature the squares are not perfectly six miles by six miles, and the rows must be offset frequently; and errors in the original surveying complicate matters still further, particularly in rugged landscapes. Figure 5.6 shows the offsetting exaggerated for a small area. Nevertheless, the PLSS remains an efficient system, and one with which many people in the Western USA and Western Canada are familiar. It is often used to specify location, particularly in managing natural resources in the oil and gas industry and in mining, and in agriculture. Systems have been built to convert PLSS locations automatically to latitude and longitude.

5.6 Measuring the Earth: latitude and longitude

The most powerful systems of georeferencing are those that provide the potential for very fine spatial resolution, that allow distance to be computed between pairs of locations, and that support other forms of spatial analysis. The system of latitude and longitude is in many ways the most comprehensive, and is often called the *geographic* system of coordinates, based on the Earth's rotation about its center of mass.

To define latitude and longitude we first identify the *axis* of the Earth's rotation. The Earth's center of mass lies on the axis, and the plane through the center of mass perpendicular to the axis defines the *Equator*. Slices through the Earth parallel to the axis, and perpendicular to the plane of the Equator, define lines of constant longitude (Figure 5.7), rather like the segments of an

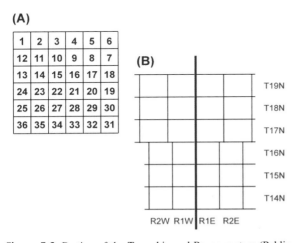

(A)

1	2	3	4	5	6
12	11	10	9	8	7
13	14	15	16	17	18
24	23	22	21	20	19
25	26	27	28	29	30
36	35	34	33	32	31

(B)

Figure 5.6 Portion of the Township and Range system (Public Lands Survey System) widely used in the western USA as the basis of land ownership (shown on the right). Townships are laid out in six-mile squares on either side of an accurately surveyed Prime Meridian. The offset shown between ranges 16 N and 17 N is needed to accommodate the Earth's curvature (shown much exaggerated). The square mile sections within each township are numbered as shown in the upper left

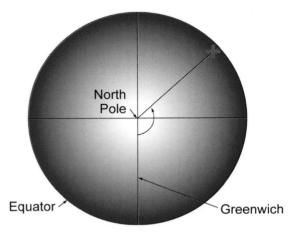

Figure 5.7 Definition of longitude. The Earth is seen here from above the North Pole, looking along the Axis, with the Equator forming the outer circle. The location of Greenwich defines the Prime Meridian. The longitude of the point at the center of the red cross is determined by drawing a plane through it and the axis, and measuring the angle between this plane and the Prime Meridian

orange. A slice through a line marked on the ground at the Royal Observatory in Greenwich, England defines zero longitude, and the angle between this slice and any other slice defines the latter's measure of longitude. Each of the 360 degrees of longitude is divided into 60 minutes and each minute into 60 seconds. But it is more conventional to refer to longitude by degrees East or West, so longitude ranges from 180 degrees West to 180 degrees East. Finally, because computers are designed to handle numbers ranging from very large and negative to very large and positive, we normally store longitude in computers as if West were negative and East were positive; and we store parts of degrees using decimals rather than minutes and seconds. A line of constant longitude is termed a *meridian*.

Longitude can be defined in this way for any rotating solid, no matter what its shape, because the axis of rotation and the center of mass are always defined. But the definition of latitude requires that we know something about the shape. The Earth is a complex shape that is only approximately spherical. A much better approximation or *figure of the Earth* is the *ellipsoid of rotation*, the figure formed by taking a mathematical ellipse and rotating it about its shorter axis (Figure 5.8). The term *spheroid* is also commonly used.

The difference between the ellipsoid and the sphere is measured by its *flattening*, or the reduction in the minor axis relative to the major axis. Flattening is defined as:

$$f = (a - b)/a$$

where *a* and *b* are the lengths of the major and minor axes respectively (we usually refer to the *semi*-axes, or half the lengths of the axes, because these are comparable to radii). The actual flattening is about 1 part in 300.

The Earth is slightly flattened, such that the distance between the Poles is about 1 part in 300 less than the diameter at the Equator.

Much effort was expended over the past 200 years in finding ellipsoids that best approximated the shape of the Earth in particular countries, so that national mapping agencies could measure position and produce accurate maps. Early ellipsoids varied significantly in their basic parameters, and were generally not centered on the Earth's center of mass. But the development of intercontinental ballistic missiles in the 1950s and the need to target them accurately, as well as new data available from satellites, drove the push to a single international standard. Without a single standard, the maps produced by different countries using different ellipsoids could never be made to fit together along their edges, and artificial steps and offsets were often necessary in moving from one country to another (navigation systems in aircraft would have to be corrected, for example).

The ellipsoid known as WGS84 (the World Geodetic System of 1984) is now widely accepted, and North American mapping is being brought into conformity with it through the adoption of the virtually identical North American Datum of 1983 (NAD83). It specifies a semi-major axis (distance from the center to the Equator) of 6378137 m, and a flattening of 1 part in 298.257. But many other ellipsoids remain in use in other parts of the world, and many older data still adhere to earlier standards, such as the North American Datum of 1927 (NAD27). Thus GIS users sometimes need to convert between datums, and functions to do that are commonly available.

We can now define latitude. Figure 5.9 shows a line drawn through a point of interest perpendicular to the ellipsoid at that location. The angle made by this line with the plane of the Equator is defined as the point's latitude, and varies from 90 South to 90 North. Again, south latitudes are usually stored as negative numbers and north latitudes as positive. Latitude is often symbolized by the Greek letter phi (ϕ) and longitude by the Greek letter lambda (λ), so the respective ranges can be expressed in mathematical shorthand as: $-180 \leq \lambda \leq 180$; $-90 \leq \phi \leq 90$. A line of constant latitude is termed a *parallel*.

It is important to have a sense of what latitude and longitude mean in terms of distances on the surface. Ignoring the flattening, two points on the same north–south line of longitude and separated by one degree

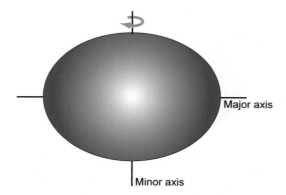

Figure 5.8 Definition of the ellipsoid, formed by rotating an ellipse about its minor axis (corresponding to the axis of the Earth's rotation)

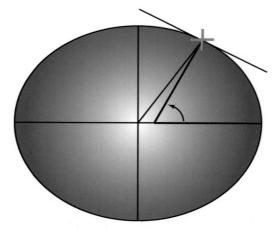

Figure 5.9 Definition of the latitude of the point marked with the red cross, as the angle between the Equator and a line drawn perpendicular to the ellipsoid

of latitude are 1/360 of the circumference of the Earth, or about 111 km, apart. One minute of latitude corresponds to 1.86 km, and also defines one nautical mile, a unit of distance that is still commonly used in navigation. One second of latitude corresponds to about 30 m. But things are more complicated in the east–west direction, and these figures only apply to east–west distances along the Equator, where lines of longitude are furthest apart. Away from the Equator the length of a line of latitude gets shorter and shorter, until it vanishes altogether at the poles. The degree of shortening is approximately equal to the cosine of latitude, or $\cos\phi$, which is 0.866 at 30 degrees North or South, 0.707 at 45 degrees, and 0.500 at 60 degrees. So a degree of longitude is only 55 km along the northern boundary of the Canadian province of Alberta (exactly 60 degrees North).

Lines of latitude and longitude are equally far apart only at the Equator; towards the Poles lines of longitude converge.

Given latitude and longitude it is possible to determine distance between any pair of points, not just pairs along lines of longitude or latitude. It is easiest to pretend for a moment that the Earth is spherical, because the flattening of the ellipsoid makes the equations much more complex. On a spherical Earth the shortest path between two points is a *great circle*, or the arc formed if the Earth is sliced through the two points and through its center (Figure 5.10; an off-center slice creates a *small circle*). The length of this arc on a spherical Earth of radius R is given by:

$$R \arccos[\sin\phi_1 \sin\phi_2 + \cos\phi_1 \cos\phi_2 \cos(\lambda_1 - \lambda_2)]$$

where the subscripts denote the two points (and see the discussion of Measurement in Section 14.3). For example, the distance from a point on the Equator at longitude

90 East (in the Indian Ocean between Sri Lanka and the Indonesian island of Sumatra) and the North Pole is found by evaluating the equation for $\phi_1 = 0$, $\lambda_1 = 90$, $\phi_2 = 90$, $\lambda_2 = 90$. It is best to work in radians (1 radian is 57.30 degrees, and 90 degrees is $\pi/2$ radians). The equation evaluates to $R \arccos 0$, or $R\,\pi/2$, or one quarter of the circumference of the Earth. Using a radius of 6378 km this comes to 10 018 km, or close to 10 000 km (not surprising, since the French originally defined the meter in the late 18th century as one ten millionth of the distance from the Equator to the Pole).

5.7 Projections and coordinates

Latitude and longitude define location on the Earth's surface in terms of angles with respect to well-defined references: the Royal Observatory at Greenwich, the center of mass, and the axis of rotation. As such, they constitute the most comprehensive system of georeferencing, and support a range of forms of analysis, including the calculation of distance between points, on the curved surface of the Earth. But many technologies for working with geographic data are inherently flat, including paper and printing, which evolved over many centuries long before the advent of digital geographic data and GIS. For various reasons, therefore, much work in GIS deals with a flattened or *projected* Earth, despite the price we pay in the distortions that are an inevitable consequence of flattening. Specifically, the Earth is often flattened because:

- paper is flat, and paper is still used as a medium for inputting data to GIS by scanning or digitizing (see Chapter 9), and for outputting data in map or image form;

- rasters are inherently flat, since it is impossible to cover a curved surface with equal squares without gaps or overlaps;

- photographic film is flat, and film cameras are still used widely to take images of the Earth from aircraft to use in GIS;

- when the Earth is seen from space, the part in the center of the image has the most detail, and detail drops off rapidly, the back of the Earth being invisible; in order to see the whole Earth with approximately equal detail it must be distorted in some way, and it is most convenient to make it flat.

The Cartesian coordinate system (Figure 5.11) assigns two coordinates to every point on a flat surface, by measuring distances from an origin parallel to two axes drawn at right angles. We often talk of the two axes as x and y, and of the associated coordinates as the x and y coordinate, respectively. Because it is common to align the y axis with North in geographic applications, the coordinates of a projection on a flat sheet are often termed *easting* and *northing*.

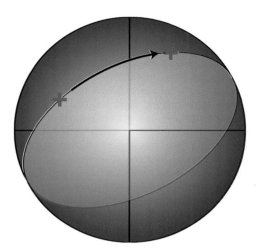

Figure 5.10 The shortest distance between two points on the sphere is an arc of a great circle, defined by slicing the sphere through the two points and the center (all lines of longitude, and the Equator, are great circles). The circle formed by a slice that does not pass through the center is a small circle (all lines of latitude except the Equator are small circles)

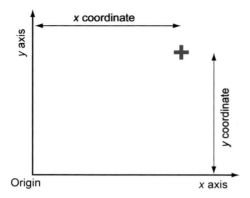

Figure 5.11 A Cartesian coordinate system, defining the location of the blue cross in terms of two measured distances from the Origin, parallel to the two axes

Although projections are not absolutely required, there are several good reasons for using them in GIS to flatten the Earth.

One way to think of a map projection, therefore, is that it transforms a position on the Earth's surface identified by latitude and longitude (ϕ, λ) into a position in Cartesian coordinates (x, y). Every recognized map projection, of which there are many, can be represented as a pair of mathematical functions:

$$x = f(\phi, \lambda)$$
$$y = g(\phi, \lambda)$$

For example, the famous Mercator projection uses the functions:

$$x = \lambda$$
$$y = \ln \tan[\phi/2 + \pi/4]$$

where ln is the natural log function. The inverse transformations that map Cartesian coordinates back to latitude and longitude are also expressible as mathematical functions: in the Mercator case they are:

$$\lambda = x$$
$$\phi = 2 \arctan e^y - \pi/2$$

where e denotes the constant 2.71828. Many of these functions have been implemented in GIS, allowing users to work with virtually any recognized projection and datum, and to convert easily between them.

Two datasets can differ in both the projection and the datum, so it is important to know both for every data set.

Projections necessarily distort the Earth, so it is impossible in principle for the scale (distance on the map compared to distance on the Earth, for a discussion of scale see Box 4.2) of any flat map to be perfectly uniform,

or for the pixel size of any raster to be perfectly constant. But projections can preserve certain properties, and two such properties are particularly important, although any projection can achieve at most one of them, not both:

- the *conformal* property, which ensures that the shapes of small features on the Earth's surface are preserved on the projection: in other words, that the scales of the projection in the x and y directions are always equal;

- the *equal area* property, which ensures that areas measured on the map are always in the same proportion to areas measured on the Earth's surface.

The conformal property is useful for navigation, because a straight line drawn on the map has a constant bearing (the technical term for such a line is a *loxodrome*). The equal area property is useful for various kinds of analysis involving areas, such as the computation of the area of someone's property.

Besides their distortion properties, another common way to classify map projections is by analogy to a physical model of how positions on the map's flat surface are related to positions on the curved Earth. There are three major classes (Figure 5.12):

- *cylindrical* projections, which are analogous to wrapping a cylinder of paper around the Earth, projecting the Earth's features onto it, and then unwrapping the cylinder;

- *azimuthal* or *planar* projections, which are analogous to touching the Earth with a sheet of flat paper; and

- *conic* projections, which are analogous to wrapping a sheet of paper around the Earth in a cone.

In each case, the projection's *aspect* defines the specific relationship, e.g., whether the paper is wrapped around the Equator, or touches at a pole. Where the paper coincides with the surface the scale of the projection is 1, and where the paper is some distance outside the surface the projected feature will be larger than it is on the Earth. *Secant* projections attempt to minimize distortion by allowing the paper to cut through the surface, so that scale can be both greater and less than 1 (Figure 5.12; projections for which the paper touches the Earth and in which scale is always 1 or greater are called *tangent*).

All three types can have either conformal or equal area properties, but of course not both. Figure 5.13 shows examples of several common projections, and shows how the lines of latitude and longitude map onto the projection, in a (distorted) grid known as a *graticule*.

The next sections describe several particularly important projections in detail, and the coordinate systems that they produce. Each is important to GIS, and users are likely to come across them frequently. The map projection (and datum) used to make a dataset is sometimes not known to the user of the dataset, so it is helpful to know enough about map projections and coordinate systems to make intelligent guesses when trying to combine such a dataset with other data. Several excellent books on map projections are listed in the References.

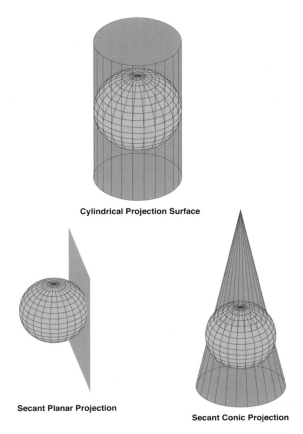

Cylindrical Projection Surface

Secant Planar Projection

Secant Conic Projection

Figure 5.12 The basis for three types of map projections – cylindrical, planar, and conic. In each case a sheet of paper is wrapped around the Earth, and positions of objects on the Earth's surface are projected onto the paper. The cylindrical projection is shown in the *tangent* case, with the paper touching the surface, but the planar and conic projections are shown in the *secant* case, where the paper cuts into the surface (Reproduced by permission of Peter H. Dana)

5.7.1 The Plate Carrée or Cylindrical Equidistant projection

The simplest of all projections simply maps longitude as *x* and latitude as *y*, and for that reason is also known informally as the *unprojected* projection. The result is a heavily distorted image of the Earth, with the poles smeared along the entire top and bottom edges of the map, and a very strangely shaped Antarctica. Nevertheless, it is the view that we most often see when images are created of the entire Earth from satellite data (for example in illustrations of sea surface temperature that show the El Niño or La Niña effects). The projection is not conformal (small shapes are distorted) and not equal area, though it does maintain the correct distance between every point and the Equator. It is normally used only for the whole Earth, and maps of large parts of the Earth, such as the USA or Canada, look distinctly odd in this projection. Figure 5.14 shows the projection applied to the world, and also shows a comparison of three familiar projections of

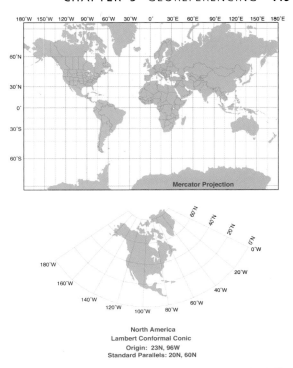

Figure 5.13 Examples of some common map projections. The Mercator projection is a tangent cylindrical type, shown here in its familiar Equatorial aspect (cylinder wrapped around the Equator). The Lambert Conformal Conic projection is a secant conic type. In this instance the cone onto which the surface was projected intersected the Earth along two lines of latitude: 20 North and 60 North (Reproduced by permission of Peter H. Dana)

the United States: the Plate Carrée, Mercator, and Lambert Conformal Conic.

When longitude is assigned to *x* and latitude to *y* a very odd-looking Earth results.

Serious problems can occur when doing analysis using this projection. Moreover, since most methods of analysis in GIS are designed to work with Cartesian coordinates rather than latitude and longitude, the same problems can arise in analysis when a dataset uses latitude and longitude, or so-called geographic coordinates. For example, a command to generate a circle of radius one unit in this projection will create a figure that is two degrees of latitude across in the north–south direction, and two degrees of longitude across in the east–west direction. On the Earth's surface this figure is not a circle at all, and at high latitudes it is a very squashed ellipse. What happens if you ask your favorite GIS to generate a circle and add it to a dataset that is in geographic coordinates? Does it recognize that you are using geographic coordinates and automatically compensate for the differences in distances east–west and north–south away from the Equator, or does it in effect operate on a Plate Carrée projection and create a figure that is an ellipse on the Earth's surface? If you ask it to compute distance between two points defined by

(A)

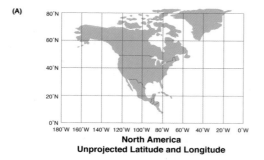

North America
Unprojected Latitude and Longitude

(B) **Three Map Projections Centered at 39 N and 96 W**

Figure 5.14 (A) The so-called *unprojected* or Plate Carrée projection, a tangent cylindrical projection formed by using longitude as *x* and latitude as *y*. (B) A comparison of three familiar projections of the USA. The Lambert Conformal Conic is the one most often encountered when the USA is projected alone, and is the only one of the three to curve the parallels of latitude, including the northern border on the 49th Parallel (Reproduced by permission of Peter H. Dana)

latitude and longitude, does it use the true shortest (great circle) distance based on the equation in Section 5.6, or the formula for distance in a Cartesian coordinate system on a distorted plane?

It is wise to be careful when using a GIS to analyze data in latitude and longitude rather than in projected coordinates, because serious distortions of distance, area, and other properties may result.

5.7.2 The Universal Transverse Mercator projection

The UTM system is often found in military applications, and in datasets with global or national coverage. It is based on the Mercator projection, but in *transverse* rather than Equatorial aspect, meaning that the projection is analogous to wrapping a cylinder around the Poles, rather than around the Equator. There are 60 zones in the system, and each zone corresponds to a half cylinder wrapped along a particular line of longitude, each zone being 6 degrees wide. Thus Zone 1 applies to longitudes from 180 W to 174 W, with the half cylinder wrapped along 177 W; Zone 10 applies to longitudes from 126 W to 120 W centered on 123 W, etc. (Figure 5.15).

The UTM system is secant, with lines of scale 1 located some distance out on both sides of the central meridian. The projection is conformal, so small features appear with the correct shape and scale is the same in all directions. Scale is 0.9996 at the central meridian and at most 1.0004 at the edges of the zone. Both parallels

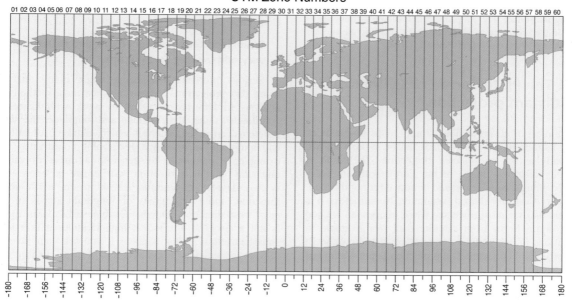

Figure 5.15 The system of zones of the Universal Transverse Mercator system. The zones are identified at the top. Each zone is six degrees of longitude in width (Reproduced by permission of Peter H. Dana)

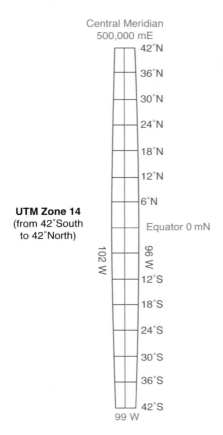

Figure 5.16 Major features of UTM Zone 14 (from 102 W to 96 W). The central meridian is at 99 W. Scale factors vary from 0.9996 at the central meridian to 1.0004 at the zone boundaries. See text for details of the coordinate system (Reproduced by permission of Peter H. Dana)

and meridians are curved on the projection, with the exception of the zone's central meridian and the Equator. Figure 5.16 shows the major features of one zone.

The coordinates of a UTM zone are defined in meters, and set up such that the central meridian's easting is always 500 000 m (a *false* easting), so easting varies from near zero to near 1 000 000 m. In the Northern Hemisphere the Equator is the origin of northing, so a point at northing 5 000 000 m is approximately 5000 km from the Equator. In the Southern Hemisphere the Equator is given a false northing of 10 000 000 m and all other northings are less than this.

> **UTM coordinates are in meters, making it easy to make accurate calculations of short distances between points.**

Because there are effectively 60 different projections in the UTM system, maps will not fit together across a zone boundary. Zones become so much of a problem at high latitudes that the UTM system is normally replaced with azimuthal projections centered on each Pole (known as the UPS or Universal Polar Stereographic system) above 80 degrees latitude. The problem is especially critical

for cities that cross zone boundaries, such as Calgary, Alberta, Canada (crosses the boundary at 114 W between Zone 11 and Zone 12). In such situations one zone can be extended to cover the entire city, but this results in distortions that are larger than normal. Another option is to define a special zone, with its own central meridian selected to pass directly through the city's center. Italy is split between Zones 32 and 33, and many Italian maps carry both sets of eastings and northings.

UTM coordinates are easy to recognize, because they commonly consist of a six-digit integer followed by a seven-digit integer (and decimal places if precision is greater than a meter), and sometimes include zone numbers and hemisphere codes. They are an excellent basis for analysis, because distances can be calculated from them for points within the same zone with no more than 0.04% error. But they are complicated enough that their use is effectively limited to professionals (the so-called 'spatially aware professionals' or SAPs defined in Section 1.4.3.2) except in applications where they can be hidden from the user. UTM grids are marked on many topographic maps, and many countries project their topographic maps using UTM, so it is easy to obtain UTM coordinates from maps for input to digital datasets, either by hand or automatically using scanning or digitizing (Chapter 9).

5.7.3 State Plane Coordinates and other local systems

Although the distortions of the UTM system are small, they are nevertheless too great for some purposes, particularly in accurate surveying. Zone boundaries also are a problem in many applications, because they follow arbitrary lines of longitude rather than boundaries between jurisdictions. In the 1930s each US state agreed to adopt its own projection and coordinate system, generally known as State Plane Coordinates (SPC), in order to support these high-accuracy applications. Projections were chosen to minimize distortion over the area of the state, so choices were often based on the state's shape. Some large states decided that distortions were still too great, and designed their SPCs with internal zones (for example, Texas has five zones based on the Lambert Conformal Conic projection, Figure 5.17, while Hawaii has five zones based on the Transverse Mercator projection). Many GIS have details of SPCs already stored, so it is easy to transform between them and UTM, or latitude and longitude. The system was revised in 1983 to accommodate the shift to the new North American Datum (NAD83).

> **All US states have adopted their own specialized coordinate systems for applications such as surveying that require very high accuracy.**

Many other countries have adopted coordinate systems of their own. For example, the UK uses a single projection and coordinate system known as the National Grid that is based on the Oblique Mercator projection (see Box 5.1)

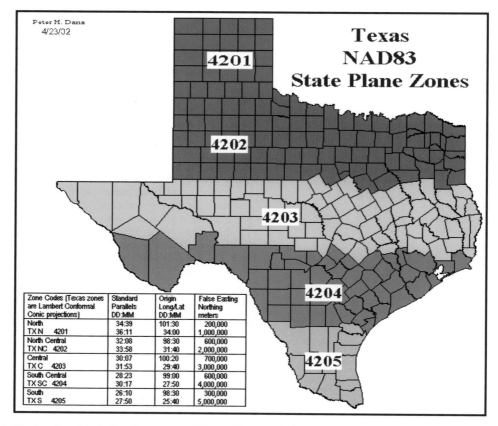

Peter H. Dana
4/23/02

Texas NAD83 State Plane Zones

Zone Codes (Texas zones are Lambert Conformal Conic projections)	Standard Parallels DD:MM	Origin Long/Lat DD:MM	False Easting Northing meters
North	34:39	101:30	200,000
TX N 4201	36:11	34:00	1,000,000
North Central	32:08	98:30	600,000
TX NC 4202	33:58	31:40	2,000,000
Central	30:07	100:20	700,000
TX C 4203	31:53	29:40	3,000,000
South Central	28:23	99:00	600,000
TX SC 4204	30:17	27:50	4,000,000
South	26:10	98:30	300,000
TX S 4205	27:50	25:40	5,000,000

Figure 5.17 The five State Plane Coordinate zones of Texas. Note that the zone boundaries are defined by counties, rather than parallels, for administrative simplicity (Reproduced by permission of Peter H. Dana)

and is marked on all topographic maps. Canada uses a uniform coordinate system based on the Lambert Conformal Conic projection, which has properties that are useful at mid to high latitudes, for applications where the multiple zones of the UTM system would be problematic.

5.8 Measuring latitude, longitude, and elevation: GPS

The Global Positioning System and its analogs (GLONASS in Russia, and the proposed Galileo system in Europe) have revolutionized the measurement of position, for the first time making it possible for people to know almost exactly where they are anywhere on the surface of the Earth. Previously, positions had to be established by a complex system of relative and absolute measurements. If one was near a point whose position was accurately known (a survey *monument*, for example), then position could be established through a series of accurate measurements of distances and directions starting from the monument. But if no monuments existed, then position had to be established through absolute measurements.

Latitude is comparatively easy to measure, based on the elevation of the sun at its highest point (local noon), or on the locations of the sun, moon, or fixed stars at precisely known times. But longitude requires an accurate method of measuring time, and the lack of accurate clocks led to massively incorrect beliefs about positions during early navigation. For example, Columbus and his contemporary explorers had no means of measuring longitude, and believed that the Earth was much smaller than it is, and that Asia was roughly as far west of Europe as the width of the Atlantic. The strength of this conviction is still reflected in the term we use for the islands of the Caribbean (the West Indies) and the first rapids on the St Lawrence in Canada (Lachine, or China). The fascinating story of the measurement of longitude is recounted by Dava Sobel.

The GPS consists of a system of 24 satellites (plus some spares), each orbiting the Earth every 12 hours on distinct orbits at a height of 20 200 km and transmitting radio pulses at very precisely timed intervals. To determine position, a receiver must make precise calculations from the signals, the known positions of the satellites, and the velocity of light. Positioning in three dimensions (latitude, longitude, and elevation) requires that at least four satellites are above the horizon, and accuracy depends on the number of such satellites and their positions (if elevation is not needed then only three

(A)

(B)

Figure 5.18 A simple GPS can provide an essential aid to wayfinding when (A) hiking or (B) driving (Reproduced by permission of David Parker, SPL, Photo Researchers)

satellites need be above the horizon). Several different versions of GPS exist, with distinct accuracies.

A simple GPS, such as one might buy in an electronics store for $100, or install as an optional addition to a laptop, cellphone, PDA (personal digital assistant, such as a Palm Pilot or iPAQ), or vehicle (Figure 5.18), has an accuracy within 10 m. This accuracy will degrade in cities with tall buildings, or under trees, and GPS signals will be lost entirely under bridges or indoors. Differential GPS (DGPS) combines GPS signals from satellites with correction signals received via radio or telephone from base stations. Networks of such stations now exist, at precisely known locations, constantly broadcasting corrections; corrections are computed by comparing each known location to its apparent location determined from GPS. With DGPS correction, accuracies improve to 1 m or better. Even greater accuracies are possible using various sophisticated techniques, or by remaining fixed and averaging measured locations over several hours.

GPS is very useful for recording ground control points when building GIS databases, for locating objects that move (for example, combine harvesters, tanks, cars, and shipping containers), and for direct capture of the locations of many types of fixed objects, such as utility assets, buildings, geological deposits, and sample points. Other applications of GPS are discussed in Chapter 11 on Distributed GIS, and by Mark Monmonier (see Box 5.2).

Some care is needed in using GPS to measure elevation. First, accuracies are typically lower, and a position determined to 10 m in the horizontal may be no better than plus or minus 50 m in the vertical. Second, a variety of reference elevations or vertical datums are in common use in different parts of the world and

by different agencies – for example, in the USA the topographic and hydrographic definitions of the vertical datum are significantly different.

5.9 Converting georeferences

GIS are particularly powerful tools for converting between projections and coordinate systems, because these transformations can be expressed as numerical operations. In fact this ability was one of the most attractive features of early systems for handling digital geographic data, and drove many early applications. But other conversions, e.g., between placenames and geographic coordinates, are much more problematic. Yet they are essential operations. Almost everyone knows their mailing address, and can identify travel destinations by name, but few are able to specify these locations in coordinates, or to interact with geographic information systems on that basis. GPS technology is attractive precisely because it allows its user to determine his or her latitude and longitude, or UTM coordinates, directly at the touch of a button.

Methods of converting between georeferences are important for:

- converting lists of customer addresses to coordinates for mapping or analysis (the task known as *geocoding*; see Box 5.3);

- combining datasets that use different systems of georeferencing;

Mark Monmonier, Cartographer

Mark Monmonier (Figure 5.19) is Distinguished Professor of Geography in the Maxwell School of Citizenship and Public Affairs at Syracuse University. He has published numerous papers on map design, automated map analysis, cartographic generalization, the history of cartography, statistical graphics, geographic demography, and mass communications. But he is best known as author of a series of widely read books on major issues in cartography, including *How to Lie with Maps* (University of Chicago Press, 1991; 2nd edition, revised and expanded, 1996) and *Rhumb Lines and Map Wars: A Social History of the Mercator Projection* (University of Chicago Press, 2004). Commenting on the power of GPS, he writes:

One of the more revolutionary aspects of geospatial technology is the ability to analyze maps without actually looking at one. Even more radical is the ability to track humans or animals around the clock by integrating a GPS fix with spatial data describing political boundaries or the street network. Social scientists recognize this kind of constant surveillance as a *panoptic gaze*, named for the Panopticon, a hypothetical prison devised by

Figure 5.19 Mark Monmonier, cartographer

Jeremy Bentham, an eighteenth-century social reformer intrigued with knowledge and power. Bentham argued that inmates aware they could be watched secretly at any time by an unseen warden were easily controlled. Location tracking can achieve similar results without walls or shutters.

GPS-based tracking can be beneficial or harmful depending on your point of view. An accident victim wants the Emergency-911 dispatcher to know where to send help. A car rental firm wants to know which client violated the rental agreement by driving out of state. Parents want to know where their children are. School principals want to know when a paroled pedophile is circling the playground. And the Orwellian thought police want to know where dissidents are gathering. Few geospatial technologies are as ambiguous and potentially threatening as location tracking.

Merge GPS, GIS, and wireless telephony, and you have the location-based services (LBS) industry (Chapter 11), useful for dispatching tow trucks, helping us find restaurants or gas stations, and letting a retailer, police detective, or stalker know where we've been. Our locational history is not only marketable but potentially invasive. How lawmakers respond to growing concern about location privacy (Figure 5.20) will determine whether we control our locational history or society lets our locational history control us.

Figure 5.20 An early edition of Mark Monmonier's government identification card. Today the ability to link such records to other information about an individual's whereabouts raises significant concerns about privacy

Geocoding: conversion of street addresses to coordinates

Geocoding is the name commonly given to the process of converting street addresses to latitude and longitude, or some similarly universal coordinate system. It is very widely used as it allows any database containing addresses, such as a company mailing list or a set of medical records, to be input to a GIS and mapped. Geocoding requires a database containing records representing the geometry of street segments between consecutive intersections, and the address ranges on each side of each segment (a *street centerline database*, see Chapter 9). Addresses are geocoded by finding the appropriate street segment record, and estimating a location based on linear interpolation within the address range. For example, 950 West Broadway in Columbia, Missouri, USA, lies on the side of the segment whose address range runs from 900 to 998, or $50/98 = 51.02\%$ of the distance from the start of the segment to the end. The segment starts at 92.3503 West longitude, 38.9519 North latitude, and ends at 92.3527 West, 38.9522 North. Simple arithmetic gives the address location as 92.3515 West, 38.9521 North. Four decimal places suggests an accuracy of about 10 m, but the estimate depends also on the accuracy of the assumption that addresses are uniformly spaced, and on the accuracy of the street centerline database.

- converting to projections that have desirable properties for analysis, e.g., no distortion of area;
- searching the Internet or other distributed data resources for data about specific locations;
- positioning GIS map displays by recentering them on places of interest that are known by name (these last two are sometimes called *locator* services).

The oldest method of converting georeferences is the *gazetteer*, the name commonly given to the index in an atlas that relates placenames to latitude and longitude, and to relevant pages in the atlas where information about that place can be found. In this form the gazetteer is a useful locator service, but it works only in one direction as a conversion between georeferences (from placename to latitude and longitude). Gazetteers have evolved substantially in the digital era, and it is now possible to obtain large databases of placenames and associated coordinates and to access services that allow such databases to be queried over the Internet (e.g., the Alexandria Digital Library gazetteer, **www.alexandria.ucsb.edu**; the US Geographic Names Information System, **geonames.usgs.gov**).

5.10 Summary

This chapter has looked in detail at the complex ways in which humans refer to specific locations on the planet, and how they measure locations. Any form of geographic information must involve some kind of georeference, and so it is important to understand the common methods, and their advantages and disadvantages. Many of the benefits of GIS rely on accurate georeferencing – the ability to link different items of information together through common geographic location; the ability to measure distances and areas on the Earth's surface, and to perform more complex forms of analysis; and the ability to communicate geographic information in forms that can be understood by others.

Georeferencing began in early societies, to deal with the need to describe locations. As humanity has progressed, we have found it more and more necessary to describe locations accurately, and over wider and wider domains, so that today our methods of georeferencing are able to locate phenomena unambiguously and to high accuracy anywhere on the Earth's surface. Today, with modern methods of measurement, it is possible to direct another person to a point on the other side of the Earth to an accuracy of a few centimeters, and this level of accuracy and referencing is achieved regularly in such areas as geophysics and civil engineering.

But georeferences can never be perfectly accurate, and it is always important to know something about spatial resolution. Questions of measurement accuracy are discussed at length in Chapter 6, together with techniques for representation of phenomena that are inherently fuzzy, such that it is impossible to say with certainty whether a given point is inside or outside the georeference.

Questions for further study

1. Visit your local map library, and determine: (1) the projections and datums used by selected maps; (2) the coordinates of your house in several common georeferencing systems.

2. Summarize the arguments for and against a single global figure of the Earth, such as WGS84.

3. How would you go about identifying the projection used by a common map source, such as the weather maps shown by a TV station or in a newspaper?

4. Chapter 14 discusses various forms of measurement in GIS. Review each of those methods, and the issues involved in performing analysis on databases that use different map projections. Identify the map projections that would be best for measurement of (1) area, (2) length, (3) shape.

Further reading

Bugayevskiy L.M. and Snyder J.P. 1995 *Map Projections: A Reference Manual*. London: Taylor and Francis.

Kennedy M. 1996 *The Global Positioning System and GIS: An Introduction*. Chelsea, Michigan: Ann Arbor Press.

Maling D.H. 1992 *Coordinate Systems and Map Projections* (2nd edn). Oxford: Pergamon.

Sobel D. 1995 *Longitude: The True Story of a Lone Genius Who Solved the Greatest Scientific Problem of His Time*. New York: Walker.

Snyder J.P. 1997 *Flattening the Earth: Two Thousand Years of Map Projections*. Chicago: University of Chicago Press.

Steede-Terry K. 2000 *Integrating GIS and the Global Positioning System*. Redlands, CA: ESRI Press.

6 *Uncertainty*

Uncertainty in geographic representation arises because, of necessity, almost all representations of the world are incomplete. As a result, data in a GIS can be subject to measurement error, out of date, excessively generalized, or just plain wrong. This chapter identifies many of the sources of geographic uncertainty and the ways in which they operate in GIS-based representations. Uncertainty arises from the way that GIS users conceive of the world, how they measure and represent it, and how they analyze their representations of it. This chapter investigates a number of conceptual issues in the creation and management of uncertainty, before reviewing the ways in which it may be measured using statistical and other methods. The propagation of uncertainty through geographical analysis is then considered. Uncertainty is an inevitable characteristic of GIS usage, and one that users must learn to live with. In these circumstances, it becomes clear that all decisions based on GIS are also subject to uncertainty.

Geographic Information Systems and Science, 2nd edition   Paul Longley, Michael Goodchild, David Maguire, and David Rhind.
© 2005 John Wiley & Sons, Ltd.   ISBNs: 0-470-87000-1 (HB); 0-470-87001-X (PB)

Learning Objectives

By the end of this chapter you will:

- Understand the concept of uncertainty, and the ways in which it arises from imperfect representation of geographic phenomena;

- Be aware of the uncertainties introduced in the three stages (conception, measurement and representation, and analysis) of database creation and use;

- Understand the concepts of vagueness and ambiguity, and the uncertainties arising from the definition of key GIS attributes;

- Understand how and why scale of geographic measurement and analysis can both create and propagate uncertainty.

6.1 Introduction

GIS-based representations of the real world are used to reconcile science with practice, concepts with applications, and analytical methods with social context. Yet, almost always, such reconciliation is imperfect, because, necessarily, representations of the world are incomplete (Section 3.4). In this chapter we will use *uncertainty* as an umbrella term to describe the problems that arise out of these imperfections. Occasionally, representations may approach perfect accuracy and precision (terms that we will define in Section 6.3.2.2) – as might be the case, for example, in the detailed site layout layer of a utility management system, in which strenuous efforts are made to reconcile fine-scale multiple measurements of built environments. Yet perfect, or nearly perfect, representations of reality are the exception rather than the rule. More usually, the inherent complexity and detail of our world makes it virtually impossible to capture every single facet, at every possible scale, in a digital representation. (Neither is this usually desirable: see the discussion of sampling in Section 4.4.) Furthermore, different individuals see the world in different ways, and in practice no single view is likely to be accepted universally as the best or to enjoy uncontested status. In this chapter we discuss how the processes and procedures of abstraction create differences between the contents of our (geographic and attribute) database and real-world phenomena. Such differences are almost inevitable and understanding of them can help us to manage uncertainty, and to live with it.

It is impossible to make a perfect representation of the world, so uncertainty about it is inevitable.

Various terms are used to describe differences between the real world and how it appears in a GIS, depending upon the context. The established scientific notion of measurement *error* focuses on differences between observers or between measuring instruments. As we saw in a previous chapter (Section 4.7), the concept of error in multivariate statistics arises in part from omission of some relevant aspects of a phenomenon – as in the failure to fully specify all of the predictor variables in a multiple regression model, for example. Similar problems arise when one or more variables are omitted from the calculation of a composite indicator – as, for example, in omitting road accessibility in an index of land value, or omitting employment status from a measure of social deprivation (see Section 16.2.1 for a discussion of indicators). More generally, the Dutch geostatistician Gerard Heuvelink (who we will introduce in Box 6.1) has defined *accuracy* as the difference between reality and *our* representation of reality. Although such differences might principally be addressed in formal mathematical terms, the use of the word *our* acknowledges the varying views that are generated by a complex, multi-scale, and inherently uncertain world.

Yet even this established framework is too simple for understanding quality or the defining standards of geographic data. The terms *ambiguity* and *vagueness* identify further considerations which need to be taken into account in assessing the *quality* of a GIS representation. Quality is an important topic in GIS, and there have been many attempts to identify its basic dimensions. The US Federal Geographic Data Committee's various standards list five components of quality: attribute accuracy, positional accuracy, logical consistency, completeness, and lineage. Definitions and other details on each of these and several more can be found on the FGDC's Web pages (**www.fgdc.gov**). Error, inaccuracy, ambiguity, and vagueness all contribute to the notion of uncertainty in the broadest sense, and uncertainty may thus be defined as a measure of the user's understanding of the difference between the contents of a dataset, and the real phenomena that the data are believed to represent. This definition implies that phenomena are real, but includes the possibility that we are unable to describe them exactly. In GIS, the term uncertainty has come to be used as the catch-all term to describe situations in which the digital representation is simply incomplete, and as a measure of the general quality of the representation.

Many geographic representations depend upon inherently vague definitions and concepts

The views outlined in the previous paragraph are themselves controversial, and a rich ground for endless philosophical discussions. Some would argue that uncertainty can be inherent in phenomena themselves, rather than just in their description. Others would argue for distinctions between *vagueness, uncertainty, fuzziness, imprecision, inaccuracy,* and many other terms that most people use as if they were essentially synonymous. Information scientist

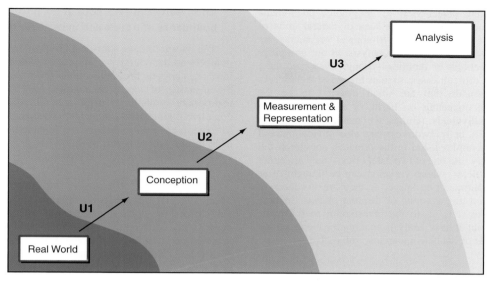

Figure 6.1 A conceptual view of uncertainty. The three filters, U1, U2, and U3 can distort the way in which the complexity of the real world is conceived, measured and represented, and analyzed in a cumulative way

Peter Fisher has provided a useful and wide-ranging discussion of these terms. We take the catch-all view here, and leave the detailed arguments to further study.

In this chapter, we will discuss some of the principal sources of uncertainty and some of the ways in which uncertainty degrades the quality of a spatial representation. The way in which we conceive of a geographic phenomenon very much prescribes the way in which we are likely to set about measuring and representing it. The measurement procedure, in turn, heavily conditions the ways in which it may be analyzed within a GIS. This chain sequence of events, in which *conception* prescribes *measurement and representation*, which in turn prescribes *analysis* is a succinct way of summarizing much of the content of this chapter, and is summarized in Figure 6.1. In this diagram, U1, U2, and U3 each denote *filters* that selectively distort or transform the representation of the real world that is stored and analyzed in GIS: a later chapter (Section 13.2.1) introduces a fourth filter that mediates interpretation of analysis, and the ways in which feedback may be accommodated through improvements in representation.

6.2 U1: Uncertainty in the conception of geographic phenomena

6.2.1 Units of analysis

Our discussion of Tobler's Law (Section 3.1) and of spatial autocorrelation (Section 4.6) established that geographic data handling is different from all other classes of non-spatial applications. A further characteristic that sets geographic information science apart from most every other science is that it is only rarely founded upon *natural* units of analysis. What is the natural unit of measurement for a soil profile? What is the spatial extent of a *pocket* of high unemployment, or a *cluster* of cancer cases? How might we delimit an environmental impact study of spillage from an oil tanker (Figure 6.2)? The questions become still more difficult in bivariate (two variable) and multivariate (more than two variable) studies. At what scale is it appropriate to investigate any relationship between background radiation and the incidence of leukemia? Or to assess any relationship between labor-force qualifications and unemployment rates?

In many cases there are no natural units of geographic analysis.

Figure 6.2 How might the spatial impact of an oil tanker spillage be delineated? We can measure the dispersion of the pollutants, but their impacts extend far beyond these narrowly defined boundaries (Reproduced by permission of Sam C. Pierson, Jr., Photo Researchers)

The discrete object view of geographic phenomena is much more reliant upon the idea of natural units of analysis than the field view. Biological organisms are almost always natural units of analysis, as are groupings such as households or families – though even here there are certainly difficult cases, such as the massive networks of fungal strands that are often claimed to be the largest living organisms on Earth, or extended families of human individuals. Things we manipulate, such as pencils, books, or screwdrivers, are also obvious natural units. The examples listed in the previous paragraph fall almost entirely into one of two categories – they are either instances of fields, where variation can be thought of as inherently continuous in space, or they are instances of poorly defined aggregations of discrete objects. In both of these cases it is up to the investigator to make the decisions about units of analysis, making the identification of the objects of analysis inherently subjective.

6.2.2 Vagueness and ambiguity

6.2.2.1 Vagueness

The frequent absence of objective geographic individual units means that, in practice, the labels that we assign to zones are often vague best guesses. What absolute or relative incidence of oak trees in a forested zone qualifies it for the label *oak woodland* (Figure 6.3)? Or, in a developing-country context in which aerial photography rather than ground enumeration is used to estimate population size, what rate of incidence of dwellings identifies a zone of *dense* population? In each of these instances, it is expedient to transform point-like events (individual trees or individual dwellings) into area objects, and pragmatic decisions must be taken in order to create a working definition of a spatial distribution. These decisions have no absolute validity, and raise two important questions:

- Is the defining boundary of a zone crisp and well-defined?
- Is our assignment of a particular label to a given zone robust and defensible?

Figure 6.3 Seeing the wood for the trees: what absolute or relative incidence rate makes it meaningful to assign the label 'oak woodland'? (Reproduced by permission of Ellan Young, Photo Researchers)

Uncertainty can exist both in the positions of the boundaries of a zone and in its attributes.

The questions have statistical implications (can we put numbers on the confidence associated with boundaries or labels?), cartographic implications (how can we convey the meaning of vague boundaries and labels through appropriate symbols on maps and GIS displays?), and cognitive implications (do people subconsciously attempt to force things into categories and boundaries to satisfy a deep need to simplify the world?).

6.2.2.2 Ambiguity

Many objects are assigned different labels by different national or cultural groups, and such groups perceive space differently. Geographic prepositions like *across, over*, and *in* (used in the Yellow Pages query in Figure 1.17) do not have simple correspondences with terms in other languages. Object names and the topological relations between them may thus be inherently *ambiguous*. Perception, behavior, language, and cognition all play a part in the conception of real-world entities and the relationships between them. GIS cannot present a value-neutral view of the world, yet it can provide a formal framework for the reconciliation of different worldviews. The geographic nature of this ambiguity may even be exploited to identify regions with shared characteristics and worldviews. To this end, Box 6.1 describes how different surnames used to describe essentially the same historic occupations provide an enduring measure in region building.

Many linguistic terms used to convey geographic information are inherently ambiguous.

Ambiguity also arises in the conception and construction of *indicators* (see also Section 16.2.1). *Direct* indicators are deemed to bear a clear correspondence with a mapped phenomenon. Detailed household income figures, for example, provide a direct indicator of the likely geography of expenditure and demand for goods and services; tree diameter at breast height can be used to estimate stand value; and field nutrient measures can be used to estimate agronomic yield. *Indirect* indicators are used when the best available measure is a perceived surrogate link with the phenomenon of interest. Thus the incidence of central heating amongst households, or rates of multiple car ownership, might provide a surrogate for (unavailable) household income data, while local atmospheric measurements of nitrogen dioxide might provide an indirect indicator of environmental health. Conception of the (direct or indirect) linkage between any indicator and the phenomenon of interest is subjective, hence ambiguous. Such measures will create (possibly systematic) errors of measurement if the correspondence between the two is imperfect. So, for example, differences in the conception of what hardship and deprivation entail can lead to specification of different composite indicators, and different geodemographic systems include different cocktails of census variables (Section 2.3.3). With regard to the natural environment, conception of critical defining properties

Historians need maps of our uncertain past

In the study of history, there are many ways in which 'spatial is special' (Section 1.1.1). For example, it is widely recognized that although what our ancestors did (their occupations) and the social groups (classes) to which they belonged were clearly important in terms of demographic behavior, location and place were of equal if not greater importance. Although population changes occur in particular socio-economic circumstances, they are also strongly influenced by the unique characteristics, or 'cultural identities', of particular places. In Great Britain today, as almost everywhere else in the world, most people still think of their nation as made up of 'regions', and their stability and defining characteristics are much debated by cultural geographers and historians.

Yet analyzing and measuring human activity by place creates particular problems for historians. Most obviously, the past was very much less data rich than the present, and few systematic data sources survive. Moreover, the geographical administrative units by which the events of the past were recorded are both complex and changing. In an ideal world, perhaps, physical and cultural boundaries would always coincide, but physical features alone rarely provide appropriate *indicators* of the limits of socio-economic conditions and cultural circumstance.

Unfortunately many mapped historical data are still presented using high-level aggregations, such as counties or regions. This achieves a measure of standardization but may depict demography in only the most arbitrary of ways. If data are forced into geographic administrative units that were delineated for other purposes, regional maps may present nothing more than misleading, or even meaningless, spatial means (see Box 1.9).

In England and in many other countries, the daily activities of most individuals historically revolved around small numbers of contiguous civil *parishes*, of which there were more than 16 000 in the 19th century. These are the smallest administrative units for which data are systematically available. They provide the best available building blocks for meaningful region building. But how can we group parishes in order to identify non-overlapping geographic territories to which people felt that they belonged? And what indicators of regional identity are likely to have survived for all individuals in the population?

Historian Kevin Schürer (Box 13.2) has investigated these questions using a historical GIS to map digital surname data from the 1881 Census of England and Wales. The motivation for the GIS arises from the observation that many surnames contain statements of regional identity, and the suggestion that distinct zones of similar surnames might be described as homogeneous regions. The digitized records of the 1881 Census for England and Wales cover some 26 million people: although some 41 000 different surnames are recorded, a fifth of the population shared just under 60 surnames, and half of the population were accounted for by some 600 surnames. Schürer suggests that these aggregate statistics conceal much that we might learn about regional identity and diversity.

Many surnames of European origin are formed from occupational titles. Occupations often have uneven regional distributions and sometimes similar occupations are described using different names in different places (at the global scale, today's 'realtors' in the US perform much the same functions as their 'estate agent' counterparts in the UK, for example). Schürer has investigated the 1881 geographical distribution of three occupational surnames – Fuller, Tucker, and Walker. These essentially refer to the same occupation; namely someone who, from around the 14th century onwards, worked in the preparation of textiles by scouring or beating cloth as a means of finishing or cleansing it. Using GIS, Schürer confirms that the geographies of these 14th century surnames remained of enduring importance in defining the regional geography of England in 1881. Figure 6.4 illustrates that in 1881 Tuckers remained concentrated in the West Country, while Fullers occurred principally in the east and Walkers resided in the Midlands and north. This map also shows that there was not much mixing of the surnames in the transition zones between names, suggesting that the maps provide a useful basis to region building.

The enduring importance of surnames as evidence of the strength and durability of regional cultures has been confirmed in an update to the work by Daryl Lloyd at University College London: Lloyd used the 2003 UK Electoral Register to map the distribution of the same three surnames (Figure 6.5) and identified persistent regional concentrations.

▶

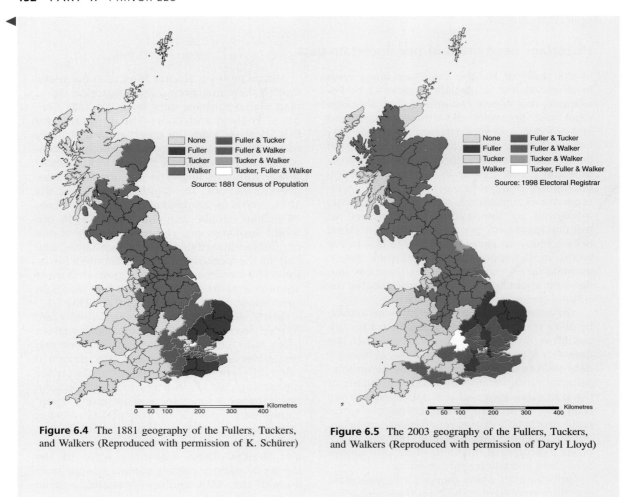

Figure 6.4 The 1881 geography of the Fullers, Tuckers, and Walkers (Reproduced with permission of K. Schürer)

Figure 6.5 The 2003 geography of the Fullers, Tuckers, and Walkers (Reproduced with permission of Daryl Lloyd)

of soils can lead to inherent ambiguity in their classification (see Section 6.2.4).

Ambiguity is introduced when imperfect indicators of phenomena are used instead of the phenomena themselves.

Fundamentally, GIS has upgraded our abilities to generalize about spatial distributions. Yet our abilities to do so may be constrained by the different taxonomies that are conceived and used by data-collecting organizations within our overall study area. A study of wetland classification in the US found no fewer than six agencies engaged in mapping the same phenomena over the same geographic areas, and each with their own definitions of wetland types (see Section 1.2). If wetland maps are to be used in regulating the use of land, as they are in many areas, then uncertainty in mapping clearly exposes regulatory agencies to potentially damaging and costly lawsuits. How might soils data classified according to the UK national classification be assimilated within a pan-European soils map, which uses a classification honed to the full range and diversity of soils found across the European continent rather than those just on an offshore

island? How might different national geodemographic classifications be combined into a form suitable for a pan-European marketing exercise? These are all variants of the question:

- How may mismatches between the categories of different classification schema be reconciled?

Differences in definitions are a major impediment to integration of geographic data over wide areas.

Like the process of pinning down the different nomenclatures developed in different cultural settings, the process of reconciling the semantics of different classification schema is an inherently *ambiguous* procedure. Ambiguity arises in data concatenation when we are unsure regarding the *meta-category* to which a particular class should be assigned.

6.2.3 Fuzzy approaches

One way of resolving the assignment process is to adopt a probabilistic interpretation. If we take a statement like 'the database indicates that this field contains wheat, but there is a 0.17 probability (or 17% chance) that it actually contains barley', there are at least two possible interpretations:

(a) If 100 randomly chosen people were asked to make independent assessments of the field on the ground, 17 would determine that it contains barley, and 83 would decide it contains wheat.

(b) Of 100 similar fields in the database, 17 actually contained barley when checked on the ground, and 83 contained wheat.

Of the two we probably find the second more acceptable because the first implies that people cannot correctly determine the crop in the field. But the important point is that, in conceptual terms, both of these interpretations are *frequentist*, because they are based on the notion that the probability of a given outcome can be defined as the proportion of times the outcome occurs in some real or imagined experiment, when the number of tests is very large. Yet while this is reasonable for classic statistical experiments, like tossing coins or drawing balls from an urn, the geographic situation is different – there is only one field with precisely these characteristics, and one observer, and in order to imagine a number of tests we have to invent more than one observer, or more than one field (the problems of imagining larger populations for some geographic samples are discussed further in Section 15.4).

In part because of this problem, many people prefer the *subjectivist* conception of probability – that it represents a judgment about relative likelihood that is not the result of any frequentist experiment, real or imagined. Subjective probability is similar in many ways to the concept of fuzzy sets, and the latter framework will be used here to emphasize the contrast with frequentist probability.

Suppose we are asked to examine an aerial photograph to determine whether a field contains wheat, and we decide that we are not sure. However, we are able to put a number on our degree of uncertainty, by putting it on a scale from 0 to 1. The more certain we are, the higher the number. Thus we might say we are 0.90 sure it is wheat, and this would reflect a greater degree of certainty than 0.80. This degree of belonging to the class *wheat* is termed the *fuzzy membership*, and it is common though not necessary to limit memberships to the range 0 to 1. In effect, we have changed our view of membership in classes, and abandoned the notion that things must either belong to classes or not belong to them – in this new world, the boundaries of classes are no longer clean and crisp, and the set of things assigned to a set can be fuzzy.

In fuzzy logic, an object's degree of belonging to a class can be partial.

One of the major attractions of fuzzy sets is that they appear to let us deal with sets that are not precisely defined, and for which it is impossible to establish membership cleanly. Many such sets or classes are found in GIS applications, including land use categories, soil types, land cover classes, and vegetation types. Classes used for maps are often fuzzy, such that two people asked to classify the same location might disagree, not because of measurement error, but because the classes themselves are not perfectly defined and because opinions vary. As such, mapping is often forced to stretch the rules of scientific repeatability, which require that two observers will always agree. Box 6.2 shows a typical extract from the legend of a soil map, and it is easy to see how two people might disagree, even though both are experts with years of experience in soil classification.

Figure 6.6 shows an example of mapping classes using the fuzzy methods developed by A-Xing Zhu of the

---

Technical Box **6.2**

Fuzziness in classification: description of a soil class

Following is the description of the Limerick series of soils from New England, USA (the type location is in Chittenden County, Vermont), as defined by the National Cooperative Soil Survey. Note the frequent use of vague terms such as 'very', 'moderate', 'about', 'typically', and 'some'. Because the definition is so loose it is possible for many distinct soils to be lumped together in this one class – and two observers may easily disagree over whether a given soil belongs to the class, even though both are experts. The definition illustrates the extreme problems of defining soil classes with sufficient rigor to satisfy the criterion of scientific repeatability.

The Limerick series consists of very deep, poorly drained soils on flood plains. They
formed in loamy alluvium. Permeability is moderate. Slope ranges from 0 to 3 percent. Mean annual precipitation is about 34 inches and mean annual temperature is about 45 degrees F.

Depth to bedrock is more than 60 inches. Reaction ranges from strongly acid to neutral in the surface layer and moderately acid to neutral in the substratum. Textures are typically silt loam or very fine sandy loam, but lenses of loamy very fine sand or very fine sand are present in some pedons. The weighted average of fine and coarser sands, in the particle-size control section, is less than 15 percent.

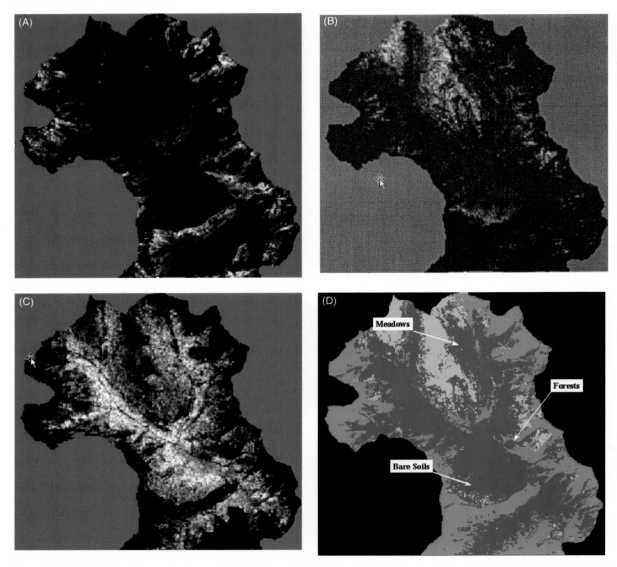

Figure 6.6 (A) Membership map for bare soils in the Upper Lake McDonald basin, Glacier National Park. High membership values are in the ridge areas where active colluvial and glacier activities prevent the establishment of vegetation. (B) Membership map for forest. High membership values are in the middle to lower slope areas where the soils are both stable and better drained. (C) Membership map for alpine meadows. High membership values are on gentle slopes at high elevation where excessive soil water and low temperature prevent the growth of trees. (D) Spatial distribution of the three cover types from hardening the membership maps. (Reproduced by permission of A-Xing Zhu)

University of Wisconsin-Madison, USA, which take both remote sensing images and the opinions of experts as inputs. There are three classes, and each map shows the fuzzy membership values in one class, ranging from 0 (darkest) to 1 (lightest). This figure also shows the result of converting to *crisp* categories, or *hardening* – to obtain Figure 6.6D, each pixel is colored according to the class with the highest membership value.

Fuzzy approaches are attractive because they capture the uncertainty that many of us feel about the assignment of places on the ground to specific categories. But researchers have struggled with the question of whether they are more *accurate*. In a sense, if we are uncertain

about which class to choose then it is more accurate to say so, in the form of a fuzzy membership, than to be forced into assigning a class without qualification. But that does not address the question of whether the fuzzy membership value is accurate. If Class A is not well defined, it is hard to see how one person's assignment of a fuzzy membership of 0.83 in Class A can be meaningful to another person, since there is no reason to believe that the two people share the same notions of what Class A means, or of what 0.83 means, as distinct from 0.91, or 0.74. So while fuzzy approaches make sense at an intuitive level, it is more difficult to see how they could be helpful in the process of communication of geographic knowledge from one person to another.

6.2.4 The scale of geographic individuals

There is a sense in which vagueness and ambiguity in the conception of *usable* (rather than *natural*) units of analysis undermines the very foundations of GIS. How, in practice, may we create a sufficiently secure base to support geographic analysis? Geographers have long grappled with the problems of defining systems of zones and have marshaled a range of deductive and inductive approaches to this end (see Section 4.9 for a discussion of what deduction and induction entail). The long-established regional geography tradition is fundamentally concerned with the delineation of zones characterized by internal homogeneity (with respect to climate, economic development, or agricultural land use, for example), within a zonal scheme which maximizes between-zone heterogeneity, such as the map illustrated in Figure 6.7. Regional geography is fundamentally about delineating *uniform* zones, and many employ multivariate statistical techniques such as cluster analysis to supplement, or post-rationalize, intuition.

Identification of homogeneous zones and spheres of influence lies at the heart of traditional regional geography as well as contemporary data analysis.

Other geographers have tried to develop *functional* zonal schemes, in which zone boundaries delineate the

breakpoints between the spheres of influence of adjacent facilities or features – as in the definition of travel-to-work areas (Figure 6.8) or the definition of a river catchment. Zones may be defined such that there is maximal interaction within zones, and minimal between zones. The scale at which uniformity or functional integrity is conceived clearly conditions the ways it is measured – in terms of the magnitude of within-zone heterogeneity that must be accommodated in the case of uniform zones, and the degree of leakage between the units of functional zones.

Scale has an effect, through the concept of spatial autocorrelation outlined in Section 4.3, upon the outcome of geographic analysis. This was demonstrated more than half a century ago in a classic paper by Yule and Kendall, where the correlation between wheat and potato yields was shown systematically to increase as English county units were amalgamated through a succession of coarser scales (Table 6.1). A succession of research papers has subsequently reaffirmed the existence of similar scale effects in multivariate analysis. However, rather discouragingly, scale effects in multivariate cases do not follow any consistent or predictable trends. This theme of dependence of results on the geographic units of analysis is pursued further in Section 6.4.3.

Relationships typically grow stronger when based on larger geographic units.

GIS appears to trivialize the task of creating composite thematic maps. Yet inappropriate conception of the scale of geographic phenomena can mean that apparent spatial

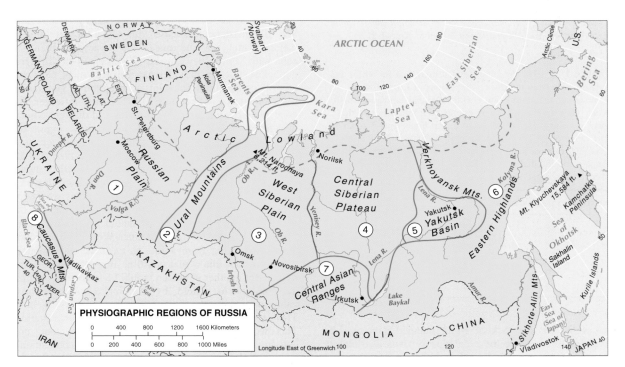

Figure 6.7 The regional geography of Russia. (*Source*: de Blij H.J. and Muller P.O. 2000 *Geography: Realms, Regions and Concepts* (9th edn) New York: Wiley, p. 113)

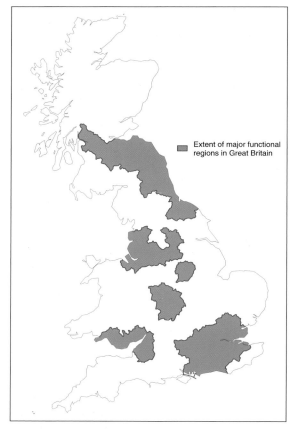

Figure 6.8 Dominant functional regions of Great Britain. (*Source*: Champion A.G., Green A.E., Owen D.W., Ellin D.J., Coombes M.G. 1987 *Changing Places: Britain's Demographic, Economic and Social Complexion*, London: Arnold, p. 9)

Table 6.1 In 1950 Yule and Kendall used data for wheat and potato yields from the (then) 48 counties of England to demonstrate that correlation coefficients tend to increase with scale. They aggregated the 48-county data into zones so that there were first 24, then 12, then 6, and finally just 3 zones. The range of their results, from near zero (no correlation) to over 0.99 (almost perfect positive correlation) demonstrates the range of results that can be obtained, although subsequent research has suggested that this range of values is atypical

No. of geographic areas	Correlation
48	0.2189
24	0.2963
12	0.5757
6	0.7649
3	0.9902

patterning (or the lack of it) in mapped data may be oversimplified, crude, or even illusory. It is also clearly inappropriate to conceive of boundaries as crisp and well-defined if significant leakage occurs across them (as happens, in practice, in the delineation of most functional

regions), or if geographic phenomena are by nature fuzzy, vague, or ambiguous.

6.3 U2: Further uncertainty in the measurement and representation of geographic phenomena

6.3.1 Measurement and representation

The conceptual models (fields and objects) that were introduced in Chapter 3 impose very different filters upon reality, and their usual corresponding representational models (raster and vector) are characterized by different uncertainties as a consequence. The vector model enables a range of powerful analytical operations to be performed (see Chapters 14 through 16), yet it also requires *a priori* conceptualization of the nature and extent of geographic individuals and the ways in which they nest together into higher-order zones. The raster model defines individual elements as square cells, with boundaries that bear no relationship at all to natural features, but nevertheless provides a convenient and (usually) efficient structure for data handling within a GIS. However, in the absence of effective automated pattern recognition techniques, human interpretation is usually required to discriminate between real-world spatial entities as they appear in a rasterized image.

Although quite different representations of reality, vector and raster data structures are both attractive in their logical consistency, the ease with which they are able to handle spatial data, and (once the software is written) the ease with which they can be implemented in GIS. But neither abstraction provides easy measurement fixes and there is no substitute for robust conception of geographic units of analysis (Section 6.2). This said, however, the conceptual distinction between fields and discrete objects is often useful in dealing with uncertainty. Figure 6.9 shows a coastline, which is often conceptualized as a discrete line object. But suppose we recognize that its position is uncertain. For example, the coastline shown on a 1:2 000 000 map is a gross generalization, in which major liberties are taken, particularly in areas where the coast is highly indented and irregular. Consequently the 1:2 000 000 version leaves substantial uncertainty about the true location of the shoreline. We might approach this by changing from a line to an area, and mapping the area where the actual coastline lies, as shown in the figure. But another approach would be to reconceptualize the coastline as a field, by mapping a variable whose value represents the probability that a point is land. This is shown in the figure as a raster representation. This would have far more information content, and consequently much more value in many applications. But at the same time it would be difficult to find an appropriate data source for the representation – perhaps a fuzzy classification of an air photo, using one of an

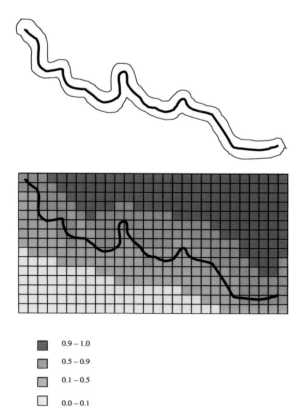

Figure 6.9 The contrast between discrete object (top) and field (bottom) conceptualizations of an uncertain coastline. In the discrete object view the line becomes an area delimiting where the true coastline might be. In the field view a continuous surface defines the probability that any point is land

0.9 – 1.0

0.5 – 0.9

0.1 – 0.5

0.0 – 0.1

increasing number of techniques designed to produce representations of the uncertainty associated with objects discovered in images.

Uncertainty can be measured differently under field and discrete object views.

Indeed, far from offering quick fixes for eliminating or reducing uncertainty, the measurement process can actually increase it. Given that the vector and raster data models impose quite different filters on reality, it is unsurprising that they can each generate additional uncertainty in rather different ways. In field-based conceptualizations, such as those that underlie remotely sensed images expressed as rasters, spatial objects are not defined *a priori*. Instead, the classification of each cell into one or other category builds together into a representation. In remote sensing, when resolution is insufficient to detect all of the detail in geographic phenomena, the term *mixel* is often used to describe raster cells that contain more than one class of land – in other words, elements in which the outcome of statistical classification suggests the occurrence of multiple land cover categories. The total area of cells classified as mixed should decrease as the resolution of the satellite sensor increases, assuming the number of categories remains constant, yet a

completely mixel-free classification is very unlikely at any level of resolution. Even where the Earth's surface is covered with perfectly homogeneous areas, such as agricultural fields growing uniform crops, the failure of real-world crop boundaries to line up with pixel edges ensures the presence of at least some mixels. Neither does higher-resolution imagery solve all problems: medium-resolution data (defined as pixel size of between 30 m × 30 m and 1000 m × 1000 m) are typically classified using between 3 and 7 bands, while high-resolution data (pixel sizes 10 × 10 m or smaller) are typically classified using between 7 and 256 bands, and this can generate much greater heterogeneity of spectral values with attendant problems for classification algorithms.

A pixel whose area is divided among more than one class is termed a mixel.

The vector data structure, by contrast, defines spatial entities and specifies explicit topological relations (see Section 3.6) between them. Yet this often entails transformations of the inherent characteristics of spatial objects (Section 14.4). In conceptual terms, for example, while the true individual members of a population might each be defined as point-like objects, they will often appear in a GIS dataset only as aggregate counts for apparently *uniform* zones. Such aggregation can be driven by the need to preserve confidentiality of individual records, or simply by the need to limit data volume. Unlike the field conceptualization of spatial phenomena, this implies that there are good reasons for partitioning space in a particular way. In practice, partitioning of space is often made on grounds that are principally pragmatic, yet are rarely completely random (see Section 6.4). In much of socio-economic GIS, for example, zones which are designed to preserve the anonymity of survey respondents may be largely *ad hoc* containers. Larger aggregations are often used for the simple reason that they permit comparisons of measures over time (see Box 6.1). They may also reflect the way that a cartographer or GIS interpolates a boundary between sampled points, as in the creation of isopleth maps (Box 4.3).

6.3.2 Statistical models of uncertainty

Scientists have developed many widely used methods for describing errors in observations and measurements, and these methods may be applicable to GIS if we are willing to think of databases as collections of measurements. For example, a digital elevation model consists of a large number of measurements of the elevation of the Earth's surface. A map of land use is also in a sense a collection of measurements, because observations of the land surface have resulted in the assignment of classes to locations. Both of these are examples of observed or measured attributes, but we can also think of location as a property that is measured.

A geographic database is a collection of measurements of phenomena on or near the Earth's surface.

Here we consider errors in nominal class assignment, such as of types of land use, and errors in continuous (interval or ratio) scales, such as elevation (see Section 3.4).

6.3.2.1 Nominal case

The values of nominal data serve only to distinguish an instance of one class from an instance of another, or to identify an object uniquely. If classes have an inherent ranking they are described as ordinal data, but for purposes of simplicity the ordinal case will be treated here as if it were nominal.

Consider a single observation of nominal data – for example, the observation that a single parcel of land is being used for agriculture (this might be designated by giving the parcel Class A as its value of the 'Land Use Class' attribute). For some reason, perhaps related to the quality of the aerial photography being used to build the database, the class may have been recorded falsely as Class G, Grassland. A certain proportion of parcels that are truly Agriculture might be similarly recorded as Grassland, and we can think of this in terms of a probability, that parcels that are truly Agriculture are falsely recorded as Grassland.

Table 6.2 shows how this might work for all of the parcels in a database. Each parcel has a true class, defined by accurate observation in the field, and a recorded class as it appears in the database. The whole table is described as a *confusion matrix*, and instances of confusion matrices are commonly encountered in applications dominated by class data, such as classifications derived from remote sensing or aerial photography. The true class might be determined by ground check, which is inherently more accurate than classification of aerial photographs, but much more expensive and time-consuming.

Ideally all of the observations in the confusion matrix should be on the principal diagonal, in the cells that correspond to agreement between true class and database class. But in practice certain classes are more easily confused than others, so certain cells off the diagonal will have substantial numbers of entries.

A useful way to think of the confusion matrix is as a set of rows, each defining a vector of values.

Table 6.2 Example of a misclassification or confusion matrix. A grand total of 304 parcels have been checked. The rows of the table correspond to the land use class of each parcel as recorded in the database, and the columns to the class as recorded in the field. The numbers appearing on the principal diagonal of the table (from top left to bottom right) reflect correct classification

	A	B	C	D	E	Total
A	80	4	0	15	7	106
B	2	17	0	9	2	30
C	12	5	9	4	8	38
D	7	8	0	65	0	80
E	3	2	1	6	38	50
Total	104	36	10	99	55	304

The vector for row i gives the proportions of cases in which what appears to be Class i is actually Class 1, 2, 3, etc. Symbolically, this can be represented as a vector $\{p_1, p_2, \ldots, p_i, \ldots, p_n\}$, where n is the number of classes, and p_i represents the proportion of cases for which what appears to be the class according to the database is actually Class i.

There are several ways of describing and summarizing the confusion matrix. If we focus on one row, then the table shows how a given class in the database falsely records what are actually different classes on the ground. For example, Row A shows that of 106 parcels recorded as Class A in the database, 80 were confirmed as Class A in the field, but 15 appeared to be truly Class D. The proportion of instances in the diagonal entries represents the proportion of correctly classified parcels, and the total of off-diagonal entries in the row is the proportion of entries in the database that appear to be of the row's class but are actually incorrectly classified. For example, there were only 9 instances of agreement between the database and the field in the case of Class D. If we look at the table's columns, the entries record the ways in which parcels that are truly of that class are actually recorded in the database. For example, of the 10 instances of Class C found in the field, 9 were recorded as such in the database and only 1 was misrecorded as Class E.

The columns have been called the *producer's* perspective, because the task of the producer of an accurate database is to minimize entries outside the diagonal cell in a given column, and the rows have been called the *consumer's* perspective, because they record what the contents of the database actually mean on the ground; in other words, the accuracy of the database's contents.

Users and producers of data look at misclassification in distinct ways.

For the table as a whole, the proportion of entries in diagonal cells is called the *percent correctly classified* (PCC), and is one possible way of summarizing the table. In this case 209/304 cases are on the diagonal, for a PCC of 68.8%. But this measure is misleading for at least two reasons. First, chance alone would produce some correct classifications, even in the worst circumstances, so it would be more meaningful if the scale were adjusted such that 0 represents chance. In this case, the number of chance hits on the diagonal in a random assignment is 76.2 (the sum of the row total times the column total divided by the grand total for each of the five diagonal cells). So the actual number of diagonal hits, 209, should be compared to this number, not 0. The more useful index of success is the *kappa index*, defined as:

$$\kappa = \frac{\sum_{i=1}^{n} c_{ii} - \sum_{i=1}^{n} c_{i.}c_{.i}/c_{..}}{c_{..} - \sum_{i=1}^{n} c_{i.}c_{.i}/c_{..}}$$

where c_{ij} denotes the entry in row i column j, the dots indicate summation (e.g., $c_{i.}$ is the summation over all columns for row i, that is, the row i total, and $c_{..}$ is the grand total), and n is the number of classes. The first term in the numerator is the sum of all the diagonal entries (entries for which the row number and the column number are the same). To compute PCC we would simply divide this term by the grand total (the first term in the denominator). For kappa, both numerator and denominator are reduced by the same amount, an estimate of the number of hits (agreements between field and database) that would occur by chance. This involves taking each diagonal cell, multiplying the row total by the column total, and dividing by the grand total. The result is summed for each diagonal cell. In this case kappa evaluates to 58.3%, a much less optimistic assessment than PCC.

The second issue with both of these measures concerns the relative abundance of different classes. In the table, Class C is much less common than Class A. The confusion matrix is a useful way of summarizing the characteristics of nominal data, but to build it there must be some source of more accurate data. Commonly this is obtained by ground observation, and in practice the confusion matrix is created by taking samples of more accurate data, by sending observers into the field to conduct spot checks. Clearly it makes no sense to visit every parcel, and instead a sample is taken. Because some classes are commoner than others, a random sample that made every parcel equally likely to be chosen would be inefficient, because too many data would be gathered on common classes, and not enough on the relatively rare ones. So, instead, samples are usually chosen such that a roughly equal number of parcels are selected in each class. Of course these decisions must be based on the class as recorded in the database, rather than the true class. This is an instance of sampling that is systematically *stratified* by class (see Section 4.4).

Sampling for accuracy assessment should pay greater attention to the classes that are rarer on the ground.

Parcels represent a relatively easy case, if it is reasonable to assume that the land use class of a parcel is uniform over the parcel, and class is recorded as a single attribute of each parcel object. But as we noted in Section 6.2, more difficult cases arise in sampling natural areas (for example in the case of vegetation cover class), where parcel boundaries do not exist. Figure 6.10 shows a typical vegetation cover class map, and is obviously highly generalized. If we were to apply the previous strategy, then we would test each area to see if its assigned vegetation cover class checks out on the ground. But unlike the parcel case, in this example the boundaries between areas are not fixed, but are themselves part of the observation process, and we need to ask whether they are correctly located. Error in this case has two forms: misallocation of an area's class and mislocation of an area's boundaries. In some cases the boundary between two areas may be fixed, because it coincides with a clearly defined line on the ground; but in other cases, the boundary's location is as much a matter of judgment as the allocation of an area's class. Peter Burrough and

Figure 6.10 An example of a vegetation cover map. Two strategies for accuracy assessment are available: to check by area (polygon), or to check by point. In the former case a strategy would be devised for field checking each area, to determine the area's correct class. In the latter, points would be sampled across the state and the correct class determined at each point

Andrew Frank have discussed many of the implications of uncertain boundaries in GIS.

Errors in land cover maps can occur in the locations of boundaries of areas, as well as in the classification of areas.

In such cases we need a different strategy, that captures the influence both of mislocated boundaries and of misallocated classes. One way to deal with this is to think of error not in terms of classes assigned to areas, but in terms of classes assigned to points. In a raster dataset, the cells of the raster are a reasonable substitute for individual points. Instead of asking whether area classes are confused, and estimating errors by sampling areas, we ask whether the classes assigned to raster cells are confused, and define the confusion matrix in terms of misclassified cells. This is often called *per-pixel* or *per-point* accuracy assessment, to distinguish it from the previous strategy of *per-polygon* accuracy assessment. As before, we would want to stratify by class, to make sure that relatively rare classes were sampled in the assessment.

6.3.2.2 Interval/ratio case

The second case addresses measurements that are made on interval or ratio scales. Here, error is best thought of not as a change of class, but as a change of value, such that the observed value x' is equal to the true value

x plus some distortion δx, where δx is hopefully small. δx might be either positive or negative, since errors are possible in both directions. For example, the measured and recorded elevation at some point might be equal to the true elevation, distorted by some small amount. If the average distortion is zero, so that positive and negative errors balance out, the observed values are said to be *unbiased*, and the average value will be true.

Error in measurement can produce a change of class, or a change of value, depending on the type of measurement.

Sometimes it is helpful to distinguish between *accuracy*, which has to do with the magnitude of δx, and *precision*. Unfortunately there are several ways of defining precision in this context, at least two of which are regularly encountered in GIS. Surveyors and others concerned with measuring instruments tend to define precision through the performance of an instrument in making repeated measurements of the same phenomenon. A measuring instrument is precise according to this definition if it repeatedly gives similar measurements, whether or not these are actually accurate. So a GPS receiver might make successive measurements of the same elevation, and if these are similar the instrument is said to be precise. Precision in this case can be measured by the variability among repeated measurements. But it is possible, for example, that all of the measurements are approximately 5 m too high, in which case the measurements are said to be biased, even though they are precise, and the instrument is said to be inaccurate. Figure 6.11 illustrates this meaning of precise, and its relationship to accuracy.

The other definition of precision is more common in science generally. It defines precision as the number of digits used to report a measurement, and again it is not necessarily related to accuracy. For example, a GPS receiver might measure elevation as 51.3456 m. But if the

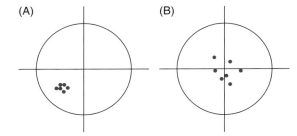

Figure 6.11 The term *precision* is often used to refer to the repeatability of measurements. In both diagrams six measurements have been taken of the same position, represented by the center of the circle. In (A) successive measurements have similar values (they are *precise*), but show a bias away from the correct value (they are *inaccurate*). In (B), precision is lower but accuracy is higher

receiver is in reality only accurate to the nearest 10 cm, three of those digits are spurious, with no real meaning. So, although the precision is one ten thousandth of a meter, the accuracy is only one tenth of a meter. Box 6.3 summarizes the rules that are used to ensure that reported measurements do not mislead by appearing to have greater accuracy than they really do.

To most scientists, precision refers to the number of significant digits used to report a measurement, but it can also refer to a measurement's repeatability.

In the interval/ratio case, the magnitude of errors is described by the *root mean square error* (RMSE), defined as the square root of the average squared error, or:

$$\left[\sum \delta x^2 / n\right]^{1/2}$$

Technical Box **6.3**

Good practice in reporting measurements

Here are some simple rules that help to ensure that people receiving measurements from others are not misled by their apparently high precision.

1. The number of digits used to report a measurement should reflect the measurement's accuracy. For example, if a measurement is accurate to 1 m then no decimal places should be reported. The measurement 14.4 m suggests accuracy to one tenth of a meter, as does 14.0, but 14 suggests accuracy to 1 m.

2. Excess digits should be removed by rounding. Fractions above one half should be rounded up, fractions below one half should be

rounded down. The following examples reflect rounding to two decimal places:

   14.57803 rounds to 14.58
   14.57397 rounds to 14.57
   14.57999 rounds to 14.58
   14.57499 rounds to 14.57

3. These rules are not effective to the left of the decimal place – for example, they give no basis for knowing whether 1400 is accurate to the nearest unit, or to the nearest hundred units.

4. If a number is known to be exactly an integer or whole number, then it is shown with no decimal point.

where the summation is over the values of δx for all of the n observations. The RMSE is similar in a number of ways to the standard deviation of observations in a sample. Although RMSE involves taking the square root of the average squared error, it is convenient to think of it as approximately equal to the average error in each observation, whether the error is positive or negative. The US Geological Survey uses RMSE as its primary measure of the accuracy of elevations in digital elevation models, and published values range up to 7 m.

Although the RMSE can be thought of as capturing the magnitude of the average error, many errors will be greater than the RMSE, and many will be less. It is useful, therefore, to know how errors are *distributed* in magnitude – how many are large, how many are small. Statisticians have developed a series of models of error distributions, of which the commonest and most important is the Gaussian distribution, otherwise known as the error function, the 'bell curve', or the Normal distribution. Figure 6.12 shows the curve's shape. If observations are unbiased, then the mean error is zero (positive and negative errors cancel each other out), and the RMSE is also the distance from the center of the distribution (zero) to the points of inflection on either side, as shown in the figure. Let us take the example of a 7 m RMSE on elevations in a USGS digital elevation model; if error follows the Gaussian distribution, this means that some errors will be more than 7 m in magnitude, some will be less, and also that the relative abundance of errors of any given size is described by the curve shown. 68% of errors will be between -1.0 and $+1.0$ RMSEs, or -7 m and $+7$ m. In practice many distributions of error do follow the Gaussian distribution, and there are good theoretical reasons why this should be so.

The Gaussian distribution predicts the relative abundances of different magnitudes of error.

To emphasize the mathematical formality of the Gaussian distribution, its equation is shown below. The symbol

σ denotes the standard deviation, μ denotes the mean (in Figure 6.12 these values are 1 and 0 respectively), and exp is the exponential function, or '2.71828 to the power of'. Scientists believe that it applies very broadly, and that many instances of measurement error adhere closely to the distribution, because it is grounded in rigorous theory. It can be shown mathematically that the distribution arises whenever a large number of random factors contribute to error, and the effects of these factors combine additively – that is, a given effect makes the same additive contribution to error whatever the specific values of the other factors. For example, error might be introduced in the use of a steel tape measure over a large number of measurements because some observers consistently pull the tape very taught, or hold it very straight, or fastidiously keep it horizontal, or keep it cool, and others do not. If the combined effects of these considerations always contributes the same amount of error (e.g., $+1$ cm, or -2 cm), then this contribution to error is said to be additive.

$$f(x) = \frac{1}{\sigma\sqrt{2\pi}} \, \exp\left[-\frac{(x-\mu)^2}{2\sigma^2}\right]$$

We can apply this idea to determine the inherent uncertainty in the locations of contours. The US Geological Survey routinely evaluates the accuracies of its digital elevation models (DEMs), by comparing the elevations recorded in the database with those at the same locations in more accurate sources, for a sample of points. The differences are summarized in a RMSE, and in this example we will assume that errors have a Gaussian distribution with zero mean and a 7 m RMSE. Consider a measurement of 350 m. According to the error model, the truth might be as high as 360 m, or as low as 340 m, and the relative frequencies of any particular error value are as predicted by the Gaussian distribution with a mean of zero and a standard deviation of 7. If we take error into account, using the Gaussian distribution with an RMSE of 7 m, it is no longer clear that a measurement of 350 m lies exactly on the 350 m contour. Instead, the truth might be 340 m, or 360 m, or 355 m. Figure 6.13 shows the implications of this in terms of the location of this contour in a real-world example. 95% of errors would put the contour within the colored zone. In areas colored red the observed value is less than 350 m, but the truth might be 350 m; in areas colored green the observed value is more than 350 m, but the truth might be 350 m. There is a 5% chance that the true location of the contour lies outside the colored zone entirely.

6.3.3 Positional error

In the case of measurements of position, it is possible for every coordinate to be subject to error. In the two-dimensional case, a measured position (x', y') would be subject to errors in both x and y; specifically, we might write $x' = x + \delta x$, $y' = y + \delta y$, and similarly in the three-dimensional case where all three coordinates are measured, $z' = z + \delta z$. The *bivariate Gaussian distribution* describes errors in the two horizontal dimensions,

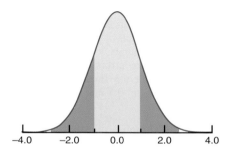

Figure 6.12 The Gaussian or Normal distribution. The height of the curve at any value of x gives the relative abundance of observations with that value of x. The area under the curve between any two values of x gives the probability that observations will fall in that range. The range between -1 standard deviation and $+1$ standard deviation is in light purple. It encloses 68% of the area under the curve, indicating that 68% of observations will fall between these limits

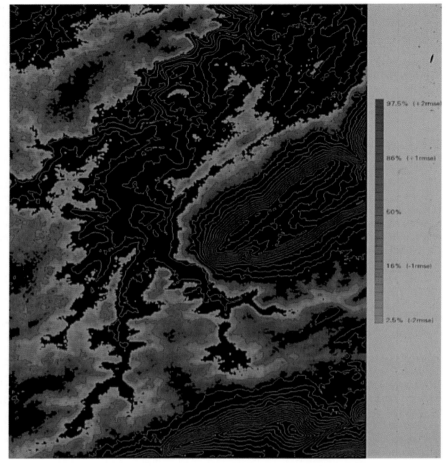

Figure 6.13 Uncertainty in the location of the 350 m contour in the area of State College, Pennsylvania, generated from a US Geological Survey DEM with an assumed RMSE of 7 m. According to the Gaussian distribution with a mean of 350 m and a standard deviation of 7 m, there is a 95% probability that the true location of the 350 m contour lies in the colored area, and a 5% probability that it lies outside (*Source*: Hunter G. J. and Goodchild M. F. 1995 'Dealing with error in spatial databases: a simple case study'. *Photogrammetric Engineering and Remote Sensing* **61**: 529–37)

and it can be generalized to the three-dimensional case. Normally, we would expect the RMSEs of x and y to be the same, but z is often subject to errors of quite different magnitude, for example in the case of determinations of position using GPS. The bivariate Gaussian distribution also allows for correlation between the errors in x and y, but normally there is little reason to expect correlations.

Because it involves two variables, the bivariate Gaussian distribution has somewhat different properties from the simple (univariate) Gaussian distribution. 68% of cases lie within one standard deviation for the univariate case (Figure 6.12). But in the bivariate case with equal standard errors in x and y, only 39% of cases lie within a circle of this radius. Similarly, 95% of cases lie within two standard deviations for the univariate distribution, but it is necessary to go to a circle of radius equal to 2.15 times the x or y standard deviations to enclose 90% of the bivariate distribution, and 2.45 times standard deviations for 95%.

National Map Accuracy Standards often prescribe the positional errors that are allowed in databases. For example, the 1947 US National Map Accuracy Standard specified that 95% of errors should fall below 1/30 inch (0.85 mm) for maps at scales of 1:20 000 and finer (more detailed), and 1/50 inch (0.51 mm) for other maps (coarser, less detailed, levels of granularity than 1:20 000). A convenient rule of thumb is that positions measured from maps are subject to errors of up to 0.5 mm at the scale of the map. Table 6.3 shows the distance on the ground corresponding to 0.5 mm for various common map scales.

A useful rule of thumb is that features on maps are positioned to an accuracy of about 0.5 mm.

6.3.4 The spatial structure of errors

The confusion matrix, or more specifically a single row of the matrix, along with the Gaussian distribution, provide convenient ways of describing the error present in a single

Table 6.3 A useful rule of thumb is that positions measured from maps are accurate to about 0.5 mm on the map. Multiplying this by the scale of the map gives the corresponding distance on the ground

Map scale	Ground distance corresponding to 0.5 mm map distance
1:1250	0.625 m
1:2500	1.25 m
1:5000	2.5 m
1:10 000	5 m
1:24 000	12 m
1:50 000	25 m
1:100 000	50 m
1:250 000	125 m
1:1 000 000	500 m
1:10 000 000	5 000 m

observation of a nominal or interval/ratio measurement respectively. When a GIS is used to respond to a simple query, such as 'tell me the class of soil at this point', or 'what is the elevation here?', then these methods are good ways of describing the uncertainty inherent in the response. For example, a GIS might respond to the first query with the information 'Class A, with a 30% probability of Class C', and to the second query with the information '350 m, with an RMSE of 7 m'. Notice how this makes it possible to describe nominal data as accurate to a percentage, but it makes no sense to describe a DEM, or any measurement on an interval/ratio scale, as accurate to a percentage. For example, we cannot meaningfully say that a DEM is '90% accurate'.

However, many GIS operations involve more than the properties of single points, and this makes the analysis of error much more complex. For example, consider the query 'how far is it from this point to that point?' Suppose the two points are both subject to error of position, because their positions have been measured using GPS units with mean distance errors of 50 m. If the two measurements were taken some time apart, with different combinations of satellites above the horizon, it is likely that the errors are independent of each other, such that one error might be 50 m in the direction of North, and the other 50 m in the direction of South. Depending on the locations of the two points, the error in distance might be as high as +100 m. On the other hand, if the two measurements were made close together in time, with the same satellites above the horizon, it is likely that the two errors would be similar, perhaps 50 m North and 40 m North, leading to an error of only 10 m in the determination of distance. The difference between these two situations can be measured in terms of the degree of *spatial autocorrelation*, or the interdependence of errors at different points in space (Section 4.6).

The spatial autocorrelation of errors can be as important as their magnitude in many GIS operations.

Spatial autocorrelation is also important in errors in nominal data. Consider a field that is known to contain a single crop, perhaps wheat. When seen from above, it is possible to confuse wheat with other crops, so there may be error in the crop type assigned to points in the field. But since the field has only one crop, we know that such errors are likely to be strongly correlated. Spatial autocorrelation is almost always present in errors to some degree, but very few efforts have been made to measure it systematically, and as a result it is difficult to make good estimates of the uncertainties associated with many GIS operations.

An easy way to visualize spatial autocorrelation and interdependence is through animation. Each frame in the animation is a single possible map, or *realization* of the error process. If a point is subject to uncertainty, each realization will show the point in a different possible location, and a sequence of images will show the point shaking around its mean position. If two points have perfectly correlated positional errors, then they will appear to shake in unison, as if they were at the ends of a stiff rod. If errors are only partially correlated, then the system behaves as if the connecting rod were somewhat elastic.

The spatial structure or autocorrelation of errors is important in many ways. DEM data are often used to estimate the slope of terrain, and this is done by comparing elevations at points a short distance apart. For example, if the elevations at two points 10 m apart are 30 m and 35 m respectively, the slope along the line between them is 5/10, or 0.5. (A somewhat more complex method is used in practice, to estimate slope at a point in the x and y directions in a DEM raster, by analyzing the elevations of nine points – the point itself and its eight neighbors. The equations in Section 14.4 detail the procedure.)

Now consider the effects of errors in these two elevation measurements on the estimate of slope. Suppose the first point (elevation 30 m) is subject to an RMSE of 2 m, and consider possible true elevations of 28 m and 32 m. Similarly the second point might have true elevations of 33 m and 37 m. We now have four possible combinations of values, and the corresponding estimates of slope range from $(33 - 32)/10 = 0.1$ to $(37 - 28)/10 = 0.9$. In other words, a relatively small amount of error in elevation can produce wildly varying slope estimates.

The spatial autocorrelation between errors in geographic databases helps to minimize their impacts on many GIS operations.

What saves us in this situation, and makes estimation of slope from DEMs a practical proposition at all, is spatial autocorrelation among the errors. In reality, although DEMs are subject to substantial errors in absolute elevation, neighboring points nevertheless tend to have similar errors, and errors tend to persist over quite large areas. Most of the sources of error in the DEM production process tend to produce this kind of persistence of error over space, including errors due to misregistration of aerial photographs. In other words, errors in DEMs exhibit strong positive spatial autocorrelation.

Another important corollary of positive spatial auto-correlation can also be illustrated using DEMs. Suppose an area of low-lying land is inundated by flooding, and our task is to estimate the area of land affected. We are asked to do this using a DEM, which is known to have an RMSE of 2 m (compare Figure 6.13). Suppose the data points in the DEM are 30 m apart, and preliminary analysis shows that 100 points have elevations below the flood line. We might conclude that the area flooded is the area represented by these 100 points, or 900×100 sq m, or 9 hectares. But because of errors, it is possible that some of this area is actually above the flood line (we will ignore the possibility that other areas outside this may also be below the flood line, also because of errors), and it is possible that *all* of the area is above. Suppose the recorded elevation for each of the 100 points is 2 m below the flood line. This is one RMSE (recall that the RMSE is equal to 2 m) below the flood line, and the Gaussian distribution tells us that the chance that the true elevation is actually above the flood line is approximately 16% (see Figure 6.12). But what is the chance that *all* 100 points are actually above the flood line?

Here again the answer depends on the degree of spatial autocorrelation among the errors. If there is none, in other words if the error at each of the 100 points is independent of the errors at its neighbors, then the answer is $(0.16)^{100}$, or 1 chance in 1 followed by roughly 70 zeroes. But if there is strong positive spatial autocorrelation, so strong that all 100 points are subject to exactly the same error, then the answer is 0.16. One way to think about this is in terms of *degrees of freedom*. If the errors are independent, they can vary in 100 independent ways, depending on the error at each point. But if they are strongly spatially autocorrelated, the effective number of degrees of freedom is much less, and may be as few as 1 if all errors behave in unison. Spatial autocorrelation has the effect of reducing the number of degrees of freedom in geographic data below what may be implied by the volume of information, in this case the number of points in the DEM.

Spatial autocorrelation acts to reduce the effective number of degrees of freedom in geographic data.

6.4 U3: Further uncertainty in the analysis of geographic phenomena

6.4.1 Internal and external validation through spatial analysis

In Chapter 1 we identified one remit of GIS as the resolution of scientific or decision-making problems through spatial analysis, which we defined in Section 1.7 as 'the process by which we turn raw spatial data into useful spatial information'. Good science needs secure foundations, yet Sections 6.2 and 6.3 have shown the conception and measurement of many geographic phenomena to be inherently uncertain. How can the outcome of spatial analysis be meaningful if it has such uncertain foundations?

Uncertainties in data lead to uncertainties in the results of analysis.

Once again, there are no easy answers to this question, although we can begin by examining the consequences of accommodating possible errors of positioning, or of aggregating clearly defined units of analysis into artificial geographic individuals (as when people are aggregated by census tracts, or disease incidences are aggregated by county). In so doing, we will illustrate how potential problems might arise, but will not present any definitive solutions – for the simple reason that the truth is inherently uncertain. The conception, measurement, and representation of geographic individuals may distort the outcome of spatial analysis by masking or accentuating apparent variation across space, or by restricting the nature and range of questions that can be asked of the GIS.

There are three ways of dealing with this risk. First, although we can only rarely tackle the *source* of distortion (we are rarely empowered to collect new, completely disaggregate data, for example), we can quantify the way in which it is likely to *operate* (or *propagates*) within the GIS, and can gauge the magnitude of its likely impacts. Second, although we may have to work with aggregated data, GIS allows us to model within-zone spatial distributions in order to ameliorate the worst effects of artificial zonation. Taken together, GIS allows us to gauge the effects of scale and aggregation through simulation of different possible outcomes. This is *internal validation* of the effects of scale, point placement, and spatial partitioning.

Because of the power of GIS to merge diverse data sources, it also provides a means of *external validation* of the effects of zonal averaging. In today's advanced GIService economy (Section 1.5.3), there may be other data sources that can be used to gauge the effects of aggregation upon our analysis. In Chapter 13 we will refine the basic model that was presented in Figure 6.1 to consider how GIS provides a medium for visualizing models of spatial distributions and patterns of homogeneity and heterogeneity.

GIS gives us maximum flexibility when working with aggregate data, and helps us to validate our data with reference to other available sources.

6.4.2 Internal validation: error propagation

The examples of Section 6.3.4 are cases of error propagation, where the objective is to measure the effects of known levels of data uncertainty on the outputs of

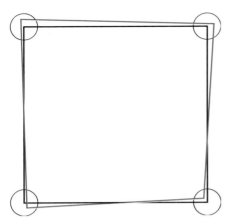

Figure 6.14 Error in the measurement of the area of a square 100 m on each side. Each of the four corner points has been surveyed; the errors are subject to bivariate Gaussian distributions with standard deviations in x and y of 1 m (dashed circles). The blue polygon shows one possible surveyed square (one *realization* of the error model)

GIS operations. We have seen how the spatial structure of errors plays a role, and how the existence of strong positive spatial autocorrelation reduces the effects of uncertainty upon estimates of properties such as slope or area. Yet the cumulative effects of error can also produce impacts that are surprisingly large, and some of the examples in this section have been chosen to illustrate the substantial uncertainties that can be produced by apparently innocuous data errors.

Error propagation measures the impacts of uncertainty in data on the results of GIS operations.

In general two strategies are available for evaluating error propagation. The examples in the previous section were instances in which it was possible to obtain a complete description of error effects based upon known measures of likely error. These enable a complete analysis of uncertainty in slope estimation, and can be applied in the DEM flooding example described in Section 6.3.4. Another example that is amenable to analysis is the calculation of the area of a polygon given knowledge of the positional uncertainties of its vertices.

For example, Figure 6.14 shows a square approximately 100 m on each side. Suppose the square has been surveyed by determining the locations of its four corner points using GPS, and suppose the circumstances of the measurements are such that there is an RMSE of 1 m in both coordinates of all four points, and that errors are independent. Suppose our task is to determine the area of the square. A GIS can do this easily, using a standard algorithm (see Figure 14.9). Computers are precise (in the sense of Section 6.3.2.2 and Box 6.3), and capable of working to many significant digits, so the calculation might be reported by printing out a number to eight digits, such as 10014.603 sq m, or even more. But the number of significant digits will have been determined by the precision of the machine, and not by the accuracy of the

determination. Box 6.3 summarized some simple rules for ensuring that the precision used to report a measurement reflects as far as possible its accuracy, and clearly those rules will have been violated if the area is reported to eight digits. But what is the appropriate precision?

In this case we can determine exactly how positional accuracy affects the estimate of area. It turns out that area has an error distribution which is Gaussian, with a standard deviation (RMSE) in this case of 200 sq m – in other words, each attempt to measure the area will give a different result, the variation between them having a standard deviation of 200 sq m. This means that the five rightmost digits in the estimate are spurious, including two digits to the left of the decimal point. So if we were to follow the rules of Box 6.3, we would print 10 000 rather than 10014.603 (note the problem with standard notation here, which does not let us omit digits to the left of the decimal point even if they are spurious, and so leaves some uncertainty about whether the tens and units digits are certain or not – and note also the danger that if the number is printed as an integer it may be interpreted as exactly the whole number). We can also turn the question around and ask how accurately the points would have to be measured to justify eight digits, and the answer is approximately 0.01 mm, far beyond the capabilities of normal surveying practice.

Analysis can be applied to many other kinds of GIS analysis, and Gerard Heuvelink (Box 6.4) discusses several further examples in his excellent text on error propagation in GIS. But analysis is a difficult strategy when spatial autocorrelation of errors is present, and many problems of error propagation in GIS are not amenable to analysis. This has led many researchers to explore a more general strategy of simulation to evaluate the impacts of uncertainty on results.

In essence, simulation requires the generation of a series of realizations, as defined earlier, and it is often called Monte Carlo simulation in reference to the realizations that occur when dice are tossed or cards are dealt in various games of chance. For example, we could simulate error in a single measurement from a DEM by generating a series of numbers with a mean equal to the measured elevation, and a standard deviation equal to the known RMSE, and a Gaussian distribution. Simulation uses everything that is known about a situation, so if any additional information is available we would incorporate it in the simulation. For example, we might know that elevations must be whole numbers of meters, and would simulate this by rounding the numbers obtained from the Gaussian distribution. With a mean of 350 m and an RMSE of 7 m the results of the simulation might be 341, 352, 356, 339, 349, 348, 355, 350, ...

Simulation is an intuitively simple way of getting the uncertainty message across.

Because of spatial autocorrelation, it is impossible in most circumstances to think of databases as decomposable into component parts, each of which can be independently disturbed to create alternative realizations, as in the previous example. Instead, we have to think of

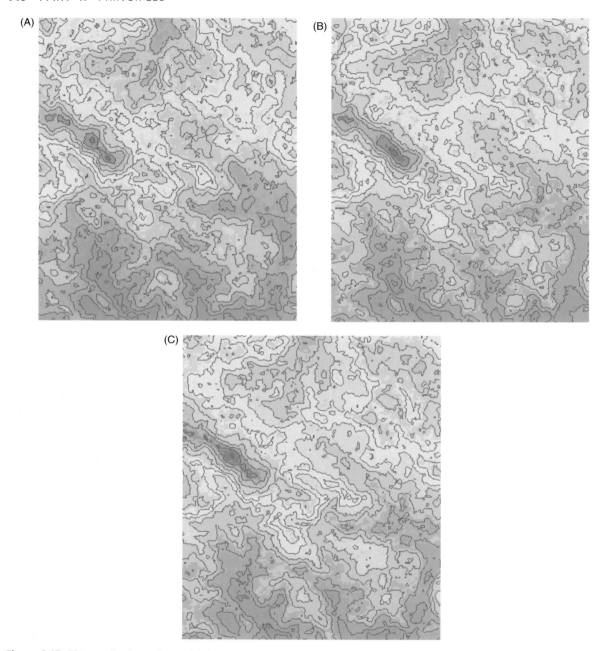

Figure 6.15 Three realizations of a model simulating the effects of error on a digital elevation model. The three datasets differ only to a degree consistent with known error. Error has been simulated using a model designed to replicate the known error properties of this dataset – the distribution of error magnitude, and the spatial autocorrelation between errors. (Reproduced by permission of Ashton Shortridge.)

the entire database as a realization, and create alternative realizations of the database's contents that preserve spatial autocorrelation. Figure 6.15 shows an example, simulating the effects of uncertainty on a digital elevation model. Each of the three realizations is a complete map, and the simulation process has faithfully replicated the strong correlations present in errors across the DEM.

6.4.3 Internal validation: aggregation and analysis

We have seen already that a fundamental difference between geography and other scientific disciplines is that the definition of its objects of study is only rarely unambiguous and, in practice, rarely precedes

Gerard Heuvelink, geostatistician

Understanding the limitations of spatial data and spatial models is essential both for managing environmental systems effectively and for encouraging safe use of GIS. Gerard Heuvelink (Figure 6.16) of the Wageningen University and Research Centre, the Netherlands, has dedicated much of his scientific career to this end, through statistical modeling of the uncertainty in spatial data and analysis of the ways in which uncertainty is propagated through GIS.

Trained as a mathematician, Gerard undertook a Ph.D. in Physical Geography working with Professor Peter Burrough of Utrecht University. His 1998 research monograph *Error Propagation in Environmental Modelling with GIS* has subsequently become the key reference in spatial uncertainty analysis. Gerard is firmly of the view that GI scientists should pay more attention to statistical validation and exploration of data, and he is actively involved in a series of symposia on 'Spatial Accuracy Assessment in Natural Resources and Environmental Sciences' (**www.spatial-accuracy.org**).

Figure 6.16 Gerard Heuvelink, geostatistician

Gerard's background in mathematics and statistics has left him with the view that spatial uncertainty analysis requires a sound statistical basis. In his view, understanding uncertainty in the position of spatial objects and in their attribute values entails use of probability distribution functions, and measuring spatial autocorrelation (Section 4.6) with uncertainties in other objects in spatial databases. He says: 'I remain disappointed with the amount of progress made in understanding the fundamental problems of uncertainty over the last fifteen years. We have moved forward in the sense that we now have a broader view of various aspects of spatial data quality. The 1980s and early 1990s were dedicated to technical topics such as uncertainty propagation in map overlay operations and the development of statistical models for representing positional uncertainty. More recently the research community has addressed a range of user-centric topics, such as visualization and communication of uncertainty, decision making under uncertainty and the development of error-aware GIS. But these developments do not hide the fact that we still do not have the statistical basics right. Until this is achieved, we run the risk of building elaborate representations on weak and uncertain foundations.'

Gerard and co-worker James Brown from the University of Amsterdam are working to contribute to filling this gap, by developing a general probabilistic framework for characterizing uncertainty in the positions and attribute values of spatial objects.

our attempts to measure their characteristics. In socio-economic GIS applications, these objects of study (geographic individuals) are usually aggregations, since the spaces that human individuals occupy are geographically unique, and confidentiality restrictions usually dictate that uniquely attributable information must be anonymized in some way. Even in natural-environment applications, the nature of sampling in the data collection process (Section 4.4) often makes it expedient to collect data pertaining to aggregations of one kind or another. Thus in socio-economic and environmental applications alike, the measurement of geographic individuals is unlikely to be determined with the end point of particular spatial-analysis applications in mind. As a consequence, we cannot be certain in ascribing even dominant characteristics *of* areas to true individuals or point locations *in* those areas. This source of uncertainty is known as the *ecological fallacy*, and has long bedevilled the analysis of spatial distributions (the opposite of ecological fallacy is

atomistic fallacy, in which the individual is considered in isolation from his or her environment). This is illustrated in Figure 6.17.

Inappropriate inference from aggregate data about the characteristics of individuals is termed the ecological fallacy.

We have also seen that the *scale* at which geographic individuals are conceived conditions our measures of association between the mosaic of zones represented within a GIS. Yet even when scale is fixed, there is a multitude of ways in which basic areal units of analysis can be *aggregated* into zones, and the requirement of spatial contiguity represents only a weak constraint upon the huge combinatorial range. This gives rise to the related *aggregation* or *zonation* problem, in which different combinations of a given number of geographic individuals into coarser-scale areal units can yield widely different results. In a classic 1984 study, the geographer Stan

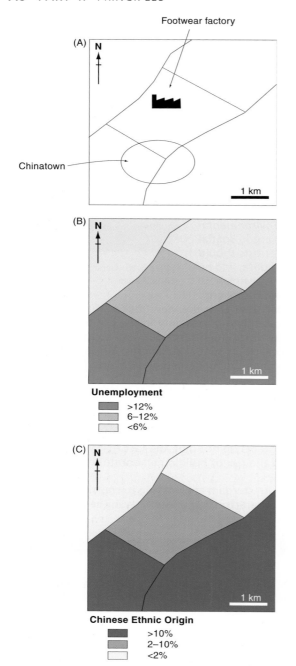

Figure 6.17 The problem of ecological fallacy. Before it closed down, the Anytown footwear factory drew its labor from blue-collar neighborhoods in its south and west sectors. Its closure led to high local unemployment, but not amongst the residents of Chinatown, who remain employed in service industries. Yet comparison of choropleth maps B and C suggests a spurious relationship between Chinese ethnicity and unemployment

Openshaw applied correlation and regression analysis to the attributes of a succession of zoning schemes. He demonstrated that the constellation of elemental zones within aggregated areal units could be used to manipulate the results of spatial analysis to a wide range of quite different prespecified outcomes. These numerical experiments have some sinister counterparts in the real world, the most notorious example of which is the political gerrymander of 1812 (see Section 14.3.2). Chance or design might therefore conspire to create apparent spatial distributions which are unrepresentative of the scale and configuration of real-world geographic phenomena. The outcome of multivariate spatial analysis is also similarly sensitive to the particular zonal scheme that is used. Taken together, the effects of scale and aggregation are generally known as the *Modifiable Areal Unit Problem* (MAUP).

The ecological fallacy and the MAUP have long been recognized as problems in applied spatial analysis and, through the concept of spatial autocorrelation (Section 4.3), they are also understood to be related problems. Increased technical capacity for numerical processing and innovations in scientific visualization have refined the quantification and mapping of these measurement effects, and have also focused interest on the effects of within-area spatial distributions upon analysis.

6.4.4 External validation: data integration and shared lineage

Goodchild and Longley (1999) use the term *concatenation* to describe the integration of two or more different data sources, such that the contents of each are accessible in the product. The polygon overlay operation that will be discussed in Section 14.4.3, and its field-view counterpart, is one simple form of concatenation. The term *conflation* is used to describe the range of functions that attempt to overcome differences between datasets, or to merge their contents (as with rubber-sheeting: see Section 9.3.2.3). Conflation thus attempts to replace two or more versions of the same information with a single version that reflects the pooling, or weighted averaging, of the sources.

The individual items of information in a single geographic dataset often share lineage, in the sense that more than one item is affected by the same error. This happens, for example, when a map or photograph is registered poorly, since all of the data derived from it will have the same error. One indicator of shared lineage is the persistence of error – because all points derived from the same misregistration will be displaced by the same, or a similar, amount. Because neighboring points are more likely to share lineage than distant points, errors tend to exhibit strong positive spatial autocorrelation.

Conflation combines the information from two data sources into a single source.

When two datasets that share no common lineage are concatenated (for example, they have not been subject to the same misregistration), then the relative positions of objects inherit the absolute positional errors of both, even over the shortest distances. While the shapes of objects in each dataset may be accurate, the relative locations of pairs of neighboring objects may be wildly inaccurate when drawn from different datasets. The anecdotal history of GIS is full of examples of datasets which were perfectly adequate for one application, but which failed completely when an application required that they be merged with some new dataset that had no common lineage. For example, merging GPS measurements of point positions with streets derived from the US Bureau of the Census TIGER files may lead to surprises where points appear on the wrong sides of streets. If the absolute positional accuracy of a dataset is 50 m, as it is with parts of the TIGER database, points located less than 50 m from the nearest street will frequently appear to be misregistered.

Datasets with different lineages often reveal unsuspected errors when overlaid.

Figure 6.18 shows an example of the consequences of overlaying data with different lineages. In this case, two datasets of streets produced by different commercial vendors using their own process fail to match in position by amounts of up to 100 m, and also fail to match in the names of many streets, and even the existence of streets.

The integrative functionality of GIS makes it an attractive possibility to generate multivariate indicators from diverse sources. Yet such data are likely to have been collected at a range of different scales, and for a range of areal units as diverse as census tracts, river catchments, land ownership parcels, travel-to-work areas, and market research surveys. Established procedures of statistical inference can only be used to reason from representative samples to the populations from which they were drawn. Yet these procedures do not regulate the assignment of inferred values to (usually smaller) zones, or their apportionment to *ad hoc* regional categorizations. There is an emergent tension within the socio-economic realm, for there is a limit to the uses of inferences drawn from conventional, scientifically valid data sources which are

frequently out-of-date, zonally coarse, and irrelevant to what is happening in modern societies. Yet the alternative of using new rich sources of marketing data may be profoundly unscientific in its inferential procedures.

6.4.5 Internal and external validation; induction and deduction

Reformulation of the MAUP into a *geocomputational* (Box 1.9 and Section 16.1) approach to zone design has been one of the key contributions of geographer Stan Openshaw. Central to this is inductive use of GIS to seek patterns through repeated scaling and aggregation experiments, alongside much better external validation, deduced using the multitude of new datasets that are a hallmark of the information age.

The Modifiable Areal Unit Problem can be investigated through simulation of large numbers of alternative zoning schemes.

Neither of these approaches, used in isolation, is likely to resolve the uncertainties inherent in spatial analysis. Zone design experiments are merely playing with the MAUP, and most of the new sources of external validation are unlikely to sustain full scientific scrutiny, particularly if they were assembled through non-rigorous survey designs. The conception and measurement of elemental zones, the geographic individuals, may be *ad hoc*, but they are rarely wholly random either. Can our recognition and understanding of the empirical effects of the MAUP help us to neutralize its effects? Not really. In measuring the distribution of all possible zonally averaged outcomes ('simple random zoning' in analogy to simple random sampling in Section 4.4), there is no tenable analogy with the established procedures of statistical inference and its concepts of precision and error. And even if there were, as we have seen in Section 4.7 there are limits to the application of classic statistical inference to spatial data.

Zoning seems similar to sampling, but its effects are very different.

The way forward seems to be to complement our new-found abilities to customize zoning schemes in GIS with external validation of data and clearer application-centered thinking about the likely degree of within-zone heterogeneity that is concealed in our aggregated data. In this view, MAUP will disappear if GIS analysts understand the particular areal units that they wish to study. There is also a sense here that resolution of the MAUP requires acknowledgment of the uniqueness of places. There is also a practical recognition that the areal objects of study are ever-changing, and our perceptions of what constitutes their appropriate definition will change. And finally, within the socio-economic realm, the act of defining zones can also be self-validating if the allocation of individuals affects the interventions they receive (be they a mail-shot about a shopping opportunity or aid under an areal policy intervention). Spatial discrimination affects

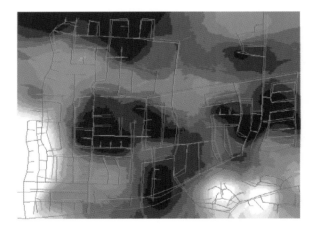

Figure 6.18 Overlay of two street databases for part of Goleta, California, USA. The red and green lines fail to match by as much as 100 m. Note also that in some cases streets in one dataset fail to appear in the other, or have different connections. The background is dark where the fit is best and white where it is poorest (it measures the average distance locally between matched intersections)

Uncertainty and town center definition

Although the locus of retail activity in many parts of the US long ago shifted to suburban locations, traditional town centers ('downtowns') remain vibrant and are cherished in most of the rest of the world. Indeed, many nations vigorously defend existing retail centers and through various planning devices seek to regulate 'out of center' development.

Therefore, many interests in the planning and retail sectors are naturally concerned with learning the precise extent of existing town centers. The pressure to devise standard definitions across its national jurisdiction led the UK's central government planning agency (the Office of the Deputy Prime Minister) to initiate a five-year research program to define and monitor changes in the shape, form, and internal geography of town centers across the nation. The work has been based at the Centre for Advanced Spatial Analysis (CASA) at University College London.

Town centers present classic examples of geographic phenomena with uncertain boundaries. Moreover the extent of any given town center is likely to change over time – for example, in response to economic fortunes consequent upon national business cycles. Candidate *indicator* variables of town centeredness might include tall buildings, pedestrian traffic, high levels of retail employment, and high retail floorspace figures.

After a consultation period with user groups (in the spirit of public participation in GIS, PPGIS: see Section 13.4) a set of the most pertinent indicators was agreed (the *conception* stage). These indicator variables measured retail and hospitality industry employment, shop and office floorspace, and retail, leisure, and service employment. The indicator measures were standardized, weighted, and summed into a summary index *measure*. This measure, mapped for all town centers, was the principal deliverable

Figure 6.19 (A) Camden Town Center, London (Reproduced by permission of Jamie G.A. Quinn); (B) a data surface representing the index of town center activity (the darker shades of red indicate greater levels of retail activity); and (C) the Camden Town Center report: Camden Town center boundary is blue, whilst the orange lines denote the retail core of Camden and of nearby town centers – the darker shades of red again indicate greater levels of retail activity

(B)

(C)

Area of Town Centre Activity: Camden Town, LB <u>Camden</u>

Employment (Persons)	
<u>Convenience retail</u>	**1,143**
<u>Comparison retail</u>	**1,008**
<u>Service retail</u>	**1,750***
<u>Offices</u>	**5,852**
<u>Civic and Public Administration</u>	**412**
<u>Restaurants & Licensed Premises</u>	**1,335**
<u>Arts, Culture and Entertainment</u>	**408**

Turnover (£000s)	
<u>Convenience retail</u>	**130,336**
<u>Comparison retail</u>	**89,475**
<u>Service retail</u>	**Disclosive**
<u>Restaurants & Licensed Premises</u>	**40,015**
<u>Arts, Culture and Entertainment</u>	**35,399**

Floorspace (Sq m)	
<u>A1</u>	**55,496**
<u>A2</u>	**6,726**
<u>A3</u>	**12,967**
<u>Retail</u>	**81,278**
<u>Offices</u>	**147,663**

<u>Back to Camden list</u>
<u>Back to Camden map</u>

<u>Printer friendly Version</u>

Production date: 21 Jun 2002

* Indicates possible unreliability in the marked data

The map background is based upon the OS map by the Office of the Deputy Prime Minister (ODPM) with permission of Ordnance Survey on behalf of The Controller of Her Majesty's Stationery Office, © Crown Copyright. All rights reserved. Unauthorised reproduction infringes Crown Copyright and may lead to prosecution or civil proceedings. Licence Number: GD272671.

Figure 6.19 (*continued*)

of the research. After further consultation, the CASA team chose to represent the 'degree of town centeredness' as a field variable. This choice reflected various priorities, including the need to maintain confidentiality of those data that were not in the public domain, an attempt to avoid the worst effects of the MAUP, and the need to communicate the rather complex concept of the 'degree of town centeredness' to an audience of 'spatially *unaware* professionals' (see Section 1.4.3.2). The datasets used in the projects each represent populations (not samples), and so kernel density estimation (Section 14.4.4.4) was used to create the

composite surfaces: the size of the kernel was subjectively set at 300 meters, because of the resolution of the data and on the basis of the empirical observation that this is the maximum distance that most shoppers are prepared to walk.

An example of the composite surface 'index of town center activity' for the town center of Camden, London (Figure 6.19A) is shown in Figure 6.19B. For reasons to be explored in our discussion of geovisualization (Chapter 13), most users prefer maps to have crisp and not graduated or uncertain boundaries. Thus crisp boundaries were subsequently interpolated as shown in Figure 6.19C.

spatial behavior, and so the principles of zone design are of much more than academic interest.

Many of issues of uncertainty in conception, measurement, representation, and analysis come together in the definition of town center boundaries (see Box 6.5).

6.5 Consolidation

Uncertainty is certainly much more than error. Just as the amount of available digital data and our abilities to process them have developed, so our understanding of the quality of digital depictions of reality has broadened. It is one of the supreme ironies of contemporary GIS that as we accrue more and better data and have more computational power at our disposal, so we seem to become more uncertain about the quality of our digital representations and the adequacy of our areal units of analysis. Richness of representation and computational power only make us more aware of the range and variety of established uncertainties, and challenge us to integrate new ones. The only way beyond this impasse is to advance hypotheses about the structure of data, in a spirit of humility rather than conviction. But this implies greater *a priori* understanding about the structure in spatial as well as attribute data. There are some general rules to guide us here and statistical measures such as spatial autocorrelation provide further structural clues (Section 4.3). The developing range of context-sensitive spatial analysis methods provides a bridge between such general statistics and methods of specifying place or local (natural) environment. Geocomputation helps too, by allowing us to assess the sensitivity of outputs to inputs, but, unaided, is unlikely to provide any unequivocal best solution. The fathoming of uncertainty requires a combination of the cumulative development of *a priori* knowledge (we should expect scientific research to be cumulative in its findings), external validation of data sources, and inductive generalization

in the fluid, eclectic data-handling environment that is contemporary GIS.

More pragmatically, here are some rules for how to live with uncertainty: First, since there can be no such thing as perfectly accurate GIS analysis, it is essential to acknowledge that uncertainty is inevitable. It is better to take a positive approach, by learning what one can about uncertainty, than to pretend that it does not exist. To behave otherwise is unconscionable, and can also be very expensive in terms of lawsuits, bad decisions, and the unintended consequences of actions (see Chapter 19).

Second, GIS analysts often have to rely on others to provide data, through government-sponsored mapping programs like those of the US Geological Survey or the UK Ordnance Survey, or commercial sources. Data should never be taken as the truth, but instead it is essential to assemble all that is known about the quality of data, and to use this knowledge to assess whether, actually, the data are fit for use. Metadata (Section 11.2.1) are designed specifically for this purpose, and will often include assessments of quality. When these are not present, it is worth spending the extra effort to contact the creators of the data, or other people who have tried to use them, for advice on quality. Never trust data that have not been assessed for quality, or data from sources that do not have good reputations for quality.

Third, the uncertainties in the outputs of GIS analysis are often much greater than one might expect given knowledge of input uncertainties, because many GIS processes are highly non-linear. Other processes dampen uncertainty, rather than enhance it. Given this, it is important to gain some impression of the impacts of input uncertainty on output.

Fourth, rely on multiple sources of data whenever you can. It may be possible to obtain maps of an area at several different scales, or to obtain several different vendors' databases. Raster and vector datasets are often complementary (e.g., combine a remotely sensed image with a topographic map). Digital elevation models can often be augmented with spot elevations, or GPS measurements.

Finally, be honest and informative in reporting the results of GIS analysis. It is safe to assume that GIS designers will have done little to help in this respect – results will have been reported to high apparent precision, with more significant digits than are justified by actual accuracy, and lines will have been drawn on maps with widths that reflect relative importance, rather than uncertainty of position. It is up to you as the user to redress this imbalance, by finding ways of communicating what you know about accuracy, rather than relying on the GIS to do so. It is wise to put plenty of caveats into reported results, so that they reflect what you believe to be true, rather than what the GIS appears to be saying. As someone once said, when it comes to influencing people 'numbers beat no numbers every time, whether or not they are right', and the same is certainly true of maps (see Chapters 12 and 13).

Questions for further study

1. What tools do GIS designers build into their products to help users deal with uncertainty? Take a look at your favorite GIS from this perspective. Does it allow you to associate metadata about data quality with datasets? Is there any support for propagation of uncertainty? How does it determine the number of significant digits when it prints numbers? What are the pros and cons of including such tools?

2. Using aggregate data for Iowa counties, Openshaw found a strong positive correlation between the proportion of people over 65 and the proportion who were registered voters for the Republican party. What if anything does this tell us about the tendency for older people to register as Republicans?

3. Find out about the five components of data quality used in GIS standards, from the information available at **www.fdgc.gov**. How are the five components applied in the case of a standard mapping agency data product, such as the US Geological Survey's Digital Orthophoto Quarter–Quadrangle program? (Search the Web for the appropriate documents.)

4. You are a senior retail analyst for Safemart, which is contemplating expansion from its home US state to three others in the Union. Assess the relative merits of your own company's store loyalty card data (which you can assume are similar to those collected by any retail chain with which you are familiar) and of data from the 2000 Census in planning this strategic initiative. Pay, particular attention to issues of survey content, the representativeness of population characteristics, and problems of scale and aggregation. Suggest ways in which the two data sources might complement one another in an integrated analysis.

Further reading

Burrough P.A. and Frank A.U. (eds) 1996 *Geographic Objects with Indeterminate Boundaries*. London: Taylor and Francis.

Fisher P.F. 2005 'Models of uncertainty in spatial data.' In Longley P.A., Goodchild M.F., Maguire D.J. and Rhind D.W. (eds) *Geographical Information Systems: Principles, Techniques, Management and Applications* (abridged edition). Hoboken, N.J.: Wiley, pp. 191–205.

Goodchild M.F. and Longley P.A. 2005 'The future of GIS and spatial analysis.' In Longley P.A., Goodchild M.F., Maguire D.J. and Rhind D.W. (eds) *Geographical Information Systems: Principles, Techniques, Management and Applications* (abridged edition). Hoboken, N.J.: Wiley, pp. 567–80.

Heuvelink G.B.M. 1998 *Error Propagation in Environmental Modelling with GIS*. London: Taylor and Francis.

Openshaw S. and Alvanides S. 2005 'Applying geocomputation to the analysis of spatial distributions'. In Longley P.A., Goodchild M.F., Maguire D.J. and Rhind D.W. (eds) *Geographical Information Systems: Principles, Techniques, Management and Applications* (abridged edition). Hoboken, N.J.: Wiley, pp. 267–282.

Zhang J.X. and Goodchild M.F. 2002 *Uncertainty in Geographical Information*. New York: Taylor and Francis.

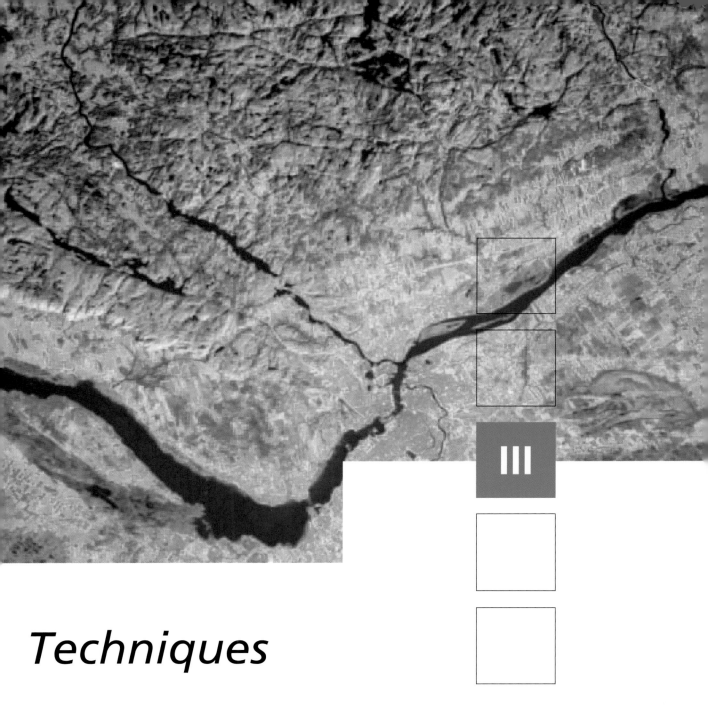

Techniques

7 GIS Software

GIS software is the processing engine and a vital component of an operational GIS. It is made up of integrated collections of computer programs that implement geographic processing functions. The three key parts of any GIS software system are the user interface, the tools (functions), and the data manager. All three parts may be located on a single computer or they may be spread over multiple machines in a departmental or enterprise configuration. Four main types of computer system architecture configurations are used to build operational GIS implementations: desktop, client-server, centralized desktop, and centralized server. There are many different types of GIS software and this chapter uses five categories to organize the discussion: desktop, server (including Internet), developer, hand-held, and other. The market leading commercial GIS software vendors are ESRI, Intergraph, Autodesk, and GE Energy (Smallworld).

Geographic Information Systems and Science, 2nd edition   Paul Longley, Michael Goodchild, David Maguire, and David Rhind.
© 2005 John Wiley & Sons, Ltd.   ISBNs: 0-470-87000-1 (HB); 0-470-87001-X (PB)

Learning Objectives

After reading this chapter you will be able to:

- Understand the architecture of GIS software systems, specifically
  - Organization by project, department, or enterprise;
  - The three-tier architecture of software systems (graphical user interface; tools, and data access);

- Describe the process of GIS customization;

- Describe the main types of commercial software
  - Desktop
  - Server
  - Developer
  - Hand-held
  - Other;

- Outline the main types of commercial GIS software products currently available.

7.1 Introduction

In Chapter 1, the four technical parts of a geographic information system were defined as the network, the hardware, the software, and the data, which together functioned with reference to people and the procedural structures within which people work (Section 1.5). This chapter is concerned with GIS software, the geographic processing engine of a complete, working GIS. The functionality or capabilities of GIS software will be discussed later in this book (especially in Chapters 12–16). The focus here is on the different ways in which these capabilities are realized in GIS software products and implemented in operational GIS.

This chapter takes a fairly narrow view of GIS software, concentrating on systems with a range of generic capabilities to collect, store, manage, query, analyze, and present geographic information. It excludes atlases, simple graphics and mapping systems, route finding software, simple location-based services, image processing systems, and spatial extensions to database management systems (DBMS), which are not true GIS as defined here. The discussion is also restricted to GIS software products – well-defined collections of software and

accompanying documentation, install scripts, etc. – that are subject to multi-versioned release control. By definition it excludes specific-purpose utilities, unsupported routines, and ephemeral codebases.

Earlier chapters, especially Chapter 3, introduced several fundamental computer concepts, including digital representations, data, and information. Two further concepts need to be introduced here. Programs are collections of instructions that are used to manipulate digital data in a computer. System software programs, such as a computer operating system, are used to support application software – the programs with which end users interact. Integrated collections of application programs are referred to as software packages or systems (or just software for short).

Software can be distributed to the market in several different ways. The dominant form of distribution is the sale of commercial-off-the-shelf (COTS) software products on hard copy media (CD/DVD). GIS software products of this type are developed with a view to providing users with a consistent and coherent model for interacting with geographic data. The product will usually comprise an integrated collection of software programs, an install script, on-line help files, sample data and maps, documentation and an associated website. Alternative distribution models that are becoming increasingly prevalent include *shareware* (usually intended for sale after a trial period), *liteware* (shareware with some capabilities disabled), *freeware* (free software but with copyright restrictions), *public domain software* (free with no restrictions), and *open source* software (where the *source code* is provided and users agree not to limit the distribution of improvements). The Internet is becoming the main medium for software distribution.

GIS software packages provide a unified approach to working with geographic information.

GIS software vendors – the companies that design, develop, and sell GIS software – build on top of basic computer operating system capabilities such as security, file management, peripheral drivers (controllers), printing, and display management. GIS software is constructed on these foundations to provide a controlled environment for geographic information collection, management, analysis, and interpretation. The unified architecture and consistent approach to representing and working with geographic information in a GIS software package aim to provide users with a standardized approach to GIS.

7.2 The evolution of GIS software

In the formative GIS years, GIS software consisted simply of collections of computer routines that a skilled programmer could use to build an operational GIS. During this period each and every GIS was unique in terms of its capabilities, and significant levels of resource were required to create a working system. As software engineering techniques advanced and the GIS market grew in the 1970s and 1980s, demand increased for higher-level applications with a standard user interface. In the late 1970s and early 1980s the standard means of communicating with a GIS was to type in command lines. User interaction with a GIS entailed typing instructions to, for example, draw a topographic map, query the attributes of a forest stand object, or summarize the length of highways in a project area. Essentially, a GIS software package was a toolbox of geoprocessing operators or commands that could be applied to datasets to create new datasets. For example, three area-based data layers *Soil*, *Slope* and *Vegetation* could be combined using an overlay function to create an *IntegratedTerrainUnit* dataset.

To make the software easier to use and more generic, there were two key developments in the late 1980s. First, command line interfaces were supplemented and eventually largely replaced by graphical user interfaces (GUIs). These menu-driven, form-based interfaces greatly simplified user interaction with a GIS. Second, a customization capability was added to allow specific-purpose applications to be created from the generic toolboxes. Software developers and advanced technical users could make calls using a high level programming language (such as Visual Basic or Java) to published application programming interfaces (APIs) that exposed key functions. Together these stimulated enormous interest in GIS, and led to much wider adoption and expansion into new areas. In particular, the ability to create custom application solutions allowed developers to build focused applications for end users in specific market areas. This led to the creation of GIS applications specifically tailored to the needs of major markets (e.g., government, utilities, military, and environment). New terms were developed to distinguish these subtypes of GIS software: planning information systems, automated mapping/facility management (AM/FM) systems, land information systems, and more recently, location-based services systems.

In the last few years a new method of software interaction has evolved that allows software systems to communicate over the Web using a Web services paradigm. A Web service is an application that exposes its functions via a well-defined published interface that can be accessed over the Web from another program or Web service. This new software interaction paradigm will allow geographically distributed GIS functions to be linked together to create complete GIS applications. For example, a market analyst who wants to determine the suitability of a particular site for locating a new store can start a small browser-based program on their desktop computer that links to remote services over the Web that provide access to the latest population census and geodemographics data, as well as analytical models. Although these data and programs are remotely hosted and maintained they can be used for site suitability analysis as though they were resident on the market analyst's desktop. Chapter 11 explores Web services in more depth in the context of distributed GIS.

The GIS software of today still embodies the same principles of an easy-to-use menu-driven interface and a customization capability, but can now be distributed over the Web.

7.3 Architecture of GIS software

7.3.1 Project, departmental, and enterprise GIS

Usually, GIS is first introduced into organizations in the context of a single, fixed-term project (Figure 7.1). The technical components (network, hardware, software, and data) of an operational GIS are assembled for the duration of the project, which may be from several months to a few years. Data are collected specifically for the project and typically little thought is given to reuse of software, data, and human knowledge. In larger organizations, multiple projects may run one after another or even in parallel. The 'one off' nature of the projects, coupled with an absence of organizational vision, often leads to duplication, as each project develops using different hardware, software, data, people, and procedures. Sharing data and experience is usually a low priority.

As interest in GIS grows, to save money and encourage sharing and resource reuse, several projects in the same department may be amalgamated. This often leads to the creation of common standards, development of a focused GIS team, and procurement of new GIS capabilities. Yet it is also quite common for different departments to have different GIS software and data standards.

As GIS becomes more pervasive, organizations learn more about it and begin to become dependent on it. This leads to the realization that GIS is a useful way to structure many of the organization's assets, processes, and workflows. Through a process of natural growth, and possibly further major procurement (e.g., purchase of upgraded hardware, software, and data), GIS gradually becomes accepted as an important enterprise-wide information system. At this point GIS standards are accepted across multiple departments, and resources to support and manage the GIS are often centrally-funded and managed. A fourth type of societal implementation has additionally been identified in which hundreds or thousands of users become engaged in GIS and connected by a network. Today there are only a few examples of societal implementations with perhaps the best being the State of Qatar in the Middle East where more than 16 government departments have joined together to create a comprehensive and nationwide GIS with thousands of users.

7.3.2 The three-tier architecture

From an information systems perspective there are three key parts to a GIS: the user interface, the tools, and

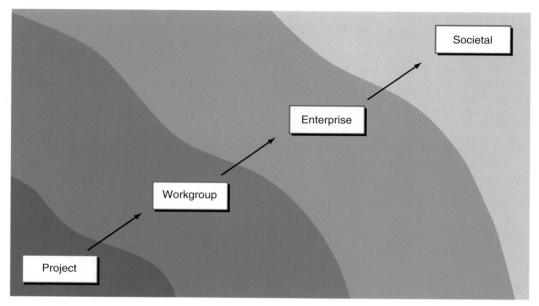

Figure 7.1 Types of GIS implementation

the data management system (Figure 7.2). The user's
interaction with the system is via a graphical user interface
(GUI), an integrated collection of menus, tool bars, and
other controls. The GUI provides access to the GIS tools.
The toolset defines the capabilities or functions that the
GIS software has available for processing geographic
data. The data are stored in files or databases organized
by data management software. In standard information
system terminology this is a three-tier architecture with
the three tiers being called: presentation, business logic,
and data server. Each of these software tiers is required
to perform different types of independent tasks. The
presentation tier must be adept at rendering (displaying)
and interacting with graphic objects. The business logic
tier is responsible for performing compute-intensive
operations such as data overlay processing and raster
analysis. It is here also that the GIS data model logic
is implemented. The data server tier must import and
export data and service requests for subsets of data
from a database or file system. In order to maximize
system performance it is useful to optimize hardware
and operating systems settings differently for each of
these types of task. For example, rendering maps requires
large amounts of memory and fast CPU clock speeds,
whereas database queries need fast disks and buses for
moving large amounts of data around. By placing each
tier on separate computers some tasks can be performed
in parallel and greater overall system scalability can
be achieved.

**GIS software systems deal with user interfaces,
tools, and data management.**

Four types of computer system architecture configurations are used to build operational GIS implementations:
desktop, client-server, centralized desktop, and centralized

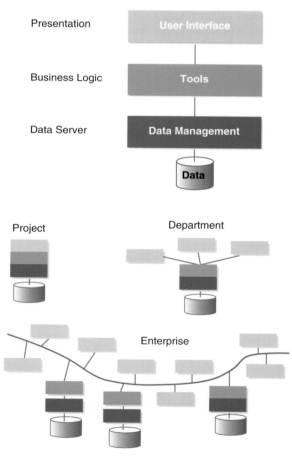

Figure 7.2 Classical three-tier architecture of a GIS software
system

server. In the simplest desktop configuration, as used in single-user project GIS, the three software tiers are all installed on a single piece of hardware (most commonly a desktop PC) in the form of a desktop GIS software package and users are usually unaware of their existence (Figure 7.3A). In a variation on this theme, data files are held on a centralized file server, but the data server functionality is still part of the desktop GIS. This means that the entire contents of any accessed file must be pulled across the network even if only a small amount of it is required (Figure 7.3B)

In larger and more advanced multiuser workgroup or departmental GIS, the three tiers can be installed on multiple machines to improve flexibility and performance (Figure 7.4). In this type of configuration, the users in a single department (for example, the planning or public works department in a typical local government organization) still interact with a desktop GIS GUI (presentation layer) on their desktop computer which also contains all the business logic, but the database management software (data server layer) and data may be located on another machine connected over a network. This type of computing architecture is usually referred to as client-server, because clients request data or processing services from servers that perform work to satisfy client requests. The data server has local processing capabilities and is able to query and process data and thus return

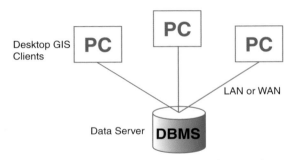

Figure 7.4 Client-server GIS: desktop GIS software and DBMS data server in a workgroup or departmental GIS configuration (LAN = Local Area Network, WAN = Wide Area Network)

part of the whole database to the client. Clients and servers can communicate over local area, wide area, or Internet networks, but given the large amount of data communication between the client and server, faster local area networks are most widely used.

In a client-server GIS, clients request data or processing services from servers that perform work to satisfy client requests.

Both the desktop and client-server architecture configurations discussed above have significant amounts of functionality on the desktop and are said to be thick clients (see also Section 7.3.4). In contrast, in the centralized desktop architecture configuration all the GUI and business logic is hosted on a centralized server, called an application (or middle tier) server (Figure 7.5). Typically this is in the form of a desktop GIS software package. An additional piece of software is also installed on the application server (Citrix or Windows Terminal Server) that allows users on remote machines full access to the software over a Local Area Network (LAN) or Wide Area Network (WAN) as though it were on the local desktop PC. Since the only application software that runs on the

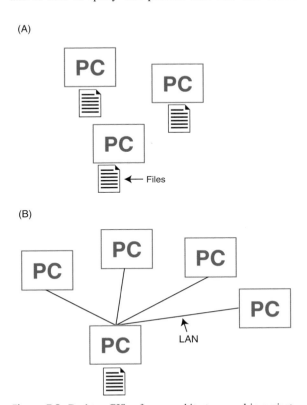

Figure 7.3 Desktop GIS software architecture used in project GIS: (A) stand alone desktop GIS on PCs each with own local files; (B) desktop GIS on PCs sharing files on a PC file server over a LAN (Local Area Network)

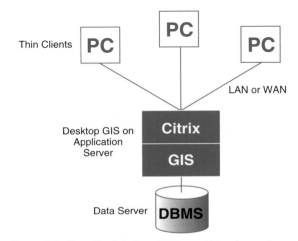

Figure 7.5 Centralized desktop GIS as used in advanced departmental and enterprise implementations

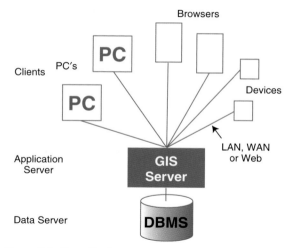

Figure 7.6 Centralized server GIS as used in advanced departmental and enterprise implementations

desktop PC is a small client library this is said to be a thin client. A data server (DBMS) is usually used for data management. This type of configuration is widely employed in large departmental and enterprise applications where high-end capabilities such as advanced editing, mapping and analysis are required.

In a more common variation of the centralized desktop implementation, the business logic is implemented as a true server system and runs on a middle tier server machine (Figure 7.6). In this configuration a range of thick and thin clients running on desktop PCs, Web browsers and specialist devices communicate with the middle tier server over a network connection. In the case of thin client access, the presentation tier (user interface) also runs on the server (although technically it is still presented on the desktop). The server machines may be connected over a local area network, but increasingly the Internet is used to connect widely distributed servers. This type of implementation is common in enterprise GIS. Large, enterprise GIS may involve more than ten servers and hundreds or even thousands of clients that are widely dispersed geographically.

Although organizations often standardize on either a project, departmental, or enterprise system, it is also common for large organizations to have project, department, and enterprise configurations all operating in parallel or as subparts of a full-scale system.

7.3.3 Software data models and customization

In addition to the three-tier model, two further topics are relevant to an understanding of software architecture: data models and customization.

GIS data models will be discussed in detail in Chapter 8, and so the discussion here will be brief. From a software perspective, a data model defines how the real world is represented in a GIS. It will also affect the type of software tools (functions or operators) that are available,

how they are organized, and their mode of operation. A software data model defines how the different tools are grouped together, how they can be used, and how they interact with data. Although such software facets are largely transparent to end users whose interaction with a GIS is via a user interface, they become very important to software developers that are interested in customizing or extending software.

Customization is the process of modifying GIS software to, for example, add new functionality to applications, embed GIS functions in other applications, or create specific-purpose applications. It can be as simple as deleting unwanted controls (for example menu choices or buttons) from a GUI, or as sophisticated as adding a major new extension to a software package for such things as network analysis, high-quality cartographic production, or land parcel management.

To facilitate customization, GIS software products must provide access to the data model and expose capabilities to use, modify, and supplement existing functions. In the late 1980s when customization capabilities were first added to GIS software products, each vendor had to provide a proprietary customization capability simply because no standard customization systems existed. Nowadays, with the widespread adoption of the Microsoft .Net and Sun Java frameworks, as well as public domain languages, a number of industry standard programming languages (such as Visual Basic, Java, and Python) are available for customizing GIS software systems.

Modern programming languages are one component of larger developer-oriented software packages called integrated development environments (IDEs). The term IDE refers to the fact that the packages combine several software development tools including a visual programming language; an editor; a debugger; and a profiler. Many of the so-called visual programming languages, such as C, C#, Visual Basic and Java, support the development of Windows-based GUIs containing forms, dialogs, buttons, and other controls. Program code can be entered and attached to the GUI elements using the integrated code editor. An interactive debugger will help identify syntactic problems in the code, for example, misspelled commands and missing instructions. Finally, there are also tools to support profiling programs. These show where resources are being consumed and how programs can be speeded up or improved in other ways.

Contemporary GIS typically use an industry-standard programming language like Visual Basic or Java for customization.

A number of mainstream COTS GIS software vendors have licensed the right to include an IDE within their GIS software package. A particularly popular choice for desktop GIS is Microsoft's Visual Basic. Figure 7.7 shows a screenshot of the customization environment within ESRI's ArcGIS 9 Desktop GIS. For server-based products both C and Java are widely used, but because of the specialist nature of this market segment they must be obtained separately from the GIS software. To support customization using open, industry-standard IDEs, a GIS vendor must expose details of the software

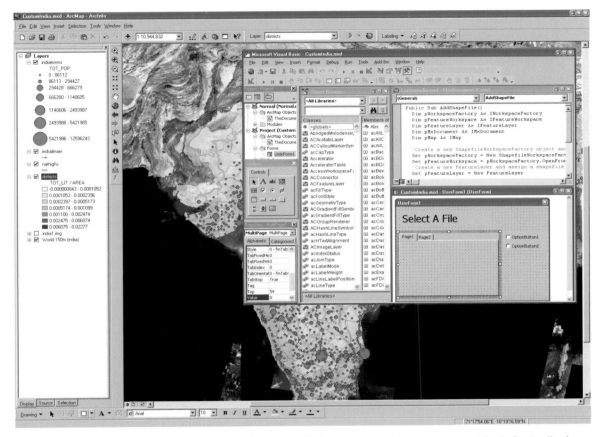

Figure 7.7 The customization capabilities of ESRI's ArcGIS 9. ESRI chose to embed Microsoft's Visual Basic for Applications as the scripting and GUI integrated development environment. The window at the front is the Visual Basic Integrated Development Environment (IDE). The window at the back is ArcMap, the main map-centric application of ArcInfo (see also Box 7.3)

package's functionality. This can be done by creating and documenting a set of application programming interfaces (APIs). These are interfaces that allow GIS functionality to be called by the programming tools in an IDE. In recent years, second-generation interfaces have been developed for accessing software functionality in the form of independent building blocks called software components.

In recent years, three technology standards have emerged for defining and reusing software functionality (components). For building interactive desktop applications, Microsoft's .Net framework is the de facto standard for high-performance, interactive applications that use fine-grained components (that is, a large number of small functionality blocks). For server-centric GIS both .Net and Sun Microsystems's Java framework are widely deployed in operational GIS applications. Although both .Net and Java work very well for building fine-grained client or server applications, they are less well suited for building applications that need to communicate over the Web. Because of the loosely-coupled, comparatively slow, heterogeneous nature of Web networks and applications, fine-grained programming models do not work well. As a consequence, coarse-grained messaging systems have been built on top of the fine-grained .Net and

Java frameworks using the XML (extensible markup language) protocol that allow applications with Web services interfaces to interact over the Web.

Components are important to software developers because they are the mechanism by which reusable, self-contained, software building blocks are created. They allow many programmers to work together to develop a large software system incrementally. The standard, open (published) format of components means that they can easily be assembled into larger systems. In addition, because the functionality within components is exposed through interfaces, developers can reuse them in many different ways, even supplementing or replacing functions if they so wish. Users also benefit from this approach because GIS software can evolve incrementally and support multiple third party extensions. In the case of GIS products this includes, for example, tools for charting, reporting, and data table management.

7.3.4 GIS on the desktop and on the Web

Today mainstream, high-end GIS users work primarily with software that runs either on the desktop or over the

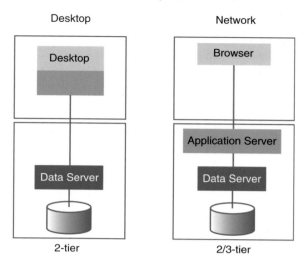

Desktop   Network

2-tier   2/3-tier

Figure 7.8 Desktop and network GIS paradigms. In the desktop top case the business logic is part of the client, but in the network case it runs on a server

Table 7.1 Comparison of desktop and network GIS

Feature	Desktop	Network
Client size	Thick	Thin
Client platform	Windows	Cross Platform Browser
Server size	Thin/thick	Thick
Server platform	Windows/Unix/ Linux	Windows/Unix/Linux
Component standard	.Net	.Net/Java
Network	LAN/WAN	LAN/WAN/Internet

Web. In the desktop case a PC (personal computer) is the main hardware platform and Microsoft Windows the operating system (Figure 7.8 and Table 7.1). In the desktop paradigm clients tend to be functionally rich and substantial in size, and are often referred to as thick or fat clients. Use of the Windows standard facilitates interoperability (interaction) with other desktop applications, such as word processors, spreadsheets, and databases. As noted earlier (Section 7.3.2), most sophisticated and mature GIS workgroups have adopted the client-server implementation approach by adding either a thin or thick server application running on the Windows, Linux or Unix operating system. The terms thin and thick are less widely used in the context of servers, but they mean essentially the same as when applied to clients. Thin servers perform relatively simple tasks, such as serving data from files or databases, whereas thick servers also offer more extensive analytical capabilities such as geocoding, routing, mapping, and spatial analysis. In desktop GIS implementations, LANs and WANs tend to be used for client-server communication. It is natural for developers to select Microsoft's .Net technology framework to build the

underlying components making up these systems given the preponderance of the Windows operating system, although other component standards could also be used. The Windows-based client-server system architecture is a good platform for hosting interactive, high-performance GIS applications. Examples of applications well suited to this platform include those involving geographic data editing, map production, 2-D and 3-D visualization, spatial analysis, and modeling. It is currently the most practical platform for general-purpose systems because of its wide availability, good performance for a given price, and common usage in business, education, and government.

GIS users are standardizing their systems on the desktop and Internet implementation models.

In the last few years there has been increasing interest in harnessing the power of the Web for GIS. Although desktop GIS have been and continue to be very successful, users are constantly looking for lower costs of ownership and improved access to geographic information. Network-based (sometimes called distributed) GIS allow previously inaccessible information resources to be made more widely available. The network GIS model intrigues many organizations because it is based on centralized software and data management, which can dramatically reduce initial implementation and ongoing support and maintenance costs. It also provides the opportunity to link nodes of distributed user and processing resources using the medium of the Internet. The continued rise in network GIS will not signal the end of desktop GIS, indeed quite the reverse, since it is likely to stimulate the demand for content and professional GIS skills in geographic database automation and administration, and application development.

In contrast to desktop GIS, network GIS can use the cross-platform Web browser to host the viewer user-interface. Currently, clients are typically very thin, often with simple display and query capabilities, although there is an increasing trend for them to become more functionally rich. Server-side functionality may be encapsulated on a single server, although in medium and large systems it is more common to have two servers, one containing the business logic (a middleware application server), the other the data manager (data server). The server applications typically contain all the business logic and are comparatively thick. The server applications may run on a Windows, Unix or Linux platform.

Recently, there has been a move to combine the best elements of the desktop and network paradigms to create so-called rich clients. These are stored and managed on the server and dynamically downloaded to a client computer according to user demand. The business logic and presentation layers run on the server, with the client hardware simply used to render the data and manage use interaction. The new software capabilities in recent editions of the .Net and Java software development kits allow the development of applications with extensive user interaction that closely emulate the user experience of working with desktop software.

7.4 Building GIS software systems

Commercial GIS software systems are built and released as GIS software products by GIS-vendor software development and product teams. Such products are subject to carefully planned versioned release cycles that incrementally enhance and extend the pool of capabilities. The key parts of a GIS software architecture – user interface, business logic (tools), data manager, data model, and customization environment – were outlined in the previous section.

GIS software vendors start with a formal design for a software system and then build each part or component separately before assembling the whole system. Typically, development will be iterative with creation of an initial prototype framework containing a small number of partially functioning parts, followed by increasing refinement of the system. Core GIS software systems are usually written in a modern programming language like Visual C++, C# or Java, with Visual Basic or Java sometimes used for operations that do not involve significant amounts of computer processing like the GUI.

As standards for software development become more widely adopted, so the prospect of reusing software components becomes a reality. A key choice that then faces all software developers or customizers is whether to design a software system by buying in components, or to build it more or less from scratch. There are advantages to both options: building components gives greater control over system capabilities and enables specific-purpose optimization; and buying components can save time and money. Examples of components which have been purchased and licensed for use in GIS software systems include: Business Objects' Crystal Reports in MapInfo Professional, Microsoft Visual Basic for Applications in ESRI ArcGIS Desktop, and Safe Software's Feature Manipulation Engine for data conversion in Autodesk Map 3D.

A key GIS implementation issue is whether to buy a system or to build one.

A modern GIS software system comprises an integrated suite of software components of three basic types: a data management system for controlling access to data (Chapters 8, 9 and 10); a mapping system for display and interaction with maps and other geographic visualizations (Chapters 12 and 13); and a spatial analysis and modeling system for transforming geographic data using operators (Chapters 14, 15 and 16). The components for these parts may reside on the same computer or can be distributed widely (Chapter 11) over a network. The work of one of GIS's leading software developers is described in Box 7.1

7.5 GIS software vendors

Daratech Inc., an IT market research and technology assessment company, produces annual estimates about the size and characteristics of the GIS market. For 2003 (Figure 7.10) they list the main players in the worldwide GIS software market as ESRI, Intergraph, Autodesk, and GE Energy. Secondary players include Leica Geosystems, IBM, and MapInfo. In order to understand the current commercial GIS software market place and the direction in which it is likely to head, it is first necessary to examine the background, current product offerings and strategy of the main players.

7.5.1 ESRI Inc.

ESRI is a privately held company founded in 1969 by Jack and Laura Dangermond. Headquartered in Redlands, California, ESRI employs over 4000 people worldwide and has annual revenues of over US $500 m. Today it serves more than 130 000 organizations and more than 1 million users. ESRI focuses solely on the GIS market, primarily as a software product company, but also generates about a quarter of its revenue from project work such as advising clients on how to implement GIS. ESRI started building commercial software products in the late 1970s. Today, ESRI's product strategy is centered on an integrated family of products called ArcGIS. The ArcGIS family is aimed at both end users and technical developers and includes products that run on hand-held devices, desktop personal computers, and servers.

ESRI is the classic high-end GIS vendor. It has a wide range of mainstream products covering all the main technical and industry markets. ESRI is a technically-led geographic company focused squarely on the needs of hard-core GIS users. Box 7.3 describes ESRI's ArcGIS product.

7.5.2 Intergraph Inc.

Like ESRI, Intergraph was also founded in 1969 as a private company. The initial focus from their Huntsville, Alabama offices was the development of computer graphics systems. After going public in 1981, Intergraph grew rapidly and diversified into a range of graphics areas including CAD and mapping software, consulting services and hardware. After a series of reorganizations in the late 1990s and early 2000s, Intergraph is today structured into four main operating units: Process, Power and Marine; Public Safety; Solutions; and Mapping and Geospatial Solutions. The latter is the main GIS focus of the company. Mapping and Geospatial Solutions accounts for more than $200 m of the annual Intergraph total revenue which exceeds $500 m.

Intergraph has a large and diverse product line. From a GIS perspective the principal product family is GeoMedia which spans the desktop and network (Internet) server markets.

Historically Intergraph has been one of the top two global GIS companies. Today it is strongest in the military, infrastructure and utility market areas. Box 7.2 describes Intergraph GeoMedia.

Scott Morehouse, software developer and father of ArcGIS

Scott Morehouse (Figure 7.9) learnt his programming and software development tradecraft at the Harvard Laboratory for Computer Graphics and Spatial Analysis in Massachusetts, in the 1970s. He was one of the lead developers of a system developed at the Lab called ODYSSEY which was the first general purpose vector-based GIS. It implemented many key GIS ideas such as digitizing, polygon overlay and choropleth mapping.

In 1981 he moved to Redlands, California to work at ESRI where he was the lead architect and developer for the initial ArcInfo release (see also Box 7.3). He was one of the programmers who implemented the first commercial vector polygon overlay algorithm which is still in use twenty-five years later. For the past two decades Scott has been the lead designer and manager for software development for all ESRI software products. He has overseen the development of ArcGIS, including ArcView, ArcGIS Engine, and ArcGIS Server.

Scott is very much a pragmatist who realized early on that successful software must not only be well designed and implemented with good algorithms and data structures, but also must be robust, well documented and widely applicable. He has always had a talent for synthesizing complex

Figure 7.9 Scott Morehouse, software architect

information and distilling out the essence so that non-experts can understand. He sees his role at ESRI as building software tools that apply geographic theory to help people solve real-world problems. Scott works alongside programmers and other product developers to design practical software architectures and robust solutions to user problems.

Although less well known in the GIS industry than ESRI's charismatic leader, Jack Dangermond, Scott is every bit as responsible for ESRI's success in the GIS software field. He has strong beliefs and a clear vision about what it takes to make things that really work. These are vital for maintaining the integrity of a large, complex system that is built by a diverse team of software engineers. He is also adept at managing the conflicting interests of supporting existing applications, while at the same time evolving GIS software products through innovation.

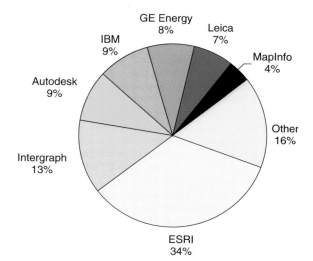

Figure 7.10 GIS 2003 software vendor market share (Courtesy Daratech)

7.5.3 Autodesk Inc.

Autodesk is a large and well known publicly traded company with headquarters in San Rafael, California. It is one of the world's leading digital design and content companies and serves customers in markets where design is critical to success: building, manufacturing, infrastructure, digital media, and location services. Autodesk is best known for its AutoCAD product family which is used worldwide by more than 4 million customers. The company was founded more than 20 years ago and has grown to become a publicly traded $1 bn entity employing over 3700 staff. The GIS division at Autodesk contributes over 10% of the company's revenue.

Autodesk's success in the GIS arena centers around three main product areas: desktop, where Autodesk Map 3D (based on AutoCad) is the flagship; an Internet server called MapGuide; and hand-held GIS, through the OnSite family of products.

Autodesk is classically thought of as a successful computer-aided design (CAD) company that has extended itself into GIS. It has been especially successful in industries that have a strong engineering and design element. Autodesk MapGuide is described in Box 7.4.

7.5.4 GE Energy

GE Energy is very different to the other major GIS players on Daratech's GIS software vendor list. GE Energy's GIS software, which is referred to as a Geospatial Asset Management Solution, is based on the Smallworld GIS. The codebase was acquired in 2000 when Smallworld was purchased by GE Power Systems (now GE Energy). Smallworld was established in the late 1980s in Cambridge, UK by several entrepreneurs with a history in the CAD industry. From the outset the technology focused on complex utility network solutions, especially in the electric and gas industries. Smallworld grew rapidly to become one of the top three GIS utility software providers.

The Smallworld product suite offers an integrated workgroup and enterprise solution that spans the desktop and Internet, and is able to integrate with other IT business systems. Increasingly the product suite is focused on specific electric, gas and telco utility design and operational system solutions where it is used for network design and operation, and asset management.

7.6 Types of GIS software systems

Over 100 commercial software systems claim to have mapping and GIS capabilities. From the previous section on the product strategy of the major GIS software vendors it can be seen that four main types of generic GIS software dominate the market: desktop; server; developer; and hand-held. In this section, these four categories of mainstream GIS software will be discussed followed by a brief summary of other types of software. Reviews of currently popular GIS software packages can be found in the various GIS magazines (see Box 1.4).

7.6.1 Desktop GIS software

Since the mid-1990s, desktop GIS has been the mainstay of the majority of GIS implementations and the most widely used category of GIS software. Desktop GIS software owes its origins to the personal computer and the Microsoft Windows operating system. Initially, the major GIS vendors ported their workstation or minicomputer GIS software to the PC, but subsequently redeveloped their software specifically for the PC platform. Desktop GIS software provides personal productivity tools for a wide variety of users across a broad cross section of industries. PCs are commonly available, relatively inexpensive and offer a large collection of user-oriented tools including databases, word processors, and spreadsheets. The

desktop GIS software category includes a range of options from simple viewers (such as ESRI ArcReader, Intergraph GeoMedia Viewer and MapInfo ProViewer) to desktop mapping and GIS software systems (such as Autodesk Map 3D, ESRI ArcView, GE Spatial Intelligence, Intergraph GeoMedia (see Box 7.2, Figure 7.11), and MapInfo Professional), and at the high-end, full-featured 'professional' editor/analysis systems (such as ESRI ArcInfo (see Box 7.3), Integraph GeoMedia Professional, and GE Energy Smallworld GIS).

Desktop GIS are the mainstream workhorses of GIS today.

In the late 1990s, a number of vendors released free GIS viewers that are able to display and query popular file formats. Today, the GIS viewer has developed into a significant product subcategory. The rationale behind the development of these products is that they help to establish market share, and can create de facto standards for specific vendor terminology and data formats. GIS users often work with viewers on a casual basis, often in conjunction with more sophisticated GIS software products. GIS viewers have limited functional capabilities restricted to display, query, and simple mapping. They do not support editing, sophisticated analysis, modeling, or customization.

With their focus on data use, rather than data creation, and their excellent tools for making maps, reports, and charts, desktop mapping and GIS software packages represent most people's experience of mainstream GIS today. The successful systems have all adopted the Microsoft standards for interoperability and user interface style. Users often see a desktop mapping and GIS package as simply a tool to enable them to do their full-time job faster, more easily, or more cheaply. Desktop mapping and GIS users work in planning, engineering, teaching, the army, marketing, and other similar professions; they are often not Spatially Aware Professionals (SAPs). Desktop GIS software prices typically range from $1000–2000 (these and other prices mentioned later typically have discounts for multiple purchases).

The term 'professional' relates to the full-featured nature of this subcategory of software. The distinctive features of professional GIS include data collection and editing, database administration, advanced geoprocessing and analysis, and other specialist tools. Professional GIS offer a superset of the capabilities of the systems in other classes. The people who use these systems are typically technically literate and think of themselves as SAPs – GIS professionals (career GIS staff) with degrees and, in many cases, advanced degrees in GIS or related disciplines. Prices for professional GIS are typically in the range $7000–20 000 per user.

Professional GIS are high-end, fully functional systems.

7.6.2 Server GIS

The last decade of GIS has been dominated by the desktop GIS architecture running on PCs. It is a fair bet that the

Desktop GIS: Intergraph GeoMedia

GeoMedia is an archetypal example of a mainstream commercial desktop GIS software product (Figure 7.11). First released in the late 1990s from a new codebase, it was built from the ground up to run on the Windows desktop operating system. Like other products in the desktop GIS category it is primarily designed with the end user in mind. It has a Windows-based graphical user interface and many tools for editing, querying, mapping and spatial analysis. Data can be stored in proprietary GeoMedia files or in DBMS such as Oracle and Microsoft Access and SQL Server. GeoMedia enables data from multiple disparate databases to be brought into a single GIS environment for viewing, analysis, and presentation. The data are read and translated on the fly directly into memory. GeoMedia provides access to all

major geospatial/CAD data file formats and to industry-standard relational databases.

GeoMedia is built as a collection of software object components. These underlying objects are exposed to developers who can customize the software using a programming language such as Visual Basic or C#.

GeoMedia offers a suite of analysis tools, including attribute and spatial query, buffer zones, spatial overlays, and thematic analysis. The product's layout composition tools provide the flexibility to design a range of types of maps that can be distributed on the Web, printed, or exported as files.

GeoMedia is the entry to a family of products. Several extensions (add-ons) offer additional functionality (e.g., image, grid and terrain analysis, and transaction management) and

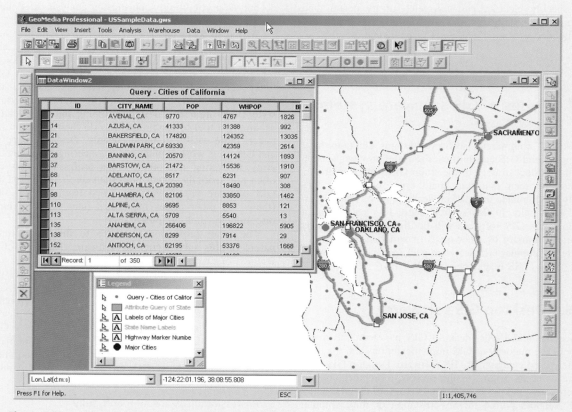

Figure 7.11 Screenshot of Intergraph GeoMedia – Desktop GIS

support for industry-specific workflows (e.g., transportation, parcel and public works). Other members of the product family include Geo-Media Viewer (free data viewer), GeoMedia Pro (high-end 'professional' functionality) and GeoMedia WebMap (Internet publishing).

All in all the GeoMedia product family offers a wide range of capabilities for core GIS activities in the mainstream markets of government, education, and private companies. It is a modern and integrated product line with strengths in the areas of data access and user productivity.

Applications Box 7.3

Desktop GIS: ESRI ArcGIS ArcInfo

ArcInfo is ESRI's full-featured professional GIS software product (Figure 7.12). It supports the full range of GIS functions including: data collection and import; editing, restructuring, and transformation; display; query; and analysis.

It is also the platform for a suite of analytic extensions for 3-D analysis, network routing, geostatistical and spatial (raster) analysis, among others. ArcInfo's strengths are the comprehensive portfolio of capabilities, the

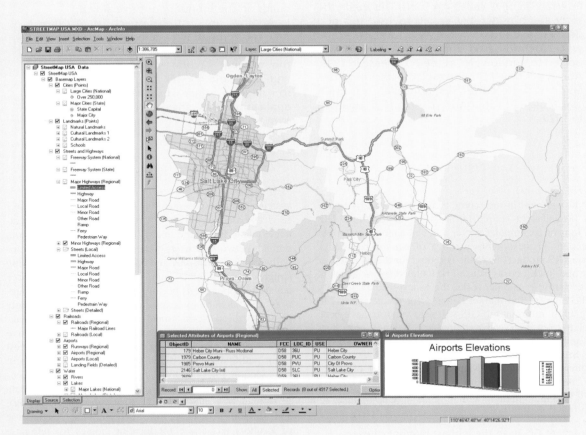

Figure 7.12 Screenshot of ESRI ArcGIS ArcInfo – Desktop GIS

sophisticated tools for data management and analysis, the customization options, and the vast array of third party tools and interfaces.

ArcInfo was originally released in 1981 on minicomputers. The early releases offered very limited functionality by today's standards and the software was basically a collection of subroutines that a programmer could use to build a working GIS software application. A major breakthrough came in 1987 when ArcInfo 4 was released with AML (Arc Macro Language), a scripting language that allowed ArcInfo to be easily customized. This release also saw the introduction of a port to Unix workstations (the software was adapted to function on this new platform) and the ability to work with data in external databases like Oracle, Informix, and Sybase. In 1991, with the release of ArcInfo 6, ESRI again re-engineered ArcInfo to take better advantage of Unix and the X-Windows windowing standard. The next major milestone was the development of a menu-driven user interface in 1993 called ArcTools. This made the software considerably easier to use and also defined a standard for how developers could write ArcInfo-based applications. ArcInfo was ported to Windows NT at the 7.1 release in 1996. About this time ESRI also took the decision to re-engineer ArcInfo from first principles. This vision was realized in the form of ArcInfo 8, released in 1999.

ArcInfo 8 was quite unlike earlier versions of the software because it was designed from the outset as a collection of reusable, self-contained software components, based on Microsoft's COM standard. ESRI used these components to create an integrated suite of menu-driven, end-user applications: ArcMap – a map-centric application supporting integrated editing and viewing; ArcCatalog – a data-centric application for browsing and managing geographic data in files and databases; and ArcToolbox – a tool-oriented application for performing geoprocessing tasks such as proximity analysis, map overlay, and data conversion. ArcInfo is customizable using either the in-built Microsoft Visual Basic for Applications (VBA) or any other COM-compliant programming language. The software is also notable because of the ability to store and manage all data (geographic and attribute) in standard commercial off-the-shelf DBMS (e.g., DB2, Informix, SQL Server, and Oracle). In 2004, ESRI released version 9 which builds on the foundations of version 8.

Another interesting aspect of ArcInfo is that for compatibility reasons since Release 8 ESRI has included a fully working version of the original ArcInfo workstation technology and applications. This has allowed ESRI users to migrate their existing databases and applications to the new version in their own time.

next decade will in turn come to be dominated by server GIS products. In simple terms, a server GIS is a GIS that runs on a computer server that can handle concurrent processing requests from a range of networked clients. Server GIS products have the potential for the largest user base and lowest cost per user. Stimulated by advances in server hardware and networks, the widespread availability of the Internet and market demand for greater access to geographic information, GIS software vendors have been quick to release server-based products. Examples of server GIS include Autodesk MapGuide, ESRI ArcGIS Server, GE Spatial Application Server, Intergraph GeoMedia Webmap, and MapInfo MapXtreme. The cost of server GIS products varies from around $5000–25 000, for small to medium-sized systems, to well beyond for large multifunction, multiuser systems.

Internet GIS have the highest number of users, although typically Internet users focus on simple display and query tasks.

Initially, such products were nothing more than ports of desktop GIS products, but second generation systems were subsequently built using a multiuser services-based architecture that allows them to run unattended and to handle many concurrent requests from remote networked users. These software systems initially focused on display and query applications – making simple things simple and cost-effective – with more advanced applications becoming available as user awareness and technology expanded. Today, it is routinely possible to perform standard operations like making maps (Microsoft Expedia has online interactive maps; **www.expediamaps.com**), routing (MapQuest offers pathfinding with directions (**www.mapquest.com**: see Section 1.4.3), publishing census data (US Census Bureau, AmericanFactFinder has online census data and maps for the whole US at **www.census.gov**), and suitability analysis (the US National Association of Realtors has a site for locating homes for purchase based on user-supplied criteria at **www.realtor.com**). A recent trend has been the development of Internet-based online data networks such as the Geography Network (**www.GeographyNetwork.com**) and ESRI ArcWeb Services and Microsoft MapPoint.Net.

Server GIS: Autodesk MapGuide

In the late 1990s, a time when desktop GIS had become dominant, a small Canadian company called Argus released an Internet GIS product called MapGuide (Figure 7.13). Subsequently purchased by Autodesk, MapGuide marked the start of another chapter in the history of GIS software. The emphasis in MapGuide and other Internet (now called server) GIS products is very much on map display and use. Indeed, these systems have few if any data editing capabilities.

MapGuide is an important innovation for the many users who have spent considerable amounts of time and money creating valuable databases, and who want to make them available to other users inside or outside their organization. Autodesk MapGuide allows users to leverage their existing GIS investment by publishing dynamic, intelligent maps at the point at which they are most valuable – in the field, at the job site, or on the desks of colleagues, clients, and the public.

There are three key components of MapGuide: the viewer – a relatively easy-to-use Web application with a browser style interface; the author – a menu-driven authoring environment used to create and publish a site for client access; and the server – the administrative software that monitors site usage, and manages requests from multiple clients, and to external databases. MapGuide works directly with Internet browsers and servers, and uses the HTTP protocol for communication. It makes good use

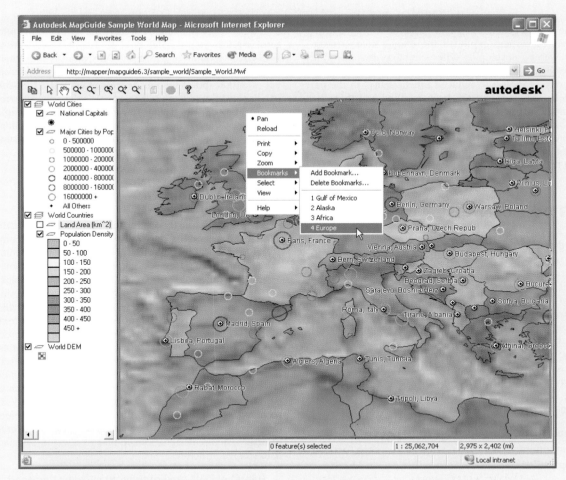

Figure 7.13 Screenshot of Autodesk MapGuide (Reproduced by permission of Autodesk MapGuide ®. Autodes, the Autodesk logo, and Autodesk MapGuide are registered trademarks of Autodesk, Inc. in the USA and/or other countries.)

of standard Internet tools like HTML (hypertext markup language) and JScript (Java Script) for building client applications and ColdFusion (an Internet site generation and management tool from Allaire Corp.) and ASP (Active Server Pages) / JSP (Java Server Pages) for managing data and queries on the server.

Typical features of MapGuide sites include the display of raster and vector maps, map navigation (pan and zoom), geographic and attribute queries, simple buffering, report generation, and printing. Like other advanced Internet server GIS, MapGuide has tools for redlining (drawing on maps) and editing geographic objects, although in MapGuide's case the editing tools are somewhat limited.

To date MapGuide has been used most widely in existing mature GIS sites that want to publish their data internally or externally, and in new sites that want a way to publish dynamic maps quickly to a widely dispersed collection of users (for example, maps showing election results, or transportation network status).

In conclusion, MapGuide and the other server GIS products are growing in importance. Their cost-effective nature, ability to be centrally managed, and focus on ease of use will help to disseminate geographic information even more widely and will introduce many new users to the field of GIS.

The second generation server GIS products were strong on architecture and exploited the unique characteristics of the Web by developing GIS technology that integrates with Web browsers and servers. Unfortunately, these gains were at the expense of reduced functionality. In the past few years this problem has been rectified and there is a new breed of true GIS server that offers 'complete' GIS functionality in a multiuser server environment. These server GIS products have functions for editing, mapping, data management, and spatial analysis, and support state of the art customization.

7.6.3 Developer GIS

With the advent of component-based software development (see Section 7.4), a number of GIS vendors have released collections of GIS software components oriented toward the needs of developers. These are really tool kits of GIS functions (components) that a reasonably knowledgeable programmer can use to build a specific-purpose GIS application. They are of interest to developers because such components can be used to create highly customized and optimized applications that can either stand alone or can be embedded within other software systems. Typically, component GIS packages offer strong display and query capabilities, but only limited editing and analysis tools.

Developer GIS products are collections of components used by developers to create focused applications.

Examples of component GIS products include Blue Marble Geographics GeoObjects, ESRI ArcGIS Engine, and MapInfo MapX. Most of the developer GIS products from mainstream vendors are built on top of Microsoft's

.Net technology standards, but there are several cross platform choices (e.g., ESRI ArcGIS Engine) and several Java-based toolkits (e.g., ObjectFX SpatialFX and Engenuity JLOOX). The typical cost for a developer GIS product is $1000–5000 for the developer kit and $100–500 per deployed application. The people who use deployed applications may not even realize that they are using a GIS, because often the run-time deployment is embedded in other applications (e.g., customer care systems, routing systems, or interactive atlases).

7.6.4 Hand-held GIS

As hardware design and miniaturization have improved dramatically over the past few years, so it has become possible to develop GIS software for mobile and personal use on hand-held systems. The development of low cost, lightweight location positioning technologies (primarily based on the Global Positioning System, GPS, see Section 5.8) and wireless networking has further stimulated this market. With capabilities similar to the desktop systems of just a few years ago, these palm and pocket devices can support many display, query, and simple analytical applications, even on displays of 320 by 240 pixels (a quarter of the VGA (640 by 480) pixel screen resolution standard). An interesting characteristic of these systems at the present time is that all programs and data are held in memory because of the lack of a hard disk. This provides fast access, but because of the cost of memory compared to disk systems, designers have had to develop compact data storage structures.

Hand-held GIS are lightweight systems designed for mobile and field use.

A very recent development is the availability of hand-held software on high-end so-called 'smartphones'.

Figure 7.14 A hand-held GIS for a smartphone

In spite of their compact size they are able to deal with comparatively large amounts of data (up to 1 GB) and surprisingly sophisticated software applications (Figure 7.14). The systems usually operate in a mixed connected/disconnected environment and so can make active use of data and software applications held on the server and accessed over a wireless telephone network.

Hand-held GIS are now available from many vendors, and include Autodesk OnSite, ESRI ArcPad, and Intergraph Intelliwhere. Many of these systems are designed to work with server GIS products (see above). Costs are typically around $400–600.

7.6.5 Other types of GIS software

The previous section has focused on mainstream GIS software from the major commercial vendors. There are many other types of commercial and non-commercial software that provide valuable GIS capabilities. This section will briefly review some of the main types of other software.

Raster-based GIS, as the name suggests, focus primarily on raster (image) data and raster analysis. Chapters 3 and 9 provide a discussion of the principles and techniques associated with raster and other data models, while Chapters 13 and 14 review their specific capabilities. Just as many vector-based systems have raster analysis extensions (for example, ESRI ArcGIS has Spatial Analyst, and GeoMedia has Image and Grid), in recent years raster systems have added vector capabilities (for example, Leica Geosystems EROAS IMAGINE and Clark Labs' Idrisi now have vector capabilities built in). The distinction between raster-based and other software system categories is becoming increasingly blurred as a consequence. The users of raster-based GIS are primarily interested in working with imagery and undertaking spatial analysis and modeling activities. The prices for raster-based GIS range from $500–10 000.

Computer-Aided Design (CAD)-based GIS are systems that started life as CAD packages and then had GIS capabilities added. Typically, a CAD system is supplemented with database, spatial analysis, and cartography capabilities. Not surprisingly, these systems appeal mainly to users whose primary focus is in typical CAD application areas such as architecture, engineering, and construction, but who also want to use geographic information and geographic analysis in their projects. The best-known examples of CAD-based GIS are Autodesk Map 3D and Bentley GeoGraphics. CAD-based GIS typically cost $3000–5000.

Many enterprise-wide GIS incorporate middleware (middle tier) GIS data and application servers. Their purpose is to manage multiple users accessing continuous geographic databases, which are stored and managed in commercial-off-the-shelf (COTS) database management systems (DBMS). GIS middleware products offer centralized management of data, the ability to process data on a server (which delivers good performance for certain types of applications), and control over database editing and update (see Chapter 11 for further details). A number of GIS vendors have developed technology that fulfills this function. Examples of GIS application servers include: Autodesk GIS Design Server, ESRI ArcSDE, and MapInfo SpatialWare. These systems typically cost $10 000–25 000 or more depending on the number of users.

To assist in managing data in standard DBMS, some vendors – notably IBM and Oracle – have developed technology to extend their DBMS servers so that they are able to store and process geographic information efficiently. Although not strictly GIS in their own right (due to the absence of editing, mapping and analysis tools) they are included here for completeness. Box 10.1. provides an overview of Oracle's Spatial DBMS extension.

In addition to the commercial-off-the-shelf GIS software products that have been the mainstay of this chapter, it is also important to acknowledge that there is a growing movement that is creating public-domain, open source and free software. In the early days such software products provided only rather simple, poorly engineered tools with no user support. Today there are several high-quality, feature rich software products. Some noteworthy examples include: GeoDa for spatial analysis and visualization (**www.csiss.org/clearinghouse/GeoDa/**), Minnesota Map Server for serving maps over the Web (**http://mapserver.gis.umn.edu/**), PostGIS for storing

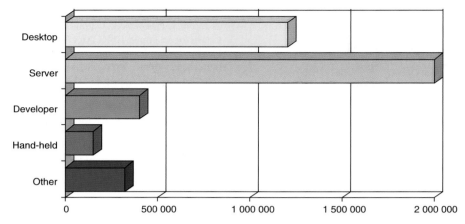

Figure 7.15 Estimated size (number of users) of the different GIS software sectors

data in a DBMS (PostgresSQL) (**http://postgis.refractions.net/**) and GRASS (**http://grass.itc.it/**) for desktop GIS tasks.

Looking forward, it is interesting to consider a new trend in GIS software. GIS functionality is now being delivered packaged along with data as seamless GIServices (Section 1.5.3). To date the most developed services have centered on simple location-based services, street mapping and routing, such as ESRI ArcWeb Services and Microsoft MapPoint, but more advanced services are now becoming available.

7.7 GIS software usage

Estimates of the size of the GIS market by type of system, based on the authors' knowledge, is shown in Figure 7.15. In 2005, the total size of the GIS market, measured in terms of the numbers of regular GIS software users (using the high-end definition adopted in this chapter), is about 4 million spread over 2 million sites. If the number of Internet GIS users is also taken into consideration then

the GIS user population rises to about 10 million in total. This excludes users of GIS products such as hard copy maps, charts, and reports.

7.8 Conclusion

GIS software is a fundamental and critical part of any operational GIS. The software employed in a GIS project has a controlling impact on the type of studies that can be undertaken and the results that can be obtained. There are also far reaching implications for user productivity and project costs. Today, there are many types of GIS software product to choose from and a number of ways to configure implementations. One of the exciting and at times unnerving characteristics of GIS software is its very rapid rate of development. This is a trend that seems set to continue as the software industry pushes ahead with significant research and development efforts. The following chapters will explore in more detail the functionality of GIS software and how it can be applied in real-world contexts.

Questions for further study

1. Design a GIS architecture that 25 users in 3 cities could use to create an inventory of recreation facilities.

2. Discuss the role of each of the three tiers of software architecture in an enterprise GIS implementation.

3. With reference to a large organization that is familiar to you, describe the ways in which its staff might use GIS, and evaluate the different types of GIS software

systems that might be implemented to fulfill these needs.

4. Go to the websites of the main GIS software vendors and compare their product strategies with open source GIS products. In what ways are they different?

   - Autodesk: **www.autodesk.com**
   - ESRI: **www.esri.com**
   - Intergraph: **www.intergraph.com**
   - GE Energy **www.gepower.com**

Further reading

Bishr Y. and Radwan M. 2000 'GDI Architectures'. In Groot R. and McLaughlin J. (eds) *Geospatial Data Infrastructure: Concepts, Cases, and Good Practice*. New York: Oxford University Press, pp. 135–50.

Coleman D.J. 2005 'GIS in networked environments'. In Longley P.A., Goodchild M.F., Maguire D.J. and Rhind D.W. (eds) *Geographical Information Systems: Principles, Techniques, Management and Applications (abridged edition)*. Hoboken, NJ: Wiley, pp. 317–29.

Elshaw Thrall S. and Thrall G.I. 2005 'Desktop GIS software'. In Longley P.A., Goodchild M.F., Maguire D.J. and Rhind D.W. (eds) *Geographical Information Systems: Principles, Techniques, Management and Applications (abridged edition)*. Hoboken, NJ: Wiley, pp. 331–45

Maguire D.J. 2005 'GIS customization'. In Longley P.A., Goodchild M.F., Maguire D.J. and Rhind D.W. (eds) *Geographical Information Systems: Principles, Techniques, Management and Applications (abridged edition)*. Hoboken, NJ: Wiley, pp. 359–69

Peng Z-H. and Tsou M-H. 2003 *Internet GIS: Distributed Geographic Information Services for the Internet and Wireless Networks*. Hoboken, NJ: Wiley.

8 *Geographic data modeling*

This chapter discusses the technical issues involved in modeling the real world in a GIS. It describes the process of data modeling and the various data models that have been used in GIS. A data model is a set of constructs for describing and representing parts of the real world in a digital computer system. Data models are vitally important to GIS because they control the way that data are stored and have a major impact on the type of analytical operations that can be performed. Early GIS were based on extended CAD, simple graphical, and image data models. In the 1980s and 1990s, the hybrid georelational model came to dominate GIS. In the last few years major software systems have been developed on more advanced and standards-based geographic object models that include elements of all earlier models.

Geographic Information Systems and Science, 2nd edition   Paul Longley, Michael Goodchild, David Maguire, and David Rhind.
© 2005 John Wiley & Sons, Ltd.   ISBNs: 0-470-87000-1 (HB); 0-470-87001-X (PB)

Learning Objectives

After reading this chapter you will be able to:

- Define what geographic data models are and discuss their importance in GIS;

- Understand how to undertake GIS data modeling;

- Outline the main geographic models used in GIS today and their strengths and weaknesses;

- Understand key topology concepts and why topology is useful for data validation, analysis, and editing;

- Read data model notation;

- Describe how to model the world and create a useful geographic database.

8.1 Introduction

This chapter builds on the material on geographic representation presented in Chapter 3. By way of introduction it should be noted that the terms representation and model as used here overlap considerably (we will return to a more detailed discussion of models in Chapter 16). For present purposes, representation can be considered to denote the conceptual and scientific issues, whereas model is used in practical and database circles. In this chapter, given the technical and practical approach, *data model* will be used to distinguish it from process models as discussed in Chapter 16. This chapter focuses on how geographic reality is modeled (abstracted or simplified) in a GIS, with particular emphasis on choosing one particular style of data model over another. A data model is an essential ingredient of any operational GIS and, as the discussion will show, has important implications for the types of operations that can be performed and the results that can be obtained.

8.1.1 Data model overview

The heart of any GIS is the data model, which is a set of constructs for representing objects and processes in the digital environment of the computer (Figure 8.1). People (GIS users) interact with operational GIS in order to undertake tasks like making maps, querying databases,

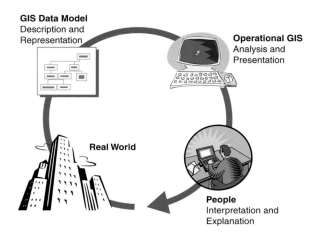

Figure 8.1 The role of a data model in GIS

and performing site suitability analyses. Because the types of analyses that can be undertaken are strongly influenced by the way the real-world is modeled, decisions about the type of data model to be adopted are vital to the success of a GIS project.

> **A data model is a set of constructs for describing and representing selected aspects of the real-world in a computer.**

As described in Chapter 3, geographic reality is continuous and infinitely complex, but computers are finite, comparatively simple, and can only deal with digital data. Therefore, difficult choices have to be made about what things are modeled in a GIS and how they are represented. Because different types of people use GIS for different purposes, and the phenomena these people study have different characteristics, there is no single type of all-encompassing GIS data model that is best for all circumstances.

8.1.2 Levels of data model abstraction

When representing the real-world in a computer, it is helpful to think in terms of four different levels of abstraction (levels of generalization or simplification) and these are shown in Figure 8.2. First, *reality* is made up of real-world phenomena (buildings, streets, wells, lakes, people, etc.), and includes all aspects that may or may not be perceived by individuals, or deemed relevant to a particular application. Second, the *conceptual model* is a human-oriented, often partially structured, model of selected objects and processes that are thought relevant to a particular problem domain. Third, the *logical model* is an implementation-oriented representation of reality that is often expressed in the form of diagrams and lists. Lastly, the *physical model* portrays the actual implementation in a GIS, and often comprises tables stored as files or databases (see Chapter 10). In terms of the discussion of uncertainty in Chapter 6 (Figure 6.1), the conceptual and logical models are found beyond the U1 filter, and the physical model is that upon which

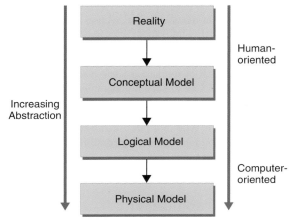

Figure 8.2 Levels of abstraction relevant to GIS data models

analysis may be performed (beyond the U2 filter). Use of the term 'physical' here is actually misleading because the models are not physical, they only exist digitally in computers, but this is the generally accepted use of the term.

In data modeling, users and system developers participate in a process that successively engages with each of these levels. The first phase of modeling begins with definition of the main types of objects to be represented in the GIS and concludes with a conceptual description of the main types of objects and relationships between them. Once this phase is complete, further work will lead to the creation of diagrams and lists describing the names of objects, their behavior, and the type of interaction between objects. This type of logical data model is very valuable for defining what a GIS will be able to do and the type of domain over which it will extend. Logical models are implementation independent, and can be created in any GIS with appropriate capabilities. The final data modeling phase involves creating a model showing how the objects under study can be digitally implemented in a

GIS. Physical models describe the exact files or database tables used to store data, the relationships between object types, and the precise operations that can be performed. For more details about the practical steps involved in data modeling see Sections 8.3 and 8.4.

A data model provides system developers and users with a common understanding and reference point. For developers, a data model is the means to represent an application domain in terms that may be translated into a design and then implemented in a system. For users, it provides a description of the structure of the system, independent of specific items of data or details of the particular application. A data model controls the types of things that can be handled by a GIS and the range of operations that can be performed on them.

The discussion of geographic representation in Chapter 3 introduced discrete objects and fields, the two fundamental conceptual models for representing things geographically in the real world. In the same chapter the raster and vector logical models were also introduced. Figure 8.3 shows two representations of the same area in a GIS, one raster and the other vector. Notice the difference in the objects represented. Major roads, in green, and areas cleared of vegetation in red, are more clearly visible in the vector representation, whereas smaller roads and built-up areas in lighter shades of gray can best be seen on the scanned raster aerial photograph. The next sections in this chapter focus on the logical and physical representation of raster, vector, and related models in GIS software systems.

8.2 GIS data models

In the past half-century, many GIS data models have been developed and deployed in GIS software systems.

(A)

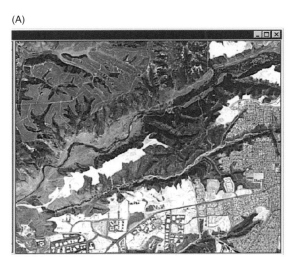

(B)

Figure 8.3 Two representations of San Diego, California (Courtesy Leica Geosystems): (A) panchromatic SPOT raster satellite image collected in 1990 at 10 m resolution; (B) vector objects digitized from the image

Table 8.1 Geographic data models used in GIS

Data model	Example applications
Computer-Aided Design (CAD)	Automating engineering design and drafting.
Graphical (non-topological)	Simple mapping.
Image	Image processing and simple grid analysis.
Raster/Grid	Spatial analysis and modeling especially in environmental and natural resources applications.
Vector/Georelational topological	Many operations on vector geometric features in cartography, socio-economic and resource analysis, and modeling.
Network	Network analysis in transportation, hydrology, and utilities.
Triangulated Irregular Network (TIN)	Surface/terrain visualization, analysis, and modeling.
Object	Many operations on all types of entities (raster/vector/TIN, etc.) in all types of applications.

The key types of geographic data models and their main areas of application are listed in Table 8.1. All are based in some way on the conceptual discrete object/field and logical vector/raster geographic data models (see Chapter 3 for more details). All GIS software systems include a core data model that is built on one or more of these GIS data models. In practice, any modern comprehensive GIS supports at least some elements of all these models. As discussed earlier, the GIS software core system data model is the means to represent geographic aspects of the real world and defines the type of geographic operations that can be performed. It is the responsibility of the GIS implementation team to populate this generic model with information about a particular problem (e.g., utility outage management, military mapping, or natural resource planning). Some GIS software packages come with a fixed data model, while others have models that can be easily extended. Those that can easily be extended are better able to model the richness of geographic domains, and in general are the easiest to use and the most productive systems.

When modeling the real world for representation inside a GIS, it is convenient to group entities of the same geometric type together (for example, all point entities such as lights, garbage cans, dumpsters, etc., might be stored together). A collection of entities of the same geometric type (dimensionality) is referred to as a class or layer. It should also be noted that the term layer is quite widely used in GIS as a general term for a specific dataset. It derived from the process of entering different types of data into a GIS from paper maps, which was undertaken one plate at a time (all entities of the same type were represented in the same color and, using printing technology, were reproduced together on film or printing plates). Grouping entities of the same geographic type

together makes the storage of geographic databases more efficient (for further discussion see Section 10.3). It also makes it much easier to implement rules for validating edit operations (for example, the addition of a new building or census administrative area) and for building relationships between entities. All of the data models discussed below use layers in some way to handle geographic entities.

A layer is a collection of geographic entities of the same geometric type (e.g., points, lines, or polygons). Grouped layers may combine layers of different geometric types.

8.2.1 CAD, graphical, and image GIS data models

The earliest GIS were based on very simple models derived from work in the fields of CAD (computer-aided design and drafting), computer cartography, and image analysis. In a CAD system, real-world entities are represented symbolically as simple point, line, and polygon vectors. This basic CAD data model never became widely popular in GIS because of three severe problems for most applications at geographic scales. First, because CAD models typically use local drawing coordinates instead of real world coordinates for representing objects, they are of little use for map-centric applications. Second, because individual objects do not have unique identifiers it is difficult to tag them with attributes. As the following discussion shows this is a key requirement for GIS applications. Third, because CAD data models are focused on graphical representation of objects they cannot store details of any relationships between objects (e.g., topology or networks), the type of information essential in many spatial analytical operations.

A second type of simple GIS geometry model was derived from work in the field of computer cartography. The main requirement for this field in the 1960s was the automated reproduction of paper topographic maps and the creation of simple thematic maps. Techniques were developed to digitize maps and store them in a computer for subsequent plotting and printing. All paper map entities were stored as points, lines, and polygons, with annotation used for placenames. Like CAD systems, there was no requirement to tag objects with attributes or to work with object relationships.

At about the same time that CAD and computer cartography systems were being developed, a third type of data model emerged in the field of image processing. Because the main data source for geographic image processing is scanned aerial photographs and digital satellite images it was natural that these systems would use rasters or grids to represent the patterning of real-world objects on the Earth's surface. The image data model is also well suited to working with pictures of real-world objects, such as photographs of water valves and scanned building floor plans that are held as attributes of geographically referenced entities in a database (Figure 8.4).

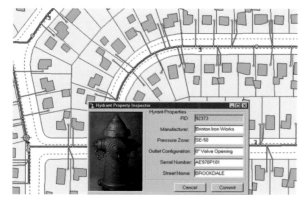

Figure 8.4 An image used as a hydrant object attribute in a water-facility system

In spite of their many limitations, GIS still exist based on these simple data models. This is partly for historical reasons – the GIS may have been built before newer, more advanced models became available – but also because of lack of knowledge about the newer approaches described below.

8.2.2 Raster data model

The raster data model uses an array of cells, or pixels, to represent real-world objects (Figure 3.7). The cells can hold any attribute values based on one of several encoding schemes including categories, and integer and floating-point numbers (see Box 3.3 for details). In the simplest case a binary representation is used (for example, presence or absence of vegetation), but in more advanced cases floating-point values are preferred (for example, height of terrain above sea level in meters). In some systems, multiple attributes can be stored for each cell in a type of value attribute table where each column is an attribute and each row either a pixel, or a pixel class (Figure 8.5).

Usually, raster data are stored as an array of grid values, with metadata (data about data: see Section 11.2.1) about the array held in a file header. Typical metadata include the geographic coordinate of the upper-left corner of the grid, the cell size, the number of row and column elements, and the projection. The data array itself is usually stored as a compressed file or record in a database management system (see Section 10.3). Techniques for compressing rasters are described in Box 8.1 (see also Section 3.6.1 for the general principles involved).

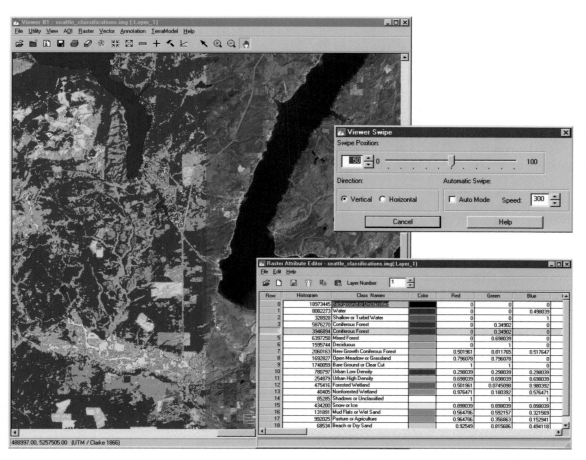

Figure 8.5 Raster data of the Olympic Peninsula, Washington State, USA, with associated value attribute table. Bands 4, 3, 2 from Landsat 5 satellite with land cover classification overlaid (Screenshot courtesy Leica Geosystems; data courtesy of US Geological Survey. Data available from US Geological Survey, EROS Data Center, Sioux Falls, SD)

Technical Box **8.1**

Raster compression techniques

Although the raster data model has many uses in GIS, one of the main operational problems associated with it is the sheer amount of raw data that must be stored. To improve storage efficiency many types of raster compression technique have been developed such as run-length encoding, block encoding, wavelet compression, and quadtrees (see Section 10.7.2.2 for another use of quadtrees as a means to index geographic data). Table 8.2 presents a comparison of file sizes and compression rates for three compression techniques based on the image in Figure 8.6. It can be seen that even the comparatively simple run-length encoding technique compresses the file size by a factor of 5. The more sophisticated

wavelet compression technique results in a compression rate of almost 40, reducing the file from 80.5 to 2.3 MB.

Table 8.2 Comparison of file sizes and compression rates for raster compression techniques (using image shown in Figure 8.6)

Compression technique	File size (MB)	Compression rate
Uncompressed original	80.5	
Run-length	17.7	5.1
Wavelet	2.3	38.3

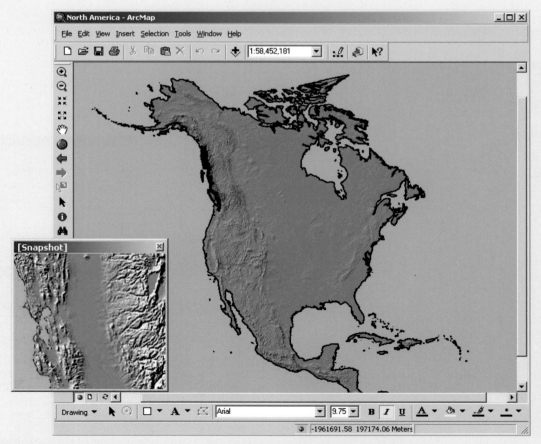

Figure 8.6 Shaded digital elevation model of North America used for comparison of image compression techniques in Table 8.2. Original image is 8726 by 10 618 pixels, 8 bits per pixel. The inset shows part of the image at a zoom factor of 1000 for the San Francisco Bay area

Run-length encoding

Run-length encoding is perhaps the simplest compression method and is very widely used. It involves encoding adjacent row cells that have the same value, with a pair of values indicating the number of cells with the same value, and the actual value.

Block encoding

Block encoding is a two-dimensional version of run-length encoding in which areas of common cell values are represented with a single value. An array is defined as a series of square blocks of the largest size possible. Recursively, the array is divided using blocks of smaller and smaller size. It is sometimes described as a quadtree data structure (see also Section 10.7.2.2).

Wavelet

Wavelet compression techniques invoke principles similar to those discussed in the treatment of fractals (Section 4.8). They remove information by recursively examining patterns in datasets at different scales, always trying to reproduce a faithful representation of the original. A useful byproduct of this for geographic applications is that wavelet-compressed raster layers can be quickly viewed at different scales with appropriate amounts of detail. MrSID (Multiresolution Seamless Image Database) from LizardTech is an example of a wavelet compression technique that is widely used in geographic applications, especially for compressing aerial photographs. Similar wavelet compression algorithms are available from other public and private sources and have been incorporated into the JPEG 2000 standard which is increasingly being used for image compression.

Run-length and block encoding both result in lossless compression of raster layers, that is, a layer can be compressed and decompressed without degradation of information. In contrast, the MrSID wavelet compression technique is lossy since information is irrevocably discarded during compression. Although MrSID compression results in very high compression ratios, because information is lost its use is limited to applications that do not need to use the raw digital numbers for processing or analysis. It is not appropriate for compressing DEMs for example, but many organizations use it to compress scanned maps and aerial photographs when access to the original data is not necessary.

Datasets encoded using the raster data model are particularly useful as a backdrop map display because they look like conventional maps and can communicate a lot of information quickly. They are also widely used for analytical applications such as disease dispersion modeling, surface water flow analysis, and store location modeling.

8.2.3 Vector data model

The raster data model discussed above is most commonly associated with the field conceptual data model. The vector data model on the other hand is closely linked with the discrete object view. Both of these conceptual perspectives were introduced in Section 3.5. The vector data model is used in GIS because of the precise nature of its representation method, its storage efficiency, the quality of its cartographic output, and the availability of functional tools for operations like map projection, overlay, and analysis.

In the vector data model each object in the real world is first classified into a geometric type: in the 2-D case point, line, or polygon (Figure 8.7). Points (e.g., wells, soil pits, and retail stores) are recoded as single coordinate pairs, lines (e.g., roads, streams, and geologic faults) as a series of ordered coordinate pairs (also called polylines – Section 3.6.2), and polygons (e.g., census tracts, soil areas, and oil license zones) as one or more line segments that close to form a polygon area. The coordinates that define the geometry of each object may have 2, 3, or 4 dimensions: 2 (x, y: row and column, or latitude and longitude), 3 (x, y, z: the addition of a height value), or 4 (x, y, z, m: the addition of another value to represent time or some other property – perhaps the offset of road signs from a road centerline, or an attribute).

For completeness, it should also be said that in some data models linear features can be represented not only as a series of ordered coordinates, but also as curves defined by a mathematical function (e.g., a spline or Bézier curve). These are particularly useful for representing built environment entities like road curbs and some buildings.

8.2.3.1 Simple features

Geographic entities encoded using the vector data model are usually called features and this will be the convention adopted here. Features of the same geometric type are stored in a geographic database as a feature class, or when speaking about the physical (database) representation the

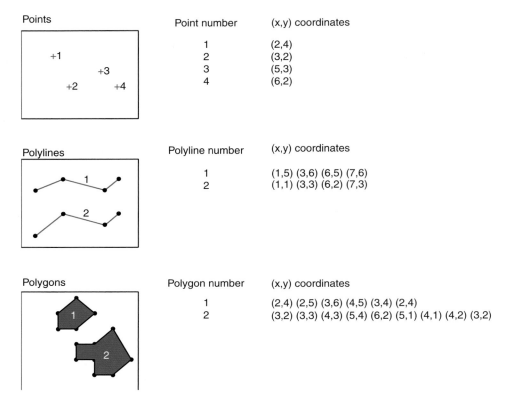

Points	Point number	(x,y) coordinates
	1	(2,4)
	2	(3,2)
	3	(5,3)
	4	(6,2)

Polylines	Polyline number	(x,y) coordinates
	1	(1,5) (3,6) (6,5) (7,6)
	2	(1,1) (3,3) (6,2) (7,3)

Polygons	Polygon number	(x,y) coordinates
	1	(2,4) (2,5) (3,6) (4,5) (3,4) (2,4)
	2	(3,2) (3,3) (4,3) (5,4) (6,2) (5,1) (4,1) (4,2) (3,2)

Figure 8.7 Representation of point, line, and polygon objects using the vector data model

term *feature table* is preferred. Here each feature occupies a row and each property of the feature occupies a column. GIS commonly deal with two types of feature: simple and topological. The structure of simple feature polyline and polygon datasets is sometimes called *spaghetti* because, like a plate of cooked spaghetti, lines (strands of spaghetti) and polygons (spaghetti hoops) can overlap and there are no relationships between any of the objects.

Features are vector objects of type point, polyline, or polygon.

Simple feature datasets are useful in GIS applications because they are easy to create and store, and because they can be retrieved and rendered on screen very quickly. On the other hand because simple features lack more advanced data structure characteristics, such as topology (see next section), operations like shortest-path network analysis and polygon adjacency cannot be performed without additional calculations.

8.2.3.2 Topological features

Topological features are essentially simple features structured using topological rules. Topology is the mathematics and science of geometrical relationships. Topological relationships are non-metric (qualitative) properties of geographic objects that remain constant when the geographic space of objects is distorted. For example, when a map is stretched properties such as distance and angle change, whereas topological properties such as adjacency

and containment do not. Topological structuring of vector layers introduces some interesting and very useful properties, especially for polyline (also called 1-cell, arc, edge, line, and link) and polygon (also called 2-cell, area, and face) data. Topological structuring of line, layers forces all line ends that are within a user-defined distance to be snapped together so that they are given exactly the same coordinate value. A node is placed wherever the ends of lines meet or cross. Following on from the earlier analogy this type of data model is sometimes referred to as spaghetti with meatballs (the nodes being the meatballs on the spaghetti lines). Topology is important in GIS because of its role in data validation, modeling integrated feature behavior, editing, and query optimization.

Topology is the science and mathematics of relationships used to validate the geometry of vector entities, and for operations such as network tracing and tests of polygon adjacency.

Data validation

Many of the geographic data collected from basic digitizing, field data collection devices, photogrammetry, and CAD systems comprise simple features of type point, polyline, or polygon, with limited structural intelligence (for example, no topology). Testing the topological integrity of a dataset is a useful way to validate the geometric quality of the data and to assess their suitability for geographic analysis. Some useful data validation topology tests include:

- Network connectivity – do all network elements connect to form a graph (i.e., are all the pipes in a water network joined to form a wastewater system)? Network elements that connect must be 'snapped' together (that is, given the same coordinate value) at junctions (intersections).

- Line intersection – are there junctions at intersecting polylines, but not at crossing polylines? It is, for example, perfectly possible for roads to cross in planimetric 2-D view, but not intersect in 3-D (for example, at a bridge or underpass; Figure 3.4).

- Overlap – do adjacent polygons overlap? In many applications (e.g., land ownership) it is important to build databases free from overlaps and gaps so that ownership is unambiguous.

- Duplicate lines – are there multiple copies of network elements or polygons? Duplicate polylines often occur during data capture. During the topological creation process it is necessary to detect and remove duplicate polylines to ensure that topology can be built for a dataset.

Modeling the integrated behavior of different feature types

In the real world many objects share common locations and partial identities. For example, water distribution areas often coincide with water catchments, electric distribution areas often share common boundaries with building sub-divisions, and multiple telecommunications fibers are often run down the same conduit. These situations can be modeled in a GIS database as either single objects with multiple geometry representations, or multiple objects with separate geometry integrated for editing, analysis, and representation. There are advantages and disadvantages to both approaches. Multiple objects with separate geometries are certainly easier to implement in commercially available databases and information systems. If one feature is moved during editing then logically both features should move. This is achieved by storing both objects separately in the database, each with their own geometry, but integrating them inside the GIS editor application so that they are treated as single features. When the geometry of one is moved the other automatically moves with it. There is further discussion of shared editing in the next section.

Editing productivity

Topology improves editor productivity by simplifying the editing process and providing additional capabilities to manipulate feature geometries. Editing requires both topological data structuring and a set of topologically aware tools. The productivity of editors can be improved in several ways:

- Topology provides editors with the ability to manipulate common, shared polylines and nodes as single geometric objects to ensure that no differences are introduced into the common geometries.

- Rubberbanding is the process of moving a node, polyline, or polygon boundary and receiving

interactive feedback on screen about the location of all topologically connected geometry.

- Snapping is a useful technique to both speed up editing and maintain a high standard of data quality.

- Auto-closure is the process of completing a polygon by snapping the last point to the first digitized point.

- Tracing is a type of network analysis technique that is used, especially in utility applications, to test the connectivity of linear features (is the newly designed service going to receive power?).

Optimized queries

There are many GIS queries that can be optimized by pre-computing and storing information about topological relationships. Some common examples include:

- Network tracing (e.g., find all connected water pipes and fittings).

- Polygon adjacency (e.g., who owns the parcels adjoining those owned by a specific owner?).

- Containment (e.g., which manholes lie within the pavement area of a given street?).

- Intersection (e.g., which census tracts intersect with a set of health areas?).

In the remainder of this section the discussion concentrates on a conceptual understanding of GIS topology focusing on the more complex polygon case. The network case is considered in the next section. The relative merits and implementations of two approaches to GIS topology are discussed later in Section 10.7.1. These two implementation approaches differ because in one case relationships are batch built and stored along with feature geometry, and in the other relationships are calculated interactively when they are needed.

Conceptually speaking, in a topologically structured polygon data layer each polygon is defined as a collection of polylines that in turn are made up of an ordered list of coordinates (vertices). Figure 8.8 shows an example of a polygon dataset comprising six polygons (including the 'outside world': polygon 1). A number in a circle identifies a polygon. The lines that make up the polygons are shown in the polygon-polyline list. For example, polygon 2 can be assembled from lines 4, 6, 7, 10, and 8. In this particular implementation example the 0 before the 8 is used to indicate that line 8 actually defines an 'island' inside polygon 2. The list of coordinates for each line is also shown in Figure 8.8. For example, line 5 begins with coordinates 7,4 and 6,3 – other coordinates have been omitted for brevity. A line may appear in the polygon-polyline list more than once (for example, line 6 is used in the definition of both polygons 2 and 5), but the actual coordinates for each polyline are only stored once in the polyline-coordinate list. Storing common boundaries between adjacent polygons avoids the potential problems of gaps (slivers) or overlaps between adjacent polygons. It has the added bonus that there are fewer coordinates in a topologically structured polygon feature layer compared with a simple feature layer representation of the same entities. The downside, however, is that drawing a polygon requires that multiple polylines must be retrieved

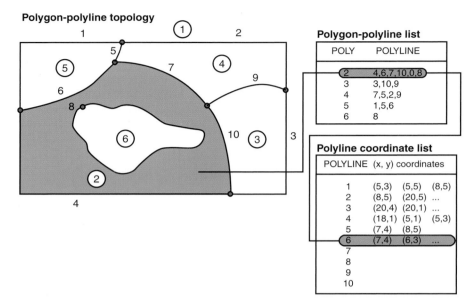

Figure 8.8 A topologically structured polygon data layer. The polygons are made up of the polylines shown in the polygon-polyline list. The lines are made up of the coordinates shown in the line coordinate list (*Source*: after ESRI 1997)

from the database and then assembled into a boundary. This process can be time consuming when repeated for each polygon in a large dataset.

Planar enforcement is a very important property of topologically structured polygons. In simple terms, planar enforcement means that all the space on a map must be filled and that any point must fall in one polygon alone, that is, polygons must not overlap. Planar enforcement implies that the phenomenon being represented is conceptualized as a field.

The contiguity (adjacency) relationship between polygons is also defined during the process of topological structuring. This information is used to define the polygons on the left- and right-hand side of each polyline, in the direction defined by the list of coordinates (Figure 8.9). In Figure 8.9, Polygon 2 is on the left of Polyline 6 and Polygon 5 is on the right. Thus we can deduce from a simple look-up operation that Polygons 2 and 5 are adjacent.

Software systems based on the vector topological data model have become popular over the years. A special case of the vector topological model is the *georelational* model. In this model derivative, the feature geometries and associated topological information are stored in regular computer files, whereas the associated attribute information is held in relational database management system (RDBMS) tables. The GIS software maintains the intimate linkage between the geometry, topology, and attribute information. This hybrid data management solution was developed to take advantage of RDBMS to store and manipulate attribute information. Geometry and topology were not placed in RDBMS because, until relatively recently, RDBMS were unable to store and retrieve geographic data efficiently. Figure 8.10 is an example of a georelational model as implemented in ESRI's ArcInfo coverage polygon dataset. It shows

file-based geometry and topology information linked to attributes in an RDBMS table. The ID (identifier) of the polygon, the label point, is linked (related or joined) to the ID column in the attribute table (see also Chapter 10). Thus, in this soils dataset polygon 3 is soil B7, of class 212, and its suitability is moderate.

The topological feature geographic data model has been extensively used in GIS applications over the last 20 years, especially in government and natural resources applications based on polygon representations (see Sections 2.3.2 and 2.3.5). Typical government applications include: cadastral management, tax assessment, parcel management, zoning, planning, and building control. In the areas of natural resources and environment, key applications include site suitability analysis, integrated land use modeling, license mapping, natural resource management, and conservation. The tax appraisal case study discussed in Section 2.3.2.2 is an example of a GIS based on the topological feature data model. The developers of this system chose this model because they wanted to avoid overlaps and gaps in tax parcels (polygons), to ensure that all parcel boundaries closed (were validated), and to store data in an efficient way. This is in spite of the fact that there is an overhead in creating and maintaining parcel topology, as well as degradation in draw and query performance for large databases.

8.2.3.3 Network data model

The network data model is really a special type of topological feature model. It is discussed here separately because it raises several new issues and has been widely applied in GIS studies.

Networks can be used to model the flow of goods and services. There are two primary types of networks: *radial* and *looped*. In radial or tree networks flow

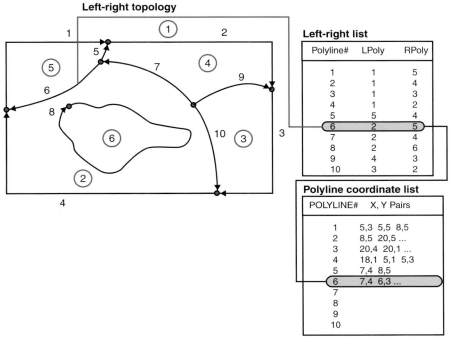

Figure 8.9 The contiguity of a topologically structured polygon data layer. For each polyline the left and right polygon is stored with the geometry data (*Source*: after ESRI 1997)

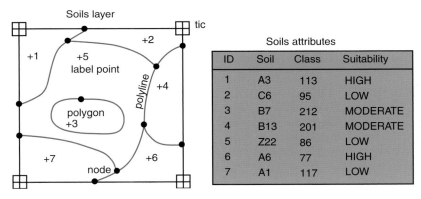

Figure 8.10 An example of a georelational polygon dataset. Each of the polygons is linked to a row in an RDBMS table. The table has multiple attributes, one in each column (*Source*: after ESRI 1997)

always has an upstream and downstream direction. Stream and storm drainage systems are examples of radial networks. In looped networks, self-intersections are common occurrences. Water distribution networks are looped by design to ensure that service interruptions affect the fewest customers.

In GIS software systems, networks are modeled as points (for example, street intersections, fuses, switches, water valves, and the confluence of stream reaches: usually referred to as nodes in topological models), and lines (for example, streets, transmission lines, pipes, and stream reaches). Network topological relationships define how lines connect with each other at nodes. For the purpose of network analysis it is also useful to define

rules about how flows can move through a network. For example, in a sewer network, flow is directional from a customer (source) to a treatment plant (sink), but in a pressurized gas network flow can be in any direction. The rate of flow is modeled as impedances (weights) on the nodes and lines. Figure 8.11 shows an example of a street network. The network comprises a collection of nodes (types of street intersection) and lines (types of street), as well as the topological relationships between them. The topological information makes it possible, for example, to trace the flow of traffic through the network and to examine the impact of street closures. The impedance on the intersections and streets determines the speed at which traffic flows. Typically, the rate of

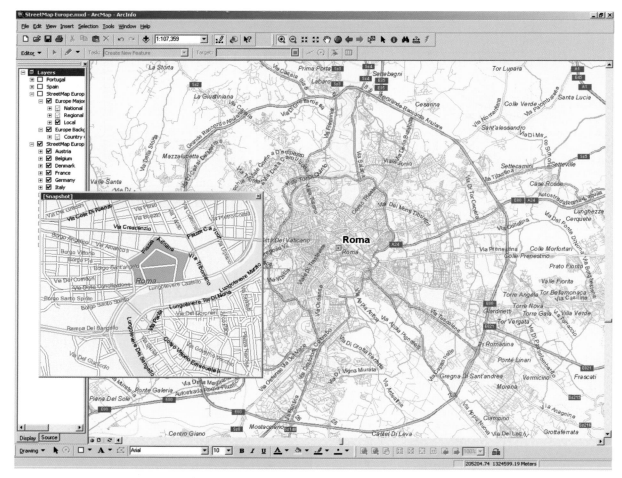

Figure 8.11 An example of a street network

flow is proportional to the street speed limit and number of lanes, and the timing of stoplights at intersections. Although this example relates to streets, the same basic principles also apply to, for example, electric, water, and railroad networks.

In georelational implementations of the topological network feature model, the geometry and topology information is typically held in ordinary computer files and the attributes in a linked database. The GIS software tools are responsible for creating and maintaining the topological information each time there is a change in the feature geometry. In more modern object models the geometry, attributes, and topology may be stored together in a DBMS, or topology may be computed on the fly.

There are many applications that utilize networks. Prominent examples include: calculating power load drops over an electricity network; routing emergency response vehicles over a street network; optimizing the route of mail deliveries over a street network; and tracing pollution upstream to a source over a stream network.

Network data models are also used to support another data model variant called *linear referencing* (Section 5.4). The basic principle of linear referencing is quite simple. Instead of recording the location of geographic entities as

explicit x, y, z coordinates, they are stored as distances along a network (called a route system) from a point of origin. This is a very efficient way of storing information such as road pavement (surface) wear characteristics (e.g., the location of pot holes and degraded asphalt), geological seismic data (e.g., shockwave measurements at sensors along seismic lines), and pipeline corrosion data. An interesting aspect of this is that a two-dimensional network is reduced to a one-dimensional linear route list. The location of each entity (often called an event) is simply a distance along the route from the origin. Offsets are also often stored to indicate the distance from a network centerline. For example, when recording the surface characteristics of a multi-carriageway road several readings may be taken for each carriageway at the same linear distance along the route. The offset value will allow the data to be related to the correct carriageway. Dynamic segmentation is a special type of linear referencing. The term derives from the fact that event data values are held separately from the actual network route in database tables (still as linear distances and offsets) and then dynamically added to the route (segmented) each time the user queries the database. This approach is especially useful in situations in which the event data change frequently and need to be stored in a database due to access from other applications (e.g., traffic volumes or rate of pipe corrosion).

8.2.3.4 TIN data model

The geographic data models discussed so far have concentrated on one- and two-dimensional data. There are several ways to model three-dimensional data, such as terrain models, sales cost surfaces, and geologic strata. The term 2.5-D is sometimes used to describe surface structure because they have dimensional properties between 2-D and 3-D (Box 3.4). A true 3-D structure will contain multiple z values at the same x, y location and thus is able to model overhangs and tunnels, and support accurate volumetric calculations like cut and fill (a term derived from civil engineering applications that describes cutting earth from high areas and placing it in low areas to construct a flat surface, as is required in, for example, railroad construction). Both grids and triangulated irregular networks (TINs) are used to create and represent surfaces in GIS. A regular grid surface is really a type of raster dataset as discussed earlier in Section 8.2.2. Each grid cell stores the height of the surface at a given location. The TIN structure, as the name suggests, represents a surface as contiguous non-overlapping triangular elements (Figure 8.12). A TIN is created from a set of points, that is, points with x, y, and z coordinate values. A key advantage of the TIN structure is that the density of sampled points, and therefore the size of triangles, can be adjusted to reflect the relief of the surface being modeled, with more points sampled in areas of variable relief (see Section 4.4). TIN surfaces can be created by performing what is called a Delaunay triangulation (Figure 8.13, Section 14.4.4.1). First, a convex hull is created for a dataset – the smallest convex polygon that contains the set of points. Next, straight lines that do not cross each other are drawn from interior points to points on the boundary of the convex hull and to each other. This divides the convex hull into a set of polygons which are then divided into triangles by drawing more lines between vertices of the polygons.

A TIN is a topological data structure that manages information about the nodes comprising each triangle and the neighbors of each triangle. Figure 8.13 shows the topology of a simple TIN. As with other topological data structures, information about a TIN may be conveniently stored in a file or database table, or computed on the fly. TINs offer many advantages for surface analysis. First, they incorporate the original sample points, providing a useful check on the accuracy of the model. Second, the variable density of triangles means that a TIN is an efficient way of storing surface representations such as terrains that have substantial variations in topography. Third, the data structure makes it easy to calculate elevation, slope, aspect, and line-of-sight between points. The combination of these factors has led to the widespread use of the TIN data structure in applications such as volumetric calculations for roadway design, drainage studies for land development, and visualization of urban forms. Figure 8.14 shows two example applications of TINs. Figure 8.14A is a shaded landslide risk TIN of the Pisa district, Italy with building objects draped on top

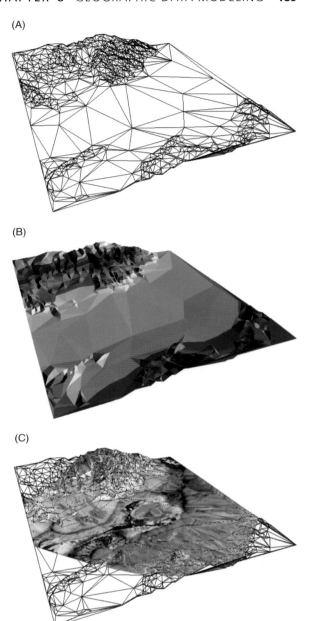

Figure 8.12 TIN surface of Death Valley, California: (A) 'wireframe' showing all triangles; (B) shaded by elevation; (C) draped with satellite image

to give a sense of landscape. Figure 8.14B is a TIN of the Yangtse River, China, greatly exaggerated in the z dimension. It shows how TINs draped with images can provide photo-realistic views of landscapes.

Like all 2.5-D and 3-D models, TINs are only as good as the input sample data. They are especially susceptible to extreme high and low values because there is no smoothing of original data. Other limitations of TINs include their inability to deal with discontinuity of slope across triangle boundaries, the difficulty of calculating

A TIN is a topologic data structure that manages
information about the nodes that comprise each triangle
and the neighbors to each triangle

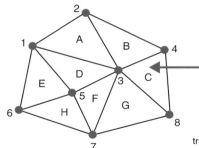

Triangle	Node list	Neighbors
A	1, 2, 3	-, B, D
B	2, 4, 3	-, C, A
C	4, 8, 3	-, G, B
D	1, 3, 5	A, F, E
E	1, 5, 6	D, H, -
F	3, 7, 5	G, H, D
G	3, 8, 7	C, -, F
H	5, 7, 6	F, -, E

Triangles always have three nodes and usually have three neighboring
triangles. Triangles on the periphery of the TIN can have one or two neighbors.

Figure 8.13 The topology of a TIN (*Source*: after Zeiler 1999)

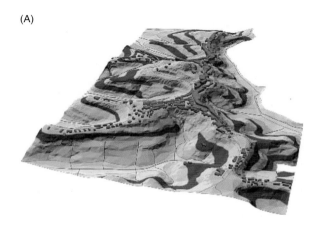

(A)

(B)

Figure 8.14 Examples of applications that use the TIN data
model: (A) Landslide risk map for Pisa, Italy (Courtesy: Earth
Science Department, University of Siena, Italy); (B) Yangtse
River, China (Courtesy: Human Settlements Research Center,
Tsinghua University, China)

optimum routes, and the need to ensure that peaks, pits,
ridges, and channels are captured if a drainage network
TIN is to be accurate.

8.2.4 Object data model

All the geographic data models described so far are
geometry-centric, that is they model the world as collections of points, lines, and areas, TINs, or rasters. Any
operations to be performed on the geometry (and, in some
cases, associated topology) are created as separate procedures (programs or scripts). Unfortunately, this approach
can present several limitations for modeling geographic
systems. All but the simplest of geographic systems contain many entities with large numbers of properties, complex relationships, and sophisticated behavior. Modeling
such entities as simple geometry types is overly simplistic
and does not easily support the sophisticated characteristics required for modern analysis. Additionally, separating
the state of an entity (attributes or properties defining what
it *is*) from the behavior of an entity (methods defining what
it *does*) makes software and database development tedious,
time-consuming, and error prone. To try to address these
problems geographic object data models were developed.
These allow the full richness of geographic systems to be
modeled in an integrated way in a GIS.

The central focus of a GIS object data model is the
collection of geographic objects and the relationships
between the objects (see Box 8.2). Each geographic object
is an integrated package of geometry, properties, and
methods. In the object data model, geometry is treated
like any other attribute of the object and not as its primary characteristic (although clearly from an application
perspective it is often the major property of interest). Geographic objects of the same type are grouped together as
object classes, with individual objects in the class referred
to as 'instances'. In many GIS software systems each
object class is stored physically as a database table, with
each row an object and each property a column. The methods that apply are attached to the object instances when
they are created in memory for use in the application.

**An object is the basic atomic unit in an object data
model and comprises all the properties that define
the state of an object, together with the methods
that define its behavior.**

Object-oriented concepts in GIS

An *object* is a self-contained package of information describing the characteristics and capabilities of an entity under study. An interaction between two objects is called a *relationship*. In a geographic object data model the real world is modeled as a collection of objects and the relationships between the objects. Each entity in the real world to be included in the GIS is an object. A collection of objects of the same type is called a *class*. In fact, classes are a more central concept than objects from the implementation point of view because many of the object-oriented characteristics are built at the class level. A class can be thought of as a template for objects. When creating an object data model the data model designer specifies classes and the relationships between classes. Only when the data model is used to create a database are objects (instances or examples of classes) actually created.

Examples of objects include oil wells, soil bodies, stream catchments, and aircraft flight paths. In the case of an oil well class, each oil well object might include *properties* defining its *state* – annual production, owner name, date of construction, and type of geometry used for representation at a given scale (perhaps a point on a small-scale map and a polygon on a large-scale one). The oil well class could have connectivity relationships with a pipeline class that represents the pipeline used to transfer oil to a refinery. There could also be a relationship defining the fact that each well must be located on a drilling platform. Finally, each oil well object might also have *methods* defining the *behavior* or what it can do. Example behavior might include how objects draw themselves on a computer screen, how objects can be created and deleted, and editing rules about how oil wells snap to pipelines.

There are three key facets of object data models that make them especially good for modeling geographic systems: encapsulation, inheritance, and polymorphism.

Encapsulation describes the fact that each object packages together a description of its state and behavior. The state of an object can be thought of as its properties or attributes (e.g., for a forest object it could be the dominant tree type, average tree age, and soil pH). The behavior is the methods or operations that can be performed on an object (for a forest object these could be create, delete, draw, query, split, and merge). For example, when splitting a forest polygon into two parts, perhaps following a part sale, it is useful to get the GIS to automatically calculate the areas of the two new parts. Combining the state and behavior of an object together in a single package is a natural way to think of geographic entities and a useful way to support the reuse of objects.

Inheritance is the ability to reuse some or all of the characteristics of one object in another object. For example, in a gas facility system a new type of gas valve could easily be created by overwriting or adding a few properties or methods to a similar existing type of valve. Inheritance provides an efficient way to create models of geographic systems by reusing objects and also a mechanism to extend models easily. New object classes can be built to reuse parts of one or more existing object classes and add some new unique properties and methods. The example described in Section 8.3 shows how inheritance and other object characteristics can be used in practice.

Polymorphism describes the process whereby each object has its own specific implementation for operations like draw, create, and delete. One example of the benefit of polymorphism is that a geographic database can have a generic object creation component that issues requests to be processed in a specific way by each type of object class. A utility system's editor software can send a generic create request to all objects (e.g., gas pipes, valves, and service lines) each of which has specific create algorithms. If a new object class is added to the system (e.g., landbase) then this mechanism will work because the new class is responsible for implementing the create method. Polymorphism is essential for isolating parts of software as self-contained components (see Chapter 10).

All geographic objects have some type of relationship to other objects in the same object class and, possibly, to objects in other object classes. Some of these relationships are inherent in the class definition (for example, some GIS remove overlapping polygons) while other interclass relationships are user-definable. Three types of relationships are commonly used in geographic object data models: topological, geographic, and general.

A class is a template for creating objects.

Generally, topological relationships are built into the class definition. For example, modeling real-world entities as a network class will cause network topology to be built for the nodes and lines participating in the network. Similarly, real-world entities modeled as topological polygon classes will be structured using the node–polyline model described in Section 8.2.3.2.

Geographic relationships between object classes are based on geographic operators (such as overlap, adjacency, inside, and touching) that determine the interaction between objects. In a model of an agricultural system, for example, it might be useful to ensure that all farm buildings are within a farm boundary using a test for geographic containment.

General relationships are useful to define other types of relationship between objects. In a tax assessment system, for example, it is advantageous to define a relationship between land parcels (polygons) and ownership data that is stored in an associated DBMS table. Similarly, an electric distribution system relating light poles (points) to text strings (called annotation) allows depiction of pole height and material of construction on a map display. This type of information is very valuable for creating work orders (requests for change) that alter the facilities. Establishing relationships between objects in this way is useful because if one object is moved then the other will move as well, or if one is deleted then the other is also deleted. This makes maintaining databases much easier and safer.

In addition to supporting relationships between objects (strictly speaking, between object classes), object data models also allow several types of rules to be defined. Rules are a valuable means of maintaining database integrity during editing tasks. The most popular types of rules used in object data models are attribute, connectivity, relationship, and geographic.

Attribute rules are used to define the possible attribute values that can be entered for any object. Both range and coded value attribute rules are widely employed. A range attribute rule defines the range of valid values that can be entered. Examples of range rules include: highway traffic speed must be in the range 25–70 miles (40–120 km) per hour; forest compartment average tree height must be in the range 0–50 meters. Coded attribute rules are used for categorical data types. Examples include: land use must be of type commercial, residential, park, or other; or pipe material must be of type steel, copper, lead, or concrete.

Connectivity rules are based on the specification of valid combinations of features, derived from the geometry, topology, and attribute properties. For example, in an electric distribution system a 28.8 kV conductor can only connect to a 14.4 kV conductor via a transformer. Similarly, in a gas distribution system it should not be possible to add pipes with free ends (that is, with no fitting or cap) to a database.

Geographic rules define what happens to the properties of objects when an editor splits or merges them (Figure 8.15). In the case of a land parcel split following the sale of part of the parcel, it is useful to define rules to determine the impact on properties like area, land use code, and owner. In this example, the original parcel area value should be divided in proportion to the size of the

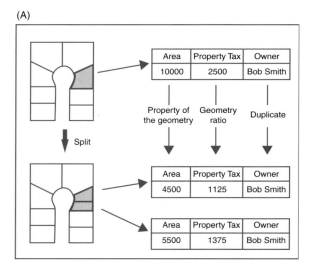

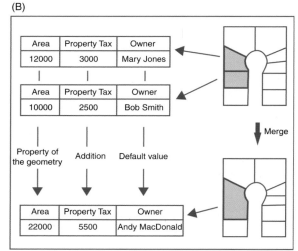

Figure 8.15 Example of split and merge rules for parcel objects: (A) split; (B) merge (*Source*: after MacDonald 1999)

two new parcels, the land use code should be transferred to both parcels, and the owner name should remain for one parcel, but a new one should be added for the part that was sold off. In the case of a merge of two adjacent water pipes, decisions need to be made about what happens to attributes like material, length, and corrosion rate. In this example, the two pipe materials should be the same, the lengths should be summed, and the new corrosion rate determined by a weighted average of both pipes.

8.3 Example of a water-facility object data model

The goal of this section is to describe an example of a geographic object model. It will discuss how many of the concepts introduced earlier in this chapter are used in practice. The example selected is that of an urban water-facility model. The types of issues raised in this example apply to all geographic object models, although of course the actual objects, object classes, and relationships under consideration will differ. The role of data modeling, as discussed in Section 8.1, is to represent the key aspects of the real world inside the digital computer for management, analysis, and display purposes.

Figure 8.16 is a diagram of part of a water distribution system, a type of pressurized network controlled by several devices. A pump is responsible for moving water through pipes (mains and laterals) connected together by fittings. Meters measure the rate of water consumption at houses. Valves and hydrants control the flow of water.

The purpose of the example object model is to support asset management, mapping, and network analysis applications. Based on this it is useful to classify the objects into two types: the landbase and the water-facilities. Landbase is a general term for objects like houses and streets that provide geographic context but are not used in network analysis. The landbase object types are Pump House, House, and Street. The water-facilities object types are: Main, Lateral (a smaller type of WaterLine), Fitting (water line connectors), Meter, Valve, and Hydrant. All of these object types need to be modeled as a network in order to support network analysis operations like network isolation traces and flow prediction. A network isolation trace is used to find all parts of a network that are unconnected (isolated). Using the topological connectivity of the network and information about whether pipes and fittings support water flow, it is possible to determine connectivity. Flow prediction is used to estimate the flow of water through the network based on network connectivity and data about water availability and consumption. Figure 8.17 shows all the main object types and the implicit geographic relationships to be incorporated into the model. The arrows indicate the direction of flow in the network. When digitizing this network using a GIS editor it will be useful to specify topological connectivity and attribute rules to control how objects can be connected (see Section 8.2.3.3 above). Before this network can be used for analysis it will also be necessary to add flow impedances to each link (for example, pipe diameter).

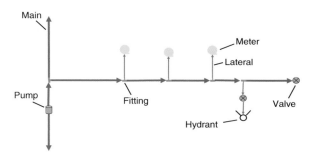

Figure 8.17 Water distribution system network

Having identified the main object types, the next step is to decide how objects relate to each other and the most efficient way to implement them. Figure 8.18 shows one possible object model that uses the Unified Modeling Language (UML) to show objects and the relationships between them. Some additional color-coding has been added to help interpret the model. In UML models each box is an object class and the lines define how one class reuses (inherits) part of the class above it in a hierarchy.

Object class names in an italic font are abstract classes; those with regular font names are used to create (instantiate) actual object instances. Abstract classes do not have instances and exist for efficiency reasons. It is sometimes useful to have a class that implements some capabilities once, so that several other classes can then be reused. For example, Main and Lateral are both types of *Line*, as is Street. Because Main and Lateral share several things in common – such as ConstructionMaterial, Diameter, and InstallDate properties, and connectivity and draw behavior – it is efficient to implement these in a separate abstract class, called *WaterLine*. The triangles indicate that one class is a type of another class. For example, Pump House and House are types of *Building*, and Street and *WaterLine* are types of *Line*. The diamonds indicate composition. For example, a network is composed of a collection of *Line* and *Node* objects. In the water-facility object model, object classes without any geometry are colored pink. The Equipment and OperationsRecord object classes have their location determined by being associated with other objects (e.g., valves and mains). The Equipment and OperationsRecord classes are useful places to store properties common to many facilities, such as EquipmentID, InstallDate, ModelNumber, and SerialNumber.

Once this logical geographic object model has been created it can be used to generate a physical data model. One way to do this is to create the model using a computer-aided software engineering (CASE) tool. A CASE tool is a software application that has graphical tools to draw and specify a logical model (Figure 8.19). A further advantage of a CASE tool is that physical models can be generated directly from the logical models, including all the database tables and much of the supporting code for implementing behavior. Once a database structure (schema) has been created, it can be populated with objects and the intended applications put into operation.

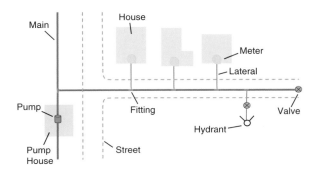

Figure 8.16 Water distribution system water-facility object types and geographic relationships

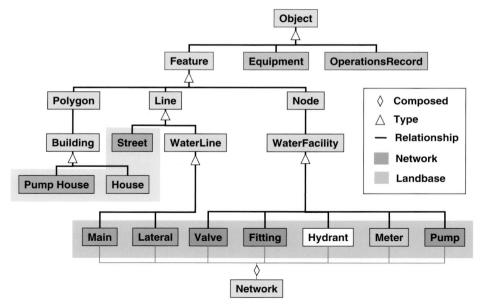

Figure 8.18 A water-facility object model

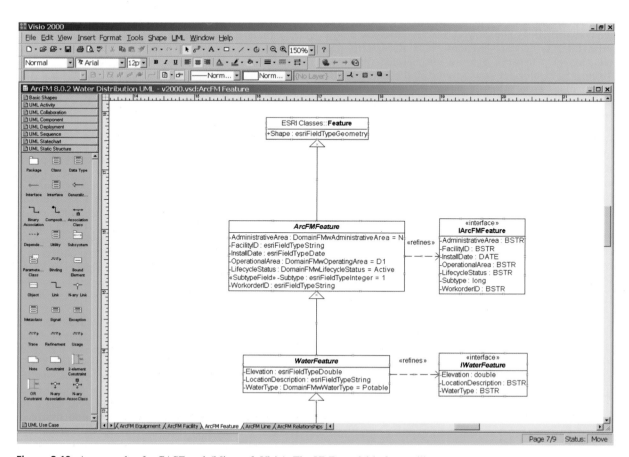

Figure 8.19 An example of a CASE tool (Microsoft Visio). The UML model is for a utility water system

8.4 Geographic data modeling in practice

Geographic analysis is only as good as the geographic database on which it is based and a geographic database is only as good as the geographic data model from which it is derived. Geographic data modeling begins with a clear definition of the project goals and progresses through an understanding of user requirements, a definition of the objects and relationships, formulation of a logical model, and then creation of a physical model. These steps are a prelude to database creation and, finally, database use. In Box 8.3 Leslie Cone of the US Bureau of Land Management describes her experience of creating a land management system with a parcel data model.

No step in data modeling is more important than understanding the purpose of the data modeling exercise. This understanding can be gained by collecting user requirements from the main users. Initially, user requirements will be vague and ill-defined, but over time they will become clearer. Project goals and user requirements should be precisely specified in a list or narrative.

Biographical Box **8.3**

Leslie M. Cone, Project Manager, BLM Land and Resources Project Office

The BLM (US Bureau of Land Management) administers some 261 million surface acres of America's public lands, located primarily in 12 Western States. The BLM mission is 'to sustain the health, diversity, and productivity of the public lands for the use and enjoyment of present and future generations'.

Those who use the Internet to access BLM's land resources and data owe many thanks to Leslie Cone, a 31-year employee of the BLM. As Project Manager for the Land and Resources Project Office, she leads the team of BLM employees, interns, and contractors who manage 24 national-level projects that include the National Integrated Land System (NILS), BLM's contribution to Geospatial One-Stop Portal, and the Automated Fluid Minerals Support System. According to Leslie 'The mission of the Land and Resources Project Office is to provide automation of BLM's land and resources data and the tools to use and manage them'.

Figure 8.20 Leslie Cone, project manager (courtesy Leslie Cone)

Addressing her contributions to the GIS, Leslie says, 'NILS was initiated to create a business solution for land managers who face an increasingly complex environment of land transactions, legal challenges, and deteriorating and difficult-to-access records'. She adds, 'This complex task was made even more challenging because the Bureau's existing land records date back over 200 years'.

The NILS project was the first step towards providing a common parcel-based solution for sharing land record information within the US government and the private sector. NILS is a joint development project of the BLM and the US Forest Service, in partnership with states, counties, and private industry. The project developed a common parcel data model and a set of software tools for the collection, management, and sharing of survey-based data, and land record information. The data model resulted from an extensive analysis of user requirements, the accumulated experience of previous-generation systems, and several prototypes. It has been crucial to the success of the project.

To government agencies, commercial businesses, students, and the general public, this provides an easy way to perform research and other tasks that were formerly a major challenge. For the public, all that is needed to access the public data is an Internet connection. The NILS GeoCommunicator application is an interactive Web-based land information portal that allows users to share, search, locate, view, and access geographic information, data, maps, and images. In addition, NILS includes proprietary applications for the government and private business sectors.

Since she became the Project Manager six years ago, the BLM Land and Resources Project Office has grown from a few individuals to approximately 60 managers, engineers, programmers, analysts, Web developers, technical writers, and student interns. Leslie Cone holds a Bachelor of Science degree in Forestry and Outdoor Recreation from Colorado State University and a Master's degree in Public Administration from the University of New Mexico.

Formulation of a logical model necessitates identification of the objects and relationships to be modeled. Both the attributes and behavior of objects are required for an object model. A useful graphic tool for creating logical data models is a CASE tool and a useful language for specifying models is UML. It is not essential that all objects and relationships be identified at the first attempt because logical models can be refined over time. The key objects and relationships for the water distribution system object model are shown in Figure 8.18.

Once an implementation-independent logical model has been created, this model can be turned into a system-dependent physical model. A physical model will result in an empty database schema – a collection of database tables and the relationships between them. Sometimes, for performance optimization reasons or because of changing requirements, it is necessary to alter the physical data model. Even at this relatively late stage in the process, flexibility is still necessary.

It is important to realize that there is no such thing as the correct geographic data model. Every problem can be represented with many possible data models. Each data model is designed with a specific purpose in mind and is sub-optimal for other purposes. A classic dilemma is whether to define a general-purpose data model that has wide applicability, but that can, potentially, be complex and inefficient, or to focus on a narrower highly optimized model. A small prototype can often help resolve some of these issues.

Geographic data modeling is both an art and a science. It requires a scientific understanding of the key geographic characteristics of real-world systems, including the state and behavior of objects, and the relationships between them. Geographic data models are of critical importance because they have a controlling influence over the type of data that can be represented and the operations that can be performed. As we have seen, object models are the best type of data model for representing rich object types and relationships in facility systems, whereas simple feature models are sufficient for elementary applications such as a map of the body. In a similar vein, so to speak, raster models are good for data represented as fields such as soils, vegetation, pollution, and population counts.

Questions for further study

1. Figure 8.21 is an oblique aerial photograph of part of the city of Kfar-Saba, Israel. Take ten minutes to list all the object classes (including their attributes and behavior) and the relationships between the classes that you can see in this picture that would be appropriate for a city information system study.

2. Why is it useful to include the conceptual, logical, and physical levels in geographic data modeling?

3. Describe, with examples, five key differences between the topological vector and raster geographic data models. It may be useful to consult Figure 8.3 and Chapter 3.

4. Review the terms encapsulation, inheritance, and polymorphism and explain with geographic examples why they make object data models superior for representing geographic systems.

Figure 8.21 Oblique aerial view of Kfar-Saba, Israel (Courtesy: ESRI)

Further reading

Arctur D. and Zeiler M. 2004 *Designing Geodatabases: Case Studies in GIS Data Modeling*. Redlands, CA: ESRI Press.

ESRI 1997 *Understanding GIS: the ArcInfo Method*. Redlands, CA: ESRI Press.

MacDonald A. 1999 *Building a Geodatabase*. Redlands, CA: ESRI Press.

Worboys M.F. and Duckham M. 2004 *GIS: A Computing Perspective* (2nd edn). Boca Raton, FL: CRC Press.

Zeiler M. 1999 *Modeling Our World: The ESRI Guide to Geodatabase Design*. Redlands, CA: ESRI Press.

9 *GIS data collection*

Data collection is one of the most time-consuming and expensive, yet important, of GIS tasks. There are many diverse sources of geographic data and many methods available to enter them into a GIS. The two main methods of data collection are data capture and data transfer. It is useful to distinguish between primary (direct measurement) and secondary (derivation from other sources) data capture for both raster and vector data types. Data transfer involves importing digital data from other sources. There are many practical issues associated with planning and executing an effective GIS data collection plan. This chapter reviews the main methods of GIS data capture and transfer and introduces key practical management issues.

Geographic Information Systems and Science, 2nd edition   Paul Longley, Michael Goodchild, David Maguire, and David Rhind.
© 2005 John Wiley & Sons, Ltd.   ISBNs: 0-470-87000-1 (HB); 0-470-87001-X (PB)

Learning Objectives

After reading this chapter you will be able to:

- Describe data collection workflows;

- Understand the primary data capture techniques in remote sensing and surveying;

- Be familiar with the secondary data capture techniques of scanning, manual digitizing, vectorization, photogrammetry, and COGO feature construction;

- Understand the principles of data transfer, sources of digital geographic data, and geographic data formats;

- Analyze practical issues associated with managing data capture projects.

9.1 Introduction

GIS can contain a wide variety of geographic data types originating from many diverse sources. Data collection activities for the purposes of organizing the material in this chapter are split into data capture (direct data input) and data transfer (input of data from other systems). From the perspective of creating geographic databases, it is convenient to classify raster and vector geographic data as primary and secondary (Table 9.1). Primary data sources are those collected in digital format specifically for use in a GIS project. Typical examples of primary GIS sources include raster SPOT and IKONOS Earth satellite images, and vector building-survey measurements captured using

Table 9.1 Classification of geographic data for data collection purposes with examples of each type

	Raster	Vector
Primary	Digital satellite remote-sensing images	GPS measurements
	Digital aerial photographs	Survey measurements
Secondary	Scanned maps or photographs	Topographic maps
	Digital elevation models from topographic map contours	Toponymy (placename) databases

a total survey station. Secondary sources are digital and analog datasets that were originally captured for another purpose and need to be converted into a suitable digital format for use in a GIS project. Typical secondary sources include raster scanned color aerial photographs of urban areas and United States Geological Survey (USGS) or Institut Géographique National, France (IGN) paper maps that can be scanned and vectorized. This classification scheme is a useful organizing framework for this chapter and, more importantly, it highlights the number of processing-stage transformations that a dataset goes through, and therefore the opportunities for errors to be introduced. However, the distinctions between primary and secondary, and raster and vector, are not always easy to determine. For example, is digital satellite remote sensing data obtained on a DVD primary or secondary? Clearly the commercial satellite sensor feeds do not run straight into GIS databases, but to ground stations where the data are pre-processed onto digital media. Here it is considered primary because usually the data has undergone only minimal transformation since being collected by the satellite sensors and because the characteristics of the data make them suitable for virtually direct use in GIS projects.

Primary geographic data sources are captured specifically for use in GIS by direct measurement. Secondary sources are those reused from earlier studies or obtained from other systems.

Both primary and secondary geographic data may be obtained in either digital or analog format (see Section 3.7 for a definition of analog). Analog data must always be digitized before being added to a geographic database. Analog to digital transformation may involve the scanning of paper maps or photographs, optical character recognition (OCR) of text describing geographic object properties, or the vectorization of selected features from an image. Depending on the format and characteristics of the digital data, considerable reformatting and restructuring may be required prior to importing into a GIS. Each of these transformations alters the original data and will introduce further uncertainty into the data (see Chapter 6 for discussion of uncertainty).

This chapter describes the data sources, techniques, and workflows involved in GIS data collection. The processes of data collection are also variously referred to as data capture, data automation, data conversion, data transfer, data translation, and digitizing. Although there are subtle differences between these terms, they essentially describe the same thing, namely, adding geographic data to a database. Data capture refers to direct entry. Data transfer is the importing of existing digital data across a network connection (Internet, wide area network (WAN), or local area network (LAN)) or from physical media such as CD ROMs, zip disks, or diskettes. This chapter focuses on the techniques of data collection; of equal, perhaps more, importance to a real-world GIS implementation are project management, cost, legal, and organization issues. These are covered briefly in Section 9.6 of this chapter as a prelude to more detailed treatment in Chapters 17 through 20.

Table 9.2 Breakdown of costs (in $1000s) for two typical client-server GIS as estimated by the authors

	10 seats		100 seats	
	$	%	$	%
Hardware	30	3.4	250	8.6
Software	25	2.8	200	6.9
Data	400	44.7	450	15.5
Staff	440	49.1	2000	69.0
Total	895	100	2900	100

Table 9.2 shows a breakdown of costs (in $1000s) for two typical client-server GIS implementations: one with 10 seats (systems) and the other with 100. The hardware costs include desktop clients and servers only (i.e., not network infrastructure). The data costs assume the purchase of a landbase (e.g., streets, parcels, and land marks) and digitizing assets such as pipes and fittings (water utility), conductors and devices (electrical utility), or land and property parcels (local government). Staff costs assume that all core GIS staff will be full-time, but that users will be part-time.

In the early days of GIS, when geographic data were very scarce, data collection was the main project task and typically it consumed the majority of the available resources. Even today data collection still remains a time-consuming, tedious, and expensive process. Typically it accounts for 15–50% of the total cost of a GIS project (Table 9.2). Data capture costs can in fact be much more significant because in many organizations (especially those that are government funded) staff costs are often assumed to be fixed and are not used in budget accounting. Furthermore, as the majority of data capture effort and expense tends to fall at the start of projects, data capture costs often receive greater scrutiny from senior managers. If staff costs are excluded from a GIS budget then in cash expenditure terms data collection can be as much as 60–85% of costs.

Data capture costs can account for up to 85% of the cost of a GIS.

After an organization has completed basic data collection tasks, the focus of a GIS project moves on to data maintenance. Over the multi-year lifetime of a GIS project, data maintenance can turn out to be a far more complex and expensive activity than initial data collection. This is because of the high volume of update transactions in many systems (for example, changes in land parcel ownership, maintenance work orders on a highway transport network, or logging military operational activities) and the need to manage multi-user access to operational databases. For more information about data maintenance, see Chapter 10.

9.1.1 Data collection workflow

In all but the simplest of projects, data collection involves a series of sequential stages (Figure 9.1). The workflow

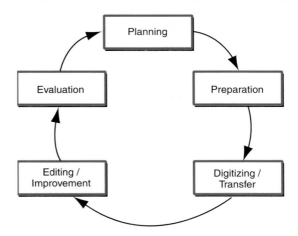

Figure 9.1 Stages in data collection projects

commences with planning, followed by preparation, digitizing/transfer (here taken to mean a range of primary and secondary techniques such as table digitizing, survey entry, scanning, and photogrammetry), editing and improvement and, finally, evaluation.

Planning is obviously important to any project and data collection is no exception. It includes establishing user requirements, garnering resources (staff, hardware, and software), and developing a project plan. Preparation is especially important in data collection projects. It involves many tasks such as obtaining data, redrafting poor-quality map sources, editing scanned map images, and removing noise (unwanted data such as speckles on a scanned map image). It may also involve setting up appropriate GIS hardware and software systems to accept data. Digitizing and transfer are the stages where the majority of the effort will be expended. It is naïve to think that data capture is really just digitizing, when in fact it involves very much more as discussed below. Editing and improvement follows digitizing/transfer. This covers many techniques designed to validate data, as well as correct errors and improve quality. Evaluation, as the name suggests, is the process of identifying project successes and failures. These may be qualitative or quantitative. Since all large data projects involve multiple stages, this workflow is iterative with earlier phases (especially a first, pilot, phase) helping to improve subsequent parts of the overall project.

9.2 Primary geographic data capture

Primary geographic capture involves the direct measurement of objects. Digital data measurements may be input directly into the GIS database, or can reside in a temporary file prior to input. Although the former is preferable as it minimizes the amount of time and the possibility of errors, close coupling of data collection devices and GIS databases is not always possible. Both raster and vector GIS primary data capture methods are available.

9.2.1 Raster data capture

Much the most popular form of primary raster data capture is remote sensing. Broadly speaking, remote sensing is a technique used to derive information about the physical, chemical, and biological properties of objects without direct physical contact (Section 3.6). Information is derived from measurements of the amount of electromagnetic radiation reflected, emitted, or scattered from objects. A variety of sensors, operating throughout the electromagnetic spectrum from visible to microwave wavelengths, are commonly employed to obtain measurements (see Section 3.6.1). Passive sensors are reliant on reflected solar radiation or emitted terrestrial radiation; active sensors (such as synthetic aperture radar) generate their own source of electromagnetic radiation. The platforms on which these instruments are mounted are similarly diverse. Although Earth-orbiting satellites and fixed-wing aircraft are by far the most common, helicopters, balloons, masts, and booms are also employed (Figure 9.2). As used here, the term *remote sensing* subsumes the fields of satellite remote sensing and aerial photography.

Remote sensing is the measurement of physical, chemical, and biological properties of objects without direct contact.

From the GIS perspective, resolution is a key physical characteristic of remote sensing systems. There are three aspects to resolution: spatial, spectral, and temporal. All sensors need to trade off spatial, spectral, and temporal properties because of storage, processing, and bandwidth considerations. For further discussion of the important topic of resolution see also Sections 3.4, 3.6.1, 4.1, 6.4.2, 7.1, and 16.1.

Three key aspects of resolution are: spatial, spectral, and temporal.

Spatial resolution refers to the size of object that can be resolved and the most usual measure is the pixel size. Satellite remote sensing systems typically provide data with pixel sizes in the range 0.5 m–1 km. The resolution of cameras used for capturing aerial photographs usually ranges from 0.1 m–5 m. Image (scene) sizes vary quite widely between sensors – typical ranges include 900 by 900 to 3000 by 3000 pixels. The total coverage of remote sensing images is usually in the range 9 by 9 to 200 by 200 km.

Spectral resolution refers to the parts of the electromagnetic spectrum that are measured. Since different objects emit and reflect different types and amounts of radiation, selecting which part of the electromagnetic spectrum to measure is critical for each application area. Figure 9.3 shows the spectral signatures of water, green vegetation, and dry soil. Remote sensing systems may capture data in one part of the spectrum (referred to as a single band) or simultaneously from several parts (multiband or multi-spectral). The radiation values are usually normalized and resampled to give a range of integers from 0–255 for each band (part of the electromagnetic spectrum measured), for each pixel, in each image. Until recently, remote sensing satellites typically measured a small number of bands, in the visible part of the spectrum. More recently a number of hyperspectral systems have come into operation that measure very large numbers of bands across a much wider part of the spectrum.

Temporal resolution, or repeat cycle, describes the frequency with which images are collected for the same area. There are essentially two types of commercial remote sensing satellite: Earth-orbiting and geostationary. Earth-orbiting satellites collect information about different parts of the Earth surface at regular intervals. To maximize utility, typically orbits are polar, at a fixed altitude and speed, and are Sun synchronous.

The French SPOT (Système Probatoire d'Observation de la Terre) 5 satellite launched in 2002, for example, passes virtually over the poles at an altitude of 822 km sensing the same location on the Earth surface during daylight every 26 days. The SPOT platform carries multiple sensors: a panchromatic sensor measuring radiation in the visible part of the electromagnetic spectrum at a spatial resolution of 2.5 by 2.5 m; a multi-spectral sensor measuring green, red, and reflected infrared radiation at a spatial resolution of 10 by 10 m; a shortwave near-infrared sensor with a resolution of 20 by 20 m; and a vegetation sensor measuring four bands at a spatial resolution of 1000 m. The SPOT system is also able to provide stereo images from which digital terrain models and 3-D measurements can be obtained. Each SPOT scene covers an area of about 60 by 60 km.

Much of the discussion so far has focused on commercial satellite remote sensing systems. Of equal importance, especially in medium- to large (coarse)-scale GIS projects, is aerial photography. Although the data products resulting from remote sensing satellites and aerial photography systems are technically very similar (i.e., they are both images) there are some significant differences in the way data are captured and can, therefore, be interpreted. The most notable difference is that aerial photographs are normally collected using analog optical cameras (although digital cameras are becoming more widely used) and then later rasterized, usually by scanning a film negative. The quality of the optics of the camera and the mechanics of the scanning process both affect the spatial and spectral characteristics of the resulting images. Most aerial photographs are collected on an *ad hoc* basis using cameras mounted in airplanes flying at low altitudes (3000–9000 m) and are either panchromatic (black and white) or color, although multi-spectral cameras/sensors operating in the non-visible parts of the electromagnetic spectrum are also used. Aerial photographs are very suitable for detailed surveying and mapping projects.

An important feature of satellite and aerial photography systems is that they can provide stereo imagery from overlapping pairs of images. These images are used to create a 3-D analog or digital model from which 3-D coordinates, contours, and digital elevation models can be created (see Section 9.3.2.4).

Satellite and aerial photograph data offer a number of advantages for GIS projects. The consistency of the data and the availability of systematic global coverage

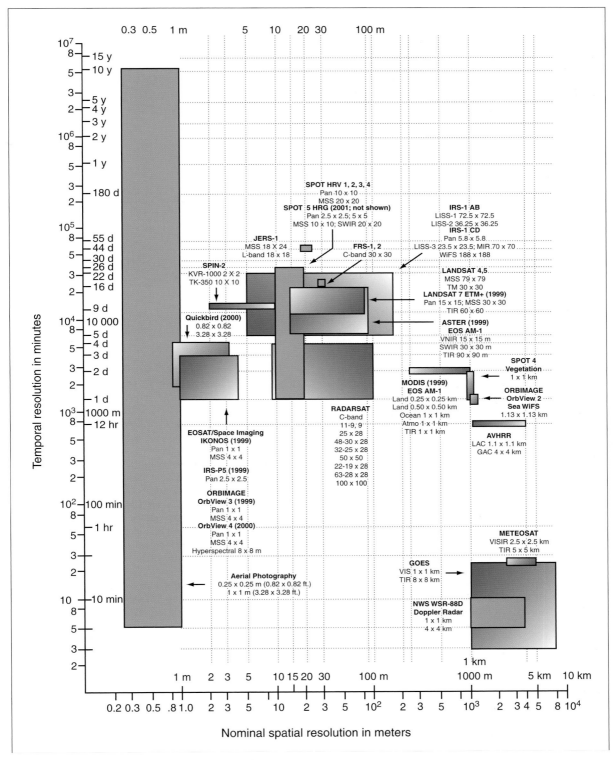

Figure 9.2 Spatial and temporal characteristics of commonly used remote sensing systems and their sensors (*Source*: after Jensen J.R. and Cowen D.C. 1999 'Remote sensing of urban/suburban infrastructure and socioeconomic attributes', *Photogrammetric Engineering and Remote Sensing* **65**, 611–622)

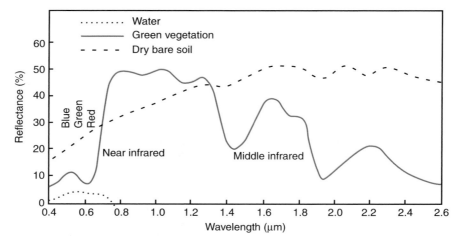

Figure 9.3 Typical reflectance signatures for water, green vegetation, and dry soil (*Source*: after Jones C. 1997 *Geographic Information Systems and Computer Cartography*. Reading, MA: Addison-Wesley Longman)

make satellite data especially useful for large-area, small-scale projects (for example, mapping landforms and geology at the river catchment-area level) and for mapping inaccessible areas. The regular repeat cycles of commercial systems and the fact that they record radiation in many parts of the spectrum make such data especially suitable for assessing the condition of vegetation (for example, the moisture stress of wheat crops). Aerial photographs in particular are very useful for detailed surveying and mapping of, for example, urban areas and archaeological sites, especially those applications requiring 3-D data (see Chapter 12).

On the other hand, the spatial resolution of commercial satellites is too coarse for many large-scale projects and the data collection capability of many sensors is restricted by cloud cover. Some of this is changing, however, as the new generation of satellite sensors now provide data at 0.6 m spatial resolution and better, and radar data can be obtained that are not affected by cloud cover. The data volumes from both satellites and aerial cameras can be very large and create storage and processing problems for all but the most modern systems. The cost of data can also be prohibitive for a single project or organization.

9.2.2 Vector data capture

Primary vector data capture is a major source of geographic data. The two main branches of vector data capture are ground surveying and GPS – which is covered in Section 5.8 – although as more surveyors use GPS routinely the distinction between the two is becoming increasingly blurred.

Ground surveying is based on the principle that the 3-D location of any point can be determined by measuring angles and distances from other known points. Surveys begin from a benchmark point. If the coordinate system of this point is known, all subsequent points can be collected in this coordinate system. If it is unknown then the survey will use a local or relative coordinate system (see Section 5.7).

Since all survey points are obtained from survey measurements, their known locations are always relative to other points. Any measurement errors need to be apportioned between multiple points in a survey. For example, when surveying a field boundary, if the last and first points are not identical in survey terms (within the tolerance employed in the survey) then errors need to be apportioned between all points that define the boundary (see Section 6.3.4). As new measurements are obtained these may change the locations of points.

Traditionally, surveyors used equipment like transits and theodolites to measure angles, and tapes and chains to measure distances. Today these have been replaced by electro-optical devices called total stations that can measure both angles and distances to an accuracy of 1 mm (Figure 9.4). Total stations automatically log data and the most sophisticated can create vector point, line, and area objects in the field, thus providing direct validation.

The basic principles of surveying have changed very little in the past 100 years, although new technology has considerably improved accuracy and productivity. Two people are usually required to perform a survey, one to operate the total station and the other to hold a reflective prism that is placed at the object being measured. On some remote-controlled systems a single person can control both the total station and the prism.

Ground survey is a very time-consuming and expensive activity, but it is still the best way to obtain highly accurate point locations. Surveying is typically used for capturing buildings, land and property boundaries, manholes, and other objects that need to be located accurately. It is also employed to obtain reference marks for use in other data capture projects. For example, large-scale aerial photographs and satellite images are frequently georeferenced using points obtained from ground survey.

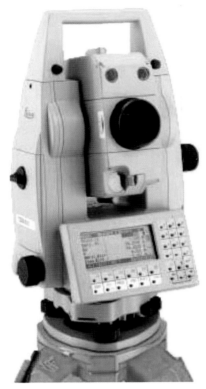

Figure 9.4 A tripod-mounted Leica TPS1100 Total Station (Courtesy: Leica Geosystems)

Figure 9.5 A large-format roll-feed image scanner (Reproduced by permission of GTCO Calcomp, Inc.)

9.3 Secondary geographic data capture

Geographic data capture from secondary sources is the process of creating raster and vector files and databases from maps, photographs, and other hard-copy documents. Scanning is used to capture raster data. Table digitizing, heads-up digitizing, stereo-photogrammetry, and COGO data entry are used for vector data.

9.3.1 Raster data capture using scanners

A scanner is a device that converts hard-copy analog media into digital images by scanning successive lines across a map or document and recording the amount of light reflected from a local data source (Figure 9.5). The differences in reflected light are normally scaled into bi-level black and white (1 bit per pixel), or multiple gray levels (8, 16, or 32 bits). Color scanners output data into 8-bit red, green, and blue color bands. The spatial resolution of scanners varies widely from as little as 200 dpi (8 dots per mm) to 2400 dpi (96 dots per mm) and beyond. Most GIS scanning is in the range 400–900 dpi (16–40 dots per mm). Depending on the type of scanner

and the resolution required, it can take from 30 seconds to 30 minutes or more to scan a map.

Scanned maps and documents are used extensively in GIS as background maps and data stores.

There are three main reasons to scan hardcopy media for use in GIS:

- Documents, such as building plans, CAD drawings, property deeds, and equipment photographs are scanned to reduced wear and tear, improve access, provide integrated database storage, and to index them geographically (e.g., building plans can be attached to building objects in geographic space).

- Film and paper maps, aerial photographs, and images are scanned and georeferenced so that they provide geographic context for other data (typically vector layers). This type of unintelligent image or background geographic wall-paper is very popular in systems that manage equipment and land and property assets (Figure 9.6).

- Maps, aerial photographs, and images are scanned prior to vectorization (see below), and sometimes as a prelude to spatial analysis.

An 8 bit (256 gray level) 400 dpi (16 dots per mm) scanner is a good choice for scanning maps for use as a background GIS reference layer. For a color aerial

Figure 9.6 An example of raster background data (black and white aerial photography) underneath vector data (land parcels)

photograph that is to be used for subsequent photo-interpretation and analysis, a color (8 bit for each of three bands) 900 dpi (40 dots per mm) scanner is more appropriate. The quality of data output from a scanner is determined by the nature of the original source material, the quality of the scanning device, and the type of preparation prior to scanning (e.g., redrafting key features or removing unwanted marks will improve output quality).

9.3.2 Vector data capture

Secondary vector data capture involves digitizing vector objects from maps and other geographic data sources. The most popular methods are manual digitizing, heads-up digitizing and vectorization, photogrammetry, and COGO data entry.

9.3.2.1 Manual digitizing

Manually operated digitizers are much the simplest, cheapest, and most commonly used means of capturing vector objects from hardcopy maps. Digitizers come in several designs, sizes, and shapes. They operate on the principle that it is possible to detect the location of a cursor or puck passed over a table inlaid with a fine mesh of wires. Digitizing table accuracies typically range from 0.0004 inch (0.01 mm) to 0.01 inch (0.25 mm). Small digitizing tablets up to 12 by 24 inches (30 by 60 cm)

(A)

(B)

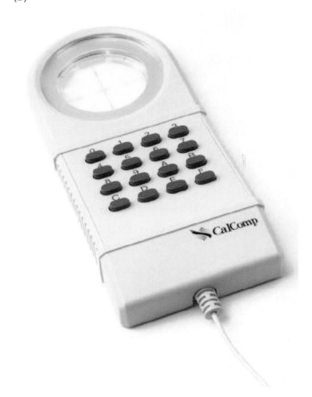

Figure 9.7 Digitizing equipment: (A) Digitizing table, (B) cursor (Reproduced by permission of GTCO Calcomp, Inc.)

are used for small tasks, but bigger (typically 44 by 60 inches (112 by 152 cm)) freestanding table digitizers are preferred for larger tasks (Figure 9.7). Both types of digitizer usually have cursors with cross hairs mounted in glass and buttons to control capture. Box 9.1 describes the process of table digitizing.

Manual digitizing is still the simplest, easiest, and cheapest method of capturing vector data from existing maps.

Manual digitizing

Manual digitizing involves five basic steps.

1. The map document is attached to the center of the digitizing table using sticky tape.

2. Because a digitizing table uses a local rectilinear coordinate system, the map and the digitizer must be registered so that vector data can be captured in real-world coordinates. This is achieved by digitizing a series of four or more well-distributed control points (also called reference points or tick marks) and then entering their real-world values. The digitizer control software (usually the GIS) will calculate a transformation and then automatically apply this to any future coordinates that are captured.

3. Before proceeding with data capture it is useful to spend some time examining a map to determine rules about which features are to be captured at what level of

generalization. This type of information is often defined in a data capture project specification.

4. Data capture involves recording the shape of vector objects using manual or stream mode digitizing as described in Section 9.3.2.1. A common rule for vector GIS is to press Button 2 on the digitizing cursor to start a line, Button 1 for each intermediate vertex, and Button 2 to finish a line. There are other similar rules to control how points and polygons are captured.

5. Finally, after all objects have been captured it is necessary to check for any errors. Easy ways to do this include using software to identify geometric errors (such as polygons that do not close or lines that do not intersect – see Figure 9.9), and producing a test plot that can be overlaid on the original document.

Vertices defining point, line, and polygon objects are captured using manual or stream digitizing methods. Manual digitizing involves placing the center point of the cursor cross hairs at the location for each object vertex and then clicking a button on the cursor to record the location of the vertex. Stream-mode digitizing partially automates this process by instructing the digitizer control to collect vertices automatically every time a distance or time threshold is crossed (e.g., every 0.02 inch (0.5 mm) or 0.25 second). Stream-mode digitizing is a much faster method, but it typically produces larger files with many redundant coordinates.

9.3.2.2 Heads-up digitizing and vectorization

One of the main reasons for scanning maps (see Section 9.3.1) is as a prelude to vectorization – the process of converting raster data into vector data. The simplest way to create vectors from raster layers is to digitize vector objects manually straight off a computer screen using a mouse or digitizing cursor. This method is called heads-up digitizing because the map is vertical and can be viewed without bending the head down. It is widely used for the selective capture of, for example, land parcels, buildings, and utility assets.

Vectorization is the process of converting raster data into vector data. The reverse is called rasterization.

A faster and more consistent approach is to use software to perform automated vectorization in either

batch or semi-interactive mode. Batch vectorization takes an entire raster file and converts it to vector objects in a single operation. Vector objects are created using software algorithms that build simple (spaghetti) line strings from the original pixel values. The lines can then be further processed to create topologically correct polygons (Figure 9.8). A typical map will take only a few minutes to vectorize using modern hardware and software systems. See Section 10.7.1 for further discussion on structuring geographic data.

Unfortunately, batch vectorization software is far from perfect and post-vectorization editing is usually required to clean up errors. To avoid large amounts of vector editing, it is useful to undertake a little raster editing of the original raster file prior to vectorization to remove unwanted noise that may affect the vectorization process. For example, text that overlaps lines should be deleted and dashed lines are best converted to solid lines. Following vectorization, topological relationships are usually created for the vector objects. This process may also highlight some previously unnoticed errors that require additional editing.

Batch vectorization is best suited to simple bi-level maps of, for example, contours, streams, and highways. For more complicated maps and where selective vec-torization is required (for example, digitizing electric conductors and devices, or water mains and fittings off topographic maps), interactive vectorization (also called semi-automatic vectorization, line following, or tracing) is preferred. In interactive vectorization, software is used to automate digitizing. The operator snaps the cursor to a pixel, indicates a direction for line following, and

(A)

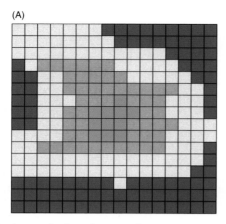

(B)

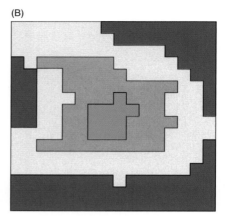

Figure 9.8 Batch vectorization of a scanned map: (A) original raster file; (B) vectorized polygons. Adjacent raster cells with the same attribute values are aggregated. Class boundaries are then created at the intersection between adjacent classes in the form of vector lines

the software then automatically digitizes lines. Typically, many parameters can be tuned to control the density of points (level of generalization), the size of gaps (blank pixels in a line) that will be jumped, and whether to pause at junctions for operator intervention or always to trace in a specific direction (most systems require that all polygons are ordered either clockwise or counterclockwise). Interactive vectorization is still quite labor intensive, but generally it results in much greater productivity than manual or heads-up digitizing. It also produces high-quality data, as software is able to represent lines more accurately and consistently than can humans. It is for these reasons that specialized data capture groups much prefer vectorization to manual digitizing.

9.3.2.3 Measurement error

Data capture, like all geographic workflows, is likely to generate errors. Because digitizing is a tedious and hence error-prone practice, it presents a source of measurement errors – as when the operator fails to position the cursor correctly, or fails to record line segments. Figure 9.9 presents some examples of human errors that are commonly introduced in the digitizing procedure. They are: overshoots and undershoots where line intersections are inexact (Figure 9.9A); invalid polygons which are topologically inconsistent because of omission of one or more lines, or omission of tag data (Figure 9.9B); and sliver polygons, in which multiple digitizing of the common boundary between adjacent polygons leads to the creation of additional polygons (Figure 9.9C).

Most GIS packages include standard software functions, which can be used to restore integrity and clean (or rather obscure, depending upon your viewpoint!) obvious measurement errors. Such operations are best carried out immediately after digitizing, in order that omissions may be easily rectified. Data cleaning operations require sensitive setting of threshold values, or else damage can be done to real-world features, as Figure 9.10 shows.

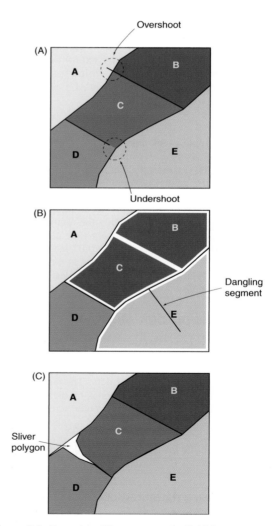

Figure 9.9 Examples of human errors in digitizing: (A) undershoots and overshoots; (B) invalid polygons; and (C) sliver polygons

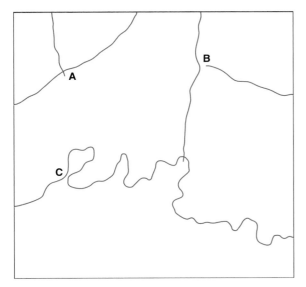

Figure 9.10 Error induced by data cleaning. If the tolerance level is set large enough to correct the errors at A and B, the loop at C will also (incorrectly) be closed

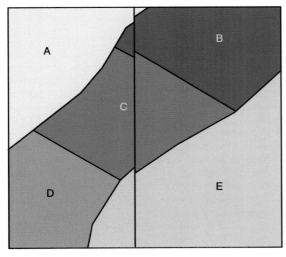

Figure 9.11 Mismatches of adjacent spatial data sources that require rubber-sheeting

Many errors in digitizing can be remedied by appropriately designed software.

Further classes of problems arise when the products of digitizing adjacent map sheets are merged together. Stretching of paper base maps, coupled with errors in rectifying them on a digitizing table, give rise to the kinds of mismatches shown in Figure 9.11. Rubber-sheeting is the term used to describe methods for removing such errors on the assumption that strong spatial autocorrelation exists among errors. If errors tend to be spatially autocorrelated up to a distance of x, say, then rubber-sheeting will be successful at removing them, at least partially, provided control points can be found that are

spaced less than x apart. For the same reason, the shapes of features that are less than x across will tend to have little distortion, while very large shapes may be badly distorted. The results of calculating areas (Section 14.3), or other geometric operations that rely only on relative position, will be accurate as long as the areas are small, but will grow rapidly with feature size. Thus it is important for the user of a GIS to know which operations depend on relative position, and over what distance; and where absolute position is important (of course, the term *absolute* simply means relative to the Earth frame, defined by the Equator and the Greenwich Meridian, or relative over a very long distance: see Section 5.6). Analogous procedures and problems characterize the rectification of raster datasets – be they scanned images of paper maps or satellite measurements of the curved Earth surface.

9.3.2.4 Photogrammetry

Photogrammetry is the science and technology of making measurements from pictures, aerial photographs, and images. Although in the strict sense it includes 2-D measurements taken from single aerial photographs, today in GIS it is almost exclusively concerned with capturing 2.5-D and 3-D measurements from models derived from stereo-pairs of photographs and images. In the case of aerial photographs, it is usual to have 60% overlap along each flight line and 30% overlap between flight lines. Similar layouts are used by remote sensing satellites. The amount of overlap defines the area for which a 3-D model can be created.

Photogrammetry is used to capture measurements from photographs and other image sources.

To obtain true georeferenced Earth coordinates from a model, it is necessary to georeference photographs using control points (the procedure is essentially analogous to that described for manual digitizing in Box 9.1). Control points can be defined by ground survey or nowadays more usually with GPS (see Section 9.2.2.1 for discussion of these techniques).

Measurements are captured from overlapping pairs of photographs using stereoplotters. These build a model and allow 3-D measurements to be captured, edited, stored, and plotted. Stereoplotters have undergone three major generations of development: analog (optical), analytic, and digital. Mechanical analog devices are seldom used today, whereas analytical (combined mechanical and digital) and digital (entirely computer-based) are much more common. It is likely that digital (soft-copy) photogrammetry will eventually replace mechanical devices entirely.

There are many ways to view stereo models, including a split screen with a simple stereoscope, and the use of special glasses to observe a red/green display or polarized light. To manipulate 3-D cursors in the x, y, and z planes, photogrammetry systems offer free-moving hand controllers, hand wheels and foot disks, and 3-D mice. The options for extracting vector objects from 3-D models are directly analogous to those available for manual digitizing as described above: namely batch, interactive, and manual (Sections 9.3.2.1 and 9.3.2.2). The obvious

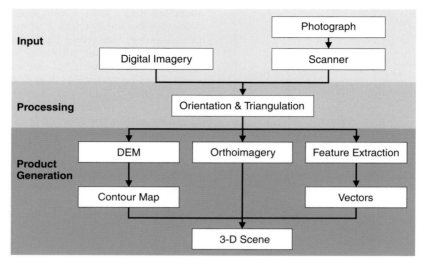

Figure 9.12 Typical photogrammetry workflow (after Tao C.V. 2002 'Digital photogrammetry: the future of spatial data collection', *GeoWorld*. **www.geoplace.com/gw/2002/0205/0205dp.asp**). (Reproduced by permission of GeoTec Media)

difference, however, is that there is a requirement for capturing *z* (elevation) values.

9.3.2.4.1 Digital photogrammetry workflow

Figure 9.12 shows a typical workflow in digital photogrammetry. There are three main parts to digital photogrammetry workflows: data input, processing, and product generation. Data can be obtained directly from sensors or by scanning secondary sources.

Orientation and triangulation are fundamental photogrammetry processing tasks. Orientation is the process of creating a stereo model suitable for viewing and extracting 3-D vector coordinates that describe geographic objects. Triangulation (also called 'block adjustment') is used to assemble a collection of images into a single model so that accurate and consistent information can be obtained from large areas.

Photogrammetry workflows yield several important product outputs including digital elevation models (DEMs), contours, orthoimages, vector features, and 3-D scenes. DEMs – regular arrays of height values – are created by 'matching' stereo image pairs together using a series of control points. Once a DEM has been created it is relatively straightforward to derive contours using a choice of algorithms. Orthoimages are images corrected for variations in terrain using a DEM. They have become popular because of their relatively low cost of creation (when compared with topographic maps) and ease of interpretation as base maps. They can also be used as accurate data sources for heads-up digitizing (see Section 9.3.2.2). Vector feature extraction is still an evolving field and there are no widely applicable fully automated methods. The most successful methods use a combination of spectral analysis and spatial rules that define context, shape, proximity, etc. Finally, 3-D scenes can be created by merging vector features with a DEM and an orthoimage (Figure 9.13).

In summary, photogrammetry is a very cost-effective data capture technique that is sometimes the only practical method of obtaining detailed topographic data about an area of interest. Unfortunately, the complexity and high cost of equipment have restricted its use to large-scale primary data capture projects and specialist data capture organizations.

9.3.2.5 COGO data entry

COGO, a contraction of the term *coordinate geometry*, is a methodology for capturing and representing geographic data. COGO uses survey-style bearings and distances to define each part of an object in much the same way as described in Section 9.2.2.1. Some examples of COGO object construction tools are shown in Figure 9.14. The Construct Along tool creates a point along a curve using a distance along the curve. The Line Construct Angle Bisector tool constructs a line that bisects an angle defined by a from-point, through-point, to-point, and a length. The Construct Fillet tool creates a circular arc tangent from two segments and a radius.

The COGO system is widely used in North America to represent land records and property parcels (also called lots). Coordinates can be obtained from COGO measurements by geometric transformation (i.e., bearings and distances are converted into *x*, *y* coordinates). Although COGO data obtained as part of a primary data capture activity are used in some projects, it is more often the case that secondary measurements are captured from hardcopy maps and documents. Source data may be in the form of legal descriptions, records of survey, tract (housing estate) maps, or similar documents.

> **COGO stands for coordinate geometry. It is a vector data structure and method of data entry.**

COGO data are very precise measurements and are often regarded as the only legally acceptable definition of land parcels. Measurements are usually very detailed and data capture is often time consuming. Furthermore, commonly occurring discrepancies in the data must be manually resolved by highly qualified individuals.

Figure 9.13 Example 3-D scene as generated from a photogrammetry workflow

Construct Along

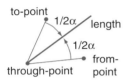

distance
(or ratio)

curve

Line Construct Angle Bisector

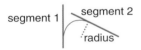

to-point

$1/2\alpha$ length

$1/2\alpha$

from-
point

through-point   point

Construct Fillet

segment 1   segment 2

radius

Figure 9.14 Example COGO construction tools used to represent geographic features

9.4 Obtaining data from external sources (data transfer)

One major decision that needs to be faced at the start of a GIS project is whether to build or buy part or all of a database. All the preceding discussion has been concerned with techniques for building databases from primary and secondary sources. This section focuses on how to import or transfer data into a GIS that has been captured by others. Some datasets are freely available, but many of them are sold as a commodity from a variety of outlets including, increasingly, Internet sites.

There are many sources and types of geographic data. Space does not permit a comprehensive review of all geographic data sources here, but a small selection of key sources is listed in Table 9.3. In any case, the characteristics and availability of datasets are constantly changing so those seeking an up-to-date list should consult one of the good online sources described below. Section 18.4.3 also discusses the characteristics of geographic information and highlights several issues to bear in mind when using data collected by others.

Table 9.3 Examples of some digital data sources that can be imported into a GIS. NMOs = National Mapping Organizations, USGS = United States Geologic Survey, NGA = US National Geospatial-Intelligence Agency, NASA = National Aeronautics and Space Administration, DEM = Digital Elevation Model, EPS = US Environmental Protection Agency, WWF = World Wildlife Fund for Nature, FEMA = Federal Emergency Management Agency, EBIS = ESRI Business Information Solutions

Type	Source	Details
Basemaps		
Geodetic framework	Many NMOs, e.g., USGS and Ordnance Survey	Definition of framework, map projections, and geodetic transformations
General topographic map data	NMOs and military agencies, e.g., NGA	Many types of data at detailed to medium scales
Elevation	NMOs, military agencies, and several commercial providers, e.g., USGS, SPOT Image, NASA	DEMs, contours at local, regional, and global levels
Transportation	National governments, and several commercial vendors, e.g., TeleAtlas and NAVTEQ	Highway/street centerline databases at national levels
Hydrology	NMOs and government agencies	National hydrological databases are available for many countries
Toponymy	NMOs, other government agencies and commercial providers	Gazetteers of placenames at global and national levels
Satellite images	Commercial and military providers, e.g., Landsat, SPOT, IRS, IKONOS, Quickbird	See Figure 9.2 for further details
Aerial photographs	Many private and public agencies	Scales vary widely, typically from 1:500–1:20 000
Environmental		
Wetlands	National agencies, e.g., US National Wetlands Inventory	Government wetlands inventory
Toxic release sites	National Environmental Protection Agencies, e.g., EPA	Details of thousands of toxic sites
World eco-regions	World Wildlife Fund for Nature (WWF)	Habitat types, threatened areas, biological distinctiveness
Flood zones	Many national and regional government agencies, e.g., FEMA	National flood risk areas
Socio-economic		
Population census	National governments, with value added by commercial providers	Typically every 10 years with annual estimates
Lifestyle classifications	Private agencies (e.g., CACI and Experian)	Derived from population censuses and other socio-economic data
Geodemographics	Private agencies (e.g., Claritas and EBIS)	Many types of data at many scales and prices
Land and property ownership	National governments	Street, property, and cadastral data
Administrative areas	National governments	Obtained from maps at scales of 1:5000–1:750 000

The best way to find geographic data is to search the Internet. Several types of resources and technologies are available to assist searching, and are described in detail in Section 11.2. These include specialist geographic data catalogs and stores, as well as the sites of specific geographic data vendors (some websites are shown in Table 9.4 and the history of one vendor is described in Box 9.2). Particularly good sites are the Data Store (**www.datastore.co.uk/**) and the AGI (Association for Geographic Information) Resource List (**www.geo.ed.ac.uk/home/giswww.html**). These sites provide access to information about the characteristics and availability of geographic data. Some also have facilities to purchase and download data directly. Probably the most useful resources for locating geographic data are the geolibraries and geoportals (see Section 11.2) that have

been created as part of national and global spatial data infrastructure initiatives (SDI).

> **The best way to find geographic data is to search the Internet using one of the specialist geolibraries or SDI geographic data geoportals.**

9.4.1 Geographic data formats

One of the biggest problems with data obtained from external sources is that they can be encoded in many different formats. There are so many different geographic data formats because no single format is appropriate for all tasks and applications. It is not possible to design a format that supports, for example, both fast rendering in police

Don Cooke, geographic data provider

Don Cooke (Figure 9.15) took a part-time job with the New Haven Census Use Study while finishing his senior year at Yale in 1967. Cooke's three years of Army artillery survey plus an introductory Fortran class gave him GIS credentials typical of most people in the field at the time. Cooke and Bill Maxfield were charged with making computer maps of census and local data. It quickly became apparent that computerized base maps linking census geometry, street addressing, and coordinates were a prerequisite to computer mapping. DIME (Dual Independent Map Encoding) was their solution, probably the first implementation of a topological data structure with redundant encoding for error correction. Cooke, Maxfield, and Jack Sweeney founded Urban Data Processing, Inc. (UDP) in 1968 to bring geocoding, computer mapping, and demographic analysis to the private sector. The Census Bureau adopted DIME which evolved into the nationwide TIGER database during the 1980s.

Figure 9.15 Don Cooke, geographic data provider

When Harte-Hanks bought UDP in 1980, Cooke founded Geographic Data Technology (GDT) to commercialize Census DIME and later TIGER files. By the late 1990s, GDT had grown to 500 employees and in 2004 was acquired by TeleAtlas. Cooke remains in his role as Founder, and the TeleAtlas North America operation (effectively a combination of ETAK and GDT) faces NAVTEQ as a competitor in GIS and Navigation markets.

Cooke served on the National Academy of Sciences Mapping Science Committee for four years, and on the Board of the Urban and Regional Information Systems Association (URISA), where he founded the first Special Interest Group (SIG) focusing on GIS. He is an active proponent of GIS and GPS technology in education at all levels. His leadership in this area helped GDT win 'School-to-Careers Company of the Year' recognition from the National Alliance of Business. He is the author of 'Fun with GPS', a GIS primer written for owners of consumer GPS receivers.

On the subject of the current state of GIS Don says: 'Suddenly it seems really easy to explain what GIS is. Most people have some contact or context; they've used MapQuest or know someone who has a GPS. People with GPS think nothing of finding mapping services through Google; they overlay their GPS tracks and points on USGS Digital Raster Graphics and Digital Ortho Quarter Quads without even knowing those terms or messing with GeoSpatial One-Stop (see Box 11.4).'

Thinking about the future, he muses: 'I like to picture a near-term future where every high-school graduate has collected GPS data for a project and mapped it with GIS; we're already there in some schools. The best thing about this is more often than not their mapping has been for a community project and they've seen through the experience how they can participate in and contribute to their community.'

command and control systems, and sophisticated topological analysis in natural resource information systems: the two are mutually incompatible. Also, given the great diversity of geographic information a single comprehensive format would simply be too large and cumbersome. The many different formats that are in use today have evolved in response to diverse user requirements.

Given the high cost of creating databases, many tools have been developed to move data between systems and to reuse data through open application programming interfaces (APIs). In the former case, the approach has been to develop software that is able to translate data (Figure 9.16), either by a direct read into memory, or via an intermediate file format. In the latter case, software developers have created open interfaces to allow access to data.

Many GIS software systems are now able to read directly AutoCAD DWG and DXF, Microstation DGN, and Shapefile, VPF, and many image formats. Unfortunately, direct read support can only easily be provided for relatively simple product-oriented formats. Complex formats, such as SDTS, were designed for exchange purposes and require more advanced processing before they can be viewed (e.g., multi-pass read and feature assembly from several parts).

Data can be transferred between systems by direct read into memory or via an intermediate file format.

More than 25 organizations are involved in the standardization of various aspects of geographic data and geoprocessing; several of them are country and domain

Table 9.4 Selected websites containing information about geographic data sources

Source	URL	Description
AGI GIS Resource List	**www.geo.ed.ac.uk/home/giswww.html**	Indexed list of several hundred sites
The Data Store	**www.data-store.co.uk/**	UK, European, and worldwide data catalog
Geospatial One-Stop	**www.geodata.gov**	Geoportal providing metadata and direct access to over 50 000 datasets
MapMart	**www.mapmart.com/**	Extensive data and imagery provider
EROS Data Center	**edc.usgs.gov/**	US government data archive
Terraserver	**www.terraserver-usa.com/**	High-resolution aerial imagery and topo maps
Geography Network	**www.GeographyNetwork.com**	Global online data and map services
National Geographic Society	**www.nationalgeographic.com**	Worldwide maps
GeoConnections	**www.connect.gc.ca/en/692-e.asp**	Canadian government's geographic data over the Web
EuroGeographics	**www.eurogeographics.org/eng/ 01_about.asp**	Coalition of European NMOs offering topographic map data
GEOWorld Data Directory	**www.geoplace.com**	List of GIS data companies
The Data Depot	**www.gisdatadepot.com**	Extensive collection of mainly free geographic data depot

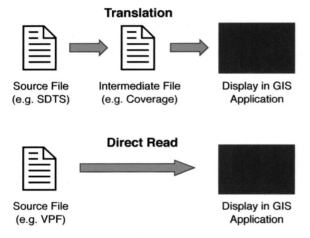

Figure 9.16 Comparison of data access by translation and direct read

specific. At the global level, the ISO (International Standards Organization) is responsible for coordinating efforts through the work of technical committees TC 211 and 287. In Europe, CEN (Comité Européen de Normalisation) is engaged in geographic standardization. At the national level, there are many complementary bodies. One other standards-forming organization of particular note is OGC (Open Geospatial Consortium: **www.opengeospatial.org**), a group of vendors, academics, and users interested in the interoperability of geographic systems (see Box 11.1). To date there have been promising OGC-coordinated efforts to standardize on simple feature access (simple geometric object types), metadata catalogs, and Web access.

The most efficient way to translate data between systems is usually via a common intermediate file format.

Having obtained a potentially useful source of geographic information the next task is to import it into a GIS database. If the data are already in the native format of the target GIS software system, or the software has a direct read capability for the format in question, then this is a relatively straightforward task. If the data are not compatible with the target GIS software then the alternatives are to ask the data supplier to convert the data to a compatible format, or to use a third-party translation software system, such as the Feature Manipulation Engine from Safe Software (**www.safe.com** lists over 60 supported geographic data formats) to convert the data. Geographic data translation software must address both syntactic and semantic translation issues. Syntactic translation involves converting specific digital symbols (letters and numbers) between systems. Semantic translation is concerned with converting the meaning inherent in geographic information. While the former is relatively simple to encode and decode, the latter is much more difficult and has seldom met with much success to date.

Although the task of translating geographic information between systems was described earlier as relatively straightforward, those that have tried this in practice will realize that things on the ground are seldom quite so simple. Any number of things can (and do!) go wrong. These range from corrupted media, to incomplete data files, wrong versions of translators, and different interpretations of a format specification, to basic user error.

There are two basic strategies used for data translation: one is direct and the other uses a neutral intermediate format. For small systems that involve the translation of a small number of formats, the first is the simplest. Directly translating data back and forth between the internal structures of two systems requires two new translators (A to B, B to A). Adding two further systems will require 12 translators to share data between all systems (A to B, A to C, A to D, B to A, B to C, B to D, C to A, C to B, C to D, D to A, D to B, and D to C). A more efficient way of solving this problem is to use the concept of a data *switchyard* and a common intermediate file format. Systems now need only to translate to and from the common format. The four systems will now need only eight translators instead of 12 (A to Neutral, B to Neutral, C to Neutral, D to Neutral, Neutral to A, Neutral to B, Neutral to C, and Neutral to D). The more systems there are the more efficient this becomes. This is one of the key principles underlying the need for common file interchange formats.

9.5 Capturing attribute data

All geographic objects have attributes of one type or another. Although attributes can be collected at the same time as vector geometry, it is usually more cost-effective to capture attributes separately. In part, this is because attribute data capture is a relatively simple task that can be undertaken by lower-cost clerical staff. It is also because attributes can be entered by direct data loggers, manual keyboard entry, optical character recognition (OCR) or, increasingly, voice recognition, which do not require expensive hardware and software systems. Much the most common method is direct keyboard data entry into a spreadsheet or database. For some projects, a custom data entry form with in-built validation is preferred. On small projects single entry is used, but for larger, more complex projects data are entered twice and then compared as a validation check.

An essential requirement for separate data entry is a common identifier (also called a key) that can be used to relate object geometry and attributes together following data capture (see Figure 10.2 for a diagrammatic explanation of relating geometry and attributes).

Metadata are a special type of non-geometric data that are increasingly being collected. Some metadata are derived automatically by the GIS software system (for example, length and area, extent of data layer, and count of features), but some must be explicitly collected (for example, owner name, quality estimate, and original source). Explicitly collected metadata can be entered in the same way as other attributes, as described above. For further information about metadata see Section 11.2.

9.6 Managing a data collection project

The subject of managing a GIS project is given extensive treatment later in this book in Chapters 17–20. The management of data capture projects is discussed briefly here both because of its critical importance and because there are several unique issues. That said, most of the general principles for any GIS project apply to data collection: the need for a clearly articulated plan, adequate resources, appropriate funding, and sufficient time.

Figure 9.17 Relationship between quality, speed, and price in data collection (*Source*: after Hohl 1997)

In any data collection project there is a fundamental tradeoff between quality, speed, and price. Collecting high-quality data quickly is possible, but it is also very expensive. If price is a key consideration then lower-quality data can be collected over a longer period (Figure 9.17).

GIS data collection projects can be carried out intensively or over a longer period. A key decision facing managers of such projects is whether to pursue a strategy of incremental or very rapid collection. Incremental data collection involves breaking the data collection project into small manageable subprojects. This allows data collection to be undertaken with lower annual resource and funding levels (although total project resource requirements may be larger). It is a good approach for inexperienced organizations that are embarking on their first data collection project because they can learn and adapt as the project proceeds. On the other hand, these longer-term projects run the risk of employee turnover and burnout, as well as changing data, technology, and organizational priorities.

Whichever approach is preferred, a pilot project carried out on part of the study area and a selection of the data types can prove to be invaluable. A pilot project can identify problems in workflow, database design, personnel, and equipment. A pilot database can also be used to test equipment and to develop procedures for quality assurance. Many projects require a test database for hardware and software acceptance tests, as well as to facilitate software customization. It is essential that project managers are prepared to discard all the data obtained during a pilot data collection project, so that the main phase can proceed unconstrained.

A further important decision is whether data collection should use in-house or external resources. It is now increasingly common to outsource geographic data collection to specialist companies that usually undertake the work in areas of the world with very low labor costs (e.g., India and Thailand). Three factors influencing this decision are: cost/schedule, quality, and long-term ramifications. Specialist external data collection agencies can often perform work faster, cheaper, with higher quality than in-house staff, but because of the need for real cash to pay external agencies this may not be possible. In the short term, project costs, quality, and time are the main considerations, but over time dependency on external groups may become a problem.

Questions for further study

1. Using the websites listed in Table 9.4 as a starting point, evaluate the suitability of free geographic data for your home region or country for use in a GIS project of your choice.

2. What are the advantages of batch vectorization over manual table digitizing?

3. What quality assurance steps would you build into a data collection project designed to construct a database of land parcels for tax assessment?

4. Why do so many geographic data formats exist? Which ones are most suitable for selling vector data?

Further reading

Hohl P. (ed) 1997 *GIS Data Conversion: Strategies, Techniques and Management*. Santa Fe, NM: OnWord Press.

Jones C. 1997 *Geographic Information Systems and Computer Cartography*. Reading, MA: Addison-Wesley Longman.

Lillesand T.M., Kiefer R.W. and Chipman R.W. 2003 *Remote Sensing and Image Interpretation* (5th edn). Hoboken, NJ: Wiley.

Paine D.P. and Kiser J.D. 2003 *Aerial Photography and Image Interpretation* (2nd edn). Hoboken, NJ: Wiley.

Walford N. 2002 *Geographical Data: Characteristics and Sources*. Hoboken, NJ: Wiley.

10 *Creating and maintaining geographic databases*

All large operational GIS are built on the foundation of a geographic database. After people, the database is arguably the most important part of a GIS because of the costs of collection and maintenance, and because the database forms the basis of all queries, analysis, and decision making. Today, virtually all large GIS implementations store data in a database management system (DBMS), a specialist piece of software designed to handle multi-user access to an integrated set of data. Extending standard DBMS to store geographic data raises several interesting challenges. Databases need to be designed with great care, and to be structured and indexed to provide efficient query and transaction performance. A comprehensive security and transactional access model is necessary to ensure that multiple users can access the database at the same time. On-going maintenance is also an essential, but very resource-intensive, activity.

Geographic Information Systems and Science, 2nd edition   Paul Longley, Michael Goodchild, David Maguire, and David Rhind.
© 2005 John Wiley & Sons, Ltd.   ISBNs: 0-470-87000-1 (HB); 0-470-87001-X (PB)

Learning Objectives

After reading this chapter you will:

- Understand the role of database management systems in GIS;

- Recognize structured query language (SQL) statements;

- Understand the key geographic database data types and functions;

- Be familiar with the stages of geographic database design;

- Understand the key techniques for structuring geographic information, specifically creating topology and indexing;

- Understand the issues associated with multi-user editing and versioning.

10.1 Introduction

A database can be thought of as an integrated set of data on a particular subject. Geographic databases are simply databases containing geographic data for a particular area and subject. It is quite common to encounter the term 'spatial' in the database world. As discussed in Section 1.1.1, 'spatial' refers to data about space at both geographic and non-geographic scales. A geographic database is a critical part of an operational GIS. This is both because of the cost of creation and maintenance, and because of the impact of a geographic database on all analysis, modeling, and decision-making activities. Databases can be physically stored in files or in specialist software programs called database management systems (DBMS). Today, most large organizations use a combination of files and DBMS for storing data assets.

A database is an integrated set of data on a particular subject.

The database approach to storing geographic data offers a number of advantages over traditional file-based datasets:

- Assembling all data at a single location reduces redundancy.

- Maintenance costs decrease because of better organization and reduced data duplication.

- Applications become data independent so that multiple applications can use the same data and can evolve separately over time.

- User knowledge can be transferred between applications more easily because the database remains constant.

- Data sharing is facilitated and a corporate view of data can be provided to all managers and users.

- Security and standards for data and data access can be established and enforced.

- DBMS are better suited to managing large numbers of concurrent users working with vast amounts of data.

On the other hand there are some disadvantages to using databases when compared to files:

- The cost of acquiring and maintaining DBMS software can be quite high.

- A DBMS adds complexity to the problem of managing data, especially in small projects

- Single user performance will often be better for files, especially for more complex data types and structures where specialist indexes and access algorithms can be implemented.

In recent years geographic databases have become increasingly large and complex. For example, AirPhoto USA's US National Image Mosaic is 25 terabytes (TB) in size, EarthSat's global Landsat mosaic at 15 m resolution is 6.5 TB, and Ordnance Survey of Great Britain has approximately 450 million vector features in its MasterMap database covering all of Britain. This chapter describes how to create and maintain geographic databases, and the concepts, tools, and techniques that are available to manage geographic data in databases. Several other chapters provide additional information that is relevant to this discussion. In particular, the nature of geographic data and how to represent them in GIS were described in Chapters 3, 4, and 5, and data modeling and data collection were discussed in Chapters 8 and 9 respectively. Later chapters introduce the tools and techniques that are available to query, model, and analyze geographic databases (Chapters 14, 15, and 16). Finally, Chapters 17 through 20 discuss the important management issues associated with creating and maintaining geographic databases.

10.2 Database management systems

A DBMS is a software application designed to organize the efficient and effective storage and access of data.

Small, simple databases that are used by a small number of people can be stored on computer disk in standard files. However, larger, more complex databases with many tens, hundreds, or thousands of users require specialist database management system (DBMS) software to ensure database integrity and longevity. A DBMS is a software application designed to organize the efficient and effective storage of and access to data. To carry out this function DBMS provide a number of important capabilities. These are introduced briefly here and are discussed further in this and other chapters. DBMS provide:

- A data model. As discussed in Section 8.4, a data model is the mechanism used to represent real-world objects digitally in a computer system. All DBMS include standard general-purpose core data models suitable for representing several types of object (e.g., integer and floating-point numbers, dates, and text). In most cases DBMS can be extended to support geographic object types.

- A data load capability. DBMS provide tools to load data into databases. Simple tools are available to load standard supported data types (e.g., character, number, and date) in well-structured formats. Other non-standard data formats can be loaded by writing custom software programs that convert the data into a structure that can be read by the standard loaders.

- Indexes. An index is a data structure used to speed up searching. All databases include tools to index standard database data types.

- A query language. One of the major advantages of DBMS is that they support a standard data query/manipulation language called SQL (Structured/Standard Query Language).

- Security. A key characteristic of DBMS is that they provide controlled access to data. This includes restricting user access to all or part of a database. For example, a casual GIS user might have read-only access to just part of a database, but a specialist user might have read and write (create, update, and delete) access to the entire database.

- Controlled update. Updates to databases are controlled through a transaction manager responsible for managing multi-user access and ensuring that updates affecting more than one part of the database are coordinated.

- Backup and recovery. It is important that the valuable data in a database are protected from system failure and incorrect (accidental or deliberate) update. Software utilities are provided to back up all or part of a database and to recover the database in the event of a problem.

- Database administration tools. The task of setting up the structure of a database (the schema), creating and maintaining indexes, tuning to improve performance, backing up and recovering, and allocating user access rights is performed by a database administrator (DBA). A specialized collection of tools and a user interface are provided for this purpose.

- Applications. Modern DBMS are equipped with standard, general-purpose tools for creating, using, and maintaining databases. These include applications for designing databases (CASE tools) and for building user interfaces for data access and presentations (forms and reports).

- Application programming interfaces (APIs). Although most DBMS have good general-purpose applications for standard use, most large, specialist applications will require further customization using a commercial off-the-shelf programming language and a DBMS programmable API.

This list of DBMS capabilities is very attractive to GIS users and so, not surprisingly, virtually all large GIS databases are based on DBMS technology. Indeed, most GIS software vendors include DBMS software within their GIS software products, or provide an interface that supports very close coupling to a DBMS. For further discussion of this see Chapter 8.

Today, virtually all large GIS use DBMS technology for data management.

10.2.1 Types of DBMS

DBMS can be classified according to the way they store and manipulate data. Three main types of DBMS are available to GIS users today: relational (RDBMS), object (ODBMS), and object-relational (ORDBMS).

A relational database comprises a set of tables, each a two-dimensional list (or array) of records containing attributes about the objects under study. This apparently simple structure has proven to be remarkably flexible and useful in a wide range of application areas, such that today over 95% of the data in DBMS are stored in RDBMS.

Object database management systems (ODBMS) were initially designed to address several of the weaknesses of RDBMS. These include the inability to store complete objects directly in the database (both object state and behavior: see Box 8.2 for an introduction to objects and object technology). Because RDBMS were focused primarily on business applications such as banking, human resource management, and stock control and inventory, they were never designed to deal with rich data types, such as geographic objects, sound, and video. A further difficulty is the poor performance of RDBMS for many types of geographic query. These problems are compounded by the difficulty of extending RDBMS to support geographic data types and processing functions, which obviously limits their adoption for geographic applications. ODBMS can store objects persistently (semi-permanently on disk or other media) and provide object-oriented query tools. A number of commercial ODBMS have been developed including GemStone/S Object Server from GemStone Systems Inc., Objectivity/DB from Objectivity Inc., ObjectStore from Progress Software, and Versant from Versant Object Technology Corp.

In spite of the technical elegance of ODBMS, they have not proven to be as commercially successful as some predicted. This is largely because of the massive installed base of RDBMS and the fact that RDBMS vendors have now added many of the important ODBMS capabilities to their standard RDBMS software systems to create hybrid object-relational DBMS (ORDBMS). An ORDBMS can be thought of as an RDBMS engine with an extensibility framework for handling objects. They can handle both the data describing what an object is (object attributes such as color, size, and age) and the behavior that determines what an object does (object methods or functions such as drawing instructions, query interfaces, and interpolation algorithms) and these can be managed and stored together as an integrated whole. Examples of ORDBMS software include IBM DB2 and Informix Dynamic Server, Microsoft SQL Server, and Oracle. As ORDBMS and the underlying relational model are so important in GIS, these topics are discussed at length in Section 10.3.

The ideal geographic ORDBMS is one that has been extended to support geographic object types and functions through the addition of the following (these topics are introduced here and discussed further later in this chapter):

- A query parser – the engine used to interpret SQL queries is extended to deal with geographic types and functions.
- A query optimizer – the software query optimizer is able to handle geographic queries efficiently. Consider a query to find all potential users of a new brand of premier wine to be marketed to wealthy households from a network of retail stores. The objective is to select all households within 3 km of a store that have an income greater than $110 000. This could be carried out in two ways:
  1. Select all households with an income greater than $110 000; from this selected set, select all households within 3 km of a store.
  2. Select all households within 3 km of a store; from this selected set select all households with an income greater than $110 000.

Selecting households with an income greater than $110 000 is an attribute query that can be performed very quickly. Selecting households within 3 km of a store is a geometric query that takes much longer. Executing the attribute query first (option 1 above) will result in fewer tests for store proximity and therefore the whole query will be completed much more quickly.

- A query language – the query language is able to handle geographic types (e.g., points and polygons) and functions (e.g., select polygons that touch each other).
- Indexing services – the standard unidimensional DBMS data index service is extended to support multidimensional (i.e., x, y, z coordinates) geographic data types.
- Storage management – the large volume of geographic records with different sizes (especially geometric and topological relationships) is accommodated through specialized storage structures.
- Transaction services – standard DBMS are designed to handle short (sub-second) transactions and are extended to deal with the long transactions common in many geographic applications.
- Replication – services for replicating databases are extended to deal with geographic types, and problems of reconciling changes made by distributed users.

10.2.2 Geographic DBMS extensions

Two of the major commercial DBMS vendors have released spatial database extensions to their standard ORDBMS products: IBM offers two solutions – DB2 Spatial Extender and Informix Spatial Datablade – and Oracle has a Spatial option (see Box 10.1).

Although there are differences in the technology, scope, and capabilities of these systems, they all provide basic functions to store, manage, and query geographic objects. This is achieved by implementing the seven key database extensions described in the previous section. It

Technical Box 10.1

Oracle Spatial

Oracle Spatial is an extension to the Oracle DBMS that provides the foundation for the management of spatial (including geographic) data inside an Oracle database. The standard types and functions in Oracle (CHAR, DATE or INTEGER, etc.) are extended with geographic equivalents. Oracle Spatial supports three basic geometric forms:

- *Points*: points can represent locations such as buildings, fire hydrants, utility poles, oil rigs, boxcars, or roaming vehicles.
- *Lines*: lines can represent things like roads, railroad lines, utility lines, or fault lines.

- *Polygons and complex polygons with holes*: polygons can represent things like outlines of cities, districts, flood plains, or oil and gas fields. A polygon with a hole might, for example, geographically represent a parcel of land surrounding a patch of wetlands.

These simple feature types can be aggregated to richer types using topology and linear referencing capabilities. Additionally, Oracle Spatial can store and manage georaster (image) data.

Oracle Spatial extends the Oracle DBMS query engine to support geographic queries. There is a set of spatial operators to perform: area-of-interest and spatial-join queries; length, area, and distance calculations; buffer and union queries; and administrative tasks. The Oracle Spatial SQL used to create a table and populate it with a single record is shown below. The characters after -- on each line are comments to describe the operations. The discussion of SQL syntax in Section 10.4 will help decode this program.

```
-- Create a table for routes (highways).
CREATE TABLE lrs_routes (
 route_id  NUMBER PRIMARY KEY,
 route_name  VARCHAR2(32),
 route_geometry  MDSYS.SDO_GEOMETRY);
 -- Populate table with just one route
    for this example.
INSERT INTO lrs_routes VALUES(
 1,
 'Route1',
MDSYS.SDO_GEOMETRY(
 3002,   -- line string, 3 dimensions:
    X,Y,M
 NULL,
 NULL,
 MDSYS.SDO_ELEM_INFO_ARRAY(1,2,1), -- one
    line string, straight segments
 MDSYS.SDO_ORDINATE_ARRAY(
  2,2,0,   -- Starting point - Exit1; 0
    is measure from start.
  2,4,2,   -- Exit2; 2 is measure from
    start.
```

```
 8,4,8,   -- Exit3; 8 is measure from
    start.
 12,4,12,  -- Exit4; 12 is measure from
    start.
 12,10,NULL, -- Not an exit; measure
    will be automatically calculated &
    filled.
 8,10,22,  -- Exit5; 22 is measure from
    start.
 5,14,27)  -- Ending point (Exit6); 27
    is measure from start.
 )
);
```

Geographic data in Oracle Spatial can be indexed using R-tree and quadtree indexing methods (these terms are defined in Section 10.7.2). There are also capabilities for managing projections and coordinate systems, as well as long transactions (see discussion in 10.9.1). Finally, there are also some tools for elementary spatial data mining (Section 15.1).

Oracle Spatial can be used with all major GIS software products and developers can create specific-purpose applications that embed SQL commands for manipulating and querying data.

As a new player in the GIS market space, Oracle has generated interest among larger IT-focused organizations. IBM has approached this market in a similar way with its Spatial Extender for the DB2 DBMS and Spatial Datablade for Informix.

is important to realize, however, that none of these is a complete GIS software system in itself. The focus of these extensions is data storage, retrieval, and management, and they have no real capabilities for geographic editing, mapping, and analysis. Consequently, they must be used in conjunction with a GIS except in the case of the simplest query-focused applications. Figure 10.1 shows how GIS and DBMS software can work together and some of the tasks best carried out by each system.

ORDBMS can support geographic data types and functions.

10.2.3 Geographic middleware extensions

An alternative to extending the DBMS software kernel to manage geographic data is to build support for spatial data types and functions into a middle-tier (or middleware) application server. Section 7.3.2 provides a description of this system architecture concept. This type of configuration offers many of the same capabilities as the core DBMS extensions, but can also support a wider range of data types and processing functions. A

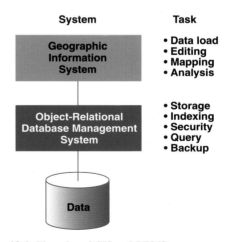

Figure 10.1 The roles of GIS and DBMS

middleware solution can also deliver better performance especially in the case of the more complex queries used in high-end GIS applications, because both the DBMS and the application server hardware resources can be used in parallel. Geographic middleware is available from ESRI in the form of ArcSDE, Intergraph in the form of

GeoMedia Transaction Server and MapInfo in the form of SpatialWare.

10.3 Storing data in DBMS tables

The lowest level of user interaction with a geographic database is usually the object class (also called a layer or feature class), which is an organized collection of data on a particular theme (e.g., all pipes in a water network, all soil polygons in a river basin, or all elevation values in a terrain surface). Object classes are stored in standard database tables. A table is a two-dimensional array of rows and columns. Each object class is stored as a single database table in a database management system (DBMS). Table rows contain objects (instances of object classes, e.g., data for a single pipe) and the columns contain object properties or attributes as they are frequently called (Figure 10.2). The data stored at individual row, column intersections are usually referred to as values. Geographic database tables are distinguished from non-geographic tables by the presence of a geometry column (often called the shape column). To save space and improve performance, the actual coordinate values may be stored in a highly compressed binary form.

Relational databases are made up of tables. Each geographic class (layer) is stored as a table.

Tables are joined together using common row/column values or keys as they are known in the database world. Figure 10.2 shows parts of tables containing data about US states. The STATES table (Figure 10.2A) contains the geometry and some basic attributes, an important one being a unique STATE_FIPS (State FIPS [Federal Information Processing Standard] code) identifier. The POPULATION table (Figure 10.2B) was created entirely independently, but also has a unique identifier column called STATE_FIPS. Using standard database tools the two tables can be joined together based on the common STATE_FIPS identifier column (the key) to create a third table, COMBINED STATES and POPULATION (Figure 10.2C). Following the join these can be treated as a single table for all GIS operations such as query, display, and analysis.

Database tables can be joined together to create new views of the database.

In a groundbreaking description of the relational model that underlies the vast majority of the world's geographic databases, In 1970 Ted Codd of IBM defined a series of rules for the efficient and effective design of database table structures. The heart of Codd's idea was that the best relational databases are made up of simple, stable tables that follow five principles:

1. Only one value is in each cell at the intersection of a row and column.

2. All values in a column are about the same subject.

3. Each row is unique (there are no duplicate records).

(A)

(B)

Figure 10.2 GIS database tables for US States: (A) STATES table; (B) POPULATION table; (C) joined table – COMBINED STATES and POPULATION

(C)

FID	Shape*	AREA	STATE_NAME	STATE_FIPS	SUB_REGION	STATE_ABBR	POP1990	POP1996
0	Polygon	67286.875	Washington	53	Pacific	WA	4866692	5629613
1	Polygon	147236.031	Montana	30	Mtn	MT	799065	885762
2	Polygon	32161.664	Maine	23	N Eng	ME	1227928	1254465
3	Polygon	70810.156	North Dakota	38	W N Cen	ND	638800	633534
4	Polygon	77193.625	South Dakota	46	W N Cen	SD	696004	721374
5	Polygon	97799.492	Wyoming	56	Mtn	WY	453588	487142
6	Polygon	56088.066	Wisconsin	55	E N Cen	WI	4891769	5144123
7	Polygon	83340.594	Idaho	16	Mtn	ID	1006749	1201327
8	Polygon	9603.218	Vermont	50	N Eng	VT	562758	587726
9	Polygon	84517.469	Minnesota	27	W N Cen	MN	4375099	4639933
10	Polygon	97070.750	Oregon	41	Pacific	OR	2842321	3203820
11	Polygon	9259.514	New Hampshire	33	N Eng	NH	1109252	1156932
12	Polygon	56257.219	Iowa	19	W N Cen	IA	2776755	2831890
13	Polygon	8172.482	Massachusetts	25	N Eng	MA	6016425	6066573
14	Polygon	77328.336	Nebraska	31	W N Cen	NE	1578385	1622272
15	Polygon	48560.578	New York	36	Mid Atl	NY	17990455	18293435
16	Polygon	45359.238	Pennsylvania	42	Mid Atl	PA	11881643	12077607
17	Polygon	4976.434	Connecticut	09	N Eng	CT	3287116	3287604
18	Polygon	1044.850	Rhode Island	44	N Eng	RI	1003464	993306
19	Polygon	7507.302	New Jersey	34	Mid Atl	NJ	7730188	7956917
20	Polygon	36399.516	Indiana	18	E N Cen	IN	5544159	5801023
21	Polygon	110667.297	Nevada	32	Mtn	NV	1201833	1532295
22	Polygon	84870.187	Utah	49	Mtn	UT	1722850	2000630
23	Polygon	157774.187	California	06	Pacific	CA	29760021	32218713
24	Polygon	41192.863	Ohio	39	E N Cen	OH	10847115	11123416
25	Polygon	56297.953	Illinois	17	E N Cen	IL	11430602	11731783
26	Polygon	66.063	District of Columbia	11	S Atl	DC	606900	550076
27	Polygon	2054.506	Delaware	10	S Atl	DE	666168	724890

Record: 0   Show: All Selected   Records (0 out of 51 Selected.)   Options

Figure 10.2 *(continued)*

4. There is no significance to the sequence of columns.

5. There is no significance to the sequence of rows.

Figure 10.3A shows a land parcel tax assessment database table that contradicts several of Codd's principles. Codd suggests a series of transformations called normal forms that successively improve the simplicity and stability, and reduce redundancy of database tables (thus reducing the risk of editing conflicts) by splitting them into sub-tables that are re-joined at query time. Unfortunately, joining large tables is computationally expensive and can result in complex database designs that are difficult to maintain. For this reason, non-normalized table designs are often used in GIS.

Figure 10.3B is a cleansed version of 10.3A that has been entered into a GIS DBMS: there is now only one value in each cell (Date and Assessed-Value are now separate columns); missing values have been added; an OBJECTID (unique system identifier) column has been added; and the potential confusion between Dave Widseler and D Widseler has been resolved. Figure 10.3C shows the same data after some normalization to make it suitable for use in a GIS tax assessment application. The database now consists of three tables that can be joined together using common keys. Figure 10.3C Attributes of Tab10_3b can be joined to Figure 10.3C Attributes of Tab10_3a

(A)

ParcelNumb	OwnerNam	OwnerAddress	PostalCode	ZoningCode	ZoningType	Date / AssessedValue
673/100	Jeff Peters	10 Railway Cuttings	114390	2	Residential	2002 220000
673-101	Joel Campbell	1115 Center Place	114390	2	Residential	2003 545500
674-100	Dave Widseler		114391	3	Commercial	99 249000
674-100		452 Diamond Plaza	114391	3	Commercial	2000 275500
674-100	D Widseler	452 Diamond Plaza	114391	3	Commercial	2001 290000
670-231	Sam Camarata	19 Big Bend Bld	114391	2	Residential	2004 450575
674-112	Chris Capelli	Hastings Barracks	114392	2	Residential	2004 350000
674-113	Sheila Sullivan	10034 Endin Mansions	114391	2	Residential	02 1005425

Figure 10.3 Tax assessment database: (A) raw data; (B) cleansed data in a GIS DBMS; (C) data partially normalized into three sub-tables; (D) joined table

(B)

OBJECTID*	ParcelNumb	OwnerNam	OwnerAddress	PostalCode	ZoningCode	ZoningType	DateAssessed	AssessedValue
1	673-100	Jeff Peters	10 Railway Cuttings	114390	2	Residential	2002	220000
2	673-101	Joel Campbell	1115 Center Place	114390	2	Residential	2003	545500
3	674-100	Dave Widseler	452 Diamond Plaza	114391	3	Commercial	1999	249000
4	674-100	Dave Widseler	452 Diamond Plaza	114391	3	Commercial	2000	275500
5	674-100	Dave Widseler	452 Diamond Plaza	114391	3	Commercial	2001	290000
6	670-231	Sam Camarata	19 Big Bend Bld	114391	2	Residential	2004	450575
7	674-112	Chris Capelli	Hastings Barracks	114392	2	Residential	2004	350000
8	674-113	Sheila Sullivan	10034 Endin Mansions	114391	2	Residential	2002	1005425

Record: 0 Show: All Selected Records (0 out of 8 Selected.)   Options ▾

(C)

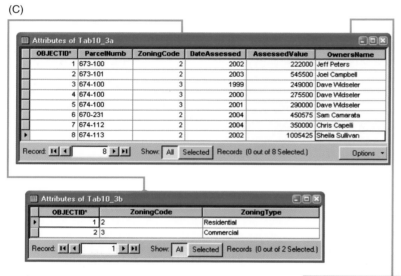

Attributes of Tab10_3a

OBJECTID*	ParcelNumb	ZoningCode	DateAssessed	AssessedValue	OwnersName
1	673-100	2	2002	222000	Jeff Peters
2	673-101	2	2003	545500	Joel Campbell
3	674-100	3	1999	249000	Dave Widseler
4	674-100	3	2000	275500	Dave Widseler
5	674-100	3	2001	290000	Dave Widseler
6	670-231	2	2004	450575	Sam Camarata
7	674-112	2	2004	350000	Chris Capelli
8	674-113	2	2002	1005425	Sheila Sullivan

Record: 8 Show: All Selected Records (0 out of 8 Selected.)   Options ▾

Attributes of Tab10_3b

OBJECTID*	ZoningCode	ZoningType
1	2	Residential
2	3	Commercial

Record: 1 Show: All Selected Records (0 out of 2 Selected.)

Attributes of Tab10_3c

OBJECTID*	OwnerName*	Address	PostalCode
2	Jeff Peters	10 Railway Cuttings	114390
3	Joel Campbell	1115 Center Place	114390
4	Dave Widseler	452 Diamond Plaza	114391
5	Sam Camarata	19 Big Bend Bld	114391
6	Chris Capelli	Hastings Barracks	114392
7	Sheila Sullivan	10034 Endin Mansions	114391

Record: 0 Show: All Selected Records (0 out of 6 Selected.)

(D)

Attributes of Tab10_3a_Tab10_3b_Tab10_3c

OBJECTID	ParcelNumb	DateAssessed	AssessedValue	ZoningType	ZoingCode	OwnerName	OwnerAddress	PostalCode
1	673-100	2002	222000	Residential	2	Jeff Peters	10 Railway Cuttings	114390
2	673-101	2003	545500	Residential	2	Joel Campbell	1115 Center Place	114390
3	674-100	1999	249000	Commercial	3	Dave Widseler	452 Diamond Plaza	114391
4	674-100	2000	275500	Commercial	3	Dave Widseler	452 Diamond Plaza	114391
5	674-102	2004	290000	Residential	2	Dave Widseler	452 Diamond Plaza	114391
6	670-231	2004	450575	Residential	2	Sam Camarata	19 Big Bend Bld	114391
7	674-112	2004	350000	Residential	2	Chris Capelli	Hastings Barracks	114392
8	674-113	2002	1005425	Residential	2	Sheila Sullivan	10034 Endin Mansions	114391

Record: 0 Show: All Selected Records (0 out of 8 Selected.)   Options ▾

Figure 10.3 *(continued)*

using the common ZoningCode column, and Figure 10.3C `Attributes of Tab10_3c` can be joined using OwnersName to create Figure 10.3D. It is now possible to execute SQL queries against these joined tables as discussed in the next section.

10.4 SQL

The standard database query language adopted by virtually all mainstream databases is SQL (Structured or Standard Query Language: ISO Standard ISO/IEC 9075). There are many good background books and system implementation manuals on SQL and so only brief details will be presented here. SQL may be used directly via an interactive command line interface; it may be compiled in a general-purpose programming language (e.g., C/C++/C#, Java, or Visual Basic); or it may be embedded in a graphical user interface (GUI). SQL is a set based, rather than a procedural (e.g., Visual Basic) or object-oriented (e.g., Java or C#), programming language designed to retrieve sets (row and column combinations) of data from tables. There are three key types of SQL statements: DDL (data definition language), DML (data manipulation language) and DCL (data control language). The third major revision of SQL (SQL 3) which came out in 2004 defines spatial types and functions as part of a multi-media extension called SQL/MM.

The data in Figure 10.3C may be queried to find parcels where the AssessedValue is greater than $300 000 and the ZoningType is Residential. This is an apparently simple query, but it requires three table joins to execute it. The SQL statements in Microsoft Access are as follows:

```
SELECT Tab10_3a.ParcelNumb, Tab10_3c.Address,
   Tab10_3a.AssessedValue
FROM (Tab10_3b INNER JOIN Tab10_3a ON
   Tab10_3b.ZoningCode =
   Tab10_3a.ZoningCode) INNER JOIN Tab10_3c
   ON Tab10_3a.OwnersName =
   Tab10_3c.OwnerName
WHERE (((Tab10_3a.AssessedValue)>300000) AND
   ((Tab10_3b.ZoningType)="Residential"));
```

The `SELECT` statement defines the columns to be displayed (the syntax is TableName.ColumnName). The `FROM` statement is used to identify and join the three tables (`INNER JOIN` is a type of join that signifies that only matching records in the two tables will be considered). The `WHERE` clause is used to select the rows from the columns using the constraints `(((Tab10_3a.AssessedValue)>300000) AND ((Tab10_3b.ZoningType)="Residential"))`. The result of this query is shown in Figure 10.4. This triplet of `SELECT, FROM, WHERE` is the staple of SQL queries.

SQL is the standard database query language. Today it has geographic capabilities.

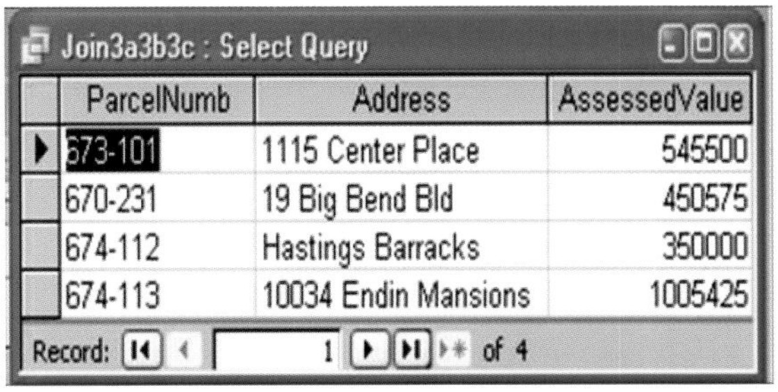
Join3a3b3c : Select Query

ParcelNumb	Address	AssessedValue
673-101	1115 Center Place	545500
670-231	19 Big Bend Bld	450575
674-112	Hastings Barracks	350000
674-113	10034 Endin Mansions	1005425

Record: 1 of 4

Figure 10.4 Results of a SQL query against the tables in Figure 10.3C (see text for query and further explanation)

In SQL, data definition language statements are used to create, alter, and delete relational database structures. The `CREATE TABLE` command is used to define a table, the attributes it will contain, and the primary key (the column used to identify records uniquely). For example, the SQL statement to create a table to store data about Countries, with two columns (name and shape (geometry)) is as follows:

```
CREATE TABLE Countries (
name            VARCHAR(200) NOT NULL PRIMARY
   KEY,
shape           POLYGON NOT NULL
CONSTRAINT   spatial reference
CHECK        (SpatialReference(shape) = 14)
)
```

This SQL statement defines several table parameters. The name column is of type `VARCHAR` (variable character) and can store values up to 200 characters. Name cannot be null (`NOT NULL`), that is, it must have a value, and it is defined as the `PRIMARY KEY`, which means that its entries must be unique. The shape column is of type `POLYGON`, and it is defined as `NOT NULL`. It has an additional spatial reference constraint (projection), meaning that a spatial reference is enforced for all shapes (Type `14` – this will vary by system, but could be Universal Transverse Mercator (UTM) – see Section 5.7.2).

Data can be inserted into this table using the SQL `INSERT` command:

```
INSERT INTO Countries
(Name, Shape) VALUES ('Kenya', Polygon('((x
   y, x y, x y, x y)) ,2))
```

Actual coordinates would need to be substituted for the x, y values. Several additions of this type would result in a table like this:

Name	Shape
Kenya	Polygon geometry
South Africa	Polygon geometry
Egypt	Polygon geometry

Data manipulation language statements are used to retrieve and manipulate data. Objects with a size greater than 11 000 can be retrieved from the countries table using a SELECT statement:

```
SELECT Countries.Name,
FROM Countries
WHERE Area(Countries.Shape) > 11000
```

In this system implementation, area is computed automatically from the shape field using a DBMS function and does not need to be stored.

Data control language statements handle authorization access. The two main DCL keywords are GRANT and REVOKE that authorize and rescind access privileges respectively.

10.5 Geographic database types and functions

There have been several attempts to define a superset of geographic data types that can represent and process geographic data in databases. Unfortunately space does not permit a review of them all. This discussion will focus on the practical aspects of this problem and will be based on the recently developed International Standards Organization (ISO) and the Open Geospatial Consortium (OGC) standards.

The GIS community working under the auspices of ISO and OGC has defined the core geographic types and functions to be used in a DBMS and accessed using the SQL language. The geometry types are shown in Figure 10.5. The Geometry class is the root class. It has an associated spatial reference (coordinate system and projection, for example, Lambert_Azimuthal_Equal_Area). The Point, Curve, Surface, and GeometryCollection classes are all subtypes of Geometry. The other classes (boxes) and relationships (lines) show how geometries of one type are aggregated from others (e.g., a LineString is a collection of Points). For further explanation of how to interpret this object model diagram see the discussion in Section 8.3.

According to this ISO/OGC standard, there are nine methods for testing spatial relationships between these geometric objects. Each takes as input two geometries (collections of one or more geometric objects) and evaluates whether the relationship is true or not. Two examples of possible relations for all point, line, and polygon combinations are shown in Figure 10.6. In the case of the point-point Contain combination (northeast square in Figure 10.6A), two comparison geometry points (big circles) are contained in the set of base geometry points (small circles). In other words, the base geometry is a superset of the comparison geometry. In the case of line-polygon Touches combination (east square in Figure 10.6B) the two lines touch the polygon because they intersect the polygon boundary. The full set of Boolean operators to test the spatial relationships between geometries is:

- Equals – are the geometries the same?

- Disjoint – do the geometries share a common point?

- Intersects – do the geometries intersect?

- Touches – do the geometries intersect at their boundaries?

- Crosses – do the geometries overlap (can be geometries of different dimensions, for example, lines and polygons)?

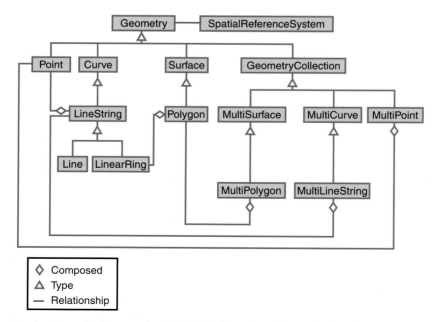

Figure 10.5 Geometry class hierarchy (*Source*: after OGC 1999) (Reproduced by permission of Open Geospatial Consortium, Inc.)

(A) **Contains**

Does the base geometry contain the comparison geometry?

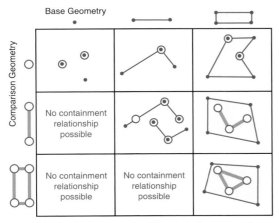

For the base geometry to contain the comparison geometry, it must be a superset of that geometry.

A geometry cannot contain another geometry of higher dimension.

(B) **Touches**

Does the base geometry touch the comparison geometry?

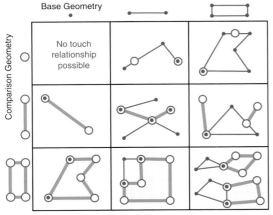

Two geometries touch when only their boundaries intersect.

Figure 10.6 Examples of possible relations for two geographic database operators: (A) Contains; and (B) Touches operators (*Source*: after Zeiler 1999)

- Within – is one geometry within another?
- Contains – does one geometry completely contain another?
- Overlaps – do the geometries overlap (must be geometries of the same dimension)?
- Relate – are there intersections between the interior, boundary, or exterior of the geometries?

Seven methods support spatial analysis on these geometries. Four examples of these methods are shown in Figure 10.7.

- Distance – determines the shortest distance between any two points in two geometries (Section 14.3.1).
- Buffer – returns a geometry that represents all the points whose distance from the geometry is less than or equal to a user-defined distance (Section 14.4.1).
- ConvexHull – returns a geometry representing the convex hull of a geometry (a convex hull is the smallest polygon that can enclose another geometry without any concave areas).
- Intersection – returns a geometry that contains just the points common to both input geometries.
- Union – returns a geometry that contains all the points in both input geometries.
- Difference – returns a geometry containing the points that are different between the two geometries.
- SymDifference – returns a geometry containing the points that are in either of the input geometries, but not both.

10.6 Geographic database design

This section is concerned with the technical aspects of logical and physical geographic database design. Chapter 8 provides an overview of these subjects and Chapters 17 to 20 discuss the organizational, strategic, and business issues associated with designing and maintaining a database.

10.6.1 The database design process

All GIS and DBMS packages have their own core data model that defines the object types and relationships that can be used in an application (Figure 10.8). The DBMS package will define and implement a model for data types and access functions, such as those in SQL discussed in Section 10.4. DBMS are capable of dealing with simple features and types (e.g., points, lines, and polygons) and relationships. A GIS can build on top of these simple feature types to create more advanced types and relationships (e.g., TINs, topologies, and feature-linked annotation geographic relationships: see Chapter 8 for a definition of these terms). The GIS types can be combined with domain data models that define specific object classes and relationships for specialist domains (e.g., water utilities, city parcel maps, and census geographies). Lastly, individual projects create specific physical data model instances that are populated with data (objects for the specified object classes). For example, a City Planning department may build a database of sewer lines that uses

(A) Buffer

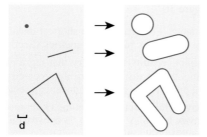

Given a geometry and a buffer distance, the buffer operator returns a polygon that covers all points whose distance from the geometry is less than or equal to the buffer distance.

(B) Convex Hull

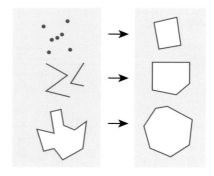

Given an input geometry, the convex hull operator returns a geometry that represents all points that are within all lines between all points in the input geometry.

A convex hull is the smallest polygon that wraps another geometry without any concave areas.

(C) Intersection

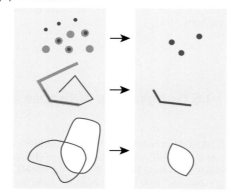

The intersect operator compares a base geometry (the object from which the operator is called) with another geometry of the same dimension and returns a geometry that contains the points that are in both the base geometry and the comparison geometry.

(D) Difference

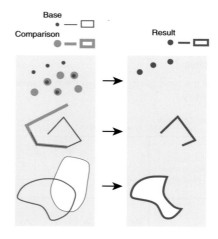

The difference operator returns a geometry that contains points that are in the base geometry and subtracts points that are in the comparison geometry.

Figure 10.7 Examples of spatial analysis methods on geometries: (A) Buffer; (B) Convex Hull; (C) Intersection; (D) Difference (*Source*: after Zeiler 1999)

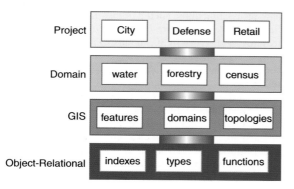

Figure 10.8 Four levels of data model available for use in GIS projects with examples of constructs used

a water/wastewater (sewer) domain data model template which is built on top of core GIS and DBMS models.

Figure 8.2 and Section 8.1.2 show three increasingly abstract stages in data modeling: conceptual, logical, and physical. The result of a data modeling exercise is a physical database design. This design will include specification of all data types and relationships, as well as the actual database configuration required to store them.

Database design involves three key stages: conceptual, logical, and physical.

Database design involves the creation of conceptual, logical, and physical models in the six practical steps shown in Figure 10.9:

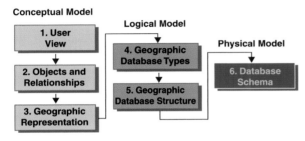

Figure 10.9 Stages in database design (*Source*: after Zeiler 1999)

10.6.1.1 Conceptual model

Model the user's view. This involves tasks such as identifying organizational functions (e.g., controlling forestry resources, finding vacant land for new building, and maintaining highways), determining the data required to support these functions, and organizing the data into groups to facilitate data management. This information can be represented in many ways – a report with accompanying tables is often used.

Define objects and their relationships. Once the functions of an organization have been defined, the object types (classes) and functions can be specified. The relationships between object types must also be described. This process usually benefits from the rigor of using object models and diagrams to describe a set of object classes and the relationships between them.

Select geographic representation. Choosing the types of geographic representation (discrete object – point, line, polygon – or field: see Section 3.6) will have profound implications for the way a database is used and so it is a critical database design task. It is, of course, possible to change between representation types, but this is computationally expensive and results in loss of information.

10.6.1.2 Logical model

Match to geographic database types. This involves matching the object types to be studied to specific data types supported by the GIS that will be used to create and maintain the database. Because the data model of the GIS is usually independent of the actual storage mechanism (i.e., it could be implemented in Oracle, Microsoft Access, or a proprietary file system), this activity is defined as a logical modeling task.

Organize geographic database structure. This includes tasks such as defining topological associations, specifying rules and relationships, and assigning coordinate systems.

10.6.1.3 Physical model

Define database schema. The final stage is definition of the actual physical database schema that will hold the database data values. This is usually created using the DBMS software's data definition language. The most popular of these is SQL with geographic extensions (see Section 10.4), although some non-standard variants also exist in older GIS/DBMS.

10.7 Structuring geographic information

Once data have been captured in a geographic database according to a schema defined in a geographic data model, it is often desirable to perform some structuring and organization in order to support efficient query, analysis, and mapping. There are two main structuring techniques relevant to geographic databases: topology creation and indexing.

10.7.1 Topology creation

The subject of building topology was covered in Section 8.2.3.2 from a conceptual data modeling perspective, and is revisited here in the context of databases where the discussion focuses on the two main approaches to structuring and storing topology in a DBMS.

Topology can be created for vector datasets using either batch or interactive techniques. Batch topology builders are required to handle CAD, survey, simple feature, and other unstructured vector data imported from non-topological systems. Creating topology is usually an iterative process because it is seldom possible to resolve all data problems during the first pass and manual editing is required to make corrections. Some typical problems that need to be fixed are shown in Figures 9.9, 9.10, and 9.11 and discussed in Section 9.3.2.3. Interactive topology creation is performed dynamically at the time objects are added to a database using GIS editing software. For example, when adding water pipes using interactive vectorization tools (see Section 9.3.2.2), before each object is committed to the database topological connectivity can be checked to see if the object is valid (that is, it conforms to some pre-established database object connectivity rules).

Two database-oriented approaches have emerged in recent years for storing and managing topology: Normalized and Physical. The Normalized Model focuses on the storage of an arc-node data structure. It is said to be normalized because each object is decomposed into individual topological primitives for storage in a database and then subsequent reassembly when a query is posed. For example, polygon objects are assembled at query time by joining together tables containing the line segment geometries and topological primitives that define topological relationships (see Section 8.2.3.2 for a conceptual description of this process). In the Physical Model topological primitives are not stored in the database and the entire geometry is stored together for each object. Topological relationships are then computed on-the-fly whenever they are required by client applications.

Figure 10.10 is a simple example of a set of database tables that store a dataset according to the Normalized Model for topology. The dataset (sketch in top left corner) comprises three feature classes (Parcels, Buildings, and Walls) and is implemented in three tables. In this example the three feature class tables have only one column

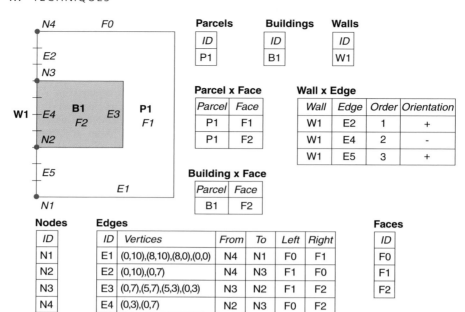

Figure 10.10 Normalized database topology model

Nodes

ID
N1
N2
N3
N4

Edges

ID	Vertices	From	To	Left	Right
E1	(0,10),(8,10),(8,0),(0,0)	N4	N1	F0	F1
E2	(0,10),(0,7)	N4	N3	F1	F0
E3	(0,7),(5,7),(5,3),(0,3)	N3	N2	F1	F2
E4	(0,3),(0,7)	N2	N3	F0	F2
E5	(0,3),(0,0)	N2	N1	F1	F0

Faces

ID
F0
F1
F2

Parcels

ID
P1

Buildings

ID
B1

Walls

ID
W1

Parcel x Face

Parcel	Face
P1	F1
P1	F2

Building x Face

Parcel	Face
B1	F2

Wall x Edge

Wall	Edge	Order	Orientation
W1	E2	1	+
W1	E4	2	-
W1	E5	3	+

(ID) and one row (a single instance of each feature in a feature class). The Nodes, Edges, and Faces tables store the points, lines, and polygons for the dataset and some of the topology (for each Edge the From-To connectivity and the Faces on the Left-Right in the direction of digitizing). Three other tables (Parcel × Face, Wall × Edge, and Building × Face) store the cross references for assembling Parcels, Buildings, and Walls from the topological primitives.

The Normalized approach offers a number of advantages to GIS users. Many people find it comforting to see topological primitives actually stored in the database. This model has many similarities to the arc-node conceptual topology model (see Section 8.2.3.2) and so it is familiar to many users and easy to understand. The geometry is only stored once thus minimizing database size and avoiding 'double digitizing' slivers (Section 9.3.2.3). Finally, the normalized approach easily lends itself to access via a SQL Application Programming Interface (API). Unfortunately, there are three main disadvantages associated with the Normalized approach to database topology: query performance, integrity checking, and update performance/complexity.

Query performance suffers because queries to retrieve features from the database (much the most common type of query) must combine data from multiple tables. For example, to fetch the geometry of Parcel P1, a query must combine data from four tables (Parcels, Parcel × Face, Faces, and Edges) using complex geometry/topology logic. The more tables that have to be visited, and especially the more that have to be joined, the longer it will take to process a query.

The standard referential integrity rules in DBMS are very simple and have no provision for complex topological relationships of the type defined here. There are many pitfalls associated with implementing topological structuring using DBMS techniques such as stored procedures (program code stored in a database) and in practice systems have resorted to implementing the business logic to manage things like topology external to the DBMS in a middle-tier application server (see Section 7.3.2).

Updates are similarly problematic because changes to a single feature will have cascading effects on many tables. This raises attendant performance (especially scalability, that is, large numbers of concurrent queries) and integrity issues. Moreover, it is uncertain how multi-user updates will be handled that entail long transactions with design alternatives (see Sections 10.9 and 10.9.1 for coverage of these two important topics).

For comparative purposes the same dataset used in the Normalized model (Figure 10.10) is implemented using the Physical model in Figure 10.11. In the physical model the three feature classes (Parcels, Buildings, and Walls) contain the same IDs, but differ significantly in that they also contain the geometry for each feature. The only other things required to be stored in the database are the specific set of topology rules that have been applied to the dataset (e.g., parcels should not overlap each other, and buildings should not overlap with each other), together with information about known errors (sometimes users defer topology clean-up and commit data with known errors to a database) and areas that have been edited, but not yet been validated (had their topology (re-)built).

The Physical model requires that an external client or middle-tier application server is responsible for validating the topological integrity of datasets. Topologically correct features are then stored in the database using a much simpler structure than the Normalized model. When compared to the Normalized model, the Physical model offers

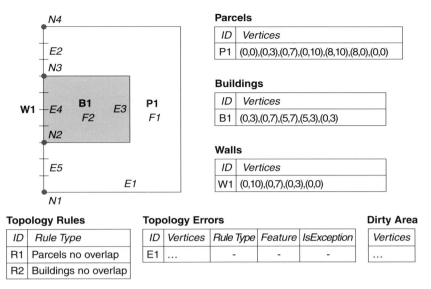

Figure 10.11 Physical database topology model

two main advantages of simplicity and performance. Since all the geometry for each feature is stored in the same table column/row value, and there is no need to store topological primitives and cross references, this is a very simple model. The biggest advantage and the reason for the appeal of this approach is query performance. Most DBMS queries do not need access to the topology primitives of features and so the overhead of visiting and joining multiple tables (four) is unnecessary. Even when topology is required it is faster to retrieve feature geometries and re-compute topology outside the database, than to retrieve geometry and topology from the database.

In summary, there are advantages and disadvantages to both the Normalized and Physical topology models. The Normalized model is implemented in Oracle Spatial and can be accessed via a SQL API making it easily available to a wide variety of users. The Physical model is implemented in ESRI ArcGIS and offers fast update and query performance for high-end GIS applications. ESRI has also implemented a long transaction and versioning model based on the physical database topology model.

10.7.2 Indexing

Geographic databases tend to be very large and geographic queries computationally expensive. Because of this, geographic queries, such as finding all the customers (points) within a store trade area (polygon), can take a very long time (perhaps 10–100 seconds or more for a 50 million customer database). The point has already been made in Section 8.2.3.2 that topological structuring can help speed up certain types of queries such as adjacency and network connectivity. A second way to speed up queries is to index a database and use the index to find data records (database table rows) more quickly. A database index is logically similar to a book index; both are special organizations that speed up searching by allowing random instead of sequential access. A database index is, conceptually speaking, an ordered list derived from the data in a table. Using an index to find data reduces the number of computational tests that have to be performed to locate a given set of records. In DBMS jargon, indexes avoid expensive full-table scans (reading every row in a table) by creating an index and storing it as a table column.

A database index is a special representation of information about objects that improves searching.

Figure 10.12 is a simple example of the standard DBMS one-dimensional B-tree (Balanced Tree) index that is found in most major commercial DBMS. Without an index a search of the original data to guarantee finding any given value will involve 16 tests/read operations (one for each data point). The B-tree index orders the data and splits the ordered list into buckets of a given size (in this example it is four and then two) and the upper value for the bucket is stored (it is not necessary to store the uppermost value). To find a specific value, such as 72, using the index involves a maximum of six tests: one at level 1 (less than or greater than 36), one at level 2 (less than or greater than 68), and a sequential read of four records at Level 3. The number of levels and buckets for each level can be optimized for each type of data. Typically, the larger the dataset the more effective indexes are in retrieving performance.

Unfortunately, creating and maintaining indexes can be quite time consuming and this is especially an issue when the data are very frequently updated. Since indexes can occupy considerable amounts of disk space, requirements can be very demanding for large datasets. As a consequence, many different types of index have been developed to try and alleviate these problems for both geographic and non-geographic data. Some indexes exploit specific characteristics of data to deliver optimum

Original Data

1
13
69
52
25
26
71
36
22
72
67
68
14
70
31
53

B-Tree Indexed Data

Level 1	Level 2	Level 3
36	22	1
		13
		14
		22
	36	25
		26
		31
		36
	68	52
		53
		67
		68
		69
		70
		71
		72

Figure 10.12 An example B-tree index

query performance, some are fast to update, and others are robust across widely different data types.

The standard indexes in DBMS, such as the B-tree, are one-dimensional and are very poor at indexing geographic objects. Many types of geographic indexing techniques have been developed. Some of these are experimental and have been highly optimized for specific types of geographic data. Research shows that even a basic spatial index will yield very significant improvements in spatial data access and that further refinements often yield only marginal improvements at the costs of simplicity, and speed of generation and update. Three main methods of general practical importance have emerged in GIS: grid indexes, quadtrees, and R-trees.

10.7.2.1 Grid index

A grid index can be thought of as a regular mesh placed over a layer of geographic objects. Figure 10.13 shows a

layer that has three features indexed using two grid levels. The highest (coarsest) grid (Index 1) splits the layer into four equal sized cells. Cell A includes parts of Features 1, 2, and 3, Cell B includes a part of Feature 3 and Cell C has part of Feature 1. There are no features on Cell D. The same process is repeated for the second level index (Index 2). A query to locate an object searches the indexed list first to find the object and then retrieves the object geometry or attributes for further analysis (e.g., tests for overlap, adjacency, or containment with other objects on the same or another layer). These two tests are often referred to as primary and secondary filters. Secondary filtering, which involves geometric processing, is much more computationally expensive. The performance of an index is clearly related to the relationship between grid and object size, and object density. If the grid size is too large relative to the size of object, too many objects will be retrieved by the primary filter, and therefore a lot of expensive secondary processing will be needed. If the grid size is too small any large objects will be spread across many grid cells, which is inefficient for draw queries (queries made to the database for the purpose of displaying objects on a screen).

For data layers that have a highly variable object density (for example, administrative areas tend to be smaller and more numerous in urban areas in order to equalize population counts) multiple levels can be used to optimize performance. Experience suggests that three grid levels are normally sufficient for good all-round performance. Grid indexes are one of the simplest and most robust indexing methods. They are fast to create and update and can handle a wide range of types and densities of data. For this reason they have been quite widely used in commercial GIS software systems.

> **Grid indexes are easy to create, can deal with a wide range of object types, and offer good performance.**

10.7.2.2 Quadtree indexes

Quadtree is a generic name for several kinds of index that are built by recursive division of space into quadrants. In many respects quadtrees are a special types of grid index. The difference here is that in quadtrees space is

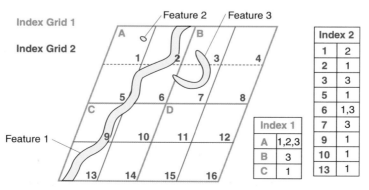

Figure 10.13 A multi-level grid geographic database index

recursively split into four quadrants based on data density. Quadtrees are data structures used for both indexing and compressing geographic database layers, although the discussion here will relate only to indexing. The many types of quadtree can be classified according to the types of data that are indexed (points, lines, areas, surfaces, or rasters), the algorithm that is used to decompose (divide) the layer being indexed, and whether fixed or variable resolution decomposition is used.

In a point quadtree, space is divided successively into four rectangles based on the location of the points (Figure 10.14). The root of the tree corresponds to the region as a whole. The rectangular region is divided into four usually irregular parts based on the x, y coordinate of the first point. Successive points subdivide each new sub-region into quadrants until all the points are indexed.

Region quadtrees are commonly used to index lines, polygons, and rasters. The quadtree index is created by successively dividing a layer into quadrants. If a quadrant cell is not completely filled by an object, then it is sub-divided again. Figure 10.15 is a quadtree of a woodland (red) and water (white) layer. Once a layer has been decomposed in this way, a linear index can be created using the search order shown in Figure 10.16. By reducing two-dimensional geographic data to a single

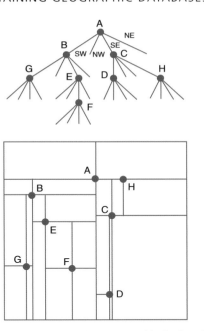

Figure 10.14 The point quadtree geographic database index (*Source*: after van Oosterom 2005) (Reproduced by permission of John Wiley & Sons Inc.)

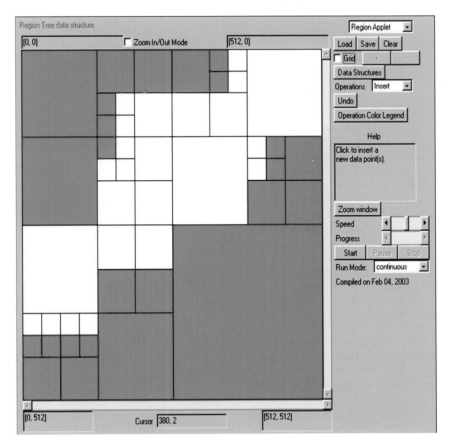

Figure 10.15 The region quadtree geographic database index (from **www.cs.umd.edu/~brabec/quadtree/**) (Reproduced by permission of Hanan Samet and Frantisek Brabec)

Original Data **Linear Quadtree Index Order**

Figure 10.16 Linear quadtree search order

linear dimension, a standard B-tree can be used to find data quickly.

> Quadtrees have found favor in GIS software systems because of their applicability to many types of data (both raster and vector), their ease of implementation, and their relatively good performance.

10.7.2.3 R-tree indexes

R-trees group objects using a rectangular approximation of their location called a minimum bounding rectangle

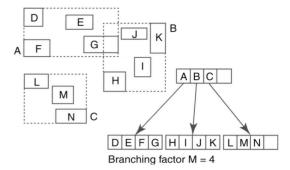

Figure 10.17 The R-tree geographic database index (*Source*: after van Oosterom 2005) (Reproduced by permission of John Wiley & Sons Inc.)

(MBR) or minimum enclosing rectangle (see Box 10.2). Groups of point, line, or polygon objects are indexed based on their MBR. Objects are added to the index by choosing the MBR that would require the least expansion to accommodate each new object. If the object causes the MBR to be expanded beyond some preset parameter then the MBR is split into two new MBR. This may also cause the parent MBR to be become too large, resulting in this also being split. The R-tree shown in Figure 10.17 has two levels. The lowest level contains three 'leaf nodes';

Technical Box **10.2**

Minimum Bounding Rectangle

Minimum Bounding Rectangles (MBR) are very useful structures that are widely implemented in GIS. An MBR essentially defines the smallest box whose sides are parallel to the axes of the coordinate system that encloses a set of one or more geographic objects. It is defined by the two coordinates at the bottom left (minimum x, minimum y) and the top right (maximum x, maximum y) as is shown in Figure 10.18.

MBR can be used to generalize a set of data by replacing the geometry of the objects in the box with two coordinates defining the box. A second use is for fast searching. For example, all the polygon objects in a database layer that are within a given study area can be found by performing a polygon on polygon contains test (see Figure 10.18) with each object and the study area boundary. If the polygon objects have complex boundaries (as is normally the case in GIS) this can be a very time-consuming task. A quicker approach is to split the task into two parts. First, screen out all the objects that are definitely in and definitely out by comparing their MBR. Because very few coordinate comparisons are required this is very

fast. Then use the full geometry outline of the remaining polygon objects to determine containment. This is computationally expensive for polygons with complex geometries.

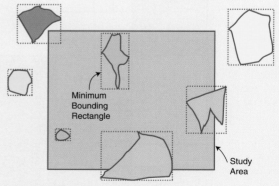

Figure 10.18 Polygon in polygon test using MBR. A MBR can be used to determine objects definitely within the study area (green) because of no overlap, definitely out (yellow), or possibly in (blue). Objects possibly in can then be analyzed further using their exact geometries. Note the purple object that is actually completely outside, although the MBR suggests it is partially within the study area

the highest has one node with pointers to the MBR of the leaf nodes. The MBR is used to reduce the number of objects that need to be examined to satisfy a query.

R-trees are popular methods of indexing geographic data because of their flexibility and excellent performance.

R-trees are suitable for a range of geographic object types and can be used for multi-dimensional data. They offer good query performance, but the speed of update is not as fast as grids and quadtrees. The Spatial Datablade extensions to the IBM Informix DBMS and Oracle Spatial both use R-tree indexes.

10.8 Editing and data maintenance

Editing is the process of making changes to a geographic database by adding new objects or changing existing objects as part of data load or database update and maintenance operations. A database update is any change to the geometry and/or attributes of one or more objects or any change to the database schema. A general-purpose geographic database will require many tools for tasks such as geometry and attribute editing, database maintenance (e.g., system administration and tuning), creating and updating indexes and topology, importing and exporting data, and georeferencing objects.

Contemporary GIS come equipped with an extensive array of tools for creating and editing geographic object geometries and attributes. These tools form workflow tasks that are exposed within the framework of a WYSIWYG (what you see is what you get) editing environment. The objects are displayed using map symbology usually in a projected coordinate system space and frequently on top of 'background' layers such as aerial photographs or satellite images, or street centerline files. Object coordinates can be digitized into a geographic database using many methods including freehand digitizing on a digitizing table, on screen heads-up vector digitizing by copying existing raster and/or vector sources, on screen semi-automatic line following, automated feature recognition, and reading survey measurements from an instrument (e.g., GPS or Total Station) file (see Sections 5.8 and 9.2.2.1). The end result is always a layer of x, y coordinates with optional z and m (height and attribute) values. Similar tools also exist for loading and editing raster data.

Data entered into the editor must be stored persistently in a file system or database and access to the database must be carefully managed to ensure continued security and quality is maintained. The mechanism for managing edits to a file or database is a called a transaction. There are many challenging issues associated with implementing multi-user access geographic data stored in a DBMS as discussed in the next section.

10.9 Multi-user editing of continuous databases

For many years, one of the most challenging problems in GIS data management was how to allow multiple users to edit the same continuous geographic database at the same time. In GIS applications the objects of interest (geographic features) do not exist in isolation and are usually closely related to surrounding objects. For example, a tax parcel will share a common boundary with an adjacent parcel and changes in one will directly affect the other; similarly connected road segments in a highway network need to be edited together to ensure continued connectivity. It is relatively easy to provide multiple users with concurrent read and query access to a continuous shared database, but more difficult to deal with conflicts and avoid potential database corruption when multiple users want write (update) access. However, solutions to both these problems have been implemented in mainstream GIS and DBMS. These solutions extend standard DBMS transaction models and provide a multi-user framework called versioning.

10.9.1 Transactions

A group of edits to a database, such as the addition of three new land parcels and changes to the attributes of a sewer line, is referred to as a 'transaction'. In order to protect the integrity of databases, transactions are atomic, that is, transactions are either completely committed to the database or they are rolled back (not committed at all). Many of the world's GIS and non-GIS databases are multi-user and transactional, that is, they have multiple users performing edit/update operations at the same time. For most types of database, transactions take a very short time (sub-second). For example, in the case of a banking system a transfer from a savings account to a checking account takes perhaps 0.01 second. It is important that the transaction is coordinated between the accounts and that it is atomic, otherwise one account might be debited and the other not credited. Multi-user access to banking and similar systems is handled simply by locking (preventing access to) affected database records (table rows) during the course of the transaction. Any attempt to write to the same record is simply postponed until the record lock is removed after the transaction is completed. Because banking transactions, like many others, take only a very short amount of time, users never even notice if a transaction is deferred.

A transaction is a group of changes that are made to a database as a coherent group. All the changes that form part of a transaction are either committed or the database is rolled back to its initial state.

Although some geographic transactions have a short duration (short transactions), many extend to hours, weeks, and months, and are called long transactions. Consider, for example, the amount of time necessary to capture all the land parcels in a city subdivision (housing estate). This might take a few hours for an efficient operator, for a small subdivision, but an inexperienced operator working on a large subdivision might take days or weeks. This may cause two multi-user update problems. First, locking the whole or even part of a database to other updates for this length of time during a long transaction is unacceptable in many types of application, especially those involving frequent maintenance changes (e.g., utilities and land administration). Second, if a system failure occurs during the editing, work may be lost unless there is a procedure for storing updates in the database. Also, unless the data are stored in the database they are not easily accessible to others that would like to use them.

10.9.2 Versioning

Short transactions use what is called a pessimistic locking concurrency strategy. That is, it is assumed that conflicts will occur in a multi-user database with concurrent users and that the only way to avoid database corruption is to lock out all but one user during an update operation. The term 'pessimistic' is used because this is a very conservative strategy assuming that update conflicts will occur and that they must be avoided at all cost. An alternative to pessimistic locking is optimistic versioning, which allows multiple users to update a database at the same time. Optimistic versioning is based on the assumption that conflicts are very unlikely to occur, but if they do occur then software can be used to resolve them.

The two strategies for providing multi-user access to geographic databases are pessimistic locking and optimistic versioning.

Versioning sets out to solve the long transaction and pessimistic locking concurrency problem described above. It also addresses a second key requirement peculiar to geographic databases – the need to support alternative representations of the same objects in the database. In many applications, it is a requirement to allow designers to create and maintain multiple object designs. For example, when designing a new subdivision, the water department manager may ask two designers to lay out alternative designs for a water system. The two designers would work concurrently to add objects to the same database layers, snapping to the same objects. At some point, they may wish to compare designs and perhaps create a third design based on parts of their two designs. While this design process is taking place, operational maintenance editors could be changing the same objects they are working with. For example, maintenance updates resulting from new service connections or repairs to broken pipes will change the database and may affect the objects used in the new designs.

Figure 10.19 compares linear short transactions and branching long transactions as implemented in a versioned database. Within a versioned database, the different

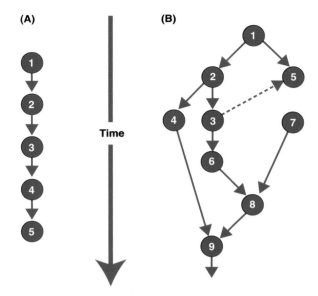

Figure 10.19 Database transactions: (A) linear short transactions; (B) branching version tree

database versions are logical copies of their parents (base tables), i.e., only the modifications (additions and deletions) are stored in the database (in version change tables). A query against a database version combines data from the base table with data in the version change tables. The process of creating two versions based on the same parent version is called branching. In Figure 10.19B, Version 4 is a branch from Version 2. Conversely, the process of combining two versions into one version is called merging. Figure 10.19B also illustrates the merging of Versions 6 and 7 into Version 8. A version can be updated at any time with any changes made in another version. Version reconciliation can be seen between Version 3 and 5. Since the edits contained within Version 5 were reconciled with 3, only the edits in Versions 6 and 7 are considered when merging to create Version 8.

There are no database restrictions or locks placed on the operations performed on each version in the database. The versioning database schema isolates changes made during the edit process. With optimistic versioning, it is possible for conflicting edits to be made within two separate versions, although normal working practice will ensure that the vast majority of edits made will not result in any conflicts (Figure 10.20). In the event that conflicts are detected, the data management software will handle them either automatically or interactively. If interactive conflict resolution is chosen, the user is directed to each feature that is in conflict and must decide how to reconcile the conflict. The GUI will provide information about the conflict and display the objects in their various states. For example, if the geometry of an object has been edited in the two versions, the user can display the geometry as it was in any of its previous states and then select the geometry to be added to the new database state.

An example will help to illustrate how versioning works in practice. In this scenario, a geographic database

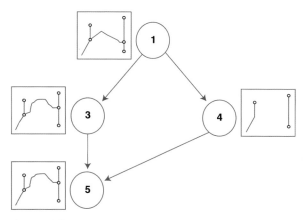

Figure 10.20 Version reconciliation. For Version 5 the user chooses via the GUI the geometry edit made in Version 3 instead of 1 or 4

of a water network similar to that shown in Figure 8.16, called *Main Plant* (Figure 10.21), is edited in parallel by four users each with different tasks. Users 1 and 2 are updating the *Main Plant* database with recent survey results (*Update 1* and *Update 2*), User 3 is performing some initial design prototype work (*Proposal 1*), and User 4 is following the progress of a construction project (*Plan A*).

The current state of the *Main Plant* database is represented by the *As Built* layer (this term is used in utility applications to denote the actual state of objects in

the real world). This version is marked as the default (base table) version of the database. Edits made to this version (1) are saved as a second version (2) which now becomes the *As Built* layer. User 4 branches the database, basing the new version (3), *Plan A*, on the *As Built* version (2). As work progresses, edits are made to *Plan A* (3) and in so doing two further states of *Plan A* are created (Versions 4 and 9). Meanwhile, User 3 creates a version (5) *Proposal 1*, which is based on User 4's *Plan A* (3). After initial design work, no further action is taken on *Proposal 1* (5), but for a historical record the version is left undeleted. Users 1 and 2 create versions of *As Built* (2) in order for their edits to be made (6 and 7). When Users 1 and 2 have completed editing, User 1 reconciles the edits in versions *Update 1* (6) and *Update 2* (7) by merging them together, so that version *Update 1* (8) contains all the edits. If any conflicts occur in the edits in the two versions then the versioning software will highlight these and support either automatic or manual conflict resolution. On completion of the work by User 4 the changes made in *Plan A* (9) are merged into the *As Built* version (10) and the intermediate versions (3, 4, and 9) can be deleted from the database (unless they are to be saved for historical reasons). Finally, the updates held in version *Update 1* (8) are merged into the *As Built* version (11).

Users performing read-only queries to the database and even simple edits (such as updating attributes) always work against the current state of the *As Built* layer. They are unaware of the other versions of *Main Plant* until they are made public, or are merged into the *Main Plant As Built* database layer. The *Main Plant* database is not directly edited by end users, conceptually each user is editing their own version of the *As Built* layer. Since the child versions of the *As Built* version are only logical copies, the system can be scaled to many concurrent users without the need for vast increases in processing and disk storage capacity. The changes to the database are actually performed with a GIS map editor using standard editing tools.

Versioning is a very useful mechanism for solving the problem of allowing multiple users to edit a shared continuous geographic database at the same time. In this section the discussion has considered the use of versioning in long transactions, design alternatives, and history (using versions to store the state of a database at specific points in time). In recent years the same versioning technology has been used to underpin replication between multiple databases. In version replication the version becomes the payload for transferring data between systems that need to be synchronized, and the versioning reconciliation framework is used to integrate changes.

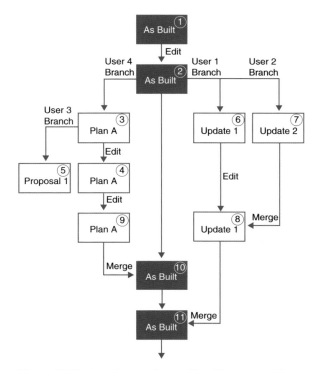

Figure 10.21 A version tree for the Main Plant geographic database. Version numbers are in circles

10.10 Conclusion

Database management systems are now a vital part of large modern operational GIS. They bring with them standardized approaches for storing and, more

Mike Worboys, mathematician and computer scientist

Mike Worboys is a graduate in mathematics and holds a masters degree in mathematical logic and the foundations of mathematics. His Ph.D. in pure mathematics developed computational approaches to very large algebraic structures arising from symmetries in high-dimensional spaces. Mike has worked as a mathematician and computer scientist and has had a long-standing interest in databases. His introduction to GIS was as a member of the newly created UK Midlands Regional Research Laboratory at Leicester in 1989. Mike's interests were in computational data modeling, and he realized that geographic data provided an interesting test-bed for the new object-oriented approaches being developed at that time. In his textbook *GIS: A Computing Perspective* (Worboys and Duckham 2004), Mike describes the database as the heart of a GIS. Much of Mike's work has been to bring more expressive approaches from the formal and computing communities to bear on the representation and management of geographic phenomena. He has subsequently been involved in research on computational models for spatio-temporal information and uncertainty in geographic information. He is presently Professor of Spatial Information Science and Engineering and a member of the National Center for Geographic Information and Analysis at the University of Maine.

Figure 10.22 Mike Worboys, mathematician and computer scientist

Reflecting on the contributions of mathematics and computing to geographic information science, Mike says: 'I increasingly believe that database technology can only take us so far along the road to effective systems. Understanding human conceptions and representations of geographic space, and translating these into formal and computational models, provide the real key. As for the future, I think the challenge is to build systems supporting truly user-centric views of information about both static and dynamic geographic phenomena.'

importantly, accessing and manipulating geographic data using the SQL query language. GIS provide the necessary tools to load, edit, query, analyze, and display geographic data. DBMS require a database administrator (DBA) to control database structure and security, and to tune the database to achieve maximum performance. Innovative work in the GIS field has extended standard DBMS to store and manage geographic data and has led to the development of long transactions and versioning that have application across several fields. Mike Worboys, a leading authority on spatial DBMS, offers his insights into the future of this important area of GIS in Box 10.3.

Questions for further study

1. Identify a geographic database with multiple layers and draw a diagram showing the tables and the relationships between them. Which are the primary keys, and which keys are used to join tables? Does the database have a good relational design?

2. What are the advantages and disadvantages of storing geographic data in a DBMS?

3. Is SQL a good language for querying geographic databases?

4. Why are there multiple methods of indexing geographic databases?

Further reading

Date C.J. 2003 *Introduction to Database Systems* (8th edn). Reading, MA: Addison-Wesley.

Hoel E., Menon S. and Morehouse S. 2003 'Building a robust relational implementation of topology.' In Hadzilacos T., Manolopoulos Y., Roddick J.F. and Theodoridis Y. (eds) *Advances in Spatial and Temporal Databases. Proceedings of 8th International Symposium, SSTD 2003 Lecture Notes in Computer Science*, Vol. 2750.

OGC 1999 OpenGIS simple features specification for SQL, Revision 1.1. Available at **www.opengis.org**

Samet H. 1990 *The Design and Analysis of Spatial Data Structures*. Reading, MA: Addison-Wesley.

van Oosterom P. 2005 'Spatial access methods.' In Longley P.A., Goodchild M.F., Maguire D.J. and Rhind D.W. (eds) *Geographic Information Systems: Principles, Techniques, Applications and Management (abridged edition)*. Hoboken, NJ: Wiley, 385–400.

Worboys M.F. and Duckham M. 2004 *GIS: A Computing Perspective* (2nd edn). Boca Raton, FL: CRC Press.

Zeiler M. 1999 *Modeling our World: The ESRI Guide to Geodatabase Design*. Redlands, CA: ESRI Press.

11 *Distributed GIS*

Until recently, the only practical way to apply GIS to a problem was to assemble all of the necessary parts in one place, on the user's desktop. But recent advances now allow all of the parts – the data and the software – to be accessed remotely, and moreover they allow the user to move away from the desktop and hence to apply GIS anywhere. Limited GI services are already available in common mobile devices such as cellphones, and are increasingly being installed in vehicles. This chapter describes current capabilities in distributed GIS, and looks to a future in which GIS is increasingly mobile and available everywhere. It is organized into three major sections, dealing with distributed data, distributed users, and distributed software.

Geographic Information Systems and Science, 2nd edition   Paul Longley, Michael Goodchild, David Maguire, and David Rhind.
© 2005 John Wiley & Sons, Ltd.   ISBNs: 0-470-87000-1 (HB); 0-470-87001-X (PB)

Learning Objectives

After reading this chapter you will understand:

■ How the parts of GIS can be distributed instead of centralized;

■ The concept of a geolibrary, and the standards and protocols that allow remotely stored data to be discovered and accessed;

■ The capabilities of mobile devices, including cellphones and wearable computers;

■ The concept of augmented reality, and its relationship to location-based services;

■ The concept of remotely invoked services, and their applications in GIS.

11.1 Introduction

Early computers were extremely expensive, forcing organizations to provide computing services centrally, from a single site, and to require users to come to the computing center to access its services. As the cost of computers fell, from millions of dollars in the 1960s, to hundreds of thousands in the late 1970s, and now to a thousand or so, it became possible for departments, small groups, and finally individuals to own computers, and to install them on their desktops (see Chapter 7). Today, of course, the average professional worker expects to have a computer in the office, and a substantial proportion of households in industrialized countries have computers in the home.

In Chapter 1 we identified the six component parts of a GIS as its hardware, software, data, users, procedures, and network. This chapter describes how the network, to which almost all computers are now connected, has enabled a new vision of *distributed* GIS, in which the component parts no longer need to be co-located. New technologies are moving us rapidly to the point where it will be possible for a GIS project to be conducted not only at the desktop but anywhere the user chooses to be, using data located anywhere on the network, and using software services provided by remote sites on the network.

In distributed GIS, the six component parts may be at different locations.

In Chapter 7 we already discussed some aspects of this new concept of distributed GIS. Many GIS vendors are now offering limited-functionality software that is capable of running in mobile devices, such as PDAs and tablet computers, which we have termed Hand-Held GIS (Section 7.6.4). Such devices are now widely used for field data collection, by statistical agencies such as the US Bureau of the Census or the US Department of Agriculture and by utilities and environmental companies. Such organizations now routinely equip their field crews with PDAs, allowing them to record data in the field and to upload the results back in the office (Figure 11.1); and similar systems are used by the field crews of utility companies to record the locations and status of transformers, poles, and switches (Figure 11.2). Vendors are also offering various forms of what we have termed Server or Internet GIS (Section 7.6.2). These allow GIS users to access processing services at remote sites, using little more than a standard Web browser, thus avoiding the cost of installing a GIS locally.

Certain concepts need to be clarified at the outset. First, there are four distinct locations of significance to distributed GIS:

■ The location of the user and the interface from which the user obtains GIS-created information, denoted by U (e.g., the doorway in Figure 11.2).

■ The location of the data being accessed by the user, denoted by D. Traditionally, data had to be moved to the user's computer before being used, but new technology is allowing data to be accessed directly from data warehouses and archives.

Figure 11.1 Using a simple GIS in the field to collect data. The device on the pole is a GPS antenna, used to georeference data as they are collected

Figure 11.2 Collecting survey data in the field, on behalf of a government statistical agency (Courtesy: Julie Dillemuth, João Hespanha, Stacy Rebich)

- The location where the data are processed, denoted by P. In Section 1.5.3 we introduced the concept of a GIService, a processing capability accessed at a remote site rather than provided locally by the user's desktop GIS.

- The area that is the focus of the GIS project, or the *subject* location, denoted by S. All GIS projects necessarily study some area, obtain data as a representation of the area, and apply GIS processes to those data.

In traditional GIS, three of these locations – U, D, and P – are the same, because the data and processing both occur at the user's desktop. The subject location could be anywhere in the world, depending on the project. But in distributed GIS there is no longer any need for D and P to be the same as U, and moreover it is possible for the user to be located in the subject area S, and able to see, touch, feel, and even smell it, rather than being in a distant office. The GIS might be held in the user's hand, or stuffed in a backpack, or mounted in a vehicle.

In distributed GIS the user location and the subject location can be the same.

Distributed GIS has many potential benefits, and these are discussed in detail in the following sections. Section 11.2 discusses the feasibility of distributing data, and the benefits of doing so, and introduces the technologies of distributed data access. Section 11.3 discusses the technologies of mobile computing, and the benefits of allowing the user to do GIS anywhere, and particularly within the subject area S. Section 11.4 reviews the status of distributed GIServices, and the benefits of obtaining processing from remote locations. It is important to distinguish at the outset between the *vision* of distributed GIS, and what is *practical* at this time, and so both perspectives are provided throughout the chapter.

Critical to distributed GIS are the standards and specifications that make it possible for devices, data, and processes to operate together, or *interoperate*. Some of these are universal, such as ASCII, the standard for

Biographical Box **11.1**

David Schell, proponent of open GIS

David Schell earned degrees in mathematics and literature from Brown University and the University of North Carolina. Initially, he worked in system development at IBM. Later, during the 1980s, David managed marketing and business development programs for various Massachusetts workstation startups. In these positions, he was actively involved in the development of Unix software libraries and the integration of productivity tools for office automation 'measurement and control' and scientific applications. In the late 1980s, while working at real-time computing pioneer Masscomp (later Concurrent Computer), David had the opportunity to work closely with the Corps of Engineers Geographic Resources Analysis Support System (GRASS) development team on technology transfer and the integration of GRASS open source with commercial GIS and imaging systems. This integration work resulted in an early prototype of the principles of 'Web mapping'. Carl Reed and John Davidson of Genasys developed an operational system in which a user could access both GenaMap and GRASS seamlessly from the same command line and where map graphics from both systems displayed into the same X-Window. This resulted in the development of some of the earliest interface-based interoperability models for geospatial processing.

Figure 11.3 David Schell, President, Open Geospatial Consortium

In 1992, with the aid of a Cooperative Research and Development Agreement with the US Army Corps of Engineers USACERL, David founded the Open GIS Foundation (OGF) to facilitate technology transfer of COE technology to the private sector as well as continuing the commercial interoperability initiatives begun with GenaMap. In 1994, responding to the need to engage geospatial technology users and providers in a formal consensus standards process, he reorganized OGF to become the Open GIS Consortium (now the Open Geospatial Consortium). With the support of both public and private sector sponsors, the OpenGIS Project began an intense and highly focused industry-wide consensus process to create an architecture to support interoperable geoprocessing. David was cited in 2003 as one of the 20 most influential innovators in the IT community by the editors of *CIO Magazine* and received the CIO 20/20 Award for visionary achievement.

Speaking about the industry's evolution, David Schell said: 'It is most important for us to realize the far-reaching cultural impact of geospatial interoperability, in particular its influence on the way we conceptualize the challenges of policy and management in the modern industrialized world. Without such a capability, we would still be wasting most of our creative energies laboring in the dark ages of time-consuming and laborious data conversion and hand-made application stove-pipes. Now it is possible for scientists and thinkers in every field to focus their energies on problems of real intellectual merit, instead of having to wrestle exhaustingly to reconcile the peculiarities of data imposed by diverse spatial constructs and vendor-limited software architectures. As our open interfaces make geospatial data and services discoverable and accessible in the context of the World Wide Web, geospatial information becomes truly more useful to more people and its potential for enabling progress and enlightenment in many domains of human activity can be more fully realized.'

coding of characters (Box 3.1), and XML, the extensible markup language that is used by many services on the Web. Others are specific to GIS, and many of these have been developed through the Open Geospatial Consortium (OGC, **www.opengeospatial.org**), an organization set up to promote openness and interoperability in GIS (see Box 11.1). Among the many successes of OGC over the past decade are the simple feature specification, which standardizes many of the terms associated with primitive elements of GIS databases (polygons, polylines, points, etc.; see also Chapter 9); Geography Markup Language (GML), a version of XML that handles geographic features and enables open-format communication of geographic data; and specifications for Web services (Web Map Service, Web Feature Service, Web Coverage Services) that allow users to request data automatically from remote servers.

Distributed GIS reinforces the notion that today's computer is not simply the device on the desk, with its hard drive, processor, and peripherals, but something more extended. The slogan 'The Network is the Computer' has provided a vision for at least one company – Sun Microsystems (**www.sun.com**) – and propels many major developments in computing. The term *cyberinfrastructure* describes a new approach to the conduct of science, relying on high-speed networks, massive processors, and distributed networks of sensors and data archives (**www.cise.nsf.gov/sci/reports/toc.cfm**). Efforts are being made to integrate the world's computers to provide the kinds of massive computing power that are needed by such projects as SETI (the Search for Extra-Terrestrial Intelligence, **www.seti.org**), which processes terabytes of data per day in the search for anomalies that might indicate life elsewhere in the universe, and makes

Figure 11.4 The power of this home computer would be wasted at night when its owner is sleeping, so instead its power has been 'harvested' by a remote server and used to process signals from radio telescopes as part of a search for extra-terrestrial intelligence

use of computer power wherever it can find it on the Internet (Figure 11.4). *Grid* computing is a generic term for such a fully integrated worldwide network of computers and data.

11.2 Distributing the data

Since its popularization in the early 1990s the Internet has had a tremendous and far-reaching impact on the accessibility of GIS data, and on the ability of GIS users to share datasets. As we saw in Chapter 9, a large and increasing number of websites offer GIS data, for free, for sale, or for temporary use, and also provide services that allow users to search for datasets satisfying certain requirements. In effect, we have gone in a period of little more than ten years from a situation in which geographic data were available only in the form of printed maps from map libraries and retailers, to one in which petabytes (Table 1.1) of information are available for download and use at electronic speed (about 1.5 million CDs would be required to store 1 petabyte). For example, the NASA-sponsored EOSDIS (Earth Observing System Data and Information System; **spsosun.gsfc.nasa.gov/New_EOSDIS.html**) archives and distributes the geographic data from the EOS series of satellites, acquiring new data at over a terabyte per day, with an accumulated total of more than 1 petabyte at this site alone.

Some GIS archives contain petabytes of data.

The vision of distributed GIS goes well beyond the ability to access and retrieve remotely located data, however, because it includes the concepts of *search*, *discovery*, and *assessment*: in the world of distributed GIS, how do users search for data, discover their existence at remote sites, and assess their fitness for use? Three concepts are important in this respect: *object-level metadata*, *geolibraries*, and *collection-level metadata*.

11.2.1 Object-level metadata

Strictly defined, metadata are data about data, and *object-level* metadata (OLM) describe the contents of a single dataset by providing essential documentation. We need information about data for many purposes, and OLM try to satisfy them all. First, we need OLM to automate the process of search and discovery over archives. In that sense OLM are similar to a library's catalog, which organizes the library's contents by author, title, and subject, and makes it easy for a user to find a book. But OLM are potentially much more powerful, because a computer is more versatile than the traditional catalog in its potential for re-sorting items by a large number of properties, going well beyond author, title, and subject, and including geographic location. Second, we need OLM to determine whether a dataset, once discovered, will satisfy the user's requirements – in other words, to assess the fitness of a dataset for a given use. Does it have sufficient spatial resolution, and acceptable quality? Such metadata may include comments provided by others who tried to use the data, or contact information for such previous users (users often comment that the most useful item of metadata is the phone number of the person who last tried to use the data). Third, OLM must provide the information needed to handle the dataset effectively. This may include technical specifications of format, or the names of software packages that are compatible with the data, along with information about the dataset's location, and its volume. Finally, OLM may

provide useful information on the dataset's contents. In the case of remotely sensed images, this may include the percentage of cloud obscuring the scene, or whether the scene contains particularly useful instances of specific phenomena, such as hurricanes.

Object-level metadata are formal descriptions of datasets that satisfy many different requirements.

OLM generalize and abstract the contents of datasets, and therefore we would expect that they would be smaller in volume than the data they describe. In reality, however, it is easy for the complete description of a dataset to generate a greater volume of information than the actual contents. OLM are also expensive to generate, because they represent a level of understanding of the data that is difficult to assemble, and require a high level of professional expertise. Generation of OLM for a geographic dataset can easily take much longer than it takes to catalog a book, particularly if it has to deal with technical issues such as the precise geographic coverage of the dataset, its projection and datum details (Chapter 5), and other properties that may not be easily accessible. Thus the cost of OLM generation, and the incentives that motivate people to provide OLM, are important issues.

For metadata to be useful, it is essential that they follow widely accepted standards. If two users are to be able to share a dataset, they must both understand the rules used to create its OLM, so that the custodian of the dataset can first create the description, and so that the potential user can understand it. The most widely used standard for OLM is the US Federal Geographic Data Committee's Content Standards for Digital Geospatial Metadata, or CSDGM, first published in 1993 and now the basis for many other standards worldwide. Box 11.2 lists some of its major features. As a *content standard* CSDGM describes the items that should be in an OLM archive, but does not prescribe exactly how they should be formatted or structured. This allows developers to implement the standard in ways that suit their own software environments, but guarantees that one implementation will be understandable to another – in other words, that the implementations will be *interoperable*. For example, ESRI's ArcGIS provides two formats for OLM, one using the widely recognized XML standard and the other using ESRI's own format.

CSGDM was devised as a system for describing geographic datasets, and most of its elements make sense only for data that are accurately georeferenced and represent the spatial variation of phenomena over the Earth's surface. As such, its designers did not attempt to place CSGDM within any wider framework. But in the past decade a number of more broadly based efforts have also been directed at the metadata problem, and at the extension of traditional library cataloging in ways that make sense in the evolving world of digital technology.

One of the best known of these is the Dublin Core (see Box 11.3), the outcome of an effort to find the minimum set of properties needed to support search and discovery for datasets in general, not only geographic datasets. Dublin Core treats both space and time as instances of a single property, coverage, and unlike CSGDM

Technical Box **11.2**

Major features of the US Federal Geographic Data Committee's Content Standards for Digital Geospatial Metadata

1. Identification Information – basic information about the dataset.

2. Data Quality Information – a general assessment of the quality of the dataset.

3. Spatial Data Organization Information – the mechanism used to represent spatial information in the dataset.

4. Spatial Reference Information – the description of the reference frame for, and the means to encode, coordinates in the dataset.

5. Entity and Attribute Information – details about the information content of the dataset, including the entity types, their attributes, and the domains from which attribute values may be assigned.

6. Distribution Information – information about the distributor of and options for obtaining the dataset.

7. Metadata Reference Information – information on the currentness of the metadata information, and the responsible party.

8. Citation Information – the recommended reference to be used for the dataset.

9. Time Period Information – information about the date and time of an event.

10. Contact Information – identity of, and means to communicate with, person(s) and organization(s) associated with the dataset.

Technical Box **11.3**

The fifteen basic elements of the Dublin Core metadata standard

1. TITLE. The name given to the resource by the CREATOR or PUBLISHER.

2. AUTHOR or CREATOR. The person(s) or organization(s) primarily responsible for the intellectual content of the resource.

3. SUBJECT or KEYWORDS. The topic of the resource, or keywords, phrases, or classification descriptors that describe the subject or content of the resource.

4. DESCRIPTION. A textual description of the content of the resource, including abstracts in the case of document-like objects or content description in the case of visual resources.

5. PUBLISHER. The entity responsible for making the resource available in its present form, such as a publisher, a university department, or a corporate entity.

6. OTHER CONTRIBUTORS. Person(s) or organization(s) in addition to those specified in the CREATOR element who have made significant intellectual contributions to the resource, but whose contribution is secondary to the individuals or entities specified in the CREATOR element.

7. DATE. The date the resource was made available in its present form.

8. RESOURCE TYPE. The category of the resource, such as home page, novel, poem, working paper, technical report, essay, dictionary.

9. FORMAT. The data representation of the resource, such as text/html, ASCII, Postscript file, executable application, or JPEG image.

10. RESOURCE IDENTIFIER. String or number used to uniquely identify the resource.

11. SOURCE. The work, either print or electronic, from which this resource is delivered, if applicable.

12. LANGUAGE. Language(s) of the intellectual content of the resource.

13. RELATION. Relationship to other resources.

14. COVERAGE. The spatial locations and temporal duration characteristic of the resource.

15. RIGHTS MANAGEMENT. The content of this element is intended to be a link (a URL or other suitable URI as appropriate) to a copyright notice, a rights-management statement, or perhaps a server that would provide such information in a dynamic way.

does not lay down how such specific properties as spatial resolution, accuracy, projection, or datum should be described.

The principle of establishing a minimum set of properties is sharply distinct from the design of CSGDM, which was oriented more toward the capture of all knowable and potentially important properties of geographic datasets. Of direct relevance here is the problem of cost, and specifically the cost of capturing a full CSGDM metadata record. While many organizations have wanted to make their data more widely available, and have been driven to create OLM for their datasets, the cost of determining the full set of CSDGM elements is often highly discouraging. There is interest therefore in a concept of *light metadata*, a limited set of properties that is both comparatively cheap to capture, and still useful to support search and discovery. Dublin Core represents this approach, and thus sits at the opposite end of a spectrum from CSGDM. Every organization must somehow determine where its needs lie on this spectrum, which ranges from light and cheap to heavy and expensive.

Light, or stripped-down, OLM provide a short but useful description of a dataset that is cheaper to create.

11.2.2 Geolibraries

The use of digital technology to support search and discovery opens up many options that were not available in the earlier world of library catalogs and bookshelves. Books must be placed in a library on a permanent basis, and there is no possibility of reordering their sequence – but in a digital catalog it is possible to reorder the sequence of holdings in a collection almost instantaneously. So while a library's shelves are traditionally sorted by subject, it would be possible to re-sort them digitally by author name or title, or by any property in the OLM catalog. Similarly the traditional card catalog allowed only three properties to be sorted – author, title, and subject – and discouraged sorting by multiple subjects. But the digital catalog can support any number of subjects.

Of particular relevance to GIS users is the possibility of sorting a collection by the coverage properties: location and time. Both the spatial and temporal dimensions are continuous, so it is impossible to capture them in a single property analogous to author that can then be sorted numerically or alphabetically. But in a digital system this is not a serious problem, and it is straightforward to capture the coverage of a dataset, and to allow the user to search for datasets that cover an area or time of interest defined by the user. Moreover, the properties of location and time are not limited to geographic datasets, since there are many types of information that are associated with specific areas on the Earth's surface, or with specific time periods. Searching based on location or time would enable users to find information about any place on the Earth's surface, or any time period – and to find reports, photographs, or even pieces of music, as long as they possessed geographical and temporal *footprints*.

The term *geolibrary* has been coined to describe digital libraries that can be searched for information about any user-defined geographic location. A US National Research Council report (see Further Reading) describes the concept and its implementation, and many instances of geolibraries can be found on the Web (see Box 11.4). Digital Earth is a vision of a geolibrary in which the user interacts with a visual representation of the surface of the Earth, rather than with a digital representation of the traditional card catalog.

A geolibrary can be searched for information about a specific geographic location.

11.2.3 Collection-level metadata

Thus far we have seen how metadata can be useful in the description of individual datasets. There are several other forms of metadata of potential interest to GIS users, and this section describes the issues associated with one of them – metadata that describe entire collections of datasets, rather than individual datasets.

Many different collections or archives of geographic datasets exist on the Internet, and Box 11.4 and Table 9.4 identify several examples. The Geospatial One-Stop is structured as a mechanism that allows custodians to make their data visible to users who access the portal, and that allows users to search across multiple servers using a simple protocol, assess the appropriateness of datasets, and access them through their own GIS. The US Geological Survey supports several major collections of geographic data, notably at the EROS Data Center in Sioux Falls, South Dakota (**edcwww.cr.usgs.gov**), and major commercial collections of data can be found at the sites of Space Imaging (**www.spaceimaging.com**), Terraserver (**www.terraserver.com**), GlobeXplorer (**www.globexplorer.com**), and at many other locations. Clearinghouses are also sponsored by many other countries and by several US states.

But this abundance of readily accessible data presents a problem – how does the user know which collection to search for a given dataset? Although the Geospatial One-Stop hopes to be a single portal for a large proportion of geographic data, its emphasis is on the US, and no site, however successful, will ever have a monopoly of all data for all parts of the world. The term *collection-level metadata* (CLM) defines the information needed to make an intelligent choice, based on general knowledge of each collection's contents. For example, the EROS Data Center is a good site to search for images and maps of the US. Microsoft's Terraserver (**terraserver.microsoft.com**) is a good source for imagery of any part of the world. In looking for maps of a specific country, it makes sense to try to find a site sponsored by that country's national mapping agency, and similarly, information about a state may exist in a clearinghouse sponsored by that state.

Collection-level metadata describe the entire collection available through a single website.

Unfortunately there has been little consistent effort to develop standards for CLM, unlike its individual dataset counterpart (OLM). Knowledge of CLM is still

Exemplar geolibraries

The Alexandria Digital Library (**www.alexandria. ucsb.edu**) provides access to over 2 million maps and images using a simple method of search based on user interaction with a world map, placenames, or coordinates. Many of the maps and images are stored in the Map and Imagery Laboratory at the University of California, Santa Barbara, but the Alexandria catalog also identifies materials in other collections. Figure 11.5 shows a screen shot of the library's Web interface.

The US Geospatial One-Stop (**www.geo-one-stop.gov**; Figure 11.6) is intended to provide a single portal (a *geoportal*) to vast and distributed resources of geographic data. Managed by the Department of the Interior, it presents a catalog of datasets residing in federal, state, local, tribal, and private archives, and allows users to search across these resources independently of their storage locations. Its benefits include reduced duplication, by encouraging users to share the data of others rather than creating them themselves, and increased collaboration between agencies. Users can volunteer their own datasets by registering ('publishing') them to the portal. Geoportals can also act as directories to remotely accessible GIServices (Section 11.4).

(A)

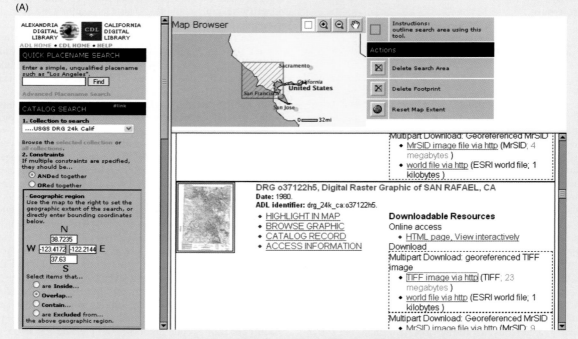

Figure 11.5 Example of a geolibrary: the Alexandria Digital Library (**alexandria.ucsb.edu**). (A) shows the interface to the library in a standard Netscape browser. The user has selected an area of California to search, and the system has returned the first 50 items in the library whose metadata match the search area footprint and the user's request for scanned images of 1:24 000 topographic maps. (B) shows one of the 'hits', part of the San Rafael quadrangle centered on Muir Woods National Monument

(B)

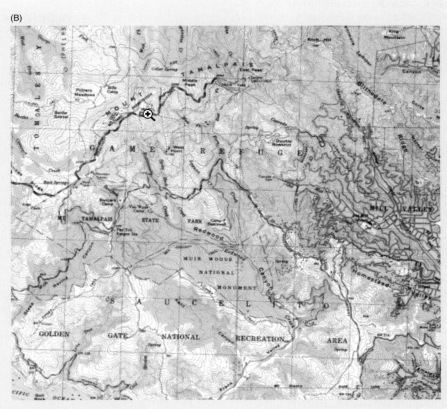

Figure 11.5 (*continued*)

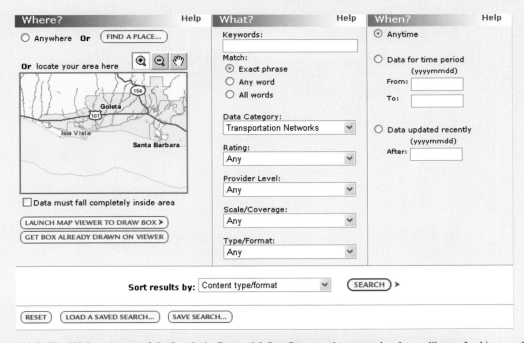

Figure 11.6 The US Department of the Interior's Geospatial One-Stop, another example of a geolibrary. In this case the user has zoomed in to the area of Goleta, California, and requested data on transportation networks. A total of 66 'hits' were identified in this search, indicating 66 possible sources of suitable information

largely unorganized, and tends to accumulate slowly in the minds of geographic information specialists (or SAPs, see Section 1.4.3.2). Knowing where to look is still largely a matter of personal knowledge and luck.

Various possible solutions to this problem can be identified. In the future, we may find the means to develop a new generation of Web search engines that are able to discover geographic datasets automatically, and catalog their OLM. Then a user needing information on a particular area could simply access the search engine's results, and be provided with a list of appropriate collections. That capability is well beyond the reach of today's generation of search engines, although several format standards, including GeoTIFF, already add the tags to geographic datasets that clever search engines could recognize. Alternatively, some scheme might be developed that authorizes a uniform system of servers, such that each server's contents are defined precisely, using geographic, thematic, or other criteria. In such a system all data about locations within the server's nominal coverage area, such as a state or a range of latitude and longitude, would be found on that server and no other. Such a system would need to be hierarchical, with very detailed data in collections with small coverage areas, and coarser data in collections with larger coverage areas. But at this time the CLM problem remains a major impediment to effective sharing of geographic data.

11.3 The mobile user

Computing has become so much a part of our lives that for many people it is difficult to imagine life without it. Increasingly, we need computers to shop, to communicate with friends, to obtain the latest news, and to entertain ourselves. In the early days, the only place one could compute was in a computing center, within a few meters of the central processor. Computing had extended to the office by the 1970s, and to the home by the 1990s. The portable computers of the 1980s opened the possibility of computing in the garden, at the beach, or 'on the road' in airports and airplanes. Wireless communication services such as WiFi (the wireless access technology based on the 802.11 family of standards) now allow broadband communication to the Internet from 'hot-spots' in many hotels, restaurants, airports, office buildings, and private homes (Figure 11.7). The range of mobile computing devices is also multiplying rapidly, from the relatively cumbersome but powerful laptop weighing several kg to the PDA, tablet computer, and the cellphone weighing a hundred or so grams. Within a few years we will likely see the convergence of these devices into a single, low-weight, and powerful mobile personal device that acts as computer, storage device, and cellphone.

In many ways the ultimate end point of this progression is the *wearable* computer, a device that is fully embedded

Figure 11.7 Map of WiFi (802.11) wireless broadband 'hotspots' within 1 mile of the White House (1600 Pennsylvania Ave, Washington, DC) and using the T-Mobile Internet provider

Figure 11.8 A wearable computer in use. The outfit consists of a processor and storage unit hung on the user's waist belt; an output unit clipped to the eyeglasses with a screen approximately 1 cm across and VGA resolution; an input device in the hand; and a GPS antenna on the shoulder. The batteries are in a jacket pocket (Courtesy: Keith Clarke)

in the user's clothing, goes everywhere, and provides ubiquitous computing service. Such devices are already obtainable, though in fairly cumbersome and expensive form. They include a small box worn on the belt and containing the processor and storage, and an output display clipped to the user's eyeglasses. Figure 11.8 shows such a device in use.

> **Computing has moved from the computer center and is now close to, even inside, the human body. The ultimate form of mobile computing is the wearable computer.**

What possibilities do such systems create for GIS? Of greatest interest is the case where the user is situated in the subject location, that is, *U* is contained within *S*, and the GIS is used to analyze the immediate surroundings. A convenient way to think about the possibilities is by comparing *virtual* reality with *augmented* reality.

11.3.1 Virtual reality and augmented reality

One of the great strengths of GIS is the window it provides on the world. A researcher in an office in Nairobi, Kenya might use GIS to obtain and analyze data on the effects of salinization in the Murray Basin of Australia, combining images from satellites with base topographic data, data on roads, soils, and population distributions, all obtained from different sources via the Internet. In doing so the researcher would build up a comprehensive picture of a part of the world he or she might never have visited, all through the medium of digital geographic data and GIS, and might learn almost as much through this GIS window

as from actually being there. The expense and time of traveling to Australia would be avoided, and the analysis could proceed almost instantaneously. Some aspects of the study area would be missing, of course – aspects of culture, for example, that can best be experienced by meeting with the local people.

Research environments such as this are termed *virtual realities*, because they replace what humans normally gather through their senses – sight, sound, touch, smell, and taste – by presenting information from a database. In most GIS applications only one of the senses, sight, is used to create this virtual reality, or VR. In principle, it is possible to record sounds and store them in GIS as attributes of features, but in practice very little use is made of any sensory channel other than vision. Moreover, in most GIS applications the view presented to the user is the view from *above*, even though our experience at looking at the world from this perspective is limited (for most of us, to times when we requested a window seat in an airplane). GIS has been criticized for what has been termed the *God's eye view* by some writers, on the basis that it distances the researcher from the real conditions experienced by people on the ground.

> **Virtual environments attempt to place the user in distant locations.**

More elaborate VR systems are capable of *immersing* the user, by presenting the contents of a database in a three-dimensional environment, using special eyeglasses or by projecting information onto walls surrounding the user, and effectively *transporting* the user into the environment represented in the database. Virtual London (Box 13.7 and see **www.casa.ucl.ac.uk/research/virtuallondon.htm**) and the Virtual Field Course (**www.geog.le.ac.uk/vfc/**) are examples of projects to build databases to support near-immersion in particular environments. Some of the most interesting of these are projects that recreate historic environments, such as those of the Cultural VR Lab at the University of California, Los Angeles (**www.cvrlab.org**), which supports roaming through three-dimensional visualizations of the Roman church of Santa Maria Maggiore (Figure 11.9) and other classical structures.

Roger Downs, Professor of Geography at Pennsylvania State University, uses Johannes Vermeer's famous painting *The Geographer* (Figure 11.10) to make an important point about virtual realities. On the table in front of the figure is a map, representing the geographer's window on a part of the world that happens to be of interest. But the subject figure is shown looking out of the window, at the real world, perhaps because he needs the information he derives from his senses to understand the world as shown on the map. The idea of combining information from a database with information derived directly through the senses is termed *augmented reality*, or AR. In terms of the locations of computing discussed earlier, AR is clearly of most value when the location of the user *U* is contained within the subject area *S*, allowing the user to augment what can be seen directly with information retrieved about the same area from a database. This might include historic information, or predictions about

Figure 11.9 Image generated from a three-dimensional digital reconstruction of the Basilica of Santa Maria Maggiore by the Virtual Reality Laboratory of the University of California, Los Angeles (**www.cvrlab.org**). The basilica was built in Rome around 440 AD and has long since been destroyed (Reproduced by permission of Regents of the University of California

Figure 11.10 Johannes Vermeer's painting *The Geographer* from 1669

(Figure 11.11), includes a differential GPS for accurate positioning, a GIS database that includes very detailed information on the immediate environment, a compass to determine the position of the user's head, and a pair of earphones. The user gives the system information to identify the desired destination, and the system then generates sufficient information to replace the normal role of sight. This might be through verbal instructions, or by generating stereo sounds that appear to come from the appropriate direction.

Augmented reality combines information from the database with information from the senses.

Steven Feiner, Professor of Computer Science at Columbia University, has demonstrated another form of AR that superimposes historic images and other information directly on the user's field of view. For example, in Figures 11.12 and 11.13, a user wearing a head-mounted device coupled to a wearable computer is seeing both the Columbia University main library building, and also an image generated from a database showing the building that occupied the library's position prior to the University's move to this site in 1896: the Bloomingdale Insane Asylum.

Figure 11.11 The system worn here by Reg Golledge (a leader of the development team) uses GIS and GPS to augment the senses of a visually impaired person navigating through a complex space such as a university campus (Reproduced by permission of Reg Golledge)

the future, or information that is for some other reason invisible to the user.

AR can be used to augment or replace the work of senses that are impaired or absent. A team led by Reginald Golledge, Professor of Geography at University of California, Santa Barbara (see Box 1.8), has been experimenting with AR as a means of helping visually impaired people to perform the simple task of navigation. The system, which is worn by the user

Figure 11.12 An augmented reality system developed at Columbia University by Prof. Steve Feiner and his group © 1999, T. Höllerer, S. Feiner & J. Pavlik, Computer Graphics & User Interfaces Lab., Columbia University. Reproduced with permission

11.3.2 Location-based services

One of the four big trends in software is location-based applications.
Bill Gates, Wireless 2000 Conference, March 2000

A location-based service (LBS) is defined as an information service provided by a device *that knows where it is* and is capable of *modifying* the information it provides based on that knowledge. Traditional computing devices, such as desktops or laptops, have no way of knowing where they are, and their functions are in no way changed when they are moved. But increasingly the essential information about a device's location is available, and is being used for a wide range of purposes, some beneficial and some distinctly questionable (see Box 5.2).

A location-based service is provided by a computing device that knows where it is.

The simplest and most obvious form of locationally enabled device is the GPS receiver, and any computer that includes a GPS capability, such as a laptop or PDA with an added PCMCIA card or chip, is capable of providing LBS. But the most ubiquitous LBS-capable device is the modern cellphone. A variety of methods exist for determining cellphone locations, including GPS embedded in the cellphone itself, measurements made by the cellphone of signals received from cellphone towers, and measurements made by the towers of signals received

Figure 11.13 The field of view of the user of Feiner's AR system, showing the Columbia University main library and the insane asylum that occupied the site in the 1800s © 1999, T. Höllerer, S. Feiner & J. Pavlik, Computer Graphics & User Interfaces Lab., Columbia University. Reproduced with permission

from the phone. As long as the phone is on, the operator of the cellphone network is able to pinpoint its location at least to the accuracy represented by the size of the cell, of the order of 10 km, and frequently much more accurately. Within a few years it is likely that the locations of the vast majority of active phones will be known to an accuracy of 10 m.

One of the strongest motives driving this process is emergency response. A large and growing proportion of emergency calls come from cellphones, and while the location of each land-line phone is likely to be recorded in a database available to the emergency responder, in a significant proportion of cases the user of a cellphone is unable to report his or her current location to sufficient accuracy to enable effective response. Several well-publicized cases have drawn attention to the problem. The magazine *Popular Science*, for example, reported a case of a woman who lost control of her car in Florida in February 2001, skidding into a canal. Although she called 911 (the emergency number standard in the US and Canada), she was unable to report her location accurately and died before the car was found.

One solution to the 911 problem is to install GPS in the vehicle, communicating location directly to the dispatcher. The Onstar system (**www.onstar.com**) is one such system, combining a GPS device and wireless communication. The system has been offered in the US to Cadillac purchasers for several years, and provides various advisory services in addition to emergency response, such as advice on local attractions and directions to local services. When the vehicle's airbags inflate, indicating a likely accident, the system automatically measures and radios its location to the dispatcher, who relays the necessary information to the emergency services.

Emergency services provide one of the strongest motivations for LBS.

There are many other examples of LBS that take advantage of locationally enabled cellphones. A *yellow-page* service responds to a user who requests information on businesses that are close to his or her current location (Where is the nearest pizza restaurant? Where is the nearest hospital? Where is the nearest WiFi hotspot?) by sending a request that includes the location to a suitable Web server. The response might consist of an ordered list presented on the cellphone screen, or a simple map centered on the user's current location (Figure 11.7). A *trip planner* gives the user the ability to find an optimum driving route from the current location to some defined destination. Similar services are now being provided by public transport operators, and in some cases these services make use of GPS transponders on buses and trains to provide information on actual, as distinct from scheduled, arrival and departure times. AT&T's Find Friends service allows a cellphone user to display a map showing the current locations of nearby friends (provided their cellphones are active and they are also registered for this service). Undercover (**www.playundercover.com**) is an example of a *location-based game* that involves the actual locations of players (see Box 11.5) moving around a real environment.

Direct determination of location, using GPS or measurement to or from towers, is only one basis on which a computing device might know its location, however. Other forms of LBS are provided by fixed devices, and rely on the determination of location when the device was installed. For example, many *point-of-sale* systems that are used by retailers record the location of the sale, combining it with other information about the buyer obtained by accessing credit-card or store-affinity-card records (Figure 11.18). In exchange for the convenience

---

Biographical Box 11.5

Antonio Câmara, designer of location-based services

Antonio Câmara (Figure 11.14) is the CEO and co-founder of YDreams (**www.ydreams.com**), a company specializing in location-based games and novel applications of computer graphics. A Professor at the New University of Lisbon in Portugal, he has been involved in GIS since 1977. Antonio Câmara was a member of the team that developed SNIG, the first spatial data infrastructure on the Internet, in 1995.

The mobile-phone-based entertainment market is expected to reach between 8 and 25 billion Euros a year in revenues by 2008. Games where the player's location is known are becoming a prominent component of that market. Undercover (**www.playundercover.com**), developed by YDreams, was the first location-based game with a visual interface. Undercover is a multi-player game where players can act solely or in a group. The purpose is to eliminate terrorists, which are either virtual or result from an infection acquired by a player. The positioning of players and terrorists can be seen on the mobile-phone screen (Figure 11.15).

Figure 11.14 Antonio Câmara, CEO and co-founder of YDreams, a company specializing in location-based games (Reproduced by permission of Antonio Câmara)

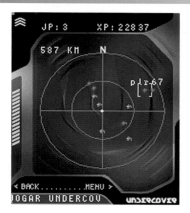

Figure 11.15 Cellphone display for a user playing Undercover: scanning for friends (blue) and foes (red) (Reproduced by permission of Antonio Câmara)

Figure 11.16 The Lex Ferrum location-based game developed by YDreams relies on the display of three-dimensional characters on the cellphone screen (Reproduced by permission of Antonio Câmara)

Figure 11.17 A Formula 1 race played on a large screen using mobile phones. The project was developed by IDEO and YDreams for Vodafone (Reproduced by permission of Antonio Câmara)

In the first version, Undercover relied on simplified maps where only main points of interest were considered. Positioning was based on Cell-ID techniques provided by the operator. Undercover's second version will use city maps in the background. Positioning may use a range of systems including GPS units communicating via Bluetooth with the mobile phones. YDreams has also developed a proximity-based location game called Lex Ferrum (Figure 11.16) for the NOKIA N-Gage. In this game, the player's location is scanned via Bluetooth or GPRS/UMTS. Local combat is done using Bluetooth. Remote combat may use the cellular network.

Such games will also benefit from the widespread deployment of large screens in cities. Mobile phones can then be used as controllers of such screens. Imagine a user visualizing a three-dimensional representation of the city, querying the underlying GIS and playing games such as an imaginary Formula 1 race (Figure 11.17).

Figure 11.18 Credit card in use (Reproduced with permission of PhotoDisc, Inc.)

of a credit card, the user effectively surrenders some degree of location privacy to the company whenever the card is used. One benefit is that it is possible for the company to analyze transactions, looking for patterns of purchase that are outside the card user's normal buying habits, perhaps because they occur in locations that the user does not normally frequent, or at anomalous times. Many of us will have experienced the embarrassment of having a credit card transaction refused in an unfamiliar city, because the techniques used by the company have flagged the transaction as an indicator the card might have been stolen. In principle, a store-affinity card gives the company access to information about buying habits and locations, in return for a modest discount.

Location is revealed every time a credit, debit, or store affinity card is used.

Another technique relies on the locations given when an Internet IP address is registered, augmented by location information that may appear on Web pages. Companies such as InfoSplit (**www.infosplit.com**; Figure 11.19) specialize in gathering and providing lists of such locations to customers. A search-engine company such as Google would be interested in knowing the locations of computers that send it queries, because it would then be possible to sort responses by distance from the computer – ranking in this way would be of commercial interest to retailers, for example, because it would make their sites more conspicuous to local customers.

11.3.3 Issues in mobile GIS

GIS in the field or 'on the road' is very different from GIS in the office. First, the location of the user is important, and directly relevant to the application. It makes good sense to center maps on the user's location, to provide the capability for maps that show the view from the user's location rather than from above, and to offer maps that are oriented to the user's direction of travel, rather than north. Second, the field environment may make certain kinds of interaction impractical, or less desirable. In a moving vehicle, for example, it would be dangerous to present the driver with visual displays, unless perhaps these are directly superimposed on the field of view (on the windshield) – instead, such systems often provide instructions through computer-generated speech, and use speech recognition to receive instructions. With wearable devices that provide output on minute screens attached to the user's eyeglasses there is no prospect of conventional point-and-click with a mouse, so again voice communication may be more appropriate. On the other hand, many environments are noisy, creating problems for voice recognition.

One of the most important limitations to mobility remains the battery, and although great strides have been made in recent years, battery technology has not advanced as rapidly as other components of mobile systems, such as processors and storage devices. Batteries typically account for the majority of a mobile system's weight, and limit its operating time.

The battery remains the major limitation to LBS and to mobile computing in general.

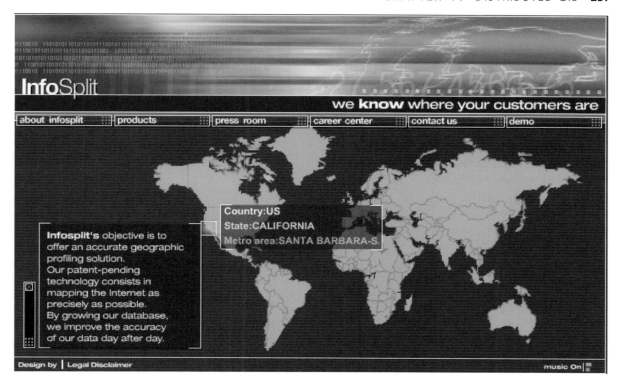

Figure 11.19 Companies such as InfoSplit (**www.infosplit.com**) specialize in determining the locations of computers from their IP addresses. In this case the server has correctly determined that the 'hit' comes from Santa Barbara, California

Although broad band wireless communication is possible using WiFi, connectivity remains a major issue. Wireless communication techniques tend to be:

- Limited in spatial coverage. WiFi hotspots are limited to a single building or perhaps a hundred meters from the router; cellphone-based techniques have wider coverage but lower bandwidth (communication speeds), and only the comparatively slow satellite-based systems approach global coverage.

- Noisy. Cellphone and WiFi communications tend to 'break up' at the edges of coverage areas, or as devices move between cells, leading to errors in communication.

- Limited in temporal coverage. A moving device is likely to lose signal from time to time, while some devices, particularly recorders installed in commercial vehicles and PDAs used for surveys, are unable to upload data until within range of a home system.

- Insecure. Wireless communications often lack adequate security to prevent unwanted intrusion. It is comparatively easy, for example, for someone to tap into a wireless session and obtain sensitive information.

Progress is being made on all of these fronts, but it will be some years before it becomes possible to 'do GIS' as efficiently, effectively, and safely anywhere in the field as it currently is in the office.

11.4 Distributing the software: GIServices

This final section addresses distributed processing, the notion that the actual operations of GIS might be provided from remote sites, rather than by the user's own computer. Despite the move to component-based software (Section 7.3.3), it is still true that almost all of the operations performed on a user's data are executed in the same computer, the one on the user's desk. Each copy of a popular GIS includes the same functions, which are replicated in every installation of that GIS around the world. When new versions are released, they must be copied and installed at each site, and because not all copies are replaced there is no guarantee that the GIS used by one user is identical to the GIS used by another, even though the vendor and product name are the same.

A GIService is defined as a program executed at a remote site that performs some specific GIS task (see also Section 7.2). The execution of the program is initiated remotely by the user, who may have supplied data, or may rely on data provided by the service, or both. A simple example of a GIService is that provided by wayfinding sites like MapQuest (**www.mapquest.com**) or Yell.com (see Section 1.4.3.1). The user's current location and desired destination are provided to the service (the current location may be entered by the user, or provided automatically from GPS), but the data representing the travel network are provided by the GIService. The results are obtained by solving for the shortest path between the origin and the destination (often a compromise between minimizing distance and minimizing expected travel time), a function that exists in many GIS but in this case is provided remotely by the GIService. Finally, the results are returned to the user in the form of driving instructions, a map, or both.

A GIService replaces a local GIS function with one provided remotely by a server.

In principle, any GIS function could be provided in this way, based on GIS server software (Section 7.6.2). In practice, however, certain functions tend to have attracted more attention than others. One obvious problem is commercial: how would a GIService pay for itself, would it charge for each transaction, and how would this compare to the normal sources of income for GIS vendors based on software sales? Some services are offered free, and generate their revenue by sales of advertising space or by offering the service as an add-on to some other service – MapQuest and Yell are good examples, generating much of their revenue through direct advertising and through embedding their service in hotel

and other sites. In general, however, the characteristics that make a GIS function suitable for offering as a service appear to be:

- Reliance on a database that must be updated frequently, and is too expensive for the average user to acquire. Both geocoding (Box 5.3) and wayfinding services fall into this category, as do gazetteer services (Section 5.9).

- Reliance on GIS operations that are complex and can be performed better by a specialized service than by a generic GIS.

The number of available GIServices is growing steadily, creating a need for directories, portals, and other mechanisms to help users find and access them, and standards and protocols for interacting with them. The Geography Network (**www.geographynetwork.com**; Figure 11.20) provides such a directory, in addition to its role as a geolibrary, as does the Geospatial One-Stop (Box 11.4). Generic standards for remote services are emerging, such as WSDL (Web Services Definition Language, **www.w3.org/TR/wsdl**) and UDDI (Universal Description, Discovery, and Integration, **www.uddi.org**), though to date these have found little application in GIS.

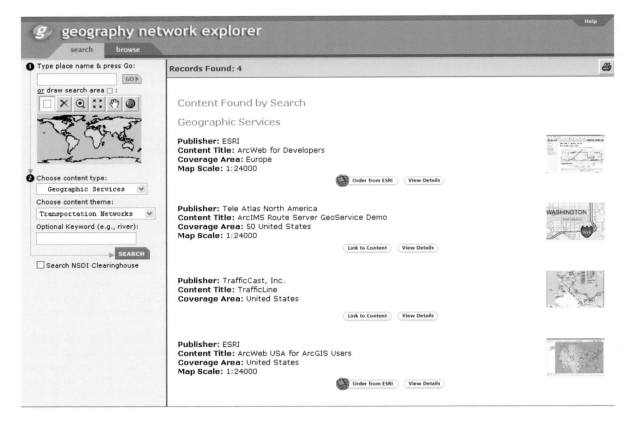

Figure 11.20 The Geography Network provides a directory of remote GIServices. In this case a search for services related to transportation networks has identified four

11.5 Prospects

Distributed GIS offers enormous advantages, in reducing duplication of effort, allowing users to access remotely located data and services through simple devices, and providing ways of combining information gathered through the senses with information provided from digital sources. Many issues continue to impede progress, however: complications resulting from the difficulties of interacting with devices in field settings; limitations placed on communication bandwidth and reliability; and limitations inherent in battery technology. Perhaps, at this time, more problematic than any of these is the difficulty of imagining the full potential of distributed GIS. We are used to associating GIS with the desktop, and conscious that we have not fully exploited its potential – so it is hard to imagine what might be possible when the GIS can be carried anywhere and its information combined with the window on the world provided by our senses.

Questions for further study

1. Design a location-based game based on GPS-enabled cellphones.

2. Find a selection of geolibraries on the Web, and identify their common characteristics. How does each of them: 1) allow the user to specify locations of interest; 2) allow the user to browse through datasets that have similar characteristics; and 3) allow the user to examine the contents of datasets before acquiring them?

3. To what extent do citizens have a right to locational privacy? What laws and regulations control the use of locational data on individuals?

4. Is an LBS a type of GIS, or is a GIS a type of LBS, or are they quite different forms of computer application? If they are different, are they likely to merge in the future, and what will the result look like?

Further reading

Câmara A.S. 2002 *Environmental Systems: A Multidimensional Approach*. New York: Oxford University Press.

National Research Council 1999 *Distributed Geolibraries: Spatial Information Resources*. Washington, DC: National Academy Press.

Peng Z.-H. and Tsou M.-H. 2003 *Internet GIS: Distributed Geographic Information Services for the Internet and Wireless Networks*. Hoboken, NJ: Wiley.

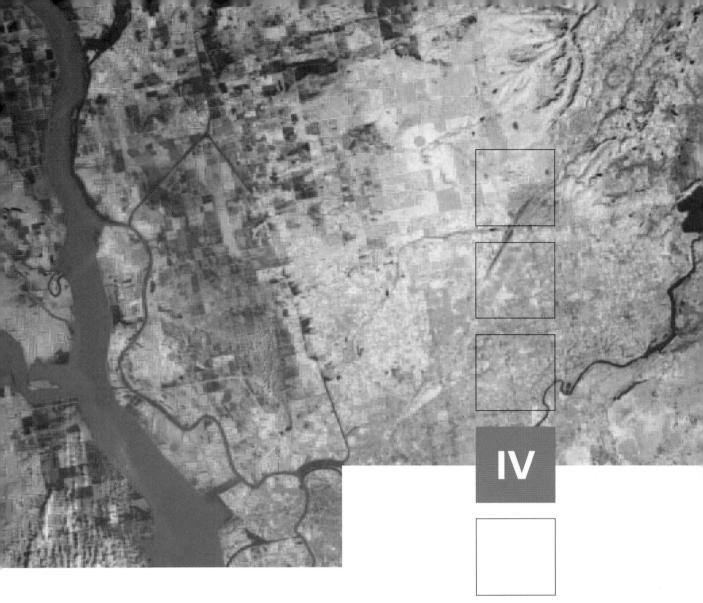

IV

Analysis

12 *Cartography and map production*

This chapter on cartography and map production is the first of a series of chapters that examine GIS output. The next, Chapter 13, deals with the closely related but distinct subject of visualization. The current chapter reviews the nature of cartography and the ways that users interact with GIS in order to produce digital and hard-copy reference and thematic maps. Standard cartographic conventions and graphic symbology are discussed, as is the range of transformations that are used in map design. Map production is reviewed in the context of creating maps for specific applications and also map series. Some specialized types of mapping are introduced that are appropriate for particular applications areas.

Geographic Information Systems and Science, 2nd edition   Paul Longley, Michael Goodchild, David Maguire, and David Rhind.
© 2005 John Wiley & Sons, Ltd.   ISBNs: 0-470-87000-1 (HB); 0-470-87001-X (PB)

Learning Objectives

At the end of this chapter you will understand:

■ The nature of maps and cartography;

■ Key map design principles;

■ Choices that are available to compose maps;

■ The many types of map symbology;

■ Concepts of map production flow-lines.

12.1 Introduction

GIS output represents the pinnacle of many GIS projects. Since the purpose of information systems is to produce results, this aspect of GIS is vitally important to many managers, technicians, and scientists. Maps are a very effective way of summarizing and communicating the results of GIS operations to a wide audience. The importance of map output is further highlighted by the fact that many consumers of geographic information only interact with GIS through their use of map products.

For the purposes of organizing the discussion in this book, it is useful to distinguish between two types of GIS output: formal maps, created according to well-established cartographic conventions, that are used as a reference or communication product (e.g., a military mapping agency 1:250 000 scale topographic map, or a geological survey 1:50 000 scale paper map; Box 12.3); and transitory map and map-like visualizations used simply to display, analyze, edit, and query geographic information (e.g., results of a database query to retrieve areas with poor health indicators viewed on a desktop computer, or routing information displayed on a PDA device Figure 15.18). Both can exist in digital form on interactive display devices (see Box 13.3) or in hard-copy form on paper and other media. In practice, this distinction is somewhat arbitrary and there is considerable overlap, but

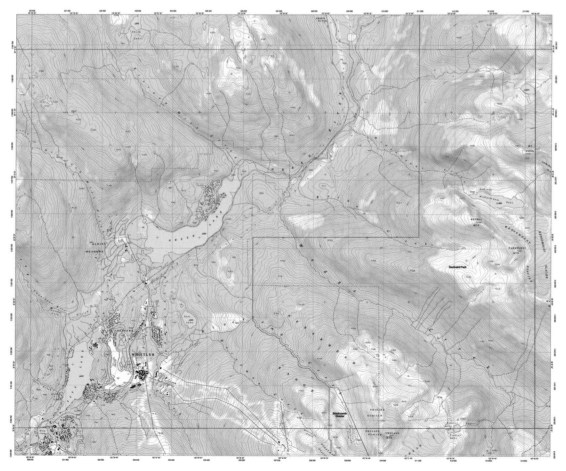

Figure 12.1 Terrestrial topographic map of Whistler, British Columbia, Canada. This is one of a collection of 7016 commercial maps at 1:20 000 scale covering the province (Courtesy: ESRI)

at the core the motivations, tools, and techniques for map production and map visualization are quite different. The current chapter will focus on maps and the next on visualization.

Cartography concerns the art, science, and techniques of making maps or charts. Conventionally, the term map is used for terrestrial areas (Figure 12.1) and chart for marine areas (Figure 12.2; see also Box 4.5), but they are both maps in the sense the word is used here. In statistical or analytical fields, charts provide the pictorial representation of statistical data, but this does not form part of the discussion in this chapter. Note, however, that statistical charts can be used on maps (e.g., Figure 12.3(B)). Cartography dates back thousands of years to a time before paper, but the main visual display principles were developed during the paper era and thus many of today's digital cartographers still use the terminology, conventions, and techniques from the paper era. Box 12.1 illustrates the importance of maps in a historic military context.

Maps are important communication and decision support tools.

Historically, the origins of many national mapping organizations can be traced to the need for mapping for 'geographical campaigns' of infantry warfare, for colonial administration, and for defense. Today such organizations fulfill a far wider range of needs of many more user types (see Chapter 19). Although the military remains a heavy user of mapping, such territorial changes as arise out of today's conflicts reflect a more subtle interplay of economic, political, and historical considerations – though, of course, the threat or actual deployment of force remains a pivotal consideration. Today, GIS-based terrestrial mapping serves a wide range of purposes – such as the support of humanitarian relief efforts (Figure 12.3) and the partitioning of territory through negotiation rather than force (e.g., GIS was used by senior decision makers in the partitioning of Bosnia

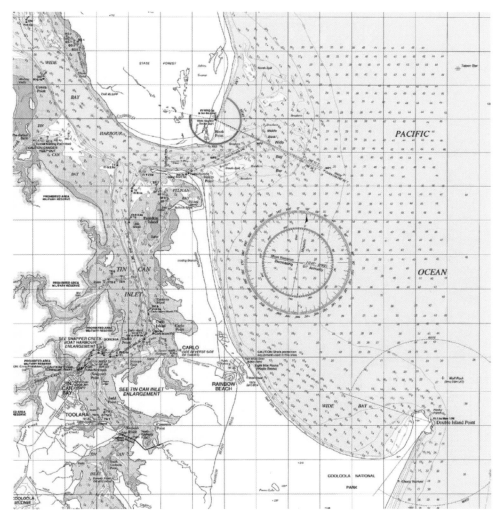

Figure 12.2 Great Sandy Strait (South), Queensland, Australia Boating Safety Chart. This marine chart conforms to international charting standards (Courtesy: ESRI) (Reproduced by permission of Maritime Safety Queensland.)

Applications Box **12.1**

Military maps in history

'Roll up that map; it will not be wanted these ten years.' British Prime Minister William Pitt the Younger made this remark after hearing of the defeat of British forces at the Battle of Austerlitz in 1805, where it became clear that his country's military campaign in Continental Europe had been thwarted for the foreseeable future. The quote illustrates the crucial historic role of mapping as a tool of decision support in warfare, in a world in which nation states were far more insular than they are today. It also identifies two other defining characteristics of the use of geographic information in 19th century society.

First, the principal, straightforward purpose of much terrestrial mapping was to further national interests by infantry warfare. Second, the timeframe over which changes in geographic activity patterns unfolded was, by today's standards, incredibly slow – Pitt envisaged that no British citizen would revisit this territory for a quarter of a (then) average lifetime!

In the 19th century, printed maps became available in many countries for local, regional, and national areas, and their uses extended to land ownership, tax assessment, and navigation, among other activities.

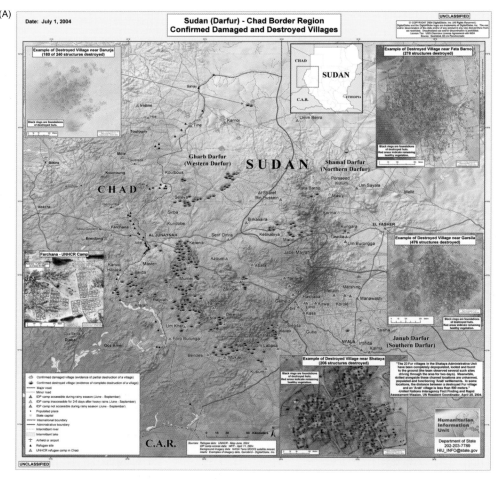

Figure 12.3 Humanitarian relief in Sudan–Chad border region: (A) damaged villages in July 2004 resulting from civil war (Courtesy Humanitarian Information Unit, Dept of State, US Government); (B) malaria vulnerability (Courtesy: World Health Organization)

(B)

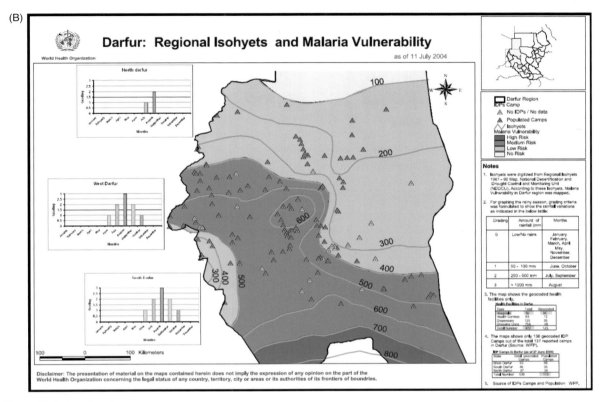

Figure 12.3 (*continued*)

in 1999). The timeframe over which events unfold is also much more rapid – it is inconceivable to think of politicians, managers, and officials being able to neglect geographic space for months, weeks, or even days, never mind years.

Paper maps remain in widespread use because of their transportability, their reliability, ease of use, and the straightforward application of printing technology that they entail. They are also amenable to conveying straightforward messages and supporting decision making. Yet the increasing detail of our understanding of the natural environment and the accelerating complexity of society mean that the messages that mapping can convey are increasingly sophisticated and, for the reasons set out in Chapter 6, uncertain. Greater democracy and accountability, coupled with the increased spatial reasoning abilities that better education brings, mean that more people than ever feel motivated and able to contribute to all kinds of spatial policy. This makes the decision support role immeasurably more challenging, varied, and demanding of visual media. Today's mapping must be capable of communicating an extensive array of messages and emulating the widest range of 'what if' scenarios (Section 16.1.1).

Both paper and digital maps have an important role to play in many economic, environmental, and social activities.

The visual medium of a given application must also be open to the widest community of users. Technology has led to the development of an enormous range of devices to bring mapping to the greatest range of users in the widest spectrum of decision environments. In-vehicle displays, palm top devices, and wearable computers are all important in this regard (Figures 11.1, 11.2 and 11.8). Most important of all, the innovation of the Internet makes 'societal representations' of space a real possibility for the first time.

12.2 Maps and cartography

There are many possible definitions of a map; here we use the term to describe digital or analog (soft- or hard-copy) output from a GIS that shows geographic information using well-established cartographic conventions. A map is the final outcome of a series of GIS data processing steps (Figure 12.4) beginning with data collection (Chapter 9), editing and maintenance, through data management (Chapter 10), analysis (Chapters 14–16) and concluding with a map. Each of these activities successively transforms a database of geographic information until it is in the form appropriate to display on a given technology.

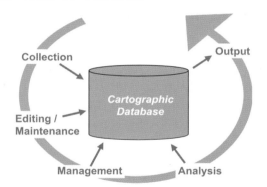

Figure 12.4 GIS processing transformations needed to create a map

Central to any GIS is the creation of a data model that defines the scope and capabilities of its operation (Chapter 8), and the management context in which it operates (Chapters 17–20). There are two basic types of map (Figure 12.5): reference maps, such as topographic maps from national mapping agencies (see also Figure 12.1), that convey general information; and thematic maps that depict specific geographic themes, such as population census statistics, soils, or climate zones (see Figures 12.7 and 12.15 for other examples).

Maps fulfill two very useful functions, acting as both storage and communication mechanisms for geographic information. The old adage 'a picture is worth a thousand words' connotes something of the efficiency of maps as a storage container. The modern equivalent of this is 'a map is worth a million bytes'. Before the advent of GIS, the paper map was the database, but a map can now be considered a single product generated from a digital database. Maps are also a mechanism to communicate information to viewers. Maps can present the results of analyses (e.g., the optimum site suitable for locating a new store, or analysis of the impact of an oil spill). They can communicate spatial relationships between phenomena across the same map, or between maps of the same or different areas. As such they can assist in the identification of spatial order and differentiation. Effective decision support requires that the message of the map is readily interpretable in the mind of the decision maker. A major function of a map is not simply to marshal and transmit known information about the world, but also to create or reinforce a particular message. Menno-Jan Kraak, an academic cartographer, speaks about his personal experiences of the development of cartography in Box 12.2

Maps are both storage and communication mechanisms.

(A)

Figure 12.5 Two maps of the Southern United States: (A) reference map showing topographic information; (B) thematic map showing population density in 1996

(B)

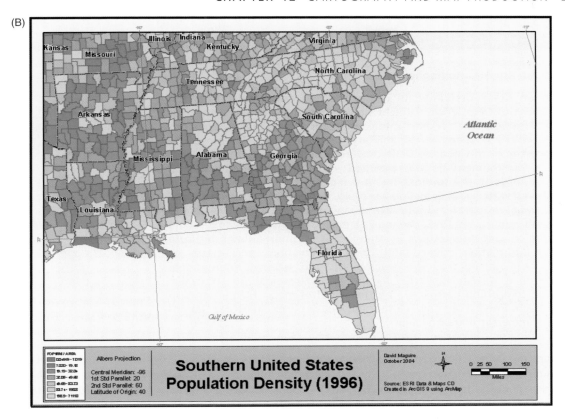

Figure 12.5 (*continued*)

Maps also have several limitations:

■ Maps can be used to miscommunicate (lie) accidentally or on purpose. For example, incorrect use of symbols can convey the wrong message to users by highlighting one type of feature at the expense of another (see Figure 12.13 for an example of different choropleth map classifications).

■ Maps are a single realization of a spatial process. If we think for a moment about maps from a statistical perspective, then each map instance represents the outcome of a sampling trial and is therefore a single occurrence generated from all possible maps on the same subject for the same area. The significance of this is that other sample maps drawn from the same population would exhibit variations and, consequently, we need to be careful in drawing inferences from a single map sample. For example, a map of soil textures is derived by interpolating soil sample texture measurements. Repeated sampling of soils will show natural variation in the texture measurements.

■ Maps are often created using complex rules, symbology, and conventions, and can be difficult to understand and interpret by the untrained viewer. This is particularly the case, for example, in multivariate statistical thematic mapping where the idiosyncrasies of classification schemes and color symbology can be challenging to comprehend.

Uncertainty pertains to maps just as it does to other geographic information.

12.2.1 Maps and media

Without question, GIS has fundamentally changed cartography and the way we create, use, and think about maps. The digital cartography of GIS frees map-makers from many of the constraints inherent in traditional (non-GIS) paper mapping (see also Section 3.7). This is because:

■ The paper map is of fixed scale. Generalization procedures (Section 3.8) can be invoked in order to maintain clarity during map creation. This detail is not recoverable, except by reference back to the data from which the map was compiled. The zoom facility of GIS can allow mapping to be viewed at a range of scales, and detail to be filtered out as appropriate at a given scale.

■ The paper map is of fixed extent and adjoining map sheets must be used if a single map sheet does not cover the entire area of interest. (An unwritten law of paper map usage is that the most important map features always lie at the intersection of four paper map sheets!) GIS, by contrast, can provide a seamless medium for viewing space, and users are able to pan across wide swathes of territory.

Biographical Box **12.2**

Menno-Jan Kraak, cartographer

Menno-Jan Kraak graduated in geography and cartography from the Department of Geosciences, Utrecht University, where he also started his academic career. After a short interruption for the obligatory conscript period, which he spent at the military geography unit, he moved to the Department of Geodesy, Delft University of Technology. In this geo-technical environment he was awarded a Ph.D. on three-dimensional mapping. In Delft, the foundations of a modern cartographic text book co-authored with Ferjan Ormeling were laid. *Cartography: Visualization of Spatial Data* was translated into several languages and is still available in its second edition. In 1996, Menno-Jan moved to Enschede to become Professor in Geovisualization at the ITC. This International Institute for Geo-Information Science and Earth Observation concentrates its activities on students from the developing world. His own international orientation is further expanded by the exotic traveling for ITC which offers an interesting perspective on the world of maps. Menno-Jan also has a keen interest in how cartography is depicted in world stamps (**www.itc.nl/personal/kraak/stamps/**).

Figure 12.6 Menno-Jan Kraak, cartographer

Reflecting on the present state of cartography, Menno-Jan says:

During my career which started as a 'real' cartographer I have been part of the process that evolved cartography via computer cartography and GIS into an integral part of what is today called the GIScience world. However, during this process GIS took on board methods and techniques from non-geo-related disciplines such as scientific and information visualization. In parallel I have witnessed the same integrative trend in the Dutch geo-community. In my early Delft years I was a stranger in an (interesting) geodetic world and we all had our own professional organizations, but trends in our professions brought the different disciplines closer together. In 2003 I was involved in the merger of the geo-societies in the Netherlands into a single new geo-information society. Another interesting observation I would like to share is that one should always be open for new methods and techniques especially from other disciplines. You should certainly try them out in your own environment, but while doing so do not forget the knowledge already obtained. It can be very tempting to concentrate on the toys instead of trying to incorporate derived new knowledge in the theory of mapping.

■ Most paper maps present a static view of the world whereas conventional paper maps and charts are not adept at portraying dynamics. GIS-based representations are able to achieve this through animation.

■ The paper map is flat and hence limited in the number of perspectives that it can offer on three-dimensional data. 3-D visualization is much more effective within GIS which can support interactive pan and zoom operations (see Figure 9.13 for a 3-D view example).

■ Paper maps provide a view of the world as essentially complete. GIS-based mapping allows the supplementation of base-map material with further data. Data layers can be turned on and off to examine data combinations.

■ Paper maps provide a single, map producer-centric, view of the world. GIS users are able to create their

own, user-centric, map images in an interactive way. Side-by-side map comparison is also possible in GIS.

GIS is a flexible medium for the production of many types of maps.

12.3 Principles of map design

Map design is a creative process during which the cartographer, or map-maker, tries to convey the message of the map's objective. Primary goals in map design are to share information, highlight patterns and processes, and illustrate results. A secondary objective is to create a pleasing and interesting picture, but this must not be

Czech Geological Survey (CGS) Map Production

In 1994, the Czech Geological Survey (CGS) began extensive use of GIS technology to meet the increasing demand for digital information about the environment. GIS in the CGS is focused on the methods of spatial data processing, unification, and dissemination. Digital processing of geological maps and the development of GIS follows standardized procedures using common geological dictionaries and graphic elements. Recently, the main objective has been to create and implement a uniform geological data model and provide the public and the scientific community with easy access to geographic data, via a Web map server (**www.geology.cz**).

CGS has a unique geographic information system, containing more than 260 000 mapped geological objects from the entire Czech Republic. The fundamental part of this geographic database is the unified national geological index (legend), which consists of four main types of information – chronostratigraphical units, regional units, lithostratigraphical units, and lithological description of rocks. The database has been under revision since 1998 leading to the creation of a seamless digital geological map of the Czech Republic. This database has already been used for land use planning by government and local authorities.

The geological map of the Krkonose–Jizera mountains shown in Figure 12.7 is a cartographic presentation of one part of the CGS database. The overview map in the bottom left corner shows the extents of all maps in the series.

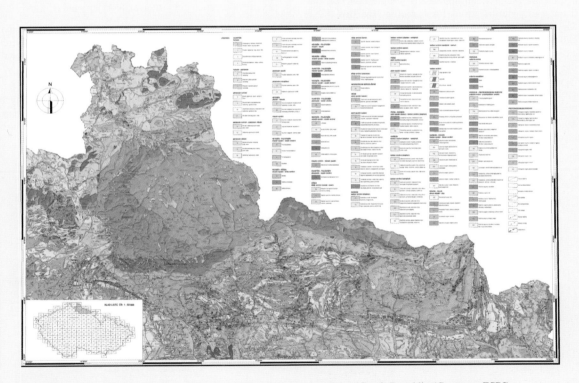

Figure 12.7 A 1:50 000 geological map of the Krkonose–Jizera mountains, Czech Republic (Courtesy: ESRI)

at the expense of fidelity to reality and meeting the primary goals. Map design is quite a complex procedure requiring the simultaneous optimization of many variables and harmonization of multiple methods. Cartographers must be prepared to compromise and balance choices.

It is difficult to define exactly what constitutes a 'good design'. The general consensus is that a good design is one that looks good, is simple and elegant, and most importantly, leads to a map that is fit for the intended purpose.

Robinson et al (1995) define seven controls on the map design process:

- *Purpose.* The purpose for which a map is being made will determine what is to be mapped and how the information is to be portrayed. Reference maps are multi-purpose, whereas thematic maps tend to be single purpose (Figure 12.5). With the digital technology of a GIS, it is easier to create maps, and many more are digital and interactive. As a consequence today maps are increasingly single purpose.

- *Reality.* The phenomena being mapped will usually impose some constraints on map design. For example, the orientation of the country – whether it be predominantly east-west (Russia) or north-south (Chile) – will determine layout in no small part.

- *Available data.* The specific characteristics of data (e.g., raster or vector, continuous or discrete, or point, line, or area) will affect the design. There are many different ways to symbolize map data of all types, as discussed in Section 12.3.2.

- *Map scale.* Scale is an apparently simple concept, but it has many ramifications for mapping (see Box 4.2 for further discussion). It will control how many data can appear in a map frame, the size of symbols, the overlap of symbols, and much more. Although one of the early promises of digital cartography and GIS was 'scale-free' databases that could be used to create multiple maps at different scales, this has never been realized because of technical complexities.

- *Audience.* Different audiences want different types of information on a map and expect to see information presented in different ways. Usually, executives (and small children!) are interested in summary information that can be assimilated quickly, whereas advanced users often want to see more information. Similarly, those with restricted eyesight find it easier to read bigger symbols.

- *Conditions of use.* The environment in which a map is to be used will impose significant constraints. Maps for outside use in poor or very bright light will need to be designed differently than maps for use indoors where the light levels are less extreme.

- *Technical limits.* The display medium, be it digital or hardcopy, will impact the design process in several ways. For example, maps to be viewed in an Internet browser, where resolution and bandwidth are limited, should be simpler and based on less data than equivalents to be displayed on a desktop PC monitor.

12.3.1 Map composition

Map composition is the process of creating a map comprising several closely interrelated elements (Figure 12.8):

- Map body. The principal focus of the map is the main map body, or in the case of comparative maps there will be two or more map bodies. It should be given space and use symbology appropriate to its significance.

- Inset/overview map. Inset and overview maps may be used to show, respectively, an area of the main map body in more detail (at a larger scale) and the general location or context of the main body.

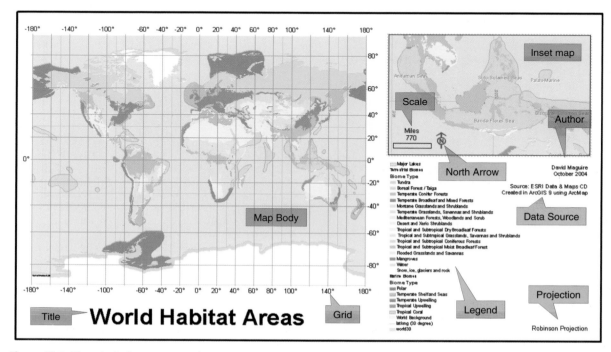

Figure 12.8 The principal components of a map composition layout

- Title. One or more map titles are used to identify the map and to inform the reader about its content.

- Legend. This lists the items represented on the map and how they are symbolized. Many different layout designs are available and there is a considerable body of information available about legend design.

- Scale. The map scale provides an indication of the size of objects and the distances between them. A paper map scale is a ratio, where one unit on the map represents some multiple of that value in the real world. The scale can be symbolized numerically (1:1000), graphically (a scalebar), or texturally ('one inch equals 1000 inches'). The scale is a representative fraction and so a 1:1000 scale is larger (finer) than a 1:100 000. A small (coarse) scale map displays a larger area than a large (fine) scale map, but with less detail. See also Box 4.2 for more details about scale.

- Direction indicator. The direction and orientation of a map can be conveyed in one of several ways including grids, graticules, and directional symbols (usually north arrows). A grid is a network of parallel and perpendicular lines superimposed on a map. A graticule is a network of longitude and latitude lines on a map that relates points on a map to their true location on the Earth.

- Map metadata. Map compositions can contain many other types of information including the map projection, date of creation, data sources, and authorship.

A key requirement for a good map is that all map elements are composed into a layout that has good visual balance. On large-scale maps, such as 1:50 000 national mapping agency topographic maps, all the contextual items (everything listed above except the map body) usually appear as marginal notations (or marginalia). In the case of map series or atlases (see Section 12.4), some of the common information may be in a separate document. On small- or medium-scale maps this information usually appears within the map border (e.g., Figure 12.7).

12.3.2 Map symbolization

The data to be displayed on a map must be classified and represented using graphic symbols that conform to well-defined and accepted conventions. The choice of symbolization is critical to the usefulness of any map. Unfortunately, the seven controls on the design process listed in Section 12.3.1 also conspire to mean that there is not a single universal symbology model applicable everywhere, but rather one for each combination of factors. Again, we see that cartographic design is a compromise reached by simultaneously optimizing several factors.

Good mapping requires that spatial objects and their attributes can be readily interpreted in applications. In Chapter 3, attributes were classified as being measured on the nominal, ordinal, interval, or ratio scales

(Box 3.3), while in Chapter 4 spatial objects were classified into points, lines, areas, and surfaces (Box 4.1). We have already seen how attribute measures that we think of as continuous are actually discretized to levels of precision imposed by measurement or design (Section 6.3.). The representation of spatial objects is similarly imposed – cities might be captured as points, areas, mixtures of points, lines, and areas (as in a street map: Figure 12.19) or 3-D 'walk-throughs' (see Section 13.4), depending on the base-scale of a representation and the importance of city objects to the application. Measurement scales and spatial object types are thus one set of conventions that are used to abstract reality. Whether using GIS or paper, mapping may entail reclassification or transformation of attribute measures.

The process of mapping attributes frequently entails further problems of classification because many spatial attributes are inherently uncertain (Chapter 6). For example, in order to create a map of occupational type, individuals' occupations will be classified first into socio-economic groups (e.g., 'factory worker') and perhaps then into super-groups, such as 'blue collar'. At every stage in the aggregation process we inevitably do injustice to many individuals who perform a mix of white and blue collar, intermediate and skilled functions by lumping them into a single group (what social class is a frogman?). In practice, the validity and usefulness of an occupational classification will have become established over repeated applications, and the task of mapping is to convey thematic variation in as efficient a way as possible.

12.3.2.1 Attribute representation and transformation

Humans are good at interpreting visual data – much more so than interpreting numbers, for example – but conventions are still necessary to convey the message that the map-maker wants to impart. Many of these conventions relate to use of symbols (such as the way highway shields denote route numbers on many US medium- and fine-scale maps; Figure 12.19) and colors (blue for rivers, green for forested areas, etc.), and have been developed over the past few hundred years. Mapping of different themes (such as vegetation cover, surface geology, and socio-economic characteristics of human populations) has a more recent history. Here too, however, mapping conventions have developed, and sometimes they are specific to particular applications.

Attribute mapping entails use of graphic symbols, which (in two dimensions) may be referenced by points (e.g., historic monuments and telecoms antennae), lines (e.g., roads and water pipes) or areas (e.g., forests and urban areas). Basic point, line, and area symbols are modified in different ways in order to communicate different types of information. The ways in which these modifications take place adhere to cognitive principles and the accumulated experience of application implementations. The nature of these modifications was first explored by Bertin in 1967, and was extended to the typology illustrated in Figure 12.9 by MacEachren. The size and orientation of point and line symbols is varied principally to

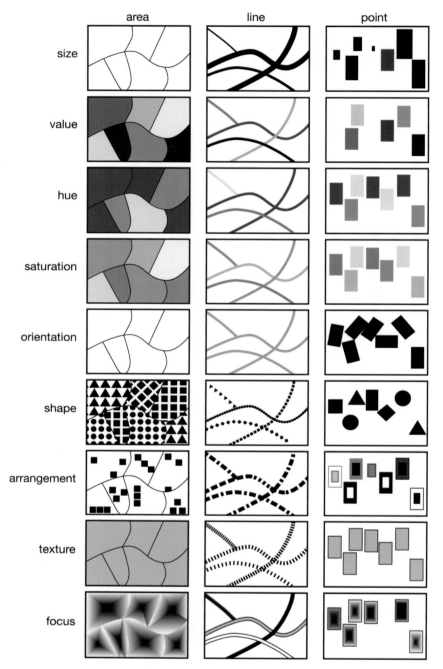

Figure 12.9 Bertin's graphic primitives, extended from seven to ten variables (the variable location is not depicted). *Source*: MacEachren 1994, from *Visualization in Geographical Information Systems,* Hearnshaw H.M. and Unwin D.J. (eds), Plate B. (Reproduced by permission of John Wiley & Sons, Ltd.)

distinguish between the values of ordinal and interval/ratio data using graduated symbols (such as the proportional pie symbols shown in Figure 12.10). Figure 12.11 illustrates how orientation and color can be used to depict the properties of locations, such as ocean current strength and direction.

Hue refers to the use of color, principally to discriminate between nominal categories, as in agricultural or urban land-use maps (Figure 12.12). Different hues may be combined with different textures or shapes (see below) if there are a large number of categories in order to avoid difficulties of interpretation. The shape of map symbols can be used either to communicate information about a spatial attribute (e.g., a viewpoint or the start of a walking trail), or its spatial location (e.g., the location of a road or boundary of a particular type: Figure 12.12), or spatial

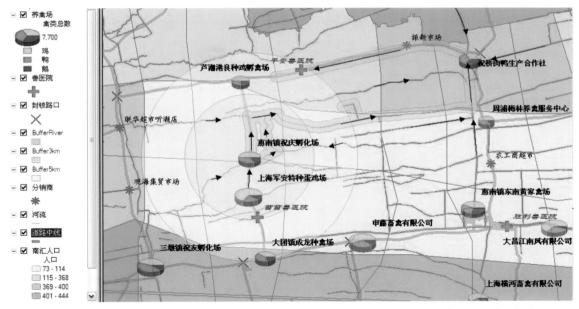

Figure 12.10 Incidence of Asian Bird Flu in Shanghai City. Pie size is proportional to the number of cases; red = duck, yellow = chicken, blue = goose

Figure 12.11 Tauranga Harbour tidal movements, Bay of Plenty, NZ. The arrows indicate speed (color) and direction (orientation) (Courtesy: ESRI)

relationships (e.g., the relationship between sub-surface topography and ocean currents). Arrangement, texture, and focus refer to within- and between-symbol properties that are used to signify pattern. A final graphic variable in the typologies of MacEachren and Bertin is location (not shown in Figure 12.9), which refers to the practice of offsetting the true coordinates of objects in order to

improve map intelligibility, or changes in map projection. We discuss this in more detail below. Some of the common ways in which these graphic variables are used to visualize spatial object types and attributes are shown in Table 12.1.

The selection of appropriate graphic variables to depict spatial locations and distributions presents one

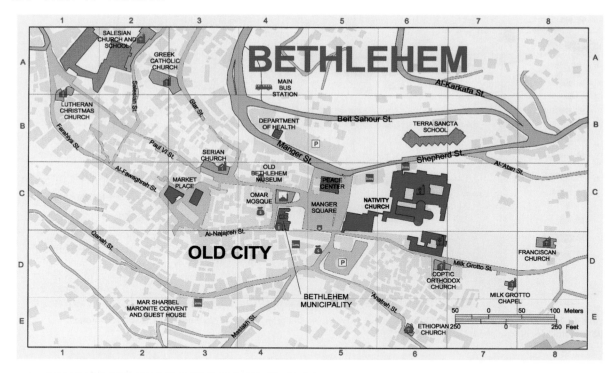

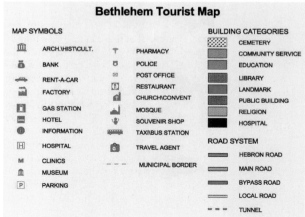

Figure 12.12 Use of hue (color) to discriminate between Bethlehem, Israel urban land-use categories, and of symbols to communicate location and other attribute information (Courtesy: ESRI)

set of problems in mapping. A related task is how best to position symbols on the map, so as to optimize map interpretability. The representation of nominal data by graphic symbols and icons is apparently trivial, although in practice automating placement presents some challenging analytical problems. Most GIS packages include generic algorithms for positioning labels and symbols in relation to geographic objects. Point labels are positioned to avoid overlap by creating a window, or mask (often invisible to the user), around text or symbols. Linear features, such as rivers, roads, and contours, are labeled by placing the text using a spline function to give a smooth even distribution, or distinguished by use of color. Area labels are assigned to

central points (Figure 12.16), using geometric algorithms similar to those used to calculate geometric centroids (Section 15.2.1). These generic algorithms are frequently customized to accommodate common conventions and rules for particular classes of application – such as topographic (Figure 12.1), utility, transportation, and seismic maps, for example.

Generic and customized algorithms also include color conventions for map symbolization and lettering.

Ordinal attribute data are assigned to point, line, and area objects in the same rule-based manner, with the ordinal property of the data accommodated through use of a hierarchy of graphic variables (symbol and lettering sizes, types, colors, intensities, etc.). As a general rule,

Table 12.1 Common methods of mapping spatial object types and attribute data with examples (2-D = two-dimensional)

Spatial object type	Attribute type		
	Nominal	Ordinal	Interval/Ratio
Point (0-D)	Symbol map (each category a different class of symbol – color, shape, orientation), and/or use of lettering: e.g., presence/absence of city (Figure 12.12)	Hierarchy of symbols or lettering (color and size): e.g., small/medium/large depots (Figure 12.22)	Graduated symbols (color and size): e.g., pollution concentrations (Figure 12.10)
Line (1-D)	Network connectivity map (color, shape, orientation): e.g., presence/absence of connection (Figure 12.20)	Graduated line symbology (color and size): e.g., road classifications (Figure 12.19)	Flow map with width or color lines proportional to flows (color and size): e.g., traffic flows (Figure 12.11)
Area (2-D)	Unique category map (color, shape, orientation, pattern): e.g., soil types (Figure 12.7)	Graduated color or shading map: e.g., timber yield low/medium/high (Figure 12.3)	Continuous hue/shading, e.g., dot-density or choropleth map: e.g., percentage of retired population (Figure 12.5B)
Surface (2.5-D)	One color per category (color, shape, orientation, pattern), e.g., relief classes: mountain/valley (Figure 6.7)	Ordered color map, e.g., areas of gentle/steep/very steep slopes	Contour map, e.g., isobars/isohyets: e.g., topography contours (Figure 12.3B)

the typical user is unable to differentiate between more than seven (plus or minus two) ordinal categories, and this provides an upper limit on the normal extent of the hierarchy.

A wide range of conventions is used to visualize interval- and ratio-scale attribute data. Proportional circles and bar charts are often used to assign interval- or ratio-scale data to point locations (Figures 12.10 and 12.14). Variable line width (with increments that correspond to the precision of the interval measure) is a standard convention for representing continuous variation in flow diagrams.

There is a variety of ways of ascribing interval or ratio scale attribute data to areal entities that are pre-defined. In practice, however, none is unproblematic. The standard method of depicting areal data is in zones (Figure 12.5B). However, as was discussed in Section 6.2, the choropleth map brings the dubious visual implication of within-zone uniformity of attribute value. Moreover, conventional choropleth mapping also allows any large (but possibly uninteresting) areas to dominate the map visually. A variant on the conventional choropleth map is the dot-density map, which uses points as a more aesthetically pleasing means of representing the relative density of zonally averaged data – but not as a means of depicting the precise locations of point events. Proportional circles provide one way around this problem; here the circle is scaled in proportion to the size of the quality being mapped and the circle can be centered on any convenient point within a zone (Figure 12.10). However, there is a tension between using circles that are of sufficient size to convey the variability in the data and the problems of overlapping circles on 'busy' areas of maps that have large numbers of symbols. Circle positioning also entails the same kind of positioning problem as that of name and symbol placement outlined above.

If the richness of the map presentation is to be equivalent to that of the representation from which it was derived, the intensity of color or shading should directly mirror the intensity or magnitude of attributes. The human eye is adept at discerning continuous variations in color and shading, and Waldo Tobler has advanced the view that continuous scales present the best means of representing geographic variation. There is no natural ordering implied by use of different colors and the common convention is to represent continuous variation on the red-green-blue (RGB) spectrum. In a similar fashion, difference in the hue, lightness, and saturation (HLS) of shading is used in color maps to represent continuous variation. International standards on intensity and shading have been formalized.

At least four basic classification schemes have been developed to divide interval and ratio data into categories (Figure 12.13):

1. Natural (Jenks) breaks, in which classes are defined according to apparently natural groupings of data values. The breaks may be imposed on the basis of break points that are known to be relevant to a particular application, such as fractions and multiples of mean income levels, or rainfall thresholds known to support different thresholds of vegetation ('arid', 'semi-arid', 'temperate', etc.). This is 'top down' or deductive assignment of breaks. Inductive ('bottom up') classification of data values may be carried out by using GIS software to look for relatively large jumps in data values.

2. Quantile breaks, in which each of a predetermined number of classes contains an equal number of observations. Quartile (four-category) classifications are widely used in statistical analysis, while quintile (five-category: Figure 12.13) classifications are well suited to the spatial display of uniformly distributed

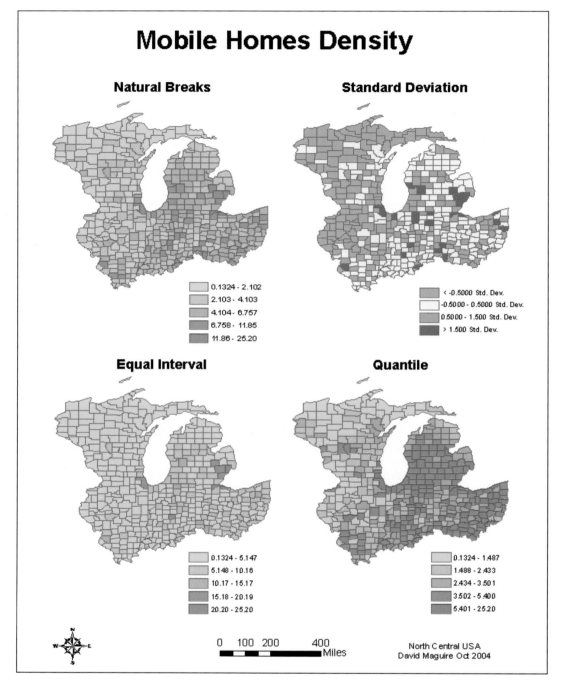

Figure 12.13 Comparison of choropleth class definition schemes: natural breaks, standard deviation, equal interval, quantile

data. Yet because the numeric size of each class is rigidly imposed, the result can be misleading. The placing of the boundaries may assign almost identical attributes to adjacent classes, or features with quite widely different values in the same class. The resulting visual distortion can be minimized by increasing the number of classes – assuming the user can assimilate the extra detail that this creates.

3. Equal interval breaks. These are best applied if the data ranges are familiar to the user of the map, such as temperature bands.

4. Standard deviation classifications show the distance of an observation from the mean. The GIS calculates the mean value and then generates class breaks in standard deviation measures above and below it. Use

of a two-color ramp helps to emphasize values above and below the mean (Figure 12.13).

Classification procedures are used in map production in order to ease user interpretation.

The choice of classification is very much the outcome of choice, convenience, and the accumulated experience of the cartographer. The automation of mapping in GIS has made it possible to evaluate different possible classifications. Looking at distributions (Figure 14.8) allows us to see if the distribution is strongly skewed – which might justify using unequal class intervals in a particular application. A study of poverty, for example, could quite happily class millionaires along with all those earning over $50 000, as they would be equally irrelevant to the study.

12.3.2.2 Multivariate mapping

Multivariate maps show two or more variables for comparative purposes. For example, Figure 12.10 shows the incidence of Asian Bird Flu as proportional symbols, and Figure 12.3B illustrates rainfall and malaria risk. Many maps are a compilation of composite maps based upon a range of constituent indicators. Climate maps, for example, are compiled from direct measures such as amount and distribution of precipitation, diurnal temperature variation, humidity, and hours of sunshine, plus indirect measures such as vegetation coverage and type. It is unlikely that there will ever be a perfect correspondence between each of these components, and historically it has been the role of the cartographer to integrate disparate data. In some instances, data may be averaged over zones which are devoid of any strong meaning, while the different components that make up a composite index may have been measured at a range of scales. Mapping can mask scale and aggregation problems where composite indicators are created at large scales using components that were only intended for use at small scales.

Figure 12.14 illustrates how multivariate data can be displayed on a shaded area map. In this soil texture map of Mexico three variables are displayed simultaneously, as indicated in the legend, using color variations to display combinations of per cent sand (base), silt (right), and clay (left).

Box 12.4 shows an innovative method for representing multivariate data on maps as clock diagrams.

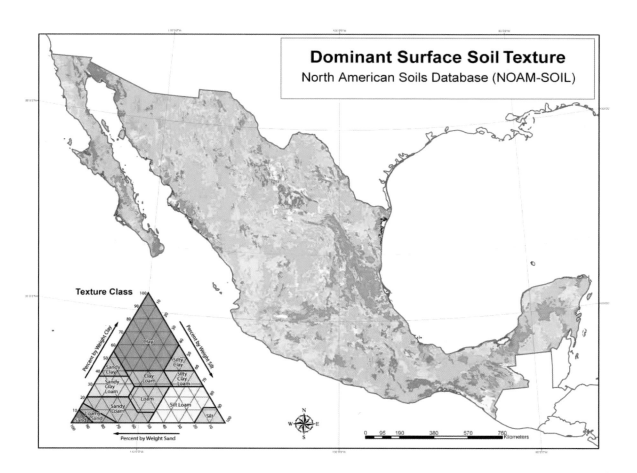

Figure 12.14 North America Soils (Mexico) – Dominant Surface Soil Texture (Courtesy: ESRI)

Applications Box **12.4**

Multivariate mapping of groundwater analyte over time

An increasing number of GIS users need to represent time dependent (temporal) data on maps in a meaningful and concise manner. The ability to analyze data for both spatial and temporal patterns in a single presentation provides a powerful incentive to develop tools for communication to map users and decision makers. Although there are many different methods to display scientific data to help discern either spatial or temporal trends, few visualization software packages allow for a single graphical presentation within a geographic context.

Chemical concentration data for many different groundwater monitoring wells is widely available for one or more chemical

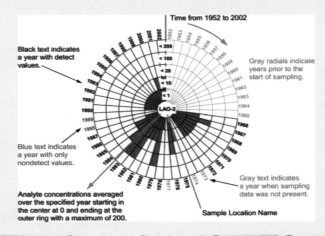

Figure 12.15 Representing temporal data as clock diagrams on maps (Courtesy: ESRI)

or analyte samples over time. Figure 12.15 shows how this complex temporal data can be represented on a map so that the movement and change in concentration of the analyte can be observed, both horizontally and vertically, and the potential for the analyte reaching the water table after many years can be assessed.

This map is innovative because it uses a clock diagram code for temporal and spatial visualization. The clock diagrams are analogous to 'rose diagrams' often used to depict wind direction in meteorological maps, or strike direction in geologic applications. The clock diagram method consists of three steps: (1) sample location data, including depth, is compiled along with attribute data such as concentration and time; (2) all data are read into the software application and clock diagram graphics are created; and (3) clock diagram graphics are placed on the map at the location of the well where samples were collected.

Although this method of representing temporal data through the use of clock diagrams is not entirely original, the use of multiple clock diagrams as GIS symbology to emphasize movement of sampled data over time is a new concept. The clock diagram code and method have been proven to be an effective means in identifying temporal and spatial trends when analyzing groundwater data.

12.4 Map series

The discussion thus far has focused on the general principles of map design that apply equally to a single map or a collection of maps for the same area or a common theme being considered over multiple areas. The real power of GIS-based digital cartography is revealed when changes need to be made to maps or when collections of similar maps need to be created. Editing or copying a map composition to create similar maps of any combination of areas and data themes is relatively straightforward with a GIS.

Many organizations use GIS to create collections or series of topographic or thematic maps. Examples include a topographic map series to cover a state or country (e.g., USGS 1:24 000 quad sheets of the United States or Eurogeographics 1:250 000 maps of Europe), an atlas of reference maps (e.g., *National Geographic Eighth Edition Atlas of the World*: Box 12.5), a series of geology maps for a country (Box 12.3), or land parcel (Figure 12.16)

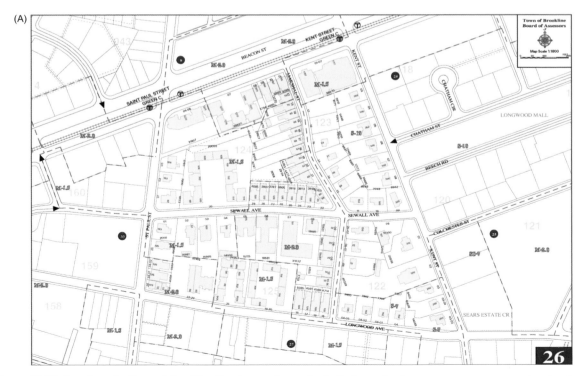

Figure 12.16 Page from the Town of Brookline Tax Assessor's map book. Each map book page uses a common template (as shown in the legend), but has a different map extent: (A) page 26; (B) legend (Courtesy: ESRI)

(B)

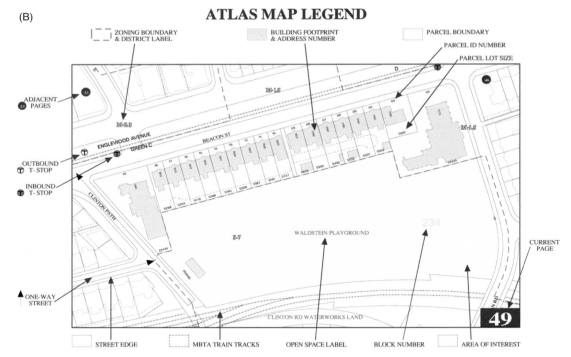

ATLAS MAP LEGEND

Figure 12.16 (*continued*)

Biographical Box **12.5**

Allen Carroll, Chief Cartographer, National Geographic Society

Allen Carroll is chief cartographer and executive vice president of National Geographic Maps. As chief cartographer, Allen presides over the editorial and creative efforts of the Society's map division, including the renowned supplement maps published in *National Geographic* magazine, the *Seventh Edition Atlas of the World*, and the *National Geographic Map Machine*, an innovative world atlas on the Internet. He has been an employee of the National Geographic Society for over twenty years, serving in a variety of positions in the map division and the art department of *National Geographic* magazine.

Figure 12.17 Allen Carroll (left), Chief Cartographer, National Geographic Society

After graduating from Connecticut College, Allen self-trained in design, illustration, and cartography. He started work as a freelance illustrator and designer, serving clients such as *The Washington Post*, Smithsonian Institute, and *Reader's Digest*. In 1991, he moved to the National Geographic Society, initially as art director of *National Geographic* magazine, producing historical, scientific, and informational artwork, then as managing director of National Geographic Maps where he presided over the shift of the unit from a division of the magazine to the Society's new subsidiary, National Geographic Ventures. During this time NG Maps expanded from its traditional role as service provider to the magazine and book divisions of the Society to become a publisher and distributor of map products. With partnerships and acquisitions, the group extended its product lines to include road atlases, road maps, and outdoor recreation maps.

Speaking about his work Allen Carroll says:

My colleagues and I feel extremely fortunate to have inherited *National Geographic*'s ninety-year tradition of superb cartography. I feel doubly blessed because I happen to be chief cartographer at the very moment that

geographic science and technology is exploding. We in National Geographic Maps are particularly aware of this explosion, having just completed preparation of our *Eighth Edition Atlas of the World*. GIS has vastly improved production of the atlas and maintenance of the data behind it – although building an atlas is still a daunting, labor-intensive task. Our new atlas features breathtaking geographic data of many types and from many sources: satellite imagery, population density, elevation, ocean dynamics, paleogeography, tectonics, protected lands, ecoregions, and more.

I'm constantly surprised and delighted by the contributions of GIS professionals throughout the public and private realm. At the same time I'm concerned about the degree to which the fruits of their labor are sequestered. So much of the GIS data that has been created for specialized uses could benefit the public by being aggregated, interpreted, and distributed to a broader audience. The new atlas, and the Map Machine interactive atlas on the Web, are small steps in the right direction. I'm hoping that we at *National Geographic*, and others throughout the GIS community, can find many other ways to show our data to the world. At the local level, publishing zoning information and other GIS data produced by municipal agencies can help spur citizen activism. Globally, the remarkable efforts of conservation organizations to map biodiversity and protect critical habitats can educate schoolchildren and enlist support for protection of our natural heritage.

In 1888 the National Geographic Society was founded 'to increase and diffuse geographic knowledge'. In the 21st century, as geographic knowledge blossoms in ways unimagined a century ago, we can – and must – diffuse that knowledge to the greatest extent possible.

or utility asset maps that might be bound into a map book. Map series by definition share a number of common elements (for example, projection, general layout, and symbology) and a number of techniques have been developed to automate the map series production process.

GIS makes it much easier to create collections of maps with common characteristics from map templates.

Figure 12.4 showed the main GIS sub-components and workflows used to create maps. In principle, this same process can be extended to build production flow-lines that result in collections of maps such as an atlas, map series, or map book. Such a GIS requires considerable financial and human resource investments, both to create and maintain it, and it will be used by many operatives over a long period of time. In an ideal system, an organization would have a single 'scale-free', continuous database, which is kept up to date by editing staff, and capable of feeding multiple flow-lines at several scales each with different content and symbology. Unfortunately, this vision remains elusive because of several significant scientific and technical issues. Nevertheless, considerable progress has been made in recent years towards this goal. The heart of map production through GIS is a geographic database covering the area and data layers of interest (see Chapter 10 for background discussion about the basic principles of creating and maintaining geographic databases). This database is built using a data model that represents interesting aspects of the real world in the GIS (see Chapter 8 for background discussion about GIS data models). Such a base cartographic data model is often referred to as a Digital Landscape Model (DLM) because its role is to represent the landscape in the GIS

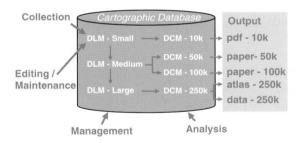

Figure 12.18 Key components and information flows in a GIS map production system

as a collection of features that is independent of any map product representation. Figure 12.18 is an extended version of Figure 12.4 that shows the key components and flows in a map production GIS.

The DLM continuous geographic database is usually stored in a Database Management System (DBMS) that is managed by the GIS. The database is created and maintained by editors that typically use sophisticated GIS desktop software. These must run carefully tailored applications that enforce strict data integrity rules (e.g., geometric connectivity, attribute domains, and topology – see Section 10.7) to maintain database quality.

In an ideal world, it will be possible to create multiple different map products from this DLM database. Example products that might be generated by a civilian or military national mapping agency include: a 1:10 000-scale topographic map in digital format (e.g., Adobe Portable Document Format (pdf)), a 1:50 000 topographic map sheet in paper format, a 1:100 000 paper map of major streets/highways, a 1:250 000 map suitable for inclusion in a digital or printed atlas, and a 1:250 000 digital database in GML (Geography Markup Language) digital format. In practice, however, for reasons of efficiency and because cartographic database generalization is still not fully automated, a series of intermediate data model layers are employed (see Section 3.8 for a discussion of generalization). The base, or fine-scale, DLM is generalized by automated and manual means to create coarser-scale DLMs (in Figure 12.18 two additional medium- and coarse-scale DLMs are shown). From each of these DLMs one or more cartographic data models (digital cartographic models (DCMs)) can be created that derive cartographic representations from real-world features. For example, for a medium-scale DCM representation a road centerline network can be simplified to remove complex junctions and information about overpasses and bridges because they are not appropriate to maps at this scale.

Once the necessary DCMs have been built the process of map creation can proceed. Each individual map will be created in the manner described in Section 12.3, but since many similar maps will be required for each series or collection, some additional work is necessary to support map creation. Many similar maps can be created efficiently from a common map template that includes any material common to all maps (for example, inset/overview maps, titles, legends, scales, direction indicators, and map metadata; see Section 12.3.1 and Figure 12.8). Once a template has been created it is simple to specify the individual map-sheet content (e.g., map data from DCM

data layers, specific title, and metadata information) for each separate sheet. The geographic extent of each map sheet is typically maintained in a separate database layer. This could be a regular grid (Figure 12.7) but commonly there is some overlap in sheets to accommodate irregular-shaped areas and to ensure that important features are not split at sheet edges. An automated batch process can then 'stamp out' each sheet from the database as and when required. Finally, it is convenient to generate an index of names (for example, places or streets), if the sheets are to form a map book or atlas, that shows on which sheet they are located.

12.5 Applications

Obviously there is a huge number and range of applications that use maps extensively. The goal here is to highlight a few examples that raise some interesting cartographic issues.

The relative importance of representing space and attributes will vary within and between different applications, as will the ability to broker improved measures of spatial distributions through integration of ancillary sources. These tensions are not new. In a topographic map, for example, the width of a single-carriageway road may be exaggerated considerably (Figure 12.19). This is done in order to enhance the legibility of features that are

Figure 12.19 Topographic map showing street centerlines from a GDT database for the area around Chicago, USA. The major roads are exaggerated with 'cased' road symbols

central to general-purpose topographic mapping. In some instances, these prevailing conventions will have evolved over long periods of time, while in others the new-found capabilities of GIS entail a distinct break with the past. As a general rule, where accuracy and precision of geo-referencing are important, the standard conventions of topographic mapping will be applied (e.g., Figure 12.1).

Utility applications use GIS that have come to be known as Automated Mapping and Facilities Management (AM/FM) systems. The prime objectives of such systems are asset management, and the production of maps that can be used in work orders that drive asset maintenance projects. Some of the maps produced from an AM/FM system use conventional cartographic representations, but for some maintenance operations schematic representations are preferred (Figure 12.20). A schematic map provides a view of the way in which a system functions, and is used for operational activities such as identifying faults during power outage and routine pipeline maintenance. Hybrid schematic and geographic maps, known as geoschematics, combine the synoptic view of the current state of the network as a schematic superimposed upon a background map with real-world coordinates.

Transportation applications use a procedure known as linear referencing to visualize point (such as street 'furniture', e.g., road signs), linear (such as parking restrictions and road surface quality measures), and continuous events (such as speed limits). In linear referencing, two-dimensional geography is collapsed into one-dimensional linear space. Data are collected as linear measures from the start of a route (a path through a network). The linear measures are added at display time, usually dynamically, and segment the route into smaller sections (hence the term 'dynamic segmentation', which is sometimes used to describe this type of model). Figure 12.21 is a map of Pitt County, USA, that shows road surface (pavement) quality rated on a 100 point scale. The data were collected as linear events by driving the roads with an in-vehicle sensor. For more information on linear referencing see Section 5.4 and 8.2.3.3.

We began this chapter with a discussion of the military uses of mapping, and such applications also have special cartographic conventions – as in the operational overlay maps used to communicate battle plans (Figure 12.22). On these maps, friendly, enemy, and neutral forces are shown in blue, red and green, respectively. The location, size, and capabilities of military units are depicted with special

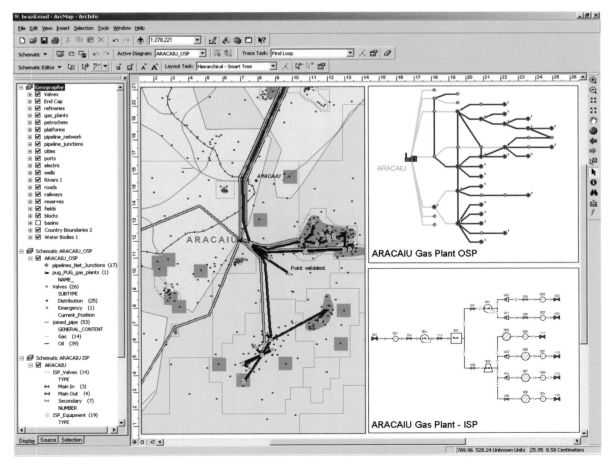

Figure 12.20 Schematic gas pipeline utility maps for Aracaiu, Brazil. Geographic view (left), Outside Plant schematic (top right), Inside Plant schematic (bottom right). Data courtesy of IHS-Energy/Tobin

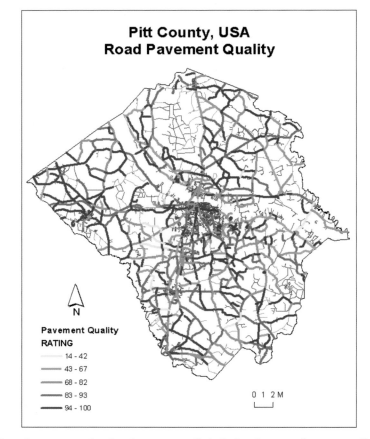

Figure 12.21 Linear referencing transportation data (pavement quality) displayed on top of street centerline files

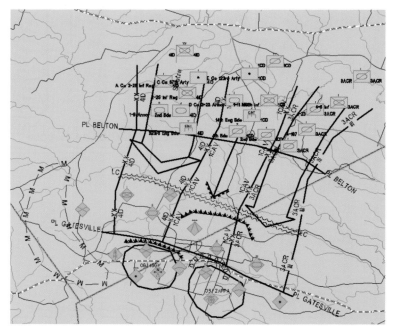

Figure 12.22 Military map showing 'overlay' graphics on top of base map (military vector map level 2) data. Blue rectangles are friendly forces units, red diamonds are enemy force units, black symbols are military operations – maneuvers, obstacles, phase lines, and unit boundaries. These 'tactical graphics' depict the battle space and the tactical environment for the operational units

multivariate symbols that have operational and tactical significance. Other subjects of significance, such as mine fields, impassable vegetation, and direction of movement, also have special symbols. Animations of such maps can be used to show the progression of a battle, including future 'what if?' scenarios.

12.6 Conclusions

Cartography is both an art and a science. The modern cartographer must also be very familiar with the application of computer technology. The very nature of cartography and map making has changed profoundly in the past few decades and will never be the same again. Nevertheless, there remains a need to understand the nature and representational characteristics of what goes into maps if they are to provide robust and defensible aids to decision making, as well as tactical and operational support tools. In cartography, there are few hard and fast rules to drive map composition, but a good map is often obvious once complete. Modern advances in GIS-based cartography make it easier than ever to create large numbers of maps very quickly using automated techniques once databases and map templates have been built. Creating databases and map templates continue to be advanced tasks requiring the services of trained professionals. The type of data that are used on maps is also changing – today's maps often reuse and recycle different datasets, obtained over the Internet, that are rich in detail but may be unsystematic in collection and incompatible in terms of scale. This all underpins the importance of metadata to evaluate datasets in terms of scale, aggregation, and representativeness prior to mapping. Collectively these changes are driving the development of new applications founded on the emerging advances in scientific visualization that will be discussed in the next chapter.

Questions for further study

1. Identify the criteria that you would have used in designing a military mapping system for use during the 2003 Iraq war. How might the system subsequently have been adapted for use in humanitarian relief?

2. Why are there so many classification schemes for area-based choropleth maps? What criteria would you use to select the best scheme?

3. Using Table 12.1 as a guide, create an equivalent table showing example maps for all the map types.

4. Using a GIS of your choice, create a map of your local area containing all the seven elements listed in Section 12.3.1. What makes this a good map?

Further reading

Dent B.D. 1999 *Cartography: Thematic Map Design* (5th edn). Dubuque, Iowa: WCB/McGraw-Hill.

Kraak M.-J. and Ormerling F. 1996 *Cartography: Visualization of Spatial Data*. Harlow: Longman.

Robinson A.H., Morrison J.L., Muehrcke P.C., Kimerling A.J. and Guptill S.C. 1995 *Elements of Cartography* (6th edn). New York: Wiley.

Slocum T.A. 1999 *Thematic Cartography and Visualization*. Upper Saddle River, NJ: Prentice-Hall Inc.

Tufte E.R. 2001 *The Visual Display of Quantitative Information*. Cheshire, CT: Graphics Press.

13 *Geovisualization*

This chapter describes a range of novel ways in which information can be presented visually to the user. Using techniques of geovisualization, GIS provides a far richer and more flexible medium for portraying attribute distributions than the paper mapping which is covered in Chapter 12. First, through techniques of spatial query, it allows users to explore, synthesize, present (communicate), and analyze the meaning of any given representation. Second, it facilitates transformation of representations using techniques such as cartograms and dasymetric mapping. Third, GIS-based geovisualization allows the user to interact with the real world from a distance, through interaction with and even immersion in artificial worlds. Together, these functions broaden the user base of GIS, and have implications for public participation in GIS (PPGIS).

Geographic Information Systems and Science, 2nd edition   Paul Longley, Michael Goodchild, David Maguire, and David Rhind.
© 2005 John Wiley & Sons, Ltd.   ISBNs: 0-470-87000-1 (HB); 0-470-87001-X (PB)

Learning Objectives

At the end of this chapter you will understand:

■ How GIS affects visual communication;

■ The ways in which good user interfaces can help to resolve spatial queries;

■ Some of the ways in which GIS-based representations may be transformed;

■ How 3-D geovisualization and virtual reality improve our ability to understand the world.

13.1 Introduction: uses, users, messages, and media

Effective decision support through GIS requires that the messages of representations (in the senses defined in Chapter 3) are readily interpretable in the minds of decision makers. Historically, many of the early uses of mapping were to support military operations (Section 12.1). The historic map shown in Figure 13.1, and the case study to which it relates (Box 13.1), illustrates how geovisualization has been used as an indirect tool of warfare. Today's GIS applications environment is immeasurably more data rich than in the past, and a major goal of GIS usage is often to make sense of the huge archive resources that are available, without creating information overload. In some respects at the opposite extreme, historical GIS applications are concerned with assembling together digital shards of evidence that may be scattered across space and time. There are commonalties of interest that link data rich and data poor applications, as Box 13.2 (and also Box 6.1) illustrates.

It is through mapping that the meaning of a spatial representation of the real world is communicated to users. Historically, the paper map was the only available interface between the map-maker and the user: it was permanent, contained a fixed array of attributes, was of predetermined and invariant scale, and rarely provided any quantitative or qualitative indications as to whether it was safe to use (Figure 13.3: see also Section 12.2). These attributes severely limit the usefulness of the paper map in today's applications environment, which can be of seemingly unfathomable complexity, and entails

Applications Box **13.1**

Geovisualization and Geopolitics

In the early 20th century the Oxford, UK, geographer Halford Mackinder proposed a theory that different national groupings organized around their own historic settlement 'heartlands'. He worked at a time when the development of railroads and roads for troop movement was opening up previously inaccessible areas of western Russia, and he envisioned this territory as a new heartland at the center of an increasingly integrated world. In *geopolitical* terms, he saw western Russia as the center of the new joined-up world, and believed that whoever occupied it would come to dominate the emerging 'world island'. Subsequently, the German geographer Friedrich Ratzel developed the concept of the 'heartland' into that of 'Lebensraum', the essence of which was that every different cultural group needed to develop its own geographic living space. In the years immediately prior to World War Two (1939–45), these related ideas of geographic heartland and Lebensraum were developed by General Haushofer, Professor of Geography at the University of Munich. The spatial extent of the Lebensraum of the German national grouping (the German Volk), he argued, had been overly constrained and dissected by the Treaty of Versailles at the end of World War One. Crudely, Haushofer mapped the spatial distribution of 'German-speaking people' (Figure 13.1) to substantiate this thesis, and implied that there was a need to extend the German cultural realm deep into Russia. This argument was subsequently used by the Nazi Party to justify German incursions into Czechoslovakia, the annexation of Austria, and the invasion of Poland that ultimately triggered World War Two. The extensive use of mapping in setting out this geopolitical thesis (termed Geopolitik) provides early and powerful evidence of the ways in which maps can be used as propaganda. Haushofer's maps portray information in order to garner support for a particular worldview and subsequent course of decision making. They illustrate the general point that the map creator is very much the arbiter and architect of what the map portrays. In Haushofer's case, the message of the map ultimately contributed to a pseudo-scientific justification of a worldview that history would rather forget – yet it is periodically rekindled in various guises by extremist groups.

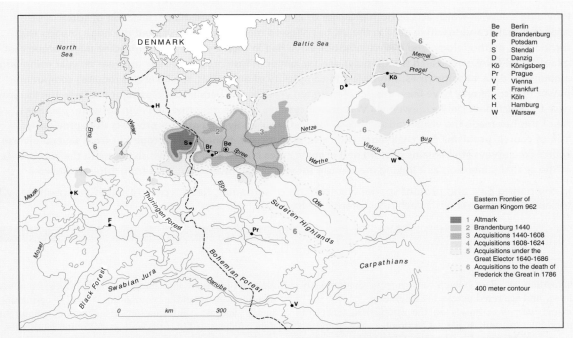

Be Berlin
Br Brandenburg
P Potsdam
S Stendal
D Danzig
Kö Königsberg
Pr Prague
V Vienna
F Frankfurt
K Köln
H Hamburg
W Warsaw

Eastern Frontier of
German Kingom 962

1 Altmark
2 Brandenburg 1440
3 Acquisitions 1440-1608
4 Acquisitions 1608-1624
5 Acquisitions under the
  Great Elector 1640-1686
6 Acquisitions to the death of
  Frederick the Great in 1786

400 meter contour

Figure 13.1 The geopolitical center of Germany. (*Source*: Parker G, 1988. Geopolitics: past, present, future. London: Pinter.)

Biographical Box **13.2**

Kevin Schürer, historical demographer and digital data archivist

Kevin Schürer has been fascinated by the past and by maps for as long as he can remember. The young child who was passionate about understanding history, especially through visiting castles and museums, was equally intrigued by atlases and images of far-off places. These dual interests led to his studying both history and geography at London and subsequently as a member of the Cambridge Group for the History of Population and Social Structure at Cambridge University, UK. Kevin recalls: 'In the 1980s the Group provided an exciting mix of historians, geographers, statisticians, computer scientists, with, from time to time, the odd anthropologist or sociologist thrown in for good measure. Each brought their own blend of research skills to problem solving. To me this was ahead of its time as a way of conducting cutting-edge inter-disciplinary team research.'

Today, Kevin is Director of the UK Data Archive (UKDA) at the University of Essex. This is the oldest and largest social science data archive in the UK. It distributes some 18 000 non-census datasets and associated metadata (Section 11.2.1) to some 2000 university-based researchers each year, as well as Census of Population data to over 10 000 registered researchers, teachers, and students. Reflecting on his role, Kevin says:

Figure 13.2 Kevin Schürer, historical demographer and digital data archivist

Providing a service which acts as both provider of and custodian to the major collection of social science research and teaching data in the UK has allowed me to witness first hand the huge impacts that technology, methodology, and geovisualization have had in recent years. Although some might argue that quantitative-focused research is unpopular and no longer in fashion, researchers are year on year using larger amounts of data. In some cases, the scale and complexity of analysis is at a level and requires computing power that could only have been dreamt of even ten years ago. A parallel development has been the increased ease by which data can be viewed via Web-based systems and be migrated between software packages. All of this has made the visualization of complex statistical analyses a greater prospect than previously. Yet, of course, problems and challenges still exist. In particular, bringing data together from several diverse sources or surveys still remains difficult and in some cases impossible. All of which points to the importance of the spatial dimension as a linking mechanism for visualizing diverse datasets from different sources.

In comparing past and present spatial distributions of surnames, Kevin's own research demonstrates why historians and data archivists each need maps (see Box 6.1).

visualization of data which are richer, continuously updated, and scattered across the Internet.

Geovisualization builds on the established tenets of map production and display. Geographer and geovisualization expert Alan MacEachren (Box 13.3) describes it as the creation and use of visual representations to facilitate thinking, understanding, and knowledge construction about human and physical environments, at geographic scales of measurement. Geovisualization is a research-led field that integrates approaches from visualization in scientific computing (ViSC), cartography, image analysis, information visualization, exploratory data analysis (EDA), as well as GIS. Its motivation is to develop theories, methods, and tools for visual exploration, analysis, synthesis, and presentation of geospatial data. The December 2000 research agenda of the International Cartographic Association's Commission on Visualization and Virtual Environments is shown in Table 13.1.

Table 13.1 Research agenda of the International Cartographic Association's Commission on Visualization and Virtual Environments (adapted from **www.geovista.psu.edu/sites/icavis/agenda/index.html**)

1. Cognitive and usability issues in geovisualization

   ■ Virtual geographic environments and 3-D geovisualization.
   ■ Dynamic representations (including animated and interactive mapping).
   ■ Improved interface design.
   ■ Understanding individual and group interaction using geovisualization.
   ■ Collaborative visualization.
   ■ Better evaluation of existing methods.

2. Representation and its relationship with cartographic visualization

   ■ Existing and potential roles of map displays.
   ■ Understanding the techniques that map producers use to visualize spatial data.
   ■ Understanding how technology might improve cartographic practice.
   ■ The relationships between user cognition, map design, and the nature of spatial data.

3. Uses of geovisualization in knowledge discovery

   ■ Visual approaches to knowledge discovery.
   ■ Visual support for knowledge discovery through geocomputation.
   ■ Use of databases and data models to support knowledge discovery.

4. User interface issues for geovisualization

   ■ Interface development and usage.
   ■ Improving usability and navigability of geovisualizations.
   ■ Improving facilities for user collaboration.

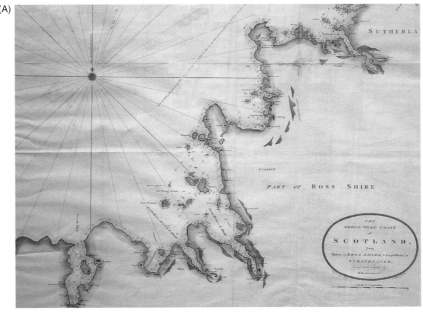

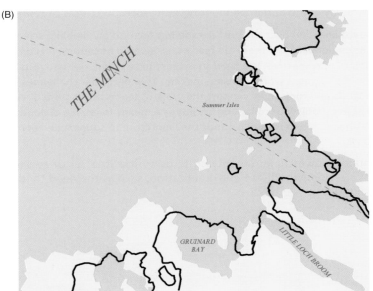

Figure 13.3 Part of the coastline of northwest Scotland (A) according to the hydrographer Murdoc Mackenzie in 1776 (orientation to magnetic north at the time) and (B) Modern day outline of the same coast with trace of Mackenzie chart overlaid in black. (Reproduced by permission of Michael J. de Smith)

As such, today's geovisualization is much more than conventional map design. It has developed into an applied area of activity that leverages geographic data resources to meet a very wide range of scientific and social needs, as well as into a research field that is developing new visual methods and tools to facilitate better user interaction with such data. In this chapter, we discuss how this is achieved through *query* and *transformation* of geographic data, and user *immersion* within them. Query and transformation are discussed in a substantially different context in Chapter 14, where they are encountered as two of six types of spatial analysis.

13.2 Geovisualization and spatial query

13.2.1 Overview

Fundamental to effective geovisualization is an understanding of how human cognition shapes GIS usage,

Biographical Box **13.3**

Alan MacEachren, geographer and geovisualization expert

Alan MacEachren (Figure 13.4) is a Professor of Geography at Pennsylvania State University. He began by studying an eclectic range of disciplines from botany through to experimental psychology at Ohio University before settling on a major in Geography, and next undertook a Masters and Ph.D. in Geography at the University of Kansas. There, he became interested in environmental perception in general and the cognitive aspects of map symbolization in particular. Following his move to Penn State in 1985, he developed interests in the relationships between cognitive science, GIS, and GIScience with an emphasis on visual representation of geographic information. Today he continues to develop a range of geovisualization applications and explores the potential of technology to enable immersive virtual environments and geocollaboration.

He acts as Director of the GeoVISTA Center at the University. Center activities have played a leading role in the development of geovisualization, allied to other developments in GIScience. The Center is avowedly interdisciplinary in addressing two goals: (a) to develop a deep understanding of human interaction with geographic information and technologies; and (b) to use that understanding to develop powerful human-centered scientific methods to solve social, environmental, and other problems. Central to these objectives is the use of computer-supported, visually-enabled exploration and analysis of an ever-expanding range of geographic data.

For Alan, the essence of geovisualization is leveraging the rapid advances in information technology to enable creative thinking. He sees better understanding of how people use dynamic visually enabled representations as key to designing and implementing more effective forms of representation. He says:

> **Most research in geovisualization over the past decade has focused on enabling science. While this remains an important application domain, and there are many unanswered questions, it is time for geovisualization research to take a broader perspective. We need to extend, or reinvent, our geovisualization methods and tools so that they can support real-world decision making. Two particular challenges stand out. First, geovisualization methods and tools must support the work of teams much more directly, with the visual display acting not just as an object of collaboration but as a means to enable dialogue and coordinate joint action. Second, geovisualization must move out of the laboratory to support decision making in the field (e.g., for crisis response).**

Alan's hobbies include birdwatching: identifying birds by jizz (the art of seeing a bird badly but still knowing what it is) involves similar skills to geovisualization, such as the need to identify phenomena that may be difficult to categorize.

(B)

(A)

Figure 13.4 Alan MacEachren: (A) geographer and geovisualization expert; and (B) birder

how people think about space and time, and how spatial environments might be better represented using computers and digital data. The conventions of map production, presented in Section 12.3, are central to the use of mapping as a decision support tool. Many of these conventions are common to paper and digital mapping, although GIS allows far greater flexibility and customization of map design. Geovisualization develops and extends these concepts in a number of new and innovative ways.

Viewed in this context, there are four principal purposes of geovisualization:

1. *Exploration*: for example, to establish whether and to what extent the general message of a dataset is sensitive to inclusion or exclusion of particular data elements.

2. *Synthesis*: to present the range, complexity, and detail of one or more datasets in ways that can be readily assimilated by users. Good geovisualization should enable the user to 'see the wood for the trees'.

3. *Presentation*: to communicate the overall message of a representation in a readily intelligible manner, and to enable the user to understand the likely overall quality of the representation.

4. *Analysis*: to provide a medium to support a range of methods and techniques of spatial analysis.

This fourth function that geovisualization supports will be discussed in detail in Chapters 14 and 15.

The most straightforward way in which reformulation and evaluation of a representation of the real world can take place is through posing *spatial queries* to ask generic spatial and temporal questions such as:

These objectives cut across the full range of different applications tasks and are pursued by users of varying expertise that desire different degrees of interaction with their data (Figure 13.5). The tasks addressed by geovisualization range from information sharing to knowledge construction (see Tables 13.1 and 1.2); user groups range from lay-public users to GIS specialists (spatially aware professionals, or SAPs: Section 1.4.3.2); and the degree of interaction that users desire ranges from low (passive) to high.

> **Geovisualization allows users to explore, synthesize, present, and analyze their data more thoroughly than was possible hitherto.**

Together these tasks may be considered in relation to the conceptual model of uncertainty that was presented in Figure 6.1, which is presented with additions in Figure 13.6. This encourages us to think of geographic analysis not as an end point, but rather as the start of an iterative process of feedbacks and 'what if?' scenario testing. The eventual best-available reformulated model is used as a decision support tool, and the real world is changed as a consequence. As such, it is appropriate to think of geovisualization as imposing a filter upon the results of analysis and on the nature of the feeds that are generated back to the conception and measurement of geographic phenomena. It can be thought of as an additional 'geovisualization filter' (U4 in Figure 13.6).

The most straightforward way in which reformulation and evaluation of a representation of the real world can take place is through posing *spatial queries* to ask generic spatial and temporal questions such as:

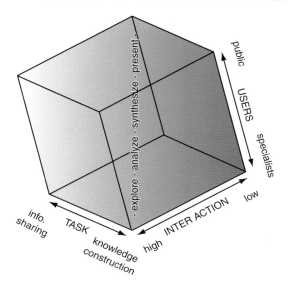

Figure 13.5 Functions of geovisualization (after MacEachren et al 2004) © IEEE 2004. Reproduced by permission.

- Where is ...?
- What is at location ...?
- What is the spatial relation between ...?
- What is similar to ...?
- Where has ... occurred?
- What has changed since ...?
- Is there a general spatial pattern, and what are the anomalies?

These questions are articulated through the graphical user interface (GUI) paradigm called a 'WIMP' interface – based upon **W**indows, **I**cons, **M**enus and **P**ointers (Figure 13.7). The familiar actions of pointing, clicking, and dragging windows and icons are the most common ways of interrogating a geographic database and summarizing results in map and tabular form. By extension, research applications increasingly use multiple displays of maps, bar charts, and scatterplots, which enable a picture of the spatial and other properties of a representation to be built up. These facilitate learning about a representation in a data-led way and are discussed in Section 14.2.

> **The WIMP interface allows spatial querying through pointing, clicking, and dragging windows and icons.**

13.2.2 Spatial query and Internet mapping

Spatial query functions are also central to many Internet GIS applications. For many users, spatial query

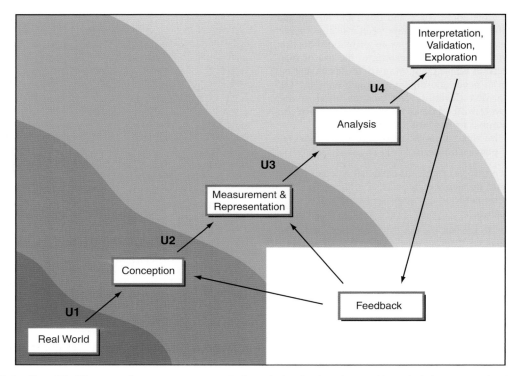

Figure 13.6 Filters U1–U4: conception, measurement, analysis, and visualization. Geographic analysis is not an end point, but rather the start of an iterative process of feedbacks and 'what if?' scenario testing. (See also Figure 6.1.)

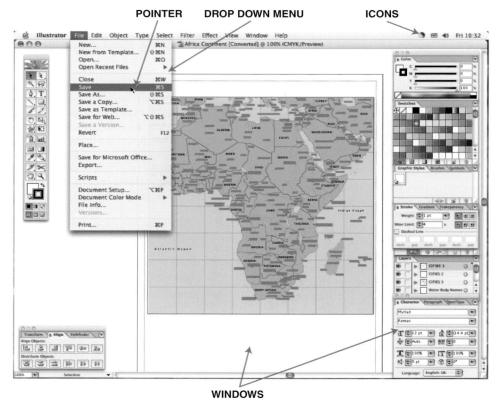

Figure 13.7 The WIMP (Windows, Icons, Menus, Pointers) interface to computing using Macintosh System 10

is the objective of a GIS application, as in queries about the taxable value of properties (Section 2.3.2.2), customer care applications about service availability (Section 2.3.3), or Internet site queries about traffic congestion (Section 2.3.4). In other applications, spatial query is a precursor to any advanced geographic analysis. Spatial query may appear routine, but entails complex operations, particularly when the objects of spatial queries are continuously updated (refreshed) in real time – as illustrated in the traffic application shown in Box 13.4 and on websites concerning weather conditions. Other common spatial queries are framed to identify the location of services, provide routing and direction information (see Figure 1.17), facilitate rapid response in disaster management (see Figure 2.16), and provide information about domestic property and neighborhoods to assist residential search (Figure 1.14).

13.3 Geovisualization and transformation

13.3.1 Overview

Chapter 12 illustrated circumstances in which it was appropriate to adjust the classification intervals of maps,

in order to highlight salient characteristics of spatial distributions. We have also seen how transformation of entire scales of attributes can assist in specifying the relationship between spatial phenomena, as in the scaling relations that characterize a fractal coastline (Box 4.6). More generally still, our discussion of the nature of geographic data (Chapter 4) illustrated some of the potentially difficult consequences arising out of the absence of natural units in geography – as when choropleth map counts are not standardized by any numerical or areal base measure. Standard map production and display functions in GIS allow us to standardize according to numerical base categories, yet large but unimportant (in attribute terms) zones continue to accrue greater visual significance in mapping than is warranted. Such problems can be addressed by using GIS to transform the shape and extent of areal units. Our use of the term 'transformation' here is thus in the cartographic sense of taking one vector space (the real world) and transforming it into another (the geovisualization). Transformation in the sense of performing analytical operations upon coordinates to assess explicit spatial properties such as adjacency and contiguity is discussed in Section 14.4.

GIS is a flexible medium for the cartographic transformation of maps.

Some ways in which example real-world phenomena of different dimensions (Box 4.1) may or may not be transformed by GIS are illustrated in Table 13.2. These examples illustrate how the measurement and representation filter of Figure 13.6 has the effect of transforming

Applications Box 13.4

Use of Internet GIS to manage traffic in Southampton, England

International port cities function as gateways to distant destinations for tourists, freight transporters, and a multitude of other travelers, and so they are often overwhelmed with unique vehicular traffic flow problems. Southampton and its surrounding area have been involved in the development and application of part of a pan-European, GIS-based traffic monitoring and analysis system called the ROad MANagement System for Europe (ROMANSE) for nearly ten years.

ROMANSE is based on Desktop GIS software linked to an Urban Traffic Control (UTC) system. The GIS acts as the graphical front end to the UTC. The GIS software has been customized to create a simplified, specific-purpose display. Basemaps originate from Great Britain's Ordnance Survey, onto which a variety of current traffic information is overlaid. The GIS creates on-the-fly status maps so that engineers at the Traffic and Travel Information Centre

(TTIC) in Southampton can analyze existing road hazards and anticipate new ones. They can then issue appropriate advice and warning messages for immediate display on digital road signs in the area or dispatch emergency services.

The information processed at the TTIC comes from a variety of sources, including closed-circuit television, roadside detectors, and satellite tracking systems, all of which are fed into the UTC. It is then processed and integrated by the GIS for instant analysis or display, providing the engineers with a good overview of what is happening on the road network. Information available to the engineers includes traffic flow per minute, traffic speeds, congestion (Figure 13.8A), and car park occupancy (Figure 13.8B).

Internet GIS has been used to provide current traffic information and maps on demand to Internet users visiting the ROMANSE Web site at **www.romanse.org.uk**. From this site, members

▶

of the public can check the current occupancy in car parks before traveling. A city-center map is displayed and by clicking on each car park it is possible to determine both the current occupancy status and what the occupancy is likely to be in the near future, based on historic records. Information for cyclists is also available at the website. Traffic information is also posted digitally on touch-screen displays at main transport interchanges, shopping centers, tourist information centers, and libraries. At bus stops, arrival times are electronically displayed and some stops feature an audio version of the information for visually impaired passengers.

This remains one of the most advanced intelligent transport information systems in the world. A key part of the success was linking GIS software to a real-time traffic control system for data collection, and to messaging systems and the Internet for data display.

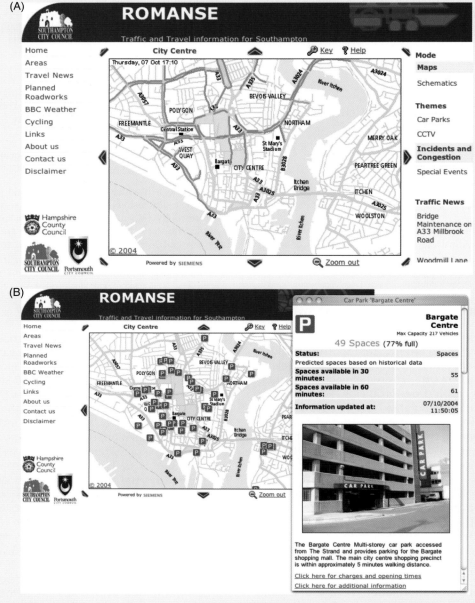

Figure 13.8 Monitoring (A) traffic flow and (B) car park occupancy in real time as part of the Web-based ROMANSE traffic management system (Base maps: Crown Copyright **southampton.romanse.org.uk/**) (Reproduced by permission of Southampton City Council)

Table 13.2 Some examples of coordinate and cartographic transformations of spatial objects of different dimensionality (based on D. Martin 1996 *Geographic Information Systems: Socioeconomic Applications (Second edition)*. London, Routledge: 65)

Dimension	0	1	2	3
Conception	Population distribution	Coastline (see Box 4.6)	Agricultural field	Land surface
	Coordinate transformation of real world arising in GIS representation/measurement			
Measurement	Imposed areal aggregation, recorded as a sequence of digitized points	Sequence of digitized points	Digitized polygon boundary	Arrangement of spot heights
	Cartographic transformation to aid interpretation of representation			
Visualization	Cartogram or dasymetric map	Generalized line	Integral/natural area	Digital elevation model

some objects, like population distributions, but not others, like agricultural fields. Further transformation may occur in order to present the information to the user in the most intelligible and parsimonious way. Thus, cartographic transformation through the U3 filter in Figure 13.6 may be the consequence of the imposition of artificial units, such as census tracts for population, or selective abstraction of data, as in the generalization of cartographic lines (Section 3.8).

The standard conventions of map production and display are not always sufficient to make the user aware of the transformations that have taken place between conception and measurement – as in choropleth mapping, for example, where mapped attributes are required to take on the proportions of the zones to which they pertain. Geovisualization techniques make it possible to manipulate the shape and form of mapped boundaries using GIS, and there are many circumstances in which it may be appropriate to do so. Indeed, where a standard mapping projection obscures the message of the attribute distribution, map transformation becomes necessary. There is nothing untoward in doing so – remember that one of the messages of Chapter 5 was that all conventional mapping of geographically extensive areas entails transformation, in order to represent the curved surface of the Earth on a flat screen or a sheet of paper. Similarly, in Section 3.8, we described how generalization procedures may be applied to line data in order to maintain their structure and character in small-scale maps. There is nothing sacrosanct about popular or conventional representations of the world, although the widely used transformations and projections do confer advantages in terms of wide user recognition, hence interpretability.

13.3.2 Cartograms

Cartograms are maps that lack planimetric correctness, and distort area or distance in the interests of some specific objective. The usual objective is to reveal patterns that might not be readily apparent from a conventional map or, more generally, to promote legibility. Thus,

the integrity of the spatial object, in terms of areal extent, location, contiguity, geometry, and/or topology, is made subservient to an emphasis upon attribute values or particular aspects of spatial relations. One of the best-known cartograms (strictly speaking a *linear cartogram*) is the London Underground map, devised in 1933 by Harry Beck to fulfill the specific purpose of helping travelers to navigate across the network. Figure 13.9A is an early map of the Underground, which shows the locations of stations in Central London in relation to the real-world pattern of streets. The much more recent central-area cartogram shown in Figure 13.9B is a descendent of Beck's 1933 map. It is a widely recognized representation of connectivity in London, and the conventions that it utilizes are well suited to the attributes of spacing, configuration, scale, and linkage of the London Underground system. The attributes of public transit systems differ between cities, as do the cultural conventions of transit users, and thus it is unsurprising that cartograms pertaining to transit systems elsewhere in the world appear quite different.

Cartograms are map transformations that distort area or distance in the interests of some specific objective.

A central tenet of Chapter 3 was that all representations are abstractions of reality. Cartograms depict transformed, hence artificial, realities, using particular exaggerations that are deliberately chosen. Figure 13.10 presents a regional map of the UK, along with an *equal population cartogram* of the same territory. The cartogram is a map projection that ensures that every area is drawn approximately in proportion to its population, and thus that every individual is accorded approximately equal weight. Wherever possible, areas that neighbor each other are kept together and the overall shape and compass orientation of the country is kept roughly correct – although the real-world pattern of zone contiguity and topology is to some extent compromised. When attributes are mapped, the variations within cities are revealed, while the variations in the countryside are reduced in size so as not to dominate the image and divert the eye away

(A)

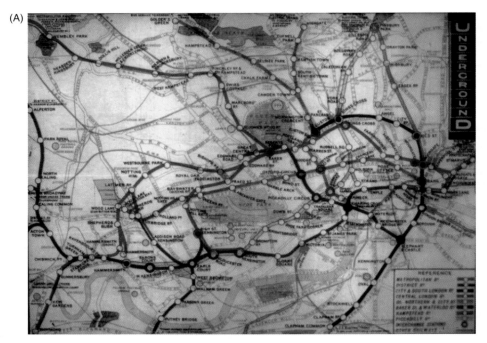

(B)

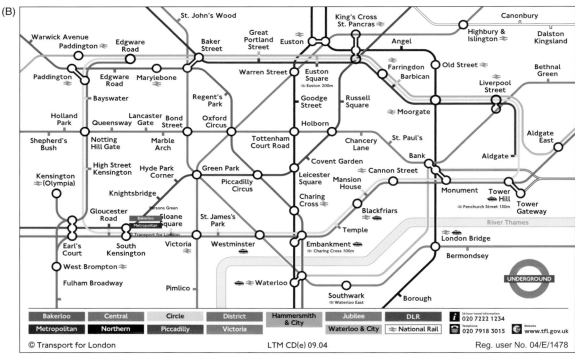

Figure 13.9 (A) A 1912 map of the London Underground and (B) its present day linear cartogram transformation

from what is happening to the majority of the population. Therefore, the transformed map of wealthy areas in the UK, shown in Figure 13.11B, depicts the considerable diversity in wealth in densely populated city areas – a phenomenon that is not apparent in the usual (scaled Transverse Mercator) projection (Figure 13.11A). There are semi-automated ways of producing cartograms,

but complicated representations such as those shown in Figures 13.10B and 13.11B almost invariably entail human judgment and design.

Throughout this book, one of the recurrent themes has been the value of GIS as a medium for data sharing and this is most easily achieved if pooled data share common coordinate systems. Yet a range of cartometric

(A)

(B)

Figure 13.10 (A) A regional map of the UK and (B) its equal population cartogram transformation (Courtesy: Daniel Dorling and Bethan Thomas: from Dorling and Thomas (2004), page (vii)) (Reproduced by permission of The Policy Press)

transformations is useful for accommodating the way that humans think about or experience space, as with the measurement of distance as travel time or travel cost. As such, the visual transformations inherent in cartograms can provide improved means of envisioning spatial structure.

13.3.3 Re-modeling spatial distributions as dasymetric maps

The cartogram offers a more radical means of transforming space, and hence restoring spatial balance, but sacrifices the familiar spatial framework valued by most users. It forces the user to make a stark choice between assigning each mapped element equal weight and being able to relate spatial objects to real locations on the Earth's surface. Yet there are also other ways in which the data-integrative power of GIS can be used to re-model spatial distributions and hence assign spatial attributes to meaningful yet recognizable spatial objects.

One example of the way in which ancillary sources of information may be used to improve the model of a spatial distribution is known as *dasymetric mapping*. Here, the intersection of two datasets is used to obtain more precise estimates of a spatial distribution. Figure 13.12A shows the census tract geography for which small area population totals are known, and Figure 13.12B shows the spatial distribution of built structures in an urban area (which might be obtained from a cadaster or very-high-resolution satellite imagery, for example). A reasonable assumption (in the absence of evidence of mixed land use or very different residential structures such as high-rise apartments and widely spaced bungalows) is that all of the built structures house resident populations at uniform density. Figure 13.12C shows how this assumption, plus an overlay of the areal extent of built structures, allows population figures to be allocated to smaller areas than census tracts, and allows calculation of indicators of residential density. The practical usefulness of dasymetric mapping is illustrated in Figure 13.13. Figures 13.13A and B show the location of an elongated census block (or enumeration district) in the City of

(A)

(B)

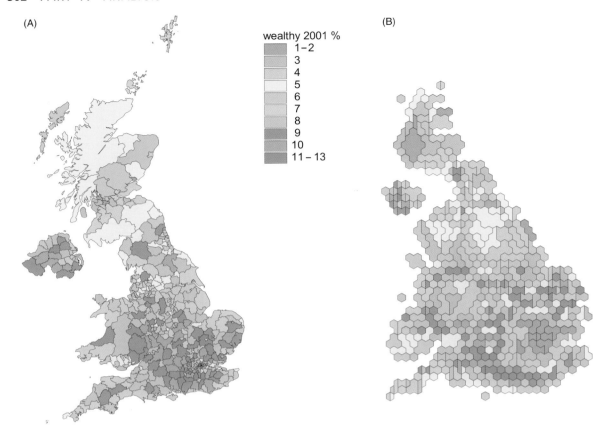

wealthy 2001 %
- 1–2
- 3
- 4
- 5
- 6
- 7
- 8
- 9
- 10
- 11 – 13

Figure 13.11 Wealthy areas of the UK, as measured by property values, property size, and levels of multiple car ownership: (A) conventional map of local authority districts and (B) its equal population cartogram transformation (Courtesy: Daniel Dorling and Bethan Thomas: from Dorling and Thomas (2004), page 16) (Reproduced by permission of The Policy Press)

Bristol, UK, which appears to have a high unemployment *rate*, but a low absolute *incidence* of unemployment. The resolution of this seeming paradox was examined in Box 4.3 but this would not help us much in deciding whether or not the zone (or part of it) should be included in an inner-city workfare program, for example. Use of GIS to overlay high-resolution aerial photography (Figure 13.13C) reveals the tract to be largely empty of population apart from a small extension to a large housing estate. It would thus appear sensible to assign this zone the same policy status as the zone to its west.

Dasymetric mapping uses the intersection of two datasets (or layers in the same dataset) to obtain more precise estimates of a spatial distribution.

Dasymetric mapping and related techniques present a window on reality that looks more convincing than conventional choropleth mapping. However, it is important to remain aware that the visualization of reality is only as good as the assumptions that are used to create it. The information about population concentration used in Figure 13.13, for example, is defined subjectively and is likely to be error prone (some built forms may be offices and shops, for example), and this will inevitably feed through into inaccuracies in visualization. Inference of

land use from land cover, in particular, is an uncertain and error-prone process, and there is a developing literature on best practice for the classification of land use using land cover information and classifying different (e.g., domestic versus non-domestic) land uses.

Inferring land use from land cover is uncertain and error prone.

13.4 Immersive interaction and PPGIS

13.4.1 Overview: scientific visualization (ViSC) and representation

Faster processing hardware and more sophisticated computer graphics, including animation, have led to the development of a new field called 'Visualization in Scientific Computing.' ViSC investigates the use of new technology

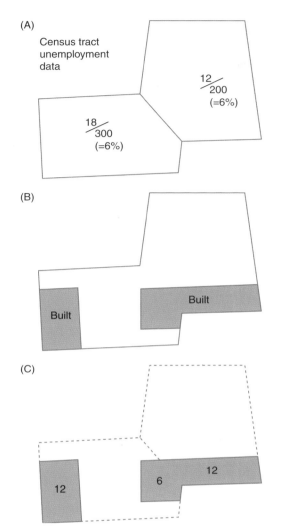

(A)

Census tract
unemployment
data

$\frac{12}{200}$
(=6%)

$\frac{18}{300}$
(=6%)

(B)

Built

Built

(C)

12

6

12

12

Figure 13.12 Modeling a spatial distribution in an urban area using dasymetric mapping: (A) zonal distribution of census population and unemployment rates; (B) distribution of built structures; and (C) overlay of (A) and (B) to obtain a more contained measure of distribution of unemployed population

Table 13.1 is to help users better understand geographic data and models, and hence improve understanding of real-world patterns and processes. Related to this is the objective of broadening the base of users that feels confident interacting with GIS. Research suggests that people first desire an overview of the totality of information, and then the ability to narrow down to information that is immediately relevant. Geodemographers (see Section 2.3.3) have investigated the ways in which individuals and different groups interact with new information and communication technologies (NICTs), and have suggested a three-stage model. Initially, people use NICTs to access information; many then graduate to using them to conduct transactions (e.g., through shopping); and user engagement culminates with people using NICTs as a gateway to participation in networked decision making. In a GIS context, this progression depends upon confidence in perceiving, manipulating, and exploring spatial phenomena through geovisualization.

Planner Mike Shiffer (Box 13.5) has described how GIS and a range of other computer media can be used to facilitate the active involvement of many different groups in the discussion and management of urban change through planning, and may act as a bulwark against officialdom or big business. Better human–computer interaction can be used to encourage public participation in (the use of) GIS, and this is often termed PPGIS. PPGIS is fundamentally about how people can perceive, manipulate, and interact with representations of the real world as manifest in GIS: its other concerns include how people evaluate options through multicriteria decision making (Section 16.4), and social issues of how GIS usage remains concentrated within networks of established interests that are often seen to control the use of technology (Section 1.7). A related theme is the use of GIS to create multiple representations – capturing and maintaining the different perspectives of stakeholders, rather than adopting only one authoritative view.

There is a range of potential uses of geovisualization in PPGIS, including:

■ Making the growing complexity of land use planning, resource use, and community development intelligible to communities and different government departments.

■ Radically transforming the planning profession through use of new tools for community design and decision making.

■ Unlocking the potential of the many digital data sources that are collected, but not used, at the local level.

■ Helping communities shift land use decisions from regulatory processes to performance-based strategies, and making the community decision making process more proactive and less reactive.

■ Improving community education about local environmental and social issues (Figure 13.14).

In each of these applications, there is a strong cognitive component to geovisualization. Users need to feel equipped and empowered to interrogate representations

and media to convey the multifaceted messages of today's mapping. ViSC provides greater flexibility, sophistication, and interaction in visualizing the world than the paper maps of the past. Some of the advanced techniques are not explicitly spatial, or are considered elsewhere: these include line generalization (Sections 3.8 and 4.8) cartographic smoothing (see the discussion of interpolation in Box 4.3), uncertainty visualization (e.g., Section 6.3.2.2), and parallel coordinate plots (Section 15.1). They may be delivered through a wide range of computer devices, from hand-held computers that help users to navigate the real world to virtual-reality systems that augment user perceptions of the real world or even present users with the artificial worlds of *virtual reality* discussed in Section 11.3.1. These activities pose many technical challenges that are interesting in themselves, but the broader motivation for many of the geovisualization research activities set out in

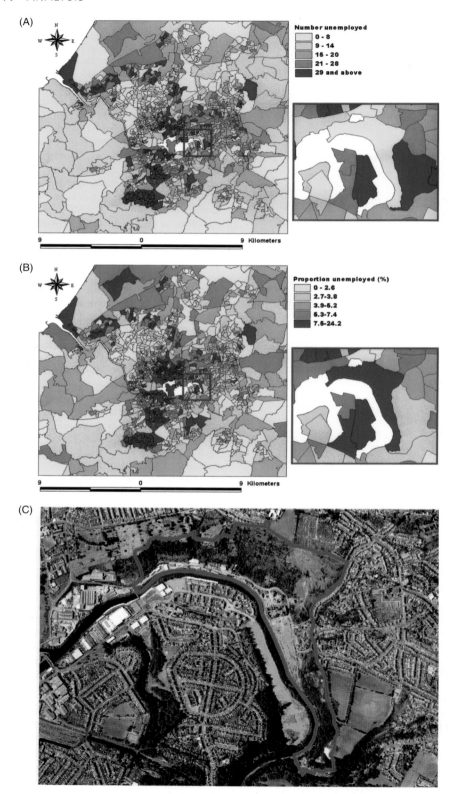

Figure 13.13 Dasymetric mapping in practice, in Bristol, UK: (A) numbers employed by census enumeration district (block); (B) proportions unemployed using the same geography (Reproduced with permission of Richard Harris); and (C) orthorectified aerial photograph of part of the area of interest. (Courtesy: Rich Harris and Cities Revealed: **www.crworld.co.uk**, reproduced by permission of The Geoinformation Group)

Figure 13.14 Field computing encourages education and public participation in GIS (PPGIS) (Courtesy: ESRI)

in order to reveal otherwise-hidden information. This requires dynamic and interactive *software* environments and *people* skills that are key to extracting meaning from a representation. GIS should allow people to use software to manipulate and represent data in multiple ways, in order to create 'what if' scenarios or to pose questions that prompt the discovery of useful relations or patterns. This is a core remit of PPGIS, where the geovisualization environment is used to support a process of knowledge construction and acquisition that is guided by the user's knowledge of the application. PPGIS research entails usability evaluations of structured tasks, using a mixture of computer-based techniques and traditional qualitative research methods, in order to identify cognitive activities and user problems associated with GIS applications.

13.4.2 Geovisualization and virtual-reality systems

Although conventional 2-D representations continue to be used in the overwhelming majority of GIS applications, increasing interaction with GIS now takes place through 3-D representations. This is closely allied with developments in computing, software, and broadband networking, which together allow:

- Users to access virtual environments and select different views of phenomena.
- Incremental changes in these perspectives to permit real time *fly-throughs* (see Figures 13.19, 13.20, and 13.21).
- Repositioning or rearrangement of the objects that make up such scenes.

- Users to be represented graphically as *avatars* – digital representations of themselves.
- Avatars to engage with others connected at different remote locations, creating a networked virtual world.
- The development, using avatars, of new kinds of representation and modeling.
- The linkage of networked virtual worlds with *virtual reality* (VR) systems. Semi-immersive and immersive VR is beginning to make the transition from the realm of computer gaming to GIS application, with new methods of spatial object manipulation and exploration using head-mounted displays and sound (see Section 11.3).

Geovisualization is enabling the creation of 3-D representations of natural and artificial (e.g., cityscapes) phenomena. Box 13.6 presents an illustration of the use of geovisualization in weather forecasting. The 3-D representation of artificial environments, specifically cities, has been aided by the increasing availability of very-high-resolution height data from airborne instruments, such as Light Detection and Ranging (LiDAR), and similar-resolution digital mapping products (Figure 13.17). Such data are achieving increasingly wide application in the extraction of discrete objects and the modeling of continuous elevation of underlying terrain; respective applications include urban planning and flood modeling. A common means of approximating extensive 3-D representations of urban environments is to use local height data to extrude the building footprints detailed on high-resolution mapping (such as GB Ordnance Survey's MasterMap 1:1250 database). Extruded block models provide a basic prismatic representation of buildings and have been augmented using computer-aided design (CAD) software to include details of (in order of complexity) roof structure, building volume, building elevation, and detailed features such as windows. CAD and GIS software also provide the facility of rendering buildings, streetscapes, and natural scenes. Many city visualizations have hitherto been restricted to small areas but, with the wide use of broadband networking, 3-D models can now be effectively 'streamed' over the Internet and make it possible to build together 3-D representations in order to visualize extensive geographic areas (see Box 13.7).

In parallel with the development of fine-scale 3-D models of cityscapes, there have been similarly impressive advances in whole-Earth global visualizations (Figure 13.18). Such software systems allow raster and vector information, including features such as buildings, trees, and automobiles, to be combined in synthetic and photo-realistic global and local displays. In global visualization systems, datasets are projected onto a global TIN-based data structure (see Section 8.2.3.4). Rapid interactive rotation and panning of the globe is enabled by caching data in an efficient in-memory data structure. Multiple levels of detail, implemented as nested TINs and reduced-resolution datasets, allow fast zooming in and out. When the observer is close to the surface of the globe, these systems allow the display angle to be tilted to provide a perspective view of the Earth's surface (Figure 13.19). The overall user experience is one of

Mike Shiffer, public participant in planning

Mike Shiffer (Figure 13.15) is Vice President responsible for planning and development at the Chicago Transit Authority (CTA). He grew up in a car-less family in Chicago where dependence on public transport led to his lifetime interest in improving urban qualities of life and mobility for those without autos. Shiffer is fond of saying that he stumbled into academia because the Chicago Transit Authority was not hiring train drivers every time he graduated (from primary school on). Shiffer received a Ph.D. in Regional Planning and Master's of Urban Planning from the University of Illinois at Urbana-Champaign, before moving in 1991 to work as an academic at the Massachusetts Institute of Technology's (MIT's) Department of Urban Studies and Planning. He returned to work in his native Chicago in 2000. Mike's academic research (he still retains a part-time academic post at the University of Illinois at Chicago) has investigated the ways in which spatial information and multimedia technologies can better inform decision making. He has developed a particular focus on how multimedia representational aids, virtual navigational aids, and broadband networking can better inform deliberation, decision making, and public debate.

At CTA, Shiffer spearheads overall strategic and operations planning efforts of the second-largest public transit system in the US, including service scheduling and facilities development. He is responsible for a staff of planners, architects, engineers, and other transit professionals in five departments. As such, his interests in how GIS technologies shape decision making are much more than academic: he has worked to reshape the CTA's planning process so that it leverages technologies and new models of human capital to create a more agile public transit system for Chicago. As our illustration (Figure 13.15) shows, he has also realized an earlier ambition: Shiffer is now a fully certified train driver.

Figure 13.15 Mike Shiffer, train enthusiast, in 1974 and 2004 (background map courtesy of Chicago Transit Authority)

Penny Tranter, weather forecaster

The UK Meteorological (Met) Office has supplied weather presenters, forecasting, and data services to the British Broadcasting Corporation (BBC) since the BBC broadcast the first British TV weather in January 1954. Today, geovisualization of the weather is required for the Corporation's established terrestrial TV channels (BBC1), its more recent all-digital TV services (including BBC4, BBC World, and BBC News 24), its interactive TV service (BBCi) and its real-time website feeds (the weather section is consistently one of the three most used on the BBC website in terms of hits: **www.bbc.co.uk/weather**).

BBC weather forecaster Penny Tranter (Figure 13.16) comes from Ayrshire in Scotland. She became interested in the weather as a teenager and, after completing a degree in Environmental Sciences at the University of East Anglia, joined the Met Office as a trainee forecaster in 1983. She now broadcasts across all BBC channels (BBC1, BBC 4, News 24, BBC World, BFBS, Radio 2, Radio 4, and Radio 5 Live).

Figure 13.16 Penny Tranter, weather forecaster (Reproduced with permission of the BBC)

BBC weather forecasters are equipped with computer-based weather information displays linked directly to the super-computers at Met Office HQ in Exeter. These powerful computer forecast models take real-time data feeds (Sections 10.8 and 10.9) from around the world and then simulate the world's atmosphere mathematically, from sea-level to the upper atmosphere. Forecasters use their scientific skills and experience as meteorologists to interpret the models and to predict what is going to happen (see Section 1.3). In consultation with the Met Office chief forecasters, they select the details they wish to use to support the weather story for that day. The BBC production team works with the forecasters to make sure that the editorial line 'fits' with the BBC's output for the day, and is consistent across all platforms.

Penny receives continuous data feeds from the Met Office in the form of digital charts showing pressure, temperature, rainfall, and cloud cover. In addition, she receives satellite pictures for every continent and hourly satellite images for the UK providing observational information. Half-hourly radar rainfall charts are also drawn up so that she can see where rain is actually falling and can compare this against the computer forecast model predictions. Over 6000 graphics products are drawn up automatically every day and Penny can also make up her own symbol or text charts using a range of background maps and images.

A state-of-the-art graphics system allows her to put together a sequence of charts and other graphics, including stills, animations, video clips, live weather cameras, and text charts. They can be in or out of vision, and can be customized for different channels or platforms. All the forecasts are ad-libbed and so can be shortened or lengthened as needed. She also prepares a summary chart for use at the end of each bulletin, and can jump to this, using a foot switch, should the bulletin have to be cut short suddenly.

Penny Tranter usually works alone in the studio. She uses an ear-piece to listen to the programme before her broadcast and to any last-minute instructions from the director in the network control room. She stands in front of a translucent screen on which is projected a faint image of the graphics. This gives her an idea of where on the map to point. The back of the screen is flooded with blue light, and an electronic system known as Colour Separation Overlay (CSO) causes any area where the camera detects blue to be replaced by a 'clean feed' of the charts from the computer, called up in sequence by the forecaster at the touch of a button. You will never see Penny dressed in blue clothing as this would merge with the graphics!

Monitors next to the camera show the output of the graphics computer and the output of the studio. In front of the camera is an autocue screen onto which is projected the service (e.g., BBC1) on which she is about to broadcast. Superimposed on this image is a countdown clock. Penny starts her forecast when she sees herself appear and stops when the countdown clock reaches zero.

A dedicated, BBC-designed weather graphics system has also been installed into each of the national and regional BBC Centres. It was radically revised in 2005, and allows each local weather presenter around the UK to prepare and transmit his or her own weather bulletin using the same information and graphics systems as the forecasters based in the BBC Weather Centre in London.

Bringing together such a range of data sources, building multiple representations of the world into a seamless presentation, and responding to the pressures of real-time delivery of forecasts is not always a straightforward task. Speaking of her work, Penny says:

> **Our state-of-the-art graphics system allows an easy-to-access, easy-to-use, and easy-to-understand approach for both the weather presenter and the viewer. Robust geovisualization conventions make our broadcasts intelligible in any part of the world. Accompanied by the BBC's graphics, they enable us to deliver our weather information – accurately and efficiently – to a wide global audience from Indian farmers to North American retailers.**

(Information courtesy of the British Broadcasting Corporation).

(A)

(B)

Figure 13.17 (A) Raw LiDAR image of football (soccer) stadium, Southampton, England, and (B) LiDAR-derived bare earth DEM draped with aerial photograph of its wider area, overlaid with 3-D buildings. (Aerial photography reproduced with permission of Ordnance Survey © Ordnance Survey. All rights reserved. (Reproduced with permission from the Environment Agency of England and Wales. Courtesy: Sarah Smith.)

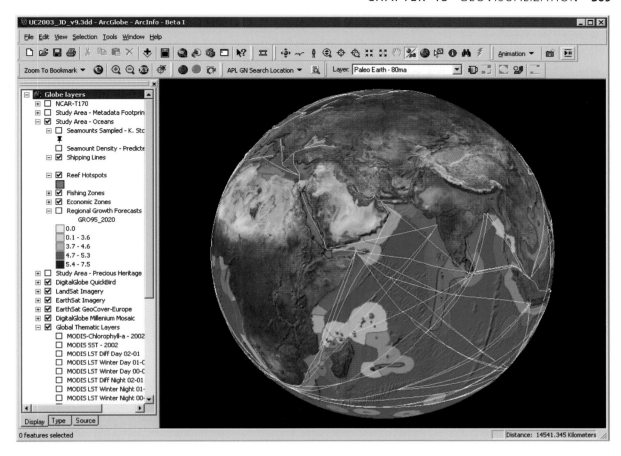

Figure 13.18 A whole-Earth visualization showing vector shipping lanes (white lines), reefs (red), and ocean fishing and economic zones, overlaid on top of raster topography (Courtesy: ESRI)

being able to roam smoothly over large volumes of global geographic information. This enables a better understanding of the distribution and abundance of globally distributed phenomena (e.g., ocean currents, atmospheric temperatures, and shipping lanes), as well as detailed examination of local features in the natural and built environment (e.g., optimum location of cellphone towers, or the impact of tree felling upon the viewscape of tourist areas).

13.4.3 Hand-held computing and geovisualization

Immersion of desk- and studio-bound users in 3-D virtual reality presents one way of promoting better remote interaction with the real world, using computers that are physically amongst the largest of the digital age. At the other end of the computer size spectrum, improved direct interaction with the real world has been made possible by the development of a range of hand-held, in-vehicle, and wearable computer devices. These are discussed, along with the geovisualization conventions they entail, in Sections 7.6.4 and 11.3.

13.5 Consolidation

Geovisualization can make a powerful contribution to decision making and can be used to simulate changes to reality. Yet, although it is governed by scientific principles, the limitations of human cognition mean that geovisualization necessarily presents a further selective filter on the reality that we seek to represent. Mapping is about seeing the detail as well as the big picture, yet the wealth of detail that is available in today's digital environments can sometimes create information overload and threaten to overwhelm the message of the representation. Geovisualization is fostering greater user interaction and participation in the use of GIS as a decision support tool. At its worst, greater sophistication of display may create information overload and obscure the inherent uncertainty in most spatial data: at its best, the media of visualization allow the user to undertake a balanced appraisal of the message of a geographic representation, and to create better description, explanation, prediction, and understanding of GI. The contribution of geovisualization to PPGIS can be viewed as developing and broadening the base of users that not

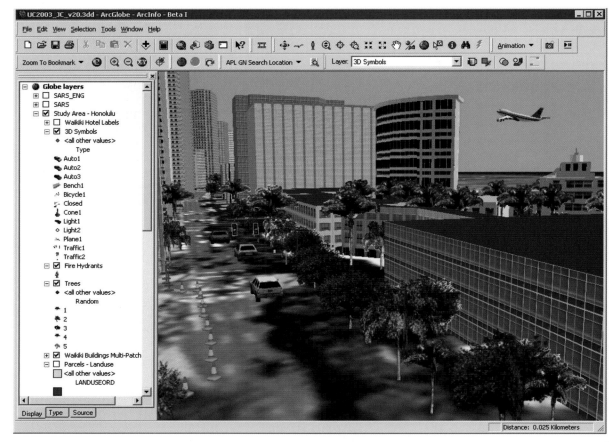

Figure 13.19 A cityscape visualization, comprising extruded buildings and specific vector objects (cars, plane, traffic cones, etc.) to create a photo-realistic simulation (Courtesy: ESRI)

Virtual London

In 2002, the UK central government announced its Electronic Government Initiative with the objective of delivering all government procurement services electronically by 2005. This has garnered political support for the vision of a 'Virtual London' and the 3-D GIS research necessary to create it.

Working at the Centre for Advanced Spatial Analysis at University College London, GIS researchers Andy Smith and Steve Evans are creating a 3-D representation of Central London (Figure 13.20). The model is being produced using GIS, CAD, and a variety of new photorealistic imaging techniques and photogrammetric methods of data capture. The core model will be distributed via broadband Internet connections, utilizing techniques to optimize sharing of large datasets and create a multi-user environment. Citizens will be able to roam around a Virtual Gallery as avatars in order to explore issues relating to London. Similar models are available for other cities (Vexcel maintains an 'off-the-shelf' library of US cities, derived principally from remotely sensed data: **www.vexcel.com**), and models of this nature have been used to encourage and enliven public participation in the planning process, particularly with focus upon urban redevelopment proposals (e.g., the Australian community spatial simulation scenarios at **www.c-s3.info**).

The model is designed for four audiences:

- Professionals, such as architects, developers, and planners who are anxious to use a full set of data query and visualization capabilities. For example, architects might place new buildings in the model in order to assess a variety of issues, such as visual impact or likely consequences for traffic and surrounding land use, and law enforcement agencies might use the model for community policing (Figure 13.21).

- Citizens interested in learning about London or in understanding the impacts of development and traffic proposals. This has obvious links to the development of e-democracy and the evaluation of 'what if' scenarios to enable digital planning by citizens.

- Virtual tourists interested in navigating and viewing scenes, or picking up links to websites of interest. This has implications for the way in which city marketing and regional development initiatives are undertaken.

- Classroom users, with diverse interests in geography, civics, and history, and university-based educators in urban planning, architecture, and computer science.

Check out the current state of Virtual London's development at **www.casa.ucl.ac.uk/ research/virtuallondon.htm**.

Figure 13.20 Scene from Virtual London, including County Hall and the London Eye (Courtesy: Andy Hudson-Smith, Steve Evans, and ESRI)

Figure 13.21 Virtual London law enforcement analysis: 3-D visualization of closed circuit television viewshed (green) and street illumination (Courtesy: Christian Castle, John Calkins, and ESRI)

only sources information through GIS, but increasingly uses GIS as a medium for information exchange and participation in decision making. The old adage that 'seeing is believing' only holds if visualization and user interaction are subservient to the underlying messages of a representation.

Questions for further study

1. Figure 13.22 is a cartogram redrawn from a newspaper feature on the costs of air travel from London in 1992. Using current advertisements in the press and on the Internet, create a similar cartogram of travel costs, in local currency, from the nearest international air hub to your place of study.

2. How can Web-based multimedia GIS tools be used to improve community participation in decision making?

3. Produce two (computer generated) maps of the Israeli West Bank security fence to illustrate opposing views of its effects. For example, the first might illustrate how it helps to preserve an Israeli 'Lebensraum' while the second might emphasize its disruptive impacts upon access between Palestinian communities. In a separate short annotative commentary, describe the structure and character of the fence at a range of spatial scales.

4. Review the common *sources* of uncertainty in geographic representation and the ways in which they can be *manifest* through geovisualization.

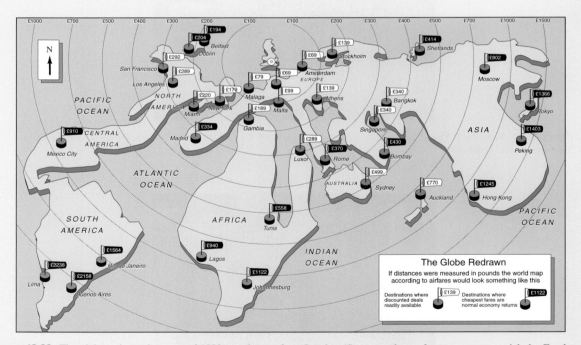

Figure 13.22 The globe redrawn in terms of 1992 travel costs from London (*Source*: redrawn from newspaper article by Frank Barrett in *The Independent* on 9 February 1992, page 6)

Further reading

Craig W.J., Harris T.M. and Weiner D. 2002 *Community Participation and Geographic Information Systems*. Boca Raton, FL: CRC Press.

Dorling D. and Thomas B. 2004 *People and Places: a 2001 Census Atlas of the UK*. Bristol, UK: Policy Press.

Egenhofer M. and Kuhn W. 2005 'Interacting with GIS'. In Longley P.A., Goodchild M.F., Maguire D.J., and Rhind D.W. (eds) *Geographical Information Systems: Principles, Techniques, Management and Applications (abridged edition)*. Hoboken, NJ: Wiley: 401–12.

MacEachren A. 1995 *How Maps Work: Representation, Visualization and Design*. New York: Guilford Press.

MacEachren A.M., Gahegan M., Pike W., Brewer I., Cai G., Lengerich E., and Hardisty F. 2004 'Geovisualization for knowledge construction and decision-support'. *Computer Graphics and Applications* **24**(1), 13–17.

Shiffer M. 2005 'Managing public discourse: towards the augmentation of GIS with multimedia'. In Longley P.A., Goodchild M.F., Maguire D.J., and Rhind D.W. (eds) *Geographical Information Systems: Principles, Techniques, Management and Applications (abridged edition)*. Hoboken, NJ: Wiley: 723–32.

Tufte E.R 2001 *The Visual Display of Quantitative Information* (2nd edn). Cheshire, CT: Graphics Press.

14 *Query, measurement, and transformation*

This chapter is the first in a set of three dealing with geographic analysis and modeling methods. The chapter begins with a review of the relevant terms, and an outline of the eight major topics covered in the three chapters: three in each of Chapters 14 and 15, and two in Chapter 16. Query methods allow users to interact with geographic databases using pointing devices and keyboards, and GIS have been designed to present data for this purpose in a number of standard views. The second area covered in this chapter, measurement, includes algorithms for determining lengths, areas, shapes, slopes, and other properties of objects. Transformations allow new information to be created through simple geometric manipulation.

Geographic Information Systems and Science, 2nd edition   Paul Longley, Michael Goodchild, David Maguire, and David Rhind.
© 2005 John Wiley & Sons, Ltd.   ISBNs: 0-470-87000-1 (HB); 0-470-87001-X (PB)

Learning Objectives

After working through this chapter you will know:

■ Definitions of geographic analysis and modeling, and tests to determine whether a method is geographic;

■ The range of queries possible with a GIS;

■ Methods for measuring length, area, shape, and other properties, and their caveats;

■ Transformations that manipulate objects to create new ones, or to determine geometric relationships between objects.

14.1 Introduction: what is spatial analysis?

The techniques covered in these three chapters are generally termed *spatial* rather than *geographic*, because they can be applied to data arrayed in any space, not only geographic space. Many of the methods might potentially be used in analysis of outer space by astronomers, or in analysis of brain scans by neuroscientists. So the term spatial is used consistently throughout these chapters.

Spatial analysis is in many ways the crux of GIS because it includes all of the transformations, manipulations, and methods that can be applied to geographic data to add value to them, to support decisions, and to reveal patterns and anomalies that are not immediately obvious – in other words, spatial analysis is the process by which we turn raw data into useful information, in pursuit of scientific discovery, or more effective decision making. If GIS is a method of communicating information about the Earth's surface from one person to another, then the transformations of spatial analysis are ways in which the sender tries to inform the receiver, by adding greater informative content and value, and by revealing things that the receiver might not otherwise see.

Some methods of spatial analysis were developed long before the advent of GIS, and carried out by hand, or by the use of measuring devices like the ruler. The term *analytical cartography* is sometimes used to refer to methods of analysis that can be applied to maps to make them more useful and informative, and spatial analysis using GIS is in many ways its logical successor.

Spatial analysis can reveal things that might otherwise be invisible – it can make what is implicit explicit.

In this and the next chapter we will look first at some definitions and basic concepts of spatial analysis. Following introductory material, the two chapters include six sections, which look at spatial analysis grouped into six more-or-less-distinct categories: queries and reasoning, measurements, transformations, descriptive summaries, optimization, and hypothesis testing. Chapter 16 is devoted to *spatial modeling*, a loosely defined term that covers a variety of more advanced and more complex techniques, and includes the use of GIS to analyze and simulate dynamic processes, in addition to analyzing static patterns. Some of the methods discussed in these two chapters were introduced in Chapter 4 as ways of describing the fundamental nature of geographic data, so references will be made to that chapter as appropriate.

Spatial analysis is the crux of GIS, the means of adding value to geographic data, and of turning data into useful information.

Methods of spatial analysis can be very sophisticated, but they can also be very simple. A large body of methods of spatial analysis has been developed over the past century or so, and some methods are highly mathematical – so much so, that it might sometimes seem that mathematical complexity is an indicator of the importance of a technique. But the human eye and brain are also very sophisticated processors of geographic data and excellent detectors of patterns and anomalies in maps and images. So the approach taken here is to regard spatial analysis as spread out along a continuum of sophistication, ranging from the simplest types that occur very quickly and intuitively when the eye and brain look at a map, to the types that require complex software and sophisticated mathematical understanding. Spatial analysis is best seen as a *collaboration* between the computer and the human, in which both play vital roles.

Effective spatial analysis requires an intelligent user, not just a powerful computer.

There is an unfortunate tendency in the GIS community to regard the making of a map using a GIS as somehow less important than the performance of a mathematically sophisticated form of spatial analysis. According to this line of thought, *real* GIS involves number crunching, and users who *just* use GIS to make maps are not serious users. But every cartographer knows that the design of a map can be very sophisticated, and that maps are excellent ways of conveying geographic information and knowledge, by revealing patterns and processes to us. We agree, and believe that map making is potentially just as important as any other application of GIS.

Spatial analysis helps us in situations when our eyes might otherwise deceive us.

There are many possible ways of defining spatial analysis, but all in one way or another express the basic idea that information on locations is essential – that analysis carried out without knowledge of locations is not spatial analysis. One fairly formal statement of this idea is:

Spatial analysis is a set of methods whose results are not invariant under changes in the locations of the objects being analyzed.

The double negative in this statement follows convention in mathematics, but for our purposes we can remove it:

Spatial analysis is a set of methods whose results change when the locations of the objects being analyzed change.

On this test the calculation of an average income for a group of people is not spatial analysis, because it in no way depends on the locations of the people. But the calculation of the center of the US population is spatial analysis because the results depend on knowing where all US residents are located. GIS is an ideal platform for

spatial analysis because its data structures accommodate the storage of object locations.

14.1.1 Examples

Spatial analysis can be used to further the aims of science, by revealing patterns that were not previously recognized, and that hint at undiscovered generalities and laws. Patterns in the occurrence of a disease may hint at the mechanisms that cause the disease and some of the most famous examples of spatial analysis are of this nature, including the work of Dr John Snow in unraveling the causes of cholera (Box 14.1).

It is interesting to speculate on what would have happened if early epidemiologists like Snow had had access to a GIS. The rules governing research today

Biographical Box **14.1**

Dr John Snow and the causes of cholera

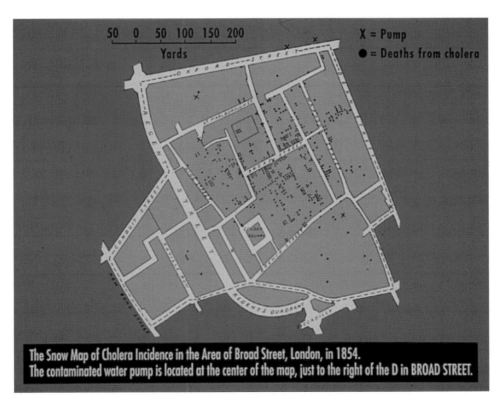

The Snow Map of Cholera Incidence in the Area of Broad Street, London, in 1854.
The contaminated water pump is located at the center of the map, just to the right of the D in BROAD STREET.

Figure 14.1 A redrafting of the map made by Dr John Snow in 1854 of the deaths that occurred in an outbreak of cholera in the Soho district of London. The existence of a public water pump in the center of the outbreak (the cross in Broad Street) convinced Snow that drinking water was the probable cause of the outbreak. Stronger evidence was obtained in support of this hypothesis when the water supply was cut off, and the outbreak subsided. (*Source*: Gilbert E. W. 1958 'Pioneer maps of health and disease in England.' *Geographical Journal*, **124**: 172–183) (Reproduced by permission of Blackwell Publishing Ltd.)

In the 1850s, cholera was very poorly understood and massive outbreaks were a common occurrence in major industrial cities (today cholera remains a significant health hazard in many parts of the world, despite progress in understanding its causes and advances in treatment). An outbreak in London in 1854 in the Soho district was typical of the time, and the deaths it caused are mapped in Figure 14.1. The map was made by Dr John Snow (Figure 14.2), who had conceived the hypothesis that cholera was transmitted through the drinking of polluted water, rather than through the air, as was commonly believed. He noticed that the outbreak appeared to be centered on a public drinking water pump in Broad Street (Figure 14.3) – and if his hypothesis was correct, the pattern shown on the map would reflect the locations of people who drank the pump's water. There appeared to be anomalies, in the sense that deaths had occurred in households that were located closer to other sources of water, but he was able to confirm that these households also drew their water from the Broad Street pump. Snow had the handle of the pump removed, and the outbreak subsided, providing direct causal evidence in favor of his hypothesis. The full story is much more complicated than this largely apocryphal version, of course; much more information is available at **www.jsi.com**.

Today, Snow is widely regarded as the father of modern epidemiology.

Figure 14.2 Dr John Snow (*Source*: John Snow Inc. **www.jsi.com**)

Figure 14.3 A modern replica of the pump that led Snow to the inference that drinking water transmitted cholera, located in what is now Broadwick Street in Soho, London (*Source*: John Snow Inc. **www.jsi.com**)

would not have allowed Snow to remove the pump handle, except after lengthy review, because the removal constituted an experiment on human subjects. To get approval, he would have had to have shown persuasive evidence in favor of his hypothesis, and it is doubtful that the map would have been sufficient because several other hypotheses might have explained the pattern equally well. First, it is conceivable that the population of Soho was inherently at risk of cholera, perhaps by being comparatively elderly, or because of poor housing conditions. The map would have been more convincing if it had shown the *rate* of incidence, relative to the population at risk. For example, if cholera was highest among the elderly, the map could have shown the number of cases in each small area of Soho as a proportion of the population aged over 50 in each area. Second, it is still conceivable that the hypothesis of transmission through the air between carriers could have produced the same observed pattern, if the first carrier happened to live in the center of the outbreak near the pump. Snow could have eliminated this alternative if he had been able to produce a sequence of maps, showing the locations of cases as the outbreak developed. Both of these options involve simple spatial analysis of the kind that is readily available today in GIS.

GIS provides tools that are far more powerful than the map at suggesting causes of disease.

Today the causal mechanisms of diseases like cholera, which results in short, concentrated outbreaks, have long since been worked out. Much more problematic are the causal mechanisms of diseases that are rare, and not sharply concentrated in space and time. The work of Stan Openshaw at the University of Leeds, using one of his Geographical Analysis Machines, illustrates the kinds of applications that make good use of the power of GIS in this contemporary context.

Figure 14.4 shows an application of one of Openshaw's techniques to a comparatively rare but devastating disease whose causal mechanisms remain largely a mystery – childhood leukemia. The study area is northern England, from the Mersey to the Tyne. The analysis begins with two datasets: one of the locations of cases of the disease, and the other of the numbers of people at risk in standard census reporting zones. Openshaw's technique then generates a large number of circles, of random sizes, and places them (*throws* them) randomly over the map. The computer generates and places the circles, and then analyzes their contents, by dividing the number of cases found in the circle by the size of the population at risk. If the ratio is anomalously high, the circle is drawn. After a large number of circles has been generated, and a small proportion have been drawn, a pattern emerges. Two large concentrations, or clusters of cases, are evident in the figure. The one on the left is located around Sellafield, the location of the British Nuclear Fuels processing plant and a site of various kinds of leaks of radioactive material. The other, in the upper right, is in the Tyneside region, and Openshaw and his colleagues discuss possible local causes.

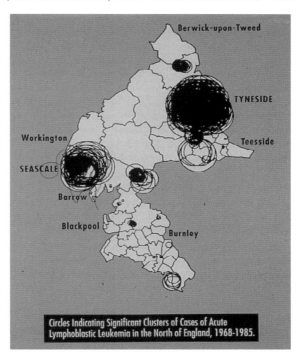

Circles Indicating Significant Clusters of Cases of Acute Lymphoblastic Leukemia in the North of England, 1968-1985.

Figure 14.4 The map made by Openshaw and colleagues by applying their Geographical Analysis Machine to the incidence of childhood leukemia in northern England. A very large number of circles of random sizes is randomly placed on the map, and a circle is drawn if the number of cases it encloses substantially exceeds the number expected in that area given the size of its population at risk (*Source*: Openshaw S., Charlton M., Wymer C., and Craft A. 1987 'A Mark I geographical analysis machine for the automated analysis of point datasets.' *International Journal of Geographical Information Systems*, **1**: 335–358. **www.tandf.co.uk/journals**)

Both of these examples are instances of the use of spatial analysis for scientific discovery and decision making. Sometimes spatial analysis is used *inductively*, to examine empirical evidence in the search for patterns that might support new theories or general principles, in this case with regard to disease causation. Other uses of spatial analysis are *deductive*, focusing on the testing of known theories or principles against data (Snow already had a theory of how cholera was transmitted, and used the map as confirmation to convince others). A third type of application is *normative*, using spatial analysis to develop or prescribe new or better designs, for the locations of new retail stores, or new roads, or new manufacturing plant. Examples of this type appear in Section 15.3.

14.1.2 Types of spatial analysis and modeling

The remaining sections of this chapter and the following one discuss methods of spatial analysis using six general headings:

Queries are the most basic of analysis operations, in which the GIS is used to answer simple questions posed by the user. No changes occur in the database, and no new data are produced. The operations vary from simple and well-defined queries like 'how many houses are found within 1 km of this point', to vaguer questions like 'which is the closest city to Los Angeles going north', where the response may depend on the system's ability to understand what the user means by 'going north' (see the discussion of vagueness in Chapter 6). Queries of databases using SQL are discussed in Section 10.4.

Measurements are simple numerical values that describe aspects of geographic data. They include measurement of simple properties of objects, like length, area, or shape, and of the relationships between pairs of objects, like distance or direction.

Transformations are simple methods of spatial analysis that change datasets, combining them or comparing them to obtain new datasets, and eventually new insights. Transformations use simple geometric, arithmetic, or logical rules, and they include operations that convert raster data into vector data, or vice versa. They may also create fields from collections of objects, or detect collections of objects in fields.

Descriptive summaries attempt to capture the essence of a dataset in one or two numbers. They are the spatial equivalent of the descriptive statistics commonly used in statistical analysis, including the mean and standard deviation.

Optimization techniques are normative in nature, designed to select ideal locations for objects given certain well-defined criteria. They are widely used in market research, in the package delivery industry, and in a host of other applications.

Hypothesis testing focuses on the process of reasoning from the results of a limited sample to make generalizations about an entire population. It allows us, for example, to determine whether a pattern of points could have arisen by chance, based on the information from a sample. Hypothesis testing is the basis of inferential statistics and lies at the core of statistical analysis, but its use with spatial data is much more problematic.

In Chapter 16, the attention turns to modeling, a very broad term that is used when the individual steps of analysis are combined into complex sequences. *Static modeling* describes the use of a sequence to attain some defined goal, such as the measurement of the suitability of locations for development, or their sensitivity to pollution or erosion. *Dynamic modeling* uses the GIS to emulate real physical or social processes operating on the geographic landscape. Time is broken up into a series of discrete steps, and the operations of the GIS are *iterated* at each time step. Dynamic modeling can be used to emulate processes of erosion, or the spread of disease, or the movement of individual vehicles in a congested street network.

14.2 Queries

In the ideal GIS it should be possible for the user to interrogate the system about any aspect of its contents, and obtain an immediate answer. Interrogation might involve pointing at a map, or typing a question, or pulling down a menu and clicking on some buttons, or sending a formal SQL request to a database (Section 10.4). The visual aspects of query are discussed in Chapter 13. Today's user interfaces are very versatile, and have very nearly reached the point where it will be possible to interrogate the system by speaking to it – this would be extremely valuable in vehicles, where the use of more conventional ways of interrogating the system through keyboards or pointing devices can be too distracting for the driver.

When a GIS is used in a moving vehicle it is impossible to rely on the normal modes of interaction through keyboards or pointing devices.

The very simplest kinds of queries involve interactions between the user and the various *views* that a GIS is capable of presenting. A *catalog* view shows the contents of a database, in the form of storage devices (hard drives, Internet sites, CDs, ZIP disks, or USB thumb-drives) with their associated folders, and the datasets contained in those folders. The catalog will likely be arranged in a hierarchy, and the user is able to expose or hide various branches of the hierarchy by clicking at appropriate points. Figure 14.5 shows a catalog view, in this case using ESRI's ArcCatalog software (a component of ArcGIS). Note how different types of datasets are symbolized using different icons so the user can tell at a glance which files contain grids, polygons, points, etc.

In contemporary software environments, such as Microsoft's Windows, the Macintosh, or Unix, many kinds of interrogation are available through simple pointing and clicking. For example, in ArcCatalog simply pointing at a dataset icon and clicking the right mouse button exposes basic statistics on the dataset when the Properties option is selected. The metadata option exposes the metadata stored with the dataset, including its projection and datum details, the names of each of its attributes, and its date of creation.

Users query a GIS database by interacting with different views.

The *map* view of a dataset shows its contents in visual form, and opens many more possibilities for querying. When the user points to any location on the screen the GIS should display the pointer's coordinates, using the units appropriate to the dataset's projection and coordinate system. For example, Figure 14.6 shows the most recent location of the pointer in the box below the map window in units of meters east and north, because the map projection in use is UTM (Section 5.7.2). If the dataset is raster the system might display a cell's row and column number, or its coordinate system if the raster is adequately georeferenced (tied to some Earth coordinate system). By

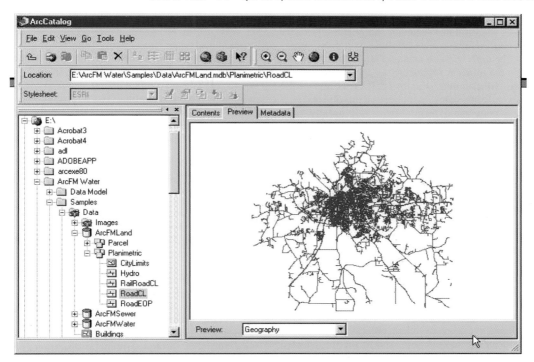

Figure 14.5 A catalog view of a GIS database (ESRI's ArcCatalog). The left window exposes the file structure of the database, and the right window provides a geographic preview of the contents of a selected file (RoadCL). Other options for the right window include a view of the file's metadata (Section 11.2.1), and its attribute table

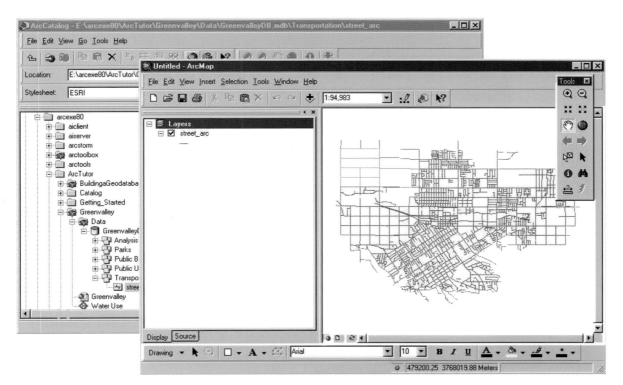

Figure 14.6 ESRI's ArcMap displays the most recent location of the pointer in a box below the map display, allowing the user to query location anywhere on the map. Location is shown in this case in meters east and north (using a UTM projection), since this is the coordinate system of the data being queried

pointing to an object the user should be able to display the values of its attributes, whether the object is a raster cell, a line representing a network link, or a polygon.

Finally, the *table* view of a dataset shows a rectangular array, with the objects organized as the rows and the attributes as the columns (Figure 14.7). This allows the user to see the attributes associated with objects at a glance, in a convenient form. There will usually be a table associated with each type of object – points, lines, areas (see Box 4.1 and Chapter 8) – in a vector database. Some systems support other views as well. In a *histogram* view, the values of a selected attribute are displayed in the form of a bar graph. In a *scatterplot* view, the values of two selected attributes are displayed plotted against each other. Scatterplots allow us to see whether relationships exist between attributes. For example, is there a tendency for the average income in a census tract to increase as the percentage of people with university education increases?

Today's GIS supports much more sophisticated forms of query than these. First, it is common for the various views to be linked, and the visual aspects of this are discussed in Chapter 13. Suppose both the map view and the table view are displayed on the screen simultaneously. Linkage allows the user to select objects in one view, perhaps by pointing and clicking, and to see the selected objects highlighted in both views, as in Figure 14.7. Linkage is often possible between other views, including the histogram and scatterplot views. For example, by linking a scatterplot with a map view, it is possible to select points in the scatterplot and see the corresponding objects highlighted on the map. This kind of linkage is very useful in examining *residuals* (Section 4.7), or cases that deviate substantially from the trend shown by a scatterplot (compare with the idealized scatterplots of Figures 4.14 and 4.15). Figure 14.8 shows four linked views of data on sudden infant death syndrome in the counties of North Carolina. The term *exploratory spatial data analysis* is sometimes used to describe these forms of interrogation, which allow the user to explore data in interesting and potentially insightful ways.

Exploratory spatial data analysis allows its users to gain insight by interacting with dynamically linked views.

Second, many methods are commonly available for interrogating the contents of tables, such as SQL (Section 10.4). Figure 14.7 shows the result of an SQL query on a simple table. The language becomes much more powerful when tables are linked, using common keys, as described in Section 10.3, and much more complex and sophisticated queries, involving multiple tables, are possible with the full language. More complex methods of table interrogation include the ability to average the values of an attribute across selected records, and to create new attributes through arithmetic operations on existing ones (e.g., create a new attribute equal to the ratio of two selected attributes).

SQL is a standard language for querying tables and relational databases.

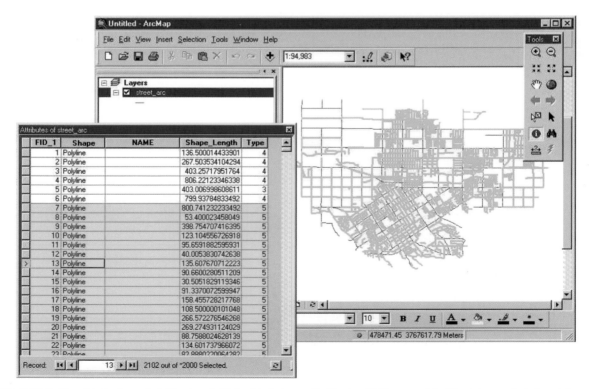

Figure 14.7 The objects shown in table view (left). In this instance all objects with Type equal to 5 have been selected, and the selected objects appear in blue in the map view (ESRI's ArcMap)

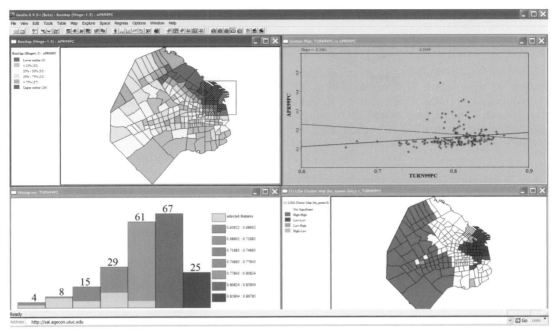

Figure 14.8 Screen shot of GeoDa, a spatial analysis package that integrates easily with a GIS and features the ability to display multiple views (clockwise from the top left a map, a scatterplot, a map showing local indicators of spatial association (LISA; Box 15.3), and a histogram – see Section 14.2 for a discussion of histograms and scatterplots) and to link them dynamically. The tracts selected by the user in the upper left are automatically highlighted in the other windows. GeoDa is available at **www.csiss.org/clearinghouse/GeoDa**. (*Source*: courtesy of Luc Anselin, University of Illinois, Urbana-Champaign)

14.3 Measurements

Many types of interrogation ask for measurements – we might want to know the total area of a parcel of land, or the distance between two points, or the length of a stretch of road – and in principle all of these measurements are obtainable by simple calculations inside a GIS. Comparable measurements by hand from maps can be very tedious and error-prone. In fact it was the ability of the computer to make accurate evaluations of area quickly that led the Canadian government to fund the development of the world's first GIS, the Canada Geographic Information System, in the mid-1960s (see the brief history of GIS in Section 1.4.1), despite the primitive state and high costs of computing at that time. Evaluation of area by hand is a messy and soul-destroying business. The *dot-counting* method uses transparent sheets on which randomly located dots have been printed – an area on the map is estimated by counting the number of dots falling within it. In the *planimeter* method a mechanical device is used to trace the area's boundary, and the required measure accumulates on a dial on the machine.

Humans have never devised good manual tools for making measurements from maps, particularly measurements of area.

By comparison, measurement of the area of a digitally represented polygon is trivial and totally reliable. The

common algorithm (see Box 14.2 for a definition of this term) calculates and sums the areas of a series of trapezia, formed by dropping perpendiculars to the x axis as shown in Figure 14.9. By making a simple change to the algorithm it is also possible to use it to compute a polygon's centroid (see Section 15.2.1 for a discussion of centroids).

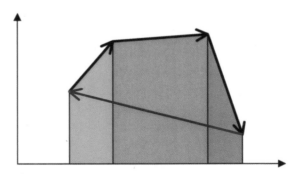

Figure 14.9 The algorithm for calculation of the area of a polygon given the coordinates of the polygon's vertices. The polygon consists of the three black arrows, plus the blue arrow forming the fourth side. Trapezia are dropped from each edge to the x axis and their areas are calculated as (difference in x) times average of y. The trapezia for the first three edges, shown in green, brown, and blue, are summed. When the fourth trapezium is formed from the blue arrow its area is negative because its start point has a larger x than its end point. When this area is subtracted from the total the result is the correct area of the polygon

Technical Box **14.2**

Definition of an algorithm

Algorithm: a procedure that provides a solution to a problem and consists of a set of unambiguous rules which specify a finite sequence of operations. Each step of an algorithm must be precisely defined and the necessary actions must be rigorously specified for each case. An algorithm should always arrive at a problem solution after a finite and reasonable number of steps. An algorithm that satisfies these requirements can be programmed as software for a digital computer.

14.3.1 Distance and length

A *metric* is a rule for the determination of distance between points in a space. Several kinds of metrics are used in GIS, depending on the application. The simplest is the rule for determining the shortest distance between two points in a flat plane, called the Pythagorean or straight-line metric. If the two points are defined by the coordinates (x_1, y_1) and (x_2, y_2), then the distance D between them is the length of the hypotenuse of a right-angled triangle (Figure 14.10), and Pythagoras's theorem tells us that the square of this length is equal to the sum of the squares of the lengths of the other two sides. So a simple formula results:

$$D = \sqrt{(x_2 - x_1)^2 + (y_2 - y_1)^2}$$

A metric is a rule for determining distance between points in space.

The Pythagorean metric gives a simple and straightforward solution for a plane, if the coordinates x and y are comparable, as they are in any coordinate system based on a projection, such as the UTM or State Plane, or National Grid (see Chapter 5). But the metric will not work for latitude and longitude, reflecting a common source of problems in GIS – the temptation to treat latitude and longitude as if they were equivalent to plane coordinates. This issue is discussed in detail in Section 5.7.1.

The assumption of a flat plane leads to significant distortion for points widely separated on the curved surface of the Earth, and distance must be measured using the metric for a spherical Earth given in Section 5.6 and based on a great circle. For some purposes even this is not sufficiently accurate because of the non-spherical nature of the Earth, and even more complex procedures must be used to estimate distance that take non-sphericity into account. Figure 14.11 shows an example of the differences that the curved surface of the Earth makes when flying long distances.

In many applications the simple rules – the Pythagorean and great circle equations – are not sufficiently accurate estimates of actual travel distance, and we are forced to resort to summing the actual lengths of travel routes. In GIS this normally means summing the lengths of links in a network representation, and many forms of GIS analysis use this approach. If a line is represented as a polyline, or a series of straight segments, then its length is simply the sum of the lengths of each segment, and each segment length can be calculated using the Pythagorean formula and the coordinates of its end points. But it is worth being aware of two problems with this simple approach.

First, a polyline is often only a rough version of the true object's geometry. A river, for example, never makes sudden changes of direction, and Figure 14.12 shows how smoothly curving streets have to be approximated by the sharp corners of a polyline. Box 4.6 discusses the case of a complex shape and the short-cutting that occurs in any generalization. Because there is a general tendency for polylines to short-cut corners, *the length of a polyline tends to be shorter than the length of the object it represents*. There are some exceptions, of course – surveyed boundaries are often truly straight between corner points and streets are often truly straight between intersections. But in general the lengths of linear objects estimated in a GIS, and this includes the lengths of the perimeters of areas represented as polygons, are often substantially shorter than their counterparts on the ground. Note that this is not similarly true of area estimates because short-cutting corners tends to produce both underestimates and overestimates of area, and these tend to cancel out (Figure 14.12; see also Box 4.6).

A GIS will almost always underestimate the true length of a geographic line.

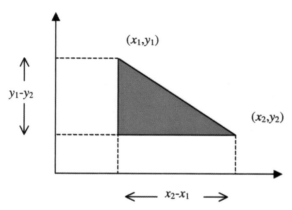

Figure 14.10 Pythagoras's Theorem and the straight-line distance between two points on a plane. The square of the length of the hypotenuse is equal to the sum of the squares of the lengths of the other two sides of the right-angled triangle

Figure 14.11 The effects of the Earth's curvature on the measurement of distance, and the choice of shortest paths. The map shows the North Atlantic on the Mercator projection. The red line shows the track made by steering a constant course of 79 degrees from Los Angeles, and is 9807 km long. The shortest path from Los Angeles to London is actually the black line, the trace of the great circle connecting them, with a length of roughly 8800 km, and this is typically the route followed by aircraft flying from London to Los Angeles. When flying in the other direction, aircraft may sometimes follow other, longer tracks, such as the red line, if by doing so they can take advantage of jetstream winds

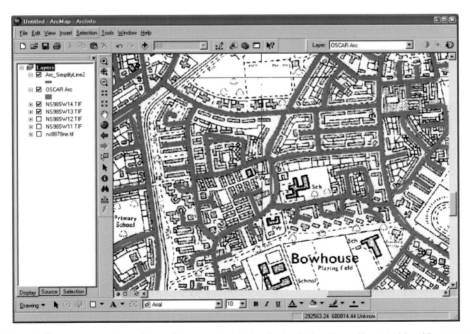

Figure 14.12 The polyline representations of smooth curves tend to be shorter in length, as illustrated by this street map (note how curves are replaced by straight-line segments). But estimates of area tend not to show systematic bias because the effects of overshoots and undershoots tend to cancel out to some extent

Second, the length of a line in a two-dimensional GIS representation will always be the length of the line's planar projection, not its true length in three dimensions, and the difference can be substantial if the line is steep (Figure 14.13). In most jurisdictions the area of a parcel of land is the area of its horizontal projection, not its true surface area. A GIS that stores the third dimension for every point is able to calculate both versions of length and area, but not a GIS that stores only the two horizontal dimensions.

14.3.2 Shape

GIS are also used to characterize the *shapes* of objects, particularly area objects. In many countries the system of political representation is based on the concept of districts or constituencies, which are used to define who will vote for each place in the legislature (Box 14.3). In the USA and the UK, and in many other countries that derived their system of representation from the UK, there is one place in the legislature for each district. It is expected that districts will be compact in shape, and the manipulation of a district's shape to achieve certain overt or covert objectives is termed gerrymandering, after an early governor of Massachusetts, Elbridge Gerry (the shape of one of the state's districts was thought to resemble a salamander, with the implication that it had been manipulated to achieve a certain outcome in the voting; Gerry was a signator both of the Declaration of Independence in 1776, and of the bill that created the offending districts in 1812). The construction of voting districts is an example of the principles of aggregation and zone design discussed in Section 6.2.

Anomalous shape is the primary means of detecting gerrymanders of political districts.

Geometric shape was the aspect that alerted Gerry's political opponents to the manipulation of districts, and today shape is measured whenever GIS is used to aid in the drawing of political district boundaries, as must occur

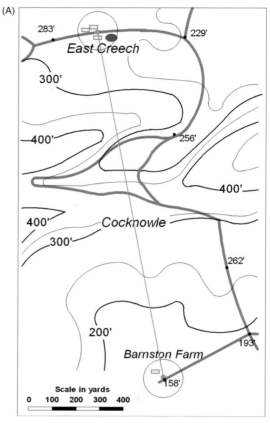

Figure 14.13 The length of a path as traveled on the Earth's surface (red line) may be substantially longer than the length of its horizontal projection as evaluated in a two-dimensional GIS. (A) shows three paths across part of Dorset in the UK. The green path is the straight route, the red path is the modern road system, and the gray path represents the route followed by the road in 1886. (B) shows the vertical profiles of all three routes, with elevation plotted against the distance traveled horizontally in each case. 1 foot = 0.3048 m, 1 yard = 0.9144 m (Courtesy: Michael De Smith)

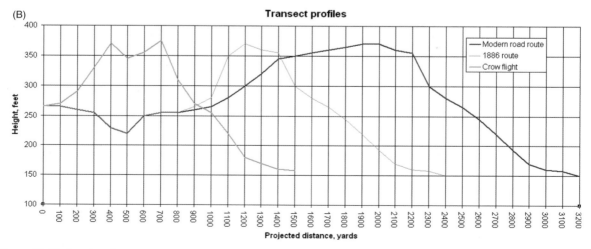

Figure 14.13 (*continued*)

Shape and the 12th Congressional District of North Carolina

In 1992, following the release of population data from the 1990 Census, new boundaries were proposed for the voting districts of North Carolina, USA (Figure 14.14). For the first time race was used as an explicit criterion, and districts were drawn that as far as possible grouped minorities (notably African Americans) into districts in which they were in the majority. The intent was to avoid the historic tendency for minorities to be thinly spread in all districts, and thus to be unable to return their own representative to Congress. African Americans were in a majority in the new 12th District, but in order to achieve this the district had to be drawn in a highly contorted shape.

The new district, and the criteria used in the redistricting, were appealed to the US Supreme Court. In striking down the new districting scheme, Chief Justice William Rehnquist wrote for the 5–4 majority that 'A generalized assertion of past discrimination in a particular industry or region is not adequate because it provides no guidance for a legislative body to determine the precise scope of the injury it seeks to remedy. Accordingly, an effort to alleviate the effects of societal discrimination is not a compelling interest.'

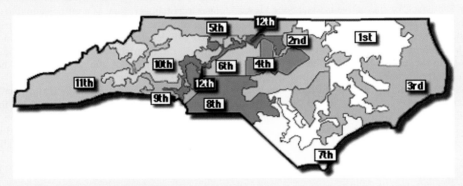

Figure 14.14 The boundaries of the 12th Congressional District of North Carolina drawn in 1992 show a very contorted shape, and were appealed to the US Supreme Court (*Source*: © Durham Herald Company, Inc., **www.herald-sun.com**)

by law in the USA after every decennial census. An easy way to define shape is by comparing the perimeter length of an area to its area measure. Normally the square root of area is used, to ensure that the numerator and denominator are both measured in the same units. A common measure of shape or compactness is:

$$S = P/3.54\sqrt{A}$$

where P is the perimeter length and A is the area. The factor 3.54 (twice the square root of π) ensures that the most compact shape, a circle, returns a shape of 1.0, and the most distended and contorted shapes return much higher values.

14.3.3 Slope and aspect

The most versatile and useful representation of terrain in GIS is the *digital elevation model*, or DEM. This is a raster representation, in which each grid cell records

the elevation of the Earth's surface, and reflects a view of terrain as a field of elevation values. The elevation recorded is often the elevation of the cell's central point, but sometimes it is the mean elevation of the cell, and other rules have been used to define the cell's elevation (the rules used to define elevation in each cell of the US Geological Survey's GTOPO30 DEM, which covers the entire Earth's surface, vary depending on the source of data). Because of this variation, it is always advisable to read the available documentation to determine exactly what is meant by the recorded elevation in the cells of any DEM.

The digital elevation model is the most useful representation of terrain in a GIS.

Knowing the exact elevation of a point above sea level is important for some applications, including prediction of the effects of global warming and rising sea levels on coastal cities, but for many applications the value of a DEM lies in its ability to produce derivative measures through transformation, specifically measures of slope and aspect, both of which are also conceptualized as fields. Imagine taking a large sheet of plywood and laying it on the Earth's surface so that it touches at the point of interest. The magnitude of steepest tilt of the sheet defines the *slope* at that point, and the direction of steepest tilt defines the *aspect* (Box 14.4).

This sounds straightforward but is complicated by a number of issues. First, what if the plywood fails to sit firmly on the surface, but instead pivots, because the point of interest happens to be a peak, or a ridge? In mathematical terms, we say that the surface at this point *lacks a well-defined tangent*, or that the surface at this point is *not differentiable*, meaning that it fails to obey the normal rules of continuous mathematical functions and differential calculus. The surface of the Earth has numerous instances of sharp breaks of slope, rocky outcrops, cliffs, canyons, and deep gullies that defy this simple mathematical approach to slope, and this is one of the issues that led Benoît Mandelbrot to develop his theory of fractals (see Section 4.8, and the additional applications described in Section 15.2.5).

A simple and satisfactory alternative is to take the view that slope must be measured at a particular resolution. To measure slope at a 30 m resolution, for example, we evaluate elevation at points 30 m apart and compute slope by comparing them (equivalent in concept to using a plywood sheet 30 m across). The value this gives is specific to the 30 m spacing, and a different spacing (or different-sized sheet of plywood) would have given a different result. In other words, *slope is a function of resolution*, and it makes no sense to talk about slope without at the same time talking about a specific resolution

or level of detail. This is convenient because slope is easily computed in this way from a DEM with the appropriate resolution.

The spatial resolution used to calculate slope and aspect should always be specified.

Second, there are several alternative *measures* of slope, and it is important to know which one is used in a particular software package and application. Slope can be measured as an *angle*, varying from 0 to 90 degrees as the surface ranges from horizontal to vertical. But it can also be measured as a percentage or ratio, defined as *rise over run*, and unfortunately there are two different ways of defining run. Figure 14.16 shows the two options, depending on whether run means the horizontal distance covered between two points, or the diagonal distance (the *adjacent* or the *hypotenuse* of the right-angled triangle respectively). In the first case (opposite over adjacent) slope as a ratio is equal to the tangent of the angle of slope, and ranges from zero (horizontal) through 1 (45 degrees) to infinity (vertical). In the second case (opposite over hypotenuse) slope as a ratio is equal to the sine of the angle of slope, and ranges from zero (horizontal) through 0.707 (45 degrees) to 1 (vertical). To avoid confusion we will use the term slope only to refer to the measurement in degrees, and call the other options tan(slope) and sin(slope) respectively.

When a GIS calculates slope and aspect from a DEM, it does so by estimating slope at each of the data points of the DEM, by comparing the elevation at that point to the elevations of surrounding points. But the number of surrounding points used in the calculation varies, as

---

Technical Box **14.4**

Calculation of slope based on the elevations of a point and its eight neighbors

The equations used are as follows (refer to Figure 14.15 for point numbering):

$$b = (z_3 + 2z_6 + z_9 - z_1 - 2z_4 - z_7)/8D$$

$$c = (z_1 + 2z_2 + z_3 - z_7 - 2z_8 - z_9)/8D$$

where b and c are tan(slope) in the x and y directions respectively, D is the grid point spacing, and z_i denotes elevation at the ith point, as shown below. These equations give the four diagonal neighbors of Point 5 only half the weight of the other four neighbors in determining slope at Point 5.

$$tan(slope) = \sqrt{b^2 + c^2}$$

where *slope* is the angle of slope in the steepest direction.

$$tan(aspect) = b/c$$

where *aspect* is the angle between the y axis and the direction of steepest slope, measured clockwise. Since *aspect* varies from 0 to 360, an additional test is necessary that adds 180 to *aspect* if c is positive.

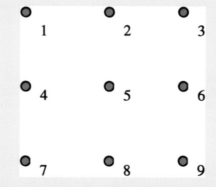

Figure 14.15 Calculation of the slope at Point 5 based on the elevation of it and its eight neighbors

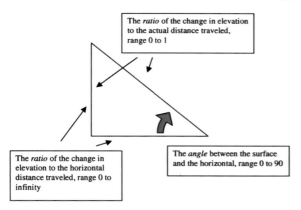

Figure 14.16 about the slope triangle shows:
The *ratio* of the change in elevation to the actual distance traveled, range 0 to 1

The *ratio* of the change in elevation to the horizontal distance traveled, range 0 to infinity

The *angle* between the surface and the horizontal, range 0 to 90

Figure 14.16 Three alternative definitions of slope. To avoid ambiguity we use the angle, which varies between 0 and 90 degrees

do the weights given to each of the surrounding points in the calculation. Box 14.4 shows this idea in practice, using one of the commonest methods, which employs eight surrounding points and gives them different weights depending on how far away they are.

Slope and aspect are the basis for many interesting and useful forms of analysis. Slope is an input to many models of the soil erosion and runoff that result from heavy storms. Slope is also an important input to analyses that find the most suitable routes across terrain for power lines, highways, and military vehicles (see Section 15.3.3).

14.4 Transformations

In this section, we look at methods that transform GIS objects and databases into more useful products, using simple rules. These operations form the basis for many applications, because they are capable of revealing aspects that are not immediately visible or obvious.

14.4.1 Buffering

One of the most important transformations available to the GIS user is the *buffer* operation (see also Section 10.5). Given any set of objects, which may include points, lines, or areas, a buffer operation builds a new object or objects by identifying all areas that are within a certain specified distance of the original objects. Figure 14.17 shows instances of a point, a line, and an area, and the results of buffering. Buffers have many uses, and they are among the most popular of GIS functions:

■ The owner of a land parcel has applied for planning permission to rebuild – the local planning authority could build a buffer around the parcel, in order to identify all homeowners who live within the legally

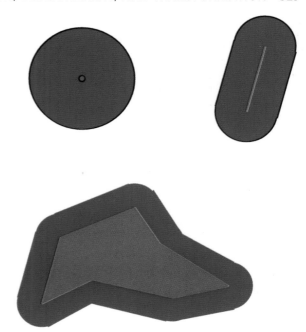

Figure 14.17 Buffers (dilations) of constant width drawn around a point, a polyline, and a polygon

mandated distance for notification of proposed redevelopments.

■ A logging company wishes to clearcut an area, but is required to avoid cutting in areas within 100 m of streams – the company could build buffers 100 m wide around all streams to identify these protected riparian areas.

■ A retailer is considering developing a new store on a site of a type that is able to draw consumers from up to 4 km away from its stores – the retailer could build a buffer around the site to identify the number of consumers living within 4 km of the site, in order to estimate the new store's potential sales (see Section 2.3.3.3).

Buffering is possible in both raster and vector GIS – in the raster case, the result is the classification of cells according to whether they lie inside or outside the buffer, while the result in the vector case is a new set of objects (Figure 14.17). But there is an additional possibility in the raster case that makes buffering more useful in some situations. Figure 14.18 shows a city; average travel speeds vary in each cell of the raster outside the city. Rather than buffer according to distance from the city, we can ask a raster GIS to *spread* outwards from the city at rates determined by the travel speed values in each cell. Where travel speeds are high the spread will extend further, so we can compute how far it is possible to go from the city in a given period of time. This idea of spreading over a variable or *friction* surface is easily implemented in raster representations, but very difficult in vector. Another form of analysis that uses a raster surface to control rate of movement is discussed in Section 15.3.3.

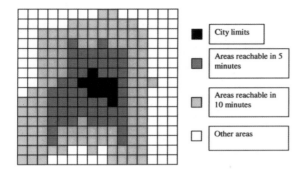

Figure 14.18 A raster generalization of the buffer function, in which spreading occurs at rates controlled by a variable (travel speed, or friction, the inverse of travel speed) whose value is recorded in every raster cell

> **Buffering is one of the most useful transformations in a GIS and is possible in both raster and vector formats.**

14.4.2 Point in polygon

In its simplest form, the point in polygon operation determines whether a given point lies inside or outside a given polygon. In more elaborate forms, there may be many polygons, and many points, and the task is to assign points to polygons. If the polygons overlap, it is possible that a given point lies in one, many, or no polygons, depending on its location. Figure 14.19 illustrates the task. The operation is popular in GIS analysis because it is the basis for answering many simple queries:

- The points represent instances of a disease in a population, and the polygons represent reporting

zones such as counties – the task is to determine how many instances of the disease occurred in each zone (in this case the zones should not overlap and each point should fall into exactly one polygon).

- The points represent the locations of transmission-line poles owned by a utility company, and the polygons are parcels of land – the task is to identify the owner of the land on which each pole lies, to verify that the company has the necessary easements and pays the necessary fees.

- The points represent the residential locations of voters, and the polygons represent voting districts – the task is to ensure that each voter receives the correct voting forms in the mail.

The point in polygon operation makes sense from both the discrete-object and the continuous-field perspectives (see Section 3.5 for a discussion of these two perspectives). From a discrete-object perspective, both points and polygons are objects and the task is simply to determine enclosure. From a continuous-field perspective, polygons representing a variable such as land ownership cannot overlap, since each polygon represents the land owned by one owner and overlap would imply that a point is owned simultaneously by two owners. Similarly from a continuous-field perspective there can be no gaps between polygons. Consequently, the result of a point in polygon operation from a continuous-field perspective must assign each point to exactly one polygon.

> **The point in polygon operation is used to determine whether a point lies inside or outside a polygon.**

The standard algorithm for the point in polygon operation is shown in Figure 14.19. In essence, it consists of drawing a line from the point to infinity, in this case parallel to the y axis, and determining the number of intersections between the line and the polygon's boundary. If the number is odd the point is inside the polygon and if it is even the point is outside. The algorithm must deal successfully with special cases – for example, if the point lies directly below a vertex (corner point) of the polygon. Some algorithms extend the task to include a third option, when the point lies exactly on the boundary. But others ignore this, on the grounds that it is never possible in practice to determine location with perfect accuracy and so never possible to determine if an infinitely small point lies on or off an infinitely thin boundary line.

14.4.3 Polygon overlay

Polygon overlay is similar to point in polygon transformation in the sense that two sets of objects are involved, but in this case both are polygons. It exists in two forms, depending on whether a continuous-field or discrete-object perspective is taken. The development of effective algorithms for polygon overlay was one of the most significant challenges of early GIS and the task remains one of the most complex and difficult to program.

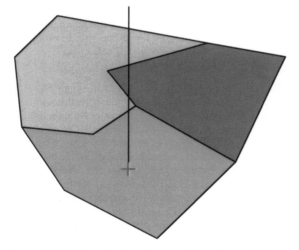

Figure 14.19 The point in polygon problem, shown in the continuous-field case (the point must by definition lie in exactly one polygon, or outside the project area). In only one instance (the pink polygon) is there an odd number of intersections between the polygon boundary and a line drawn vertically upward from the point

The complexity of computing a polygon overlay was one of the greatest barriers to the development of vector GIS.

From the discrete-object perspective, the task is to determine whether two area objects overlap, to determine the area of overlap, and to define the area formed by the overlap as one or more new area objects (the overlay of two polygons can produce a large number of distinct area objects, see Figure 14.20). This operation is useful to determine answers to such queries as:

- How much of this proposed clearcut lies in this riparian zone?

- How much of the projected catchment area of this proposed retail store lies in the catchment of this other existing store in the same chain?

- How much of this land parcel is affected by this easement?

- What proportion of the land area of the USA lies in areas managed by the Bureau of Land Management?

From the continuous-field perspective the task is somewhat different. Figure 14.21 shows two datasets, both representations of fields – one differentiates areas according to land ownership and the other differentiates the same region according to land cover class. In the terminology of ESRI's ArcGIS, both datasets are instances of *area coverages*, or fields of nominal variables represented by non-overlapping polygons. The methods discussed earlier in this chapter could be used to interrogate either dataset separately, but there are numerous queries that require simultaneous access to both datasets – for example:

- What is the land cover class and who is the owner of the point indicated by the user?

- What is the total area of land owned by X and with land cover class A?

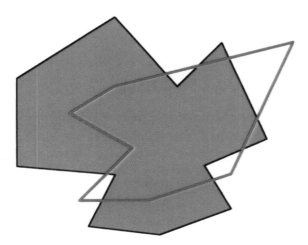

Figure 14.20 Polygon overlay, in the discrete-object case. Here, the overlay of two polygons produces nine distinct polygons. One has the properties of both polygons, four have the properties of the brown polygon but not the green polygon, and four are outside the brown polygon but inside the green polygon

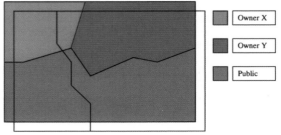

Figure 14.21 Polygon overlay in the continuous-field case. Here, a dataset representing two types of land cover (A on the left, B on the right) is overlaid on a dataset representing three types of ownership (the two datasets have been offset slightly for clarity). The result will be a single dataset in which every point is identified with one land cover type and one ownership type. It will have five polygons, since land cover A intersects with two ownership types and land cover B intersects with three

- Where are the areas that lie on publicly owned land and have land cover class B?

None of these queries can be answered by interrogating one of the datasets alone – the datasets must somehow be combined so that interrogation can be directed simultaneously at both of them.

The continuous-field version of polygon overlay does this by first computing a new dataset in which the region is partitioned into smaller areas that have uniform characteristics on both variables. Each area in the new dataset will have two sets of attributes – those obtained from one of the input datasets and those obtained from the other. All of the boundaries will be retained, but they will be broken into shorter fragments by the intersections that occur between boundaries in one input dataset and boundaries in the other. Note the unusual characteristics of the new dataset shown in Figure 14.21. Unlike the two input datasets, where boundaries meet in junctions of three lines, the new map contains a new junction of four lines, formed by the new intersection discovered during the overlay process. Because the results of overlay are distinct in this way it is almost always possible to discover whether a GIS dataset was formed by overlaying two earlier datasets.

Polygon overlay has different meanings from the continuous-field and discrete-object perspectives.

With a single dataset that combines both inputs, it is an easy matter to answer all of the queries listed above through simple interrogation. It is also easy to reverse the overlay process – if neighboring areas that share the same land cover class are merged, for example, the result is the land ownership map, and vice versa.

Polygon overlay is a computationally complex operation and as noted earlier much work has gone into developing algorithms that function efficiently for large datasets. One of the issues that must be tackled by a practically useful algorithm is known as the *spurious polygon* or *coastline weave* problem. It is almost inevitable that there will be instances in any practical application where the same line on the ground occurs in both datasets. This happens, for example, when a coastal region is being analyzed, because the coastline is almost certain to appear in every dataset of the region. Rivers and roads often form boundaries in many different datasets – a river may function both as a land cover class boundary and as a land ownership boundary, for example. But although the same line is represented in both datasets, its representations will almost certainly not be the same. They may have been digitized from different maps, digitized using different numbers of points, subjected to different manipulations, obtained from entirely different sources (an air photograph and a topographic map, for example), and subjected to different measurement errors. When overlaid, the result is a series of small slivers. Paradoxically, the more care one takes in digitizing or processing, the worse the problem appears, as the result is simply more slivers, albeit smaller in size.

> **In two vector datasets of the same area there will almost certainly be instances where lines in each dataset represent the same feature on the ground.**

Table 14.1 shows an example of the consequences of slivers and how a GIS can be rapidly overwhelmed if it fails to anticipate and deal with them adequately. Today, a GIS will offer various methods for dealing with the problem, the most common of which is the specification of a *tolerance*. If two lines fall within this distance of each other, the GIS will treat them as a single line and not create slivers. The resulting overlay contains just one version of the line, not two. But at least one of the input lines has been moved, and if the tolerance is set too high the movement can be substantial and lead to problems later.

Overlay in raster is an altogether simpler operation, and this has often been cited as a good reason to adopt raster rather than vector structures. When two raster layers are overlaid, the attributes of each cell are combined according to a set of rules. For example, suppose the task

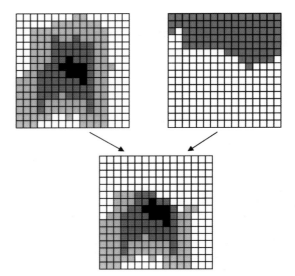

Figure 14.22 The raster overlay case, resulting in a new dataset that applies a set of rules to the input datasets. The two input datasets are maps of (top left) travel time from the urban area (see Figure 14.18) and (top right) county (red indicates County X, white indicates County Y). The output map identifies travel time to areas in County Y and might be used to compute average travel time to points in that county in a subsequent step. This operation is not reversible

is to find all areas that belong to owner A and have land use class B. Areas with these characteristics would be assigned a value, perhaps 1, and all other areas would be assigned a value of 0. Note the important difference between raster and vector overlay – in vector overlay there is no rule for combination, and instead the result of overlay contains all of the input information, rearranged and combined so that it can be used to respond to queries and can be subjected to analysis. Figure 14.22 shows an example of raster overlay.

Table 14.1 Numbers of polygons resulting from an overlay of five datasets, illustrating the spurious polygon problem. The datasets come from the Canada Geographic Information System discussed in Section 1.4.1, and all are representations of continuous fields. Dataset 1 is a representation of a map of soil capability for agriculture (Figure 15.13 shows a map of this type), datasets 2 through 4 are land-use maps of the same area at different times (the probability of finding the same real boundary in more than one such map is very high), and dataset 5 is a map of land capability for recreation. The final three columns show the numbers of polygons in overlays of three, four, and five of the input datasets

Acres	1	2	3	4	5	1 + 2 + 5	1 + 2 + 3 + 5	1 + 2 + 3 + 4 + 5
0–1	0	0	0	1	2	2640	27566	77346
1–5	0	165	182	131	31	2195	7521	7330
5–10	5	498	515	408	10	1421	2108	2201
10–25	1	784	775	688	38	1590	2106	2129
25–50	4	353	373	382	61	801	853	827
50–100	9	238	249	232	64	462	462	413
100–200	12	155	152	158	72	248	208	197
200–500	21	71	83	89	92	133	105	99
500–1000	9	32	31	33	56	39	34	34
1000–5000	19	25	27	21	50	27	24	22
>5000	8	6	7	6	11	2	1	1
Totals	88	2327	2394	2149	487	9558	39188	90599

Raster overlay is simpler, but it produces a fundamentally different kind of result.

14.4.4 Spatial interpolation

Spatial interpolation is a pervasive operation in GIS. Although it is often used explicitly in analysis, it is also used implicitly, in various operations such as the preparation of a contour map display, where spatial interpolation is invoked without the user's direct involvement. Spatial interpolation is a process of intelligent guesswork, in which the investigator (and the GIS) attempt to make a reasonable estimate of the value of a continuous field at places where the field has not actually been measured. Spatial interpolation is an operation that makes sense only from the continuous-field perspective. The principles of spatial interpolation are discussed in Section 4.5; here the emphasis is on practical applications of the technique and commonly used implementations of the principles.

Spatial interpolation finds applications in many areas:

- In estimating rainfall, temperature, and other attributes at places that are not weather stations and where no direct measurements of these variables are available.

- In estimating the elevation of the surface between the measured locations of a DEM.

- In *resampling* rasters, the operation that must take place whenever raster data must be transformed to another grid.

- In contouring, when it is necessary to guess where to place contours between measured locations.

In all of these instances, spatial interpolation calls for intelligent guesswork, and the one principle that underlies all spatial interpolation is the Tobler Law (Section 3.1) – 'all places are related but nearby places are more related than distant places'. In other words, the best guess as to the value of a field at some point is the value measured at the closest observation points – the rainfall *here* is likely to be more similar to the rainfall recorded at the nearest weather stations than to the rainfall recorded at more distant weather stations. A corollary of this same principle is that in the absence of better information, it is reasonable to assume that any continuous field exhibits relatively smooth variation – fields tend to vary slowly and to exhibit strong positive spatial autocorrelation, a property of geographic data discussed in Section 4.6.

Spatial interpolation is the GIS version of intelligent guesswork.

In this section three methods of spatial interpolation are discussed: Thiessen polygons; inverse-distance weighting (IDW), which is the simplest commonly used method; and Kriging, a popular statistical method that is grounded in the theory of regionalized variables and falls within the field of *geostatistics*.

14.4.4.1 Thiessen polygons

Thiessen polygons were suggested by Thiessen as a way of interpolating rainfall estimates from a few rain gauges to obtain estimates at other locations where rainfall had not been measured. The method is very simple: to estimate rainfall at any point take the rainfall measured at the closest gauge. This leads to a map in which rainfall is constant within polygons surrounding each gauge, and changes sharply as polygon boundaries are crossed. Although many GIS users associate polygons defined in this way with Thiessen, they are also known as Voronoi and Dirichlet polygons (Box 14.5). They have many other uses besides spatial interpolation:

- Thiessen polygons can be used to estimate the trade areas of each of a set of retail stores or shopping centers.

- They are used internally in the GIS as a means of speeding up certain geometric operations, such as search for nearest neighbor.

- They are the basis of some of the more powerful methods for generalizing vector databases.

As a method of spatial interpolation they leave something to be desired, however, because the sharp change in interpolated values at polygon boundaries is often implausible.

Figure 14.23 shows a typical set of Thiessen polygons. If each pair of points that share a Thiessen polygon boundary is connected, the result is a network of irregular triangles. These are named after Delaunay, and frequently used as the basis for the triangles of a TIN representation of terrain (Chapter 8).

14.4.4.2 Inverse-distance weighting

IDW is the workhorse of spatial interpolation, the method that is most often used by GIS analysts. It employs the Tobler Law by estimating unknown measurements as weighted averages over the known measurements

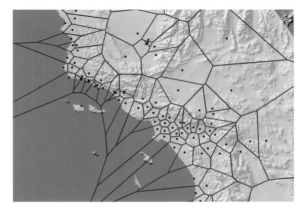

Figure 14.23 Thiessen polygons drawn around each station in part of the Southern California ozone-monitoring network. Note how the polygons, which enclose the area closest to each point, in theory extend off the map to infinity, so must be truncated by the GIS at the edge of the map

Atsu Okabe, expert on Thiessen polygons

Atsuyuki Okabe received his Ph.D. from the University of Pennsylvania in 1975 and the degree of Doctor of Engineering from the University of Tokyo in 1977. He is Director of the Center for Spatial Information Science (1998–2005) and Professor of the Department of Urban Engineering, University of Tokyo. His research interests include computational spatial analysis, spatial statistics, spatial optimization, and spatial psychology. He has published many papers in journals, books, and conference proceedings on these topics. He is a co-author (with B. Boots, K. Sugihara, and S. N. Chiu) of *Spatial Tessellations: Concepts and Applications of Voronoi Diagrams* (published by John Wiley). He currently leads the team developing a free GIS-based toolbox for spatial analysis on a network, called SANET.

Asked what had surprised him most about the development of GIS, he wrote: 'It is amazing to consider the remarkable change in spatial analysis brought about by GIS. In the early 1970s when I began to study spatial analysis, it remained conceptual. Digital data were scarce and researchers had to strive to assemble datasets. There were few efficient

Figure 14.24 Atsu Okabe, expert on Thiessen polygons

algorithms for geometrical calculations, and computation often took more than a week. Consequently, methods were difficult to test. I could, for example, consider the concepts of Voronoi diagrams, but it was hard to implement them. Try to construct a Voronoi diagram of five generators by hand, for instance! The first breakthrough was made in the mid 1980s, when 'Computational Geometry' (named by M.I. Shamos in 1975) became established through the book by Preparata and Shamos. The second development occurred in the late 1980s when GIS software became readily available. Surprisingly, almost all modern GIS software packages handle Voronoi diagrams, which we can now construct easily. Furthermore, recent progress in the design of GIS-based toolboxes has changed spatial analysis from an exclusive technique used only in academic research to a routine process used in a wide variety of professional activities. In the same way that Excel has made statistics available for general application, I hope that GIS-based spatial analysis will not only enhance studies in the humanities and social/natural sciences, but also come to be useful in everyday life.'

at nearby points, giving the greatest weight to the nearest points.

More specifically, denote the point of interest as **x**, and the points where measurements were taken as $\mathbf{x}_i$, where i runs from 1 to n, if there are n data points. Denote the unknown value as $z(\mathbf{x})$ and the known measurements as z_i. Give each of these points a weight d_i, which will be evaluated based on the distance from $\mathbf{x}_i$ to **x**. Figure 14.25 explains this notation with a diagram. Then the weighted average computed at **x** is:

$$z(\mathbf{x}) = \sum_i w_i z_i \Big/ \sum_i w_i$$

In other words, the interpolated value is an average over the observed values, weighted by the ws.

There are various ways of defining the weights, but the option most often employed is to compute them as

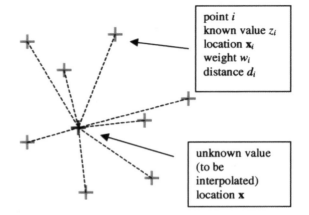

point i
known value z_i
location $\mathbf{x}_i$
weight w_i
distance d_i

unknown value (to be interpolated) location **x**

Figure 14.25 Notation used in the equations defining spatial interpolation

the inverse squares of distances, in other words (compare the options discussed in Section 4.5):

$$w_i = 1/d_i^2$$

This means that the weight given to a point drops by a factor of 4 when the distance to the point doubles (or by a factor of 9 when the distance triples). In addition, most software gives the user the option of ignoring altogether points that are further than some specified distance away, or of limiting the average to a specified number of nearest points, or of averaging over the closest points in each of a number of direction sectors. But if these values are not specified the software will assign default values to them.

IDW provides a simple way of guessing the values of a continuous field at locations where no measurement is available.

IDW achieves the desired objective of creating a smooth surface whose value at any point is more like the values at nearby points than the values at distant points. If it is used to determine z at a location where z has already been measured it will return the measured value, because the weight assigned to a point at zero distance is infinite, and for this reason IDW is described as an *exact* method of interpolation because its interpolated results honor the data points exactly. (An *approximate* method is allowed to deviate from the measured values in the interests of

greater smoothness, a property which is often useful if deviations are interpreted as indicating possible errors of measurement or local deviations that are to be separated from the general trend of the surface.)

But because IDW is an average it suffers from certain specific characteristics that are generally undesirable. A weighted average that uses weights that are never negative must always return a value that is between the limits of the measured values – no point on the interpolated surface can have an interpolated z that is more than the largest measured z, or less than the smallest measured z. Imagine an elevation surface with some peaks and pits where the peaks and pits have not actually been measured but are merely indicated by the values of the measured points. Figure 14.26 shows a cross-section of such a surface. Instead of interpolating peaks and pits as one might expect, IDW produces the kind of result shown in the figure – small pits where there should be peaks, and small peaks where there should be pits. This behavior is often obvious in GIS output that has been generated using IDW. A related problem concerns extrapolation – if a trend is indicated by the data, as shown in Figure 14.26, IDW will inappropriately indicate a regression to the mean outside the area of the data points.

IDW interpolation may produce counter-intuitive results in areas of peaks and pits, and outside the area covered by the data points.

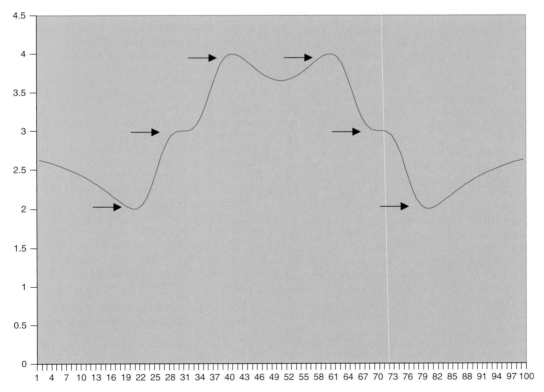

Figure 14.26 Potentially undesirable characteristics of IDW interpolation. Data points located at 20, 30, 40, 60, 70, and 80 have measured values of 2, 3, 4, 4, 3, and 2 respectively. The interpolated profile shows a pit between the two highest values, and regression to the overall mean value of 3 outside the area covered by the data

In short, the results of IDW are not always what one would want. There are many better methods of spatial interpolation that address the problems that were just identified, but the ease of programming of IDW and its conceptual simplicity make it among the most popular. Users should simply beware, and take care to examine the results of interpolation to ensure that they make good sense.

14.4.4.3 Kriging

Of the common methods of spatial interpolation, Kriging makes the most convincing claim to be grounded in good theoretical principles. The basic idea is to discover something about the general properties of the surface, as revealed by the measured values, and then to apply these properties in estimating the missing parts of the surface. Smoothness is the most important property (note the inherent conflict between this and the properties of fractals, Section 4.8), and it is operationalized in Kriging in a statistically meaningful way. There are many forms of Kriging, and the overview provided here is very brief. Further reading is identified at the end of the chapter.

There are many forms of Kriging, but all are firmly grounded in theory.

Suppose we take a point $\mathbf{x}$ as a reference and start comparing the values of the field there with the values at other locations at increasing distances from the reference point. If the field is smooth (if the Tobler Law is true, that is, if there is positive spatial autocorrelation) the values nearby will not be very different – $z(\mathbf{x})$ will not be very different from $z(\mathbf{x}_i)$. To measure the amount, we take the difference and square it, since the sign of the difference is not important: $(z(\mathbf{x}) - z(\mathbf{x}_i))^2$. We could do this with any pair of points in the area.

As distance increases, this measure will likely increase also, and in general a monotonic (consistent) increase in squared difference with distance is observed for most geographic fields (note that z must be measured on a scale that is at least interval, though *indicator Kriging* has been developed to deal with the analysis of nominal fields). In Figure 14.27, each point represents one pair of values drawn from the total set of data points at which measurements have been taken. The vertical axis represents one half of the squared difference (one half is taken for mathematical reasons), and the graph is known as the *semivariogram* (or *variogram* for short – the difference of a factor of two is often overlooked in practice, though it is important mathematically). To express its contents in summary form, the distance axis is divided into a number of ranges or *bins*, as shown, and points within each range are averaged to define the heavy points shown in the figure.

This semivariogram has been drawn without regard to the *directions* between points in a pair. As such it is said to be an *isotropic* variogram. Sometimes there is sharp variation in the behavior in different directions, and *anisotropic* semivariograms are created for different ranges of direction (e.g., for pairs in each 90 degree sector).

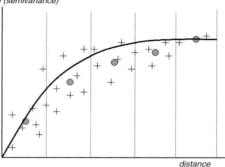

One half the mean squared difference (semivariance)

distance

Figure 14.27 A semivariogram. Each cross represents a pair of points. The solid circles are obtained by averaging within the ranges or *buckets* of the distance axis. The solid line is the best fit to these five points, using one of a small number of standard mathematical functions

An anisotropic variogram asks how spatial dependence changes in different directions.

Note how the points of this typical variogram show a steady increase in squared difference up to a certain limit, and how that increase then slackens off and virtually ceases. Again, this pattern is widely observed for fields, and it indicates that difference in value tends to increase up to a certain limit, but then to increase no further. In effect, there is a distance beyond which there are no more geographic surprises. This distance is known as the *range*, and the value of semivariance at this distance as the *sill*.

Note also what happens at the other, lower end of the distance range. As distance shrinks, corresponding to pairs of points that are closer and closer together, the semivariance falls, but there is a suggestion that it never quite falls to zero, even at zero distance. In other words, if two points were sampled a vanishingly small distance apart they would give different values. This is known as the *nugget* of the semivariogram. A non-zero nugget occurs when there is substantial error in the measuring instrument, such that measurements taken a very small distance apart would be different due to error, or when there is some other source of local noise that prevents the surface being truly smooth. Accurate estimation of a nugget depends on whether there are pairs of data points sufficiently close together. In practice the sample points may have been located at some time in the past, outside the user's control, or may have been spread out to capture the overall variation in the surface, so it is often difficult to make a good estimate of the nugget.

The nugget can be interpreted as the variation among repeated measurements at the same point.

To make estimates using Kriging, we need to reduce the semivariogram to a mathematical function, so that semivariance can be evaluated at any distance, not just at the midpoints of buckets as shown in Figure 14.27. In practice, this means selecting one from a set of standard functional forms and fitting that form to the observed data points to get the best possible fit. This is shown in the figure. The user of a Kriging function in a GIS will have control over the selection of distance ranges and functional forms, and whether a nugget is allowed.

Finally, the fitted semivariogram is used to estimate the values of the field at points of interest. As with IDW, the estimate is obtained as a weighted combination of neighboring values, but the estimate is designed to be the best possible given the evidence of the semivariogram. In general, nearby values are given greater weight, but unlike IDW direction is also important – a point can be *shielded* from influence if it lies behind another point, since the latter's greater proximity suggests greater importance in determining the estimated value, whereas relative direction is unimportant in an IDW estimate. The process of maximizing the quality of the estimate is carried out mathematically, using the precise measures available in the semivariogram.

Kriging responds both to the proximity of sample points and to their directions.

Unlike IDW, Kriging has a solid theoretical foundation, but it also includes a number of options (e.g., the choice of the mathematical function for the semivariogram) that require attention from the user. In that sense it is definitely not a *black box* that can be executed blindly and automatically (Section 2.3.5.5), but instead forces the user to become directly involved in the estimation process. For that reason GIS software designers will likely continue to offer several different methods, depending on whether the user wants something that is quick, despite its obvious faults, or better but more demanding of the user.

14.4.4.4 Density estimation

Density estimation is in many ways the logical twin of spatial interpolation – it begins with points and ends with a surface. But conceptually the two approaches could

not be more different, because one seeks to estimate the missing parts of a continuous field from samples of the field taken at data points, while the other creates a continuous field from discrete objects.

Figure 14.28 illustrates this difference. The dataset can be interpreted in two sharply different ways. In the first, it is interpreted as sample measurements from a continuous field, and in the second as a collection of discrete objects. In the discrete-object view there is nothing between the objects but empty space – no missing field to be filled in through spatial interpolation. It would make no sense at all to apply spatial interpolation to a collection of discrete

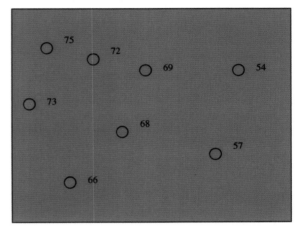

Figure 14.28 A dataset with two possible interpretations: first, a continuous field of atmospheric temperature measured at eight irregularly spaced sample points, and second, eight discrete objects representing cities, with associated populations in thousands. Spatial interpolation makes sense only for the former, and density estimation only for the latter

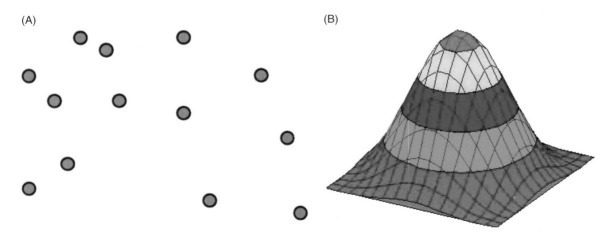

Figure 14.29 (A) A collection of point objects, and (B) a kernel function. The kernel's shape depends on a distance parameter – increasing the value of the parameter results in a broader and lower kernel and reducing it results in a narrower and sharper kernel. When each point is replaced by a kernel and the kernels are added the result is a density surface whose smoothness depends on the value of the distance parameter

objects – and no sense at all to apply density estimation to samples of a field.

> **Density estimation makes sense only from the discrete-object perspective, and spatial interpolation only from the field perspective.**

Although density estimation could be applied to any type of discrete object, it is most often applied to the estimation of point density, and that is the focus here. The most obvious example is the estimation of population density, and that example is used in this discussion, but it could be equally well applied to the density of different kinds of diseases, or animals, or any other set of well-defined points.

Consider the mainland of Australia. One way of defining its population density is to take the entire population and divide by the total area – on this basis the 1996 population density was roughly 2.38 per sq km. But we know that Australia's settlement pattern is very non-uniform, with most of the population concentrated in five coastal cities (Brisbane, Sydney, Melbourne, Adelaide, and Perth). So if we looked at the landscape in smaller pieces, such as circles 10 km in radius, and computed population density by dividing the number of people in each circle by the circle's area, we would get very different results depending on where the circle was centered. So, in general, population density at a location, and at a spatial resolution of d, might be defined by centering a circle at the location and dividing the total population within the circle by its area. Using this definition there are an infinite number of possible population density maps of Australia, depending on the value selected for d. And it follows that there is no such thing as *population density*, only population density *at a spatial resolution of d*. Note the similarity between this idea and the previous discussion of slope – in general, many geographic themes can only be defined rigorously if spatial resolution is made explicit, and much confusion results in GIS because of our willingness to talk about themes without at the same time specifying spatial resolution.

> **Density estimation with a kernel allows the spatial resolution of a field of population density to be made explicit.**

The theory of density estimation formalizes these ideas. Consider a collection of point objects, such as those shown in Figure 14.29. The surface shown in the figure is an example of a *kernel function*, the central idea in density estimation. Any kernel function has an associated length measure, and in the case of the function shown, which is a Gaussian distribution, the length measure is a parameter of the distribution – we can generate Gaussian distributions with any value of this parameter and they become flatter and wider as the value increases. In density estimation, each point is replaced by its kernel function and the various kernel functions are added to obtain an aggregate surface, or continuous field of density. If one thinks of each kernel as a pile of sand, then each pile has the same total weight of one unit. The total weight of all

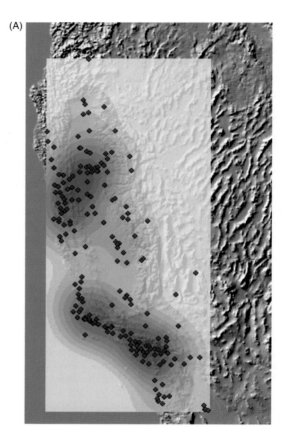

(A)

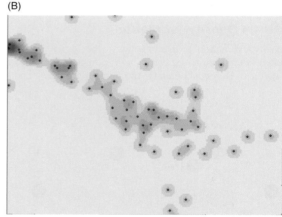

(B)

Figure 14.30 Density estimation using two different distance parameters in the respective kernel functions: (A) the surface shows the density of ozone-monitoring stations in California, using a kernel radius of 150 km; (B) zoomed to an area of Southern California, this shows that a kernel radius of 16 km is clearly too small for this dataset, as it leaves each kernel isolated from its neighbors

piles of sand is equal to the number of points and the total weight of sand within a given area, such as the area shown in the figure, is an estimate of the total population in that area. Mathematically, if the population density is represented by a field $\rho(x, y)$, then the total population within area A is the integral of the field function over that area, that is:

$$P = \int_A \rho dA$$

A variety of kernel functions are used in density estimation, but the form shown in Figure 14.29 is perhaps the commonest. This is the traditional bell curve or Gaussian distribution of statistics and is encountered elsewhere in this book in connection with errors in the measurement of position in two dimensions (Section 6.3.2.2). By adjusting the width of the bell it is possible to produce a range of density surfaces of different amounts of smoothness. Figure 14.30 contrasts two density estimations from the same data, one using a comparatively narrow bell to produce a complex surface, and the other using a broader bell to produce a smoother surface.

14.5 Conclusion

This chapter has discussed the first three types of spatial analysis; Chapter 15 discusses the remaining three, and Chapter 16 examines spatial modeling. Several general issues of scale and resolution, and accuracy and uncertainty, have been raised throughout the discussion: these are discussed in greater detail in Chapter 6.

Questions for further study

1. Did Dr John Snow actually make his inference strictly from looking at his map? What information can you find on the Web on this issue? (try **www.jsi.com**)

2. You are given a map showing the home locations of the customers of an insurance agent and asked to construct a map showing the agent's market area. Would spatial interpolation or density estimation be more appropriate, and why?

3. What is conditional simulation, and how does it differ from Kriging? Under what circumstances might it be useful (a possible source on this question is the chapter by Englund in Goodchild, Steyaert and Parks, *Environmental Modeling with GIS*, Oxford University Press, 1993).

4. What are the most important characteristics of the three methods of spatial interpolation discussed in this chapter? Using a test dataset of your own choosing, compute and describe the major features of the surfaces interpolated by each method.

Further reading

Bailey T.C. and Gatrell A.C. 1995 *Interactive Spatial Data Analysis*. Harlow, UK: Longman Scientific and Technical.

Burrough P.A. and McDonnell R.A. 1998 *Principles of Geographical Information Systems*. New York: Oxford University Press.

Isaaks E.H. and Srivastava R.M. 1989 *Applied Geostatistics*. New York: Oxford University Press.

O'Sullivan D. and Unwin D.J. 2003 *Geographic Information Analysis*. Hoboken, NJ: Wiley.

Silverman B.W. 1986 *Density Estimation for Statistics and Data Analysis*. New York: Chapman and Hall.

15 *Descriptive summary, design, and inference*

This is the second of two chapters on spatial analysis and focuses on three areas: descriptive summaries, the use of analysis in design decisions, and statistical inference. These methods are all conceptually more complex than those in Chapter 14, but are equally important in practical applications of GIS.

The chapter begins with a brief discussion of data mining. Descriptive summaries attempt to capture the nature of geographic distributions, patterns, and phenomena in simple statistics that can be compared through time, across themes, and between geographic areas. Optimization techniques apply much of the same thinking, and extend it, to help users who must select the best locations for services, or find the best routes for vehicles, or a host of similar tasks. Hypothesis testing addresses the basic scientific need to be able to generalize results from a small study to a much larger context, perhaps the entire world.

Geographic Information Systems and Science, 2nd edition   Paul Longley, Michael Goodchild, David Maguire, and David Rhind.
© 2005 John Wiley & Sons, Ltd.   ISBNs: 0-470-87000-1 (HB); 0-470-87001-X (PB)

Learning Objectives

After reading this chapter you will know about:

■ Data mining, a new form of analysis that is enabled by GIS and by vast new supplies of data;

■ The concept of summarizing a pattern in a few simple statistics;

■ Methods that support decisions by enlisting GIS to search automatically across thousands or millions of options;

■ The concept of a hypothesis, and how to make inferences from small samples to larger populations.

15.1 More spatial analysis

The methods introduced in this chapter use more sophisticated conceptual frameworks than those of Chapter 14. That chapter began with simple queries and measurement, but in this chapter the discussion turns to summaries, optimization methods, and inferential methods. Many date from well before the era of GIS, and the advent of geographic databases and cheap computing technology has turned what were once esoteric methods buried in the academic literature into practical ways of solving everyday problems.

The advent of large databases and fast computing has also led to new ways of thinking about spatial analysis and many of these are captured in the term *data mining*. Surely, the thinking goes, there must be interesting patterns, clusters, anomalies, and truths buried in the masses of data being collected and archived every day by humanity's vast new investment in information technology. Cameras now capture images of the traffic on freeways that are precise enough to read license plates. As noted in Section 11.3, every use of a credit card creates a record in a database that includes the location of use, along with information on what was bought, and having the name of the user allows that information to be linked to many other records. If we only had suitable computer methods, the thinking goes, we could program software to scan through these vast databases, looking for things of interest. Many of these might be regarded as invasions of privacy, but others could be very useful in spotting criminal activity, for example. Data mining has become a major area of application in business, where it is used to detect anomalies in spending patterns that

might be indicative of a stolen credit card – for example, a card owned by a resident of California, USA is suddenly used for a series of large purchases in a short period of time in the New York area, in types of stores not normally frequented by the card owner (Figure 15.1). The *geographic* aspect of this evidence – the expenditures in a small, unusual area – is one of the most significant.

Data mining is used to detect anomalies and patterns in vast archives of digital data.

Data mining techniques could be used to watch for anomalous patterns in the records of disease diagnosis, or to provide advanced warnings of new outbreaks of virulent forms of influenza. In all of these examples the objective is to find *patterns*, and *anomalies* that stand out from the normal in an area, or stand out at a particular point in time when compared to long-term averages. In recent years surprising patterns in digital data have led to the discovery of the ozone hole over the Antarctic, the recall of millions of tires made by the Firestone-Bridgestone Company, and numerous outbreaks of disease. Figure 15.2 shows an example of a *parallel coordinate plot*, one of many methods developed in recent years to display large multi-dimensional datasets. Note that location plays no part in this plot, which therefore

Figure 15.1 Massive geographic databases can be built from records on the use of credit cards

Figure 15.2 The search for patterns in massive datasets is known as data mining. This example of a *parallel coordinate plot* from the GeoDa software (see Figure 14.8) shows data on all 3141 US counties. Each horizontal line corresponds to one variable. Each county is represented by a polyline connecting its scores on each of the variables. In this illustration the user has selected all counties with a very high proportion of people over 65 (the bottom line), and the plot shows that such counties (the yellow polylines) tend also to have low population densities, average rates of owner occupancy, low median values of housing, and average percentages of males

fails to qualify as a form of spatial analysis according to the definitions given in Section 14.1.

15.2 Descriptive summaries

15.2.1 Centers

The topics of generalization and abstraction were discussed earlier in Section 3.8 as ways of reducing the complexity of data. This section reviews a related topic, that of numerical summaries. If we want to describe the nature of summer weather in an area we cite the *average* or *mean*, knowing that there is substantial variation around this value, but that it nevertheless gives a reasonable *expectation* about what the weather will be like on any given day. The mean (the more formal term) is one of a number of measures of *central tendency*, all of which attempt to create a summary description of a series of numbers in the form of a single number. Another is the *median*, the value such that one half of the numbers are larger, and one half are smaller. Although the mean can be computed only for numbers measured on interval or ratio scales, the median can be computed for ordinal data. For nominal data the

appropriate measure of central tendency is the *mode*, or the commonest value. For definitions of nominal, ordinal, interval, and ratio see Box 3.3. Special methods must be used to measure central tendency for cyclic data – they are discussed in texts on directional data, for example by Mardia and Jupp.

The spatial equivalent of the mean would be some kind of center, calculated to summarize the positions of a number of points. Early in US history, the Bureau of the Census adopted a practice of calculating a center for the US population. As agricultural settlement advanced across the West in the 19th century, the repositioning of the center every ten years captured the popular imagination. Today, the movement west has slowed and shifted more to the south (Figure 15.3), and by the next census may even have reversed.

Centers are the two-dimensional equivalent of the mean.

The mean of a set of numbers has several properties. First, it is calculated by summing the numbers and dividing by the number of numbers. Second, if we take any value d and sum the squares of the differences between the numbers and d, then when d is set equal to the mean this sum is minimized (Figure 15.4). Third, the mean is the point about which the set of numbers would balance if we made a physical model such as the one shown in Figure 15.5, and suspended it.

These properties extend easily into two dimensions. Figure 15.6 shows a set of points on a flat plane, each

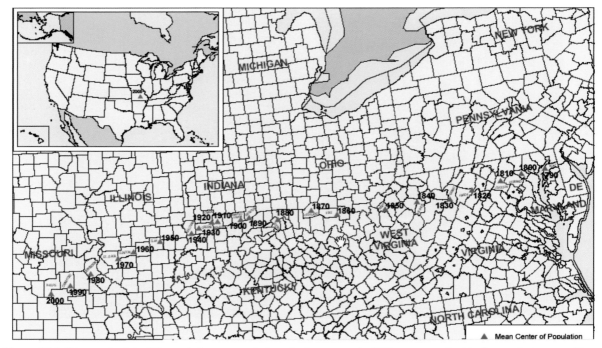

Figure 15.3 The march of the US population westward since the first census in 1790, as summarized in the population centroid (*Source*: US Bureau of the Census)

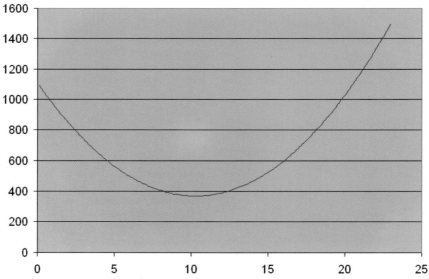

Figure 15.4 Seven points are distributed along a line at coordinates 1, 3, 5, 11, 12, 18, and 22. The curve shows the sum of distances squared from these points and how it is minimized at the mean [(1 + 3 + 5 + 11 + 12 + 18 + 22)/7 = 10.3]

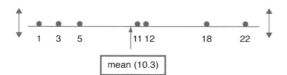

Figure 15.5 The mean is also the balance point, the point about which the distribution would balance if it were modeled as a set of equal weights on a weightless, rigid rod

one located at a point (x_i, y_i) and with weight w_i. The *centroid* or *mean center* is found by taking the weighted average of the x and y coordinates:

$$\bar{x} = \sum_i w_i x_i \Big/ \sum_i w_i$$

$$\bar{y} = \sum_i w_i y_i \Big/ \sum_i w_i$$

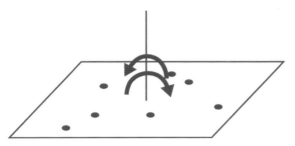

Figure 15.6 The centroid or mean center replicates the balance-point property in two dimensions – the point about which the two-dimensional pattern would balance if it were transferred to a weightless, rigid plane and suspended

It is also the point that minimizes the sum of squared distances, and it is the balance point.

Just like the mean, the centroid is a useful summary of a distribution of points. Although any single centroid may not be very interesting, a comparison of centroids for different sets of points or for different times can provide useful insights.

The centroid is the most convenient way of summarizing the locations of a set of points.

The property of minimizing functions of distance (the square of distance in the case of the centroid) makes

centers useful for different reasons. Of particular interest is the location that minimizes the sum of distances, rather than the sum of squared distances, since this could be the most effective location for any service that is intended to serve a dispersed population. The point that minimizes total straight-line distance is known as the *point of minimum aggregate travel* or MAT. Historically there have been numerous instances of confusion between the MAT and the centroid, and the US Bureau of the Census used to claim that the Center of Population had the MAT property, even though their calculations were based on the equations for the centroid.

The centroid is often confused with the point of minimum aggregate travel.

There is no simple mathematical expression for the MAT, and instead its location must be found by a trial-and-error process of *iteration*, in which an initial guess is successively improved using a suitable algorithm. There is also an interesting way of computing the MAT using a physical analog, an experiment known as the *Varignon Frame* (see Box 15.1). There is an intriguing paradox associated with the experiment. Imagine that one of the points must be relocated further away from the MAT but in the same direction. The force pulling on the string remains the same, in the same direction, so the position of the knot does not move. But how could the solution to a problem of minimizing distance not be affected by a change in distance? It is straightforward

---

The Varignon Frame experiment

This experiment provides a way of finding the point of minimum aggregate travel (MAT). Because it uses a physical model to solve the problem rather than a digital representation we can think of it as an *analog* computer (see Section 16.1 for a more extensive discussion of analog models, and see Section 3.7 for a discussion of the paper map as an analog representation). Suppose we want to find the best location for a steel mill, to minimize the total costs of shipping raw materials to the plant (iron ore, limestone, and coal) and the total costs of shipping the product to markets. Each source of raw materials or market is represented by a point, with a weight equal to the cost per km of shipping the required amount of material. Then the total shipping cost is the product of these weights times distances to the plant, summed over all sources and markets.

Take a piece of plywood, and sketch a map on it, locating the sources and markets (see Figure 15.7). At each of these points drill a hole,

and hang a weight on a string through the hole, using the appropriate weight (proportional to the shipping cost of the required amount of material per km). Bring all of the strings together into a knot. When the knot is released to find its own position the result is the MAT, though friction in the holes will create some error in the solution.

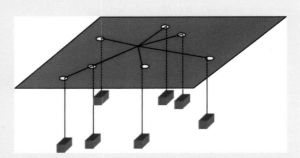

Figure 15.7 The Varignon Frame experiment, an analog computer for the point of minimum aggregate travel (MAT)

to show with a little application of calculus that the experiment is indeed correct, despite what appears to be a counter-intuitive result – that the location of the MAT does indeed depend only on the weights and the directions of pull, not on the distances. The same property occurs in the one-dimensional case: the point that minimizes distance along a line is the median, and the median's location is not affected by moving points as long as no point is relocated to the other side of the current median.

This example serves to make a very general point about spatial analysis and its role in human–computer interaction. In the introduction to Chapter 14 the point was made that all of these methods of analysis work best in the context of a collaboration between human and machine, and that one benefit of the machine is that it sometimes serves to correct the misleading aspects of human intuition. The MAT and more complex problems that are based on it illustrate this principle very well. Intuition would have suggested that moving one of the weights in the Varignon Frame experiment would have moved the knot, even though the movement was in the same direction. But theory says otherwise. Humans can be very poor at guessing the answers to optimization problems in space and the results of machine analysis can often be sharply different from what we expect.

The MAT is one instance of a type of spatial analysis in which the objective is not insight so much as *design* – it is a *normative* or *prescriptive* method because it imposes a norm (minimizing distance) or prescribes a solution. In the 1920s, the Soviet Union supported the Mendeleev Centrographical Laboratory whose objective was to identify the centers of various kinds of economic activity in the USSR as the most suitable sites for the state's central planning functions. For example, the center of piano manufacture would be the best place to locate the state planning office for piano production.

Many methods of spatial analysis are used to make design decisions.

In Section 14.3.1 it was noted that simple metrics of distance, such as the length of a straight line, have the disadvantage that they ignore the effects of networks on travel. The approach to the MAT described here also assumes straight-line travel. Comparable approaches based on networks are often better suited to practical problem solving and some examples of these are discussed in Section 15.3.

All of the methods described in this and the following section are based on plane geometry, and simple x,y coordinate systems. If the curvature of the Earth's surface is taken into account, the centroid of a set of points must be calculated in three dimensions, and will always lie under the surface. More useful perhaps are versions of the MAT and centroid that minimize distance over the curved surface (the minimum total distance and the minimum total of squared distances, respectively, using great circle distances, Sections 5.6 and 14.3.1).

15.2.2 Dispersion

Central tendency is the obvious choice if a set of numbers must be summarized in a single value, but what if there is the opportunity for a second summary value? Here, the measure of choice for numbers with interval or ratio properties is the *standard deviation*, or the square root of the mean squared difference from the mean:

$$s = \sqrt{\sum_i (x_i - \overline{x})^2 \Big/ n}$$

where n is the number of numbers, s is the standard deviation, x_i refers to the ith observation, and $\overline{x}$ is the mean of the observations. In weighted form the equation becomes:

$$s = \sqrt{\sum_i w_i (x_i - \overline{x})^2 \Big/ \sum_i w_i}$$

where w_i is the weight given to the ith observation. The *variance*, or the square of the standard deviation (the mean squared difference from the mean), is often encountered but it is not as convenient a measure for descriptive purposes. Standard deviation and variance are considered more appropriate measures of dispersion than the *range* (difference between the highest and lowest numbers) because as averages they are less sensitive to the specific values of the extremes.

The standard deviation has also been encountered in Section 6.3.2.2 in a different guise, as the root mean squared error (RMSE), a measure of dispersion of observations about a true value. Just as in that instance, the Gaussian distribution provides a basis for generalizing about the contents of a sample of numbers, using the mean and standard deviation as the parameters of a simple bell curve. If data follow a Gaussian distribution, then approximately 68% of values lie within one standard deviation of the mean; and approximately 5% of values lie outside two standard deviations.

These ideas convert very easily to the two-dimensional case. A simple measure of dispersion in two dimensions is the *mean distance from the centroid*. In some applications it may be desirable to give greater weight to more distant points. For example, if a school is being located, then students living at distant locations are comparatively disadvantaged. They can be given greater weight if each distance is squared, such that a student twice as far away receives four times the weight. This property is minimized by locating the school at the centroid.

Mean distance from the centroid is a useful summary of dispersion.

Measures of dispersion can be found in many areas of GIS. The breadth of the kernel function of density estimation (Section 14.4.5) is a measure of how broadly the pile of sand associated with each point is dispersed.

RMSE is a measure of the dispersion inherent in positional errors (Section 6.3.2.2).

15.2.3 Measures of pattern: unlabeled points

A very large number of techniques exist for summarizing the general properties of patterns. Most of these have been developed for points, and their extension to patterns of lines and areas is generally not straightforward, so the focus here will be on points only. The discussion in this section centers on points without distinguishing attributes – such as patterns of incidence of a disease, or sightings of a bird species, or customers of a retail store. The subsequent section looks at methods for describing how attributes are distributed among points, rather than at the locations of the points *per se*.

One of the questions most commonly asked about distributions of points is whether they display a random pattern, in the sense that all locations are equally likely to contain a point, or whether some locations are more likely than others – and particularly, whether the presence of one point makes other points either more or less likely in its immediate neighborhood. This leads to three possibilities:

- the pattern is *random* (points are located independently, and all locations are equally likely);
- the pattern is *clustered* (some locations are more likely than others, and the presence of one point may attract others to its vicinity);
- the pattern is *dispersed* (the presence of one point may make others less likely in its vicinity).

Establishing the existence of clusters is often of great interest, since it may point to possible causal factors, as for example with the case of childhood leukemia studied by Stan Openshaw (Figure 14.4). Dispersed patterns are the typical result of competition for space, as each point establishes its own territory and excludes others – so such patterns are commonly found among organisms that exhibit territorial behavior, as well as among market towns in rural areas, and among retail outlets.

Point patterns can be identified as clustered, dispersed, or random.

It is helpful to distinguish two kinds of processes responsible for point patterns. *First-order* processes involve points being located independently, but may still result in clusters because of varying point density. For example, the drinking-water hypothesis investigated by Dr John Snow and described in Box 14.1 led to a higher density of points around the pump, because of greater access; and similarly the density of organisms of a particular species may vary over an area because of varying suitability of habitat. *Second-order* processes involve interaction between points, and lead to clusters when the interactions are attractive in nature, and

dispersion when they are competitive or repulsive. In the cholera case, the contagion hypothesis rejected by Snow is a second-order process and results in clustering even in situations when all other density-controlling factors are perfectly uniform. Unfortunately, as argued in Section 14.1, Snow's evidence did not allow him to resolve with complete confidence between first-order and second-order processes, and in general it is not possible to determine whether a given clustered point pattern was created by varying density factors, or by interactions. On the other hand, dispersed patterns can only be created by second-order processes.

Clustering can be produced by two distinct mechanisms, identified as first-order and second-order.

There are many available descriptive statistics of pattern and some excellent books on the subject. To illustrate the method only one will be discussed in this section. This is the K function, and unlike many such statistics it provides an analysis of clustering and dispersion over a range of scales. In the interests of brevity the technical details will be omitted, but they can be found in the texts listed at the end of this chapter. They include procedures for dealing with the effects of the study area boundary, which is an important distorting factor for many pattern statistics.

$K(d)$ is defined as the expected number of points within a distance d of an arbitrarily chosen point, divided by the density of points per unit area. When the pattern is random this number is πd^2, so the normal practice is to plot the function:

$$\hat{L}(d) = \sqrt{K(d)/\pi}$$

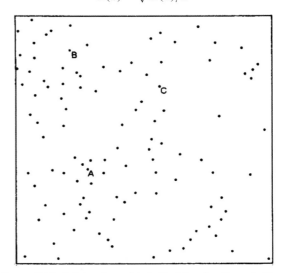

Figure 15.8 Point pattern of individual tree locations. A, B, and C identify the individual trees analyzed in Figure 15.9 (*Source*: Getis A. and Franklin J. 1987 'Second-order neighborhood analysis of mapped point patterns.' *Ecology* **68**(3): 473–477)

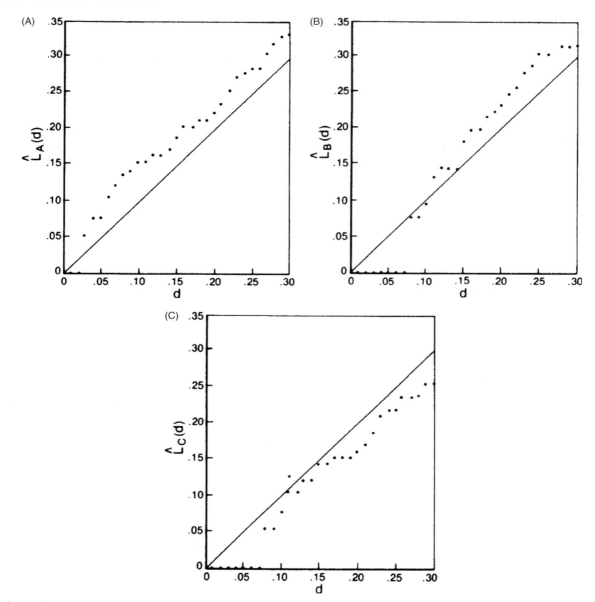

Figure 15.9 Analysis of the local distribution of trees around three reference trees in Figure 15.8 (see text for discussion) (*Source*: Getis A. and Franklin J. 1987 'Second-order neighborhood analysis of mapped point patterns.' *Ecology* **68**(3): 473–477)

since $\hat{L}(d)$ will equal d for all d in a random pattern, and a plot of $\hat{L}(d)$ against d will be a straight line with a slope of 1. Clustering at certain distances is indicated by departures of $\hat{L}$ above the line and dispersion by departures below the line.

Figures 15.8 and 15.9 show a simple example, used to discover how trees are spaced relative to each other in a forest. The locations of the trees are shown in Figure 15.8. In Figure 15.9(A) locations are analyzed in relation to Tree A. At very short distances there are fewer trees than would occur in a random pattern, but for most of the range up to a distance of 30% of the width (and height) of the study area there are *more* trees than would be expected, showing a degree of clustering.

Tree B (Figure 15.9(B)) has no nearby neighbors, but shows a degree of clustering at longer distances. Tree C (Figure 15.9(C)) shows fewer trees than expected in a random pattern over most distances, and it is evident in Figure 15.8 that C is in a comparatively widely spaced area of forest.

15.2.4 Measures of pattern: labeled features

When features are differentiated, such as by values of some attribute, the typical question concerns whether the values are randomly distributed over the features, or whether high extreme values tend to cluster: high values surrounded by high values and low values surrounded by low values. In such investigations the processes that determined the locations of features, and were the major concern in the previous section, tend to be ignored and may have nothing to do with the processes that created the pattern of labels. For example, the concern might be with some attribute of counties – their average house value, or percent married – and hypotheses about the processes that lead to counties having different values on these indicators. The processes that led to the locations and shapes of counties, which were political processes operating perhaps 100 years ago, would be of no interest.

The Moran statistic (Section 4.3) is designed precisely for this purpose, to indicate general properties of the pattern of attributes. It distinguishes between positively autocorrelated patterns, in which high values tend to be surrounded by high values, and low values by low values; random patterns, in which neighboring values are independent of each other; and dispersed patterns, in which high values tend to be surrounded by low, and vice versa.

The Moran statistic looks for patterns among the attributes assigned to features.

In recent years there has been much interest in going beyond these global measures of spatial dependence to identify dependences locally. Is it possible, for example, to identify *hot spots*, areas where high values are surrounded by high values, and *cold spots*, where low values are surrounded by low? Is it possible to identify anomalies, where high values are surrounded by low or vice versa? Local versions of the Moran statistic are among this group and along with several others now form a useful resource that is easily implemented in GIS.

Figure 15.10 shows an example, using the Local Moran statistic to differentiate states according to their roles in the pattern of one variable, the median value of housing. The weights (see Box 4.4) have been defined by adjacency, such that pairs of states that share a common

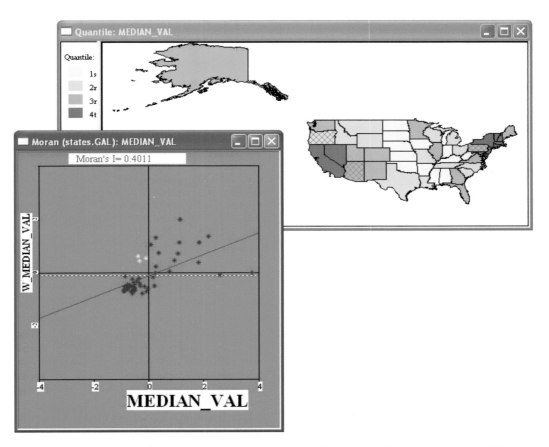

Figure 15.10 The Local Moran statistic, applied to describe local aspects of the pattern of housing value among US states. In the map window the states are colored according to median value, with the darker shades corresponding to more expensive housing. In the scatterplot window the median value appears on the x axis, while on the y axis is the weighted average of neighboring states. The three points colored yellow are instances where a state of below-average housing value is surrounded by states of above-average value. The windows are linked (see Figure 14.8 for details of this GeoDa software), and the three points are identified as Oregon, Arizona, and Pennsylvania. The global Moran statistic is also shown (+0.4011, indicating a general tendency for clustering of similar values)

Comparing attributes when objects do not coincide spatially

GIS users often encounter situations where attributes must be compared, but for different sets of objects. Figure 15.11 shows a study area, with two sets of points, where one indicates the locations where levels of ambient sound were measured using recorders mounted on telephone poles and the other shows the locations of interviews conducted with local residents to determine attitudes to noise. We would like to know about the relationship between sound and attitudes, but the locations and numbers of cases (the *spatial support*) are different. A simple expedient is to use spatial interpolation (Section 14.4.4) to map both variables as complete fields. Then it would be possible to conduct the analysis in any of three ways:

- by interpolating the second dataset to the locations at which the first attribute was measured;
- by the reverse – interpolating the first dataset to the locations at which the second attribute was measured;
- by interpolating both datasets to a common geometric base, such as a raster.

In the third case, note that it is possible to create a vast amount of data by using a sufficiently detailed raster. In essence, all of these options involve manufacturing information and the results will depend to some extent on the nature and suitability of the method used to do the spatial interpolation. We normally think of a relationship that is demonstrated with a large number of observations as stronger and more convincing than one based on a small number of observations. But the ability to manufacture data upsets this standard view.

Figure 15.11 shows a possible solution using this third option. The number of grid points has been determined by the smaller number of cases – in this case, approximately 10 (12 are shown) – to minimize concerns about manufacturing information.

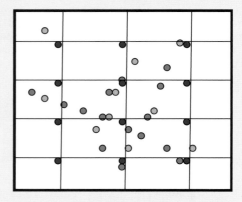

Figure 15.11 Coping with the comparison of two sets of attributes when the respective objects do not coincide. In this instance, attitudes regarding ambient noise have been obtained through a household survey of 15 residents (green dots) and are to be compared to ambient noise levels measured at 10 observation points (brown dots). The solution is to interpolate both sets of data to 12 comparison points (blue dots), using the methods discussed in Section 14.4.4

---

boundary are given a weight of 1 and all other pairs a weight of zero.

15.2.5 Fragmentation and fractional dimension

Figure 15.13 shows a type of dataset that occurs frequently in GIS. It might represent the classes of land cover in an area, but in fact it is a map of soils, and each patch corresponds to an area of uniform class and is bounded by areas of different class. A landscape ecologist or forester might be interested in the degree to which the landscape is fragmented – is it broken up into small or large patches, are the patches compact or contorted, and how long are the common boundaries between the patches? Fragmented landscapes are less favorable for wildlife because habitat is broken into small patches with potentially dangerous land between. Fragmentation statistics have been designed to provide the numerical basis for these kinds of questions, by measuring relevant properties of datasets.

An interesting application of this kind of technique is to deforestation, in areas like the Amazon Basin. Figure 15.14 shows a sequence of images of part of the basin from space, and the impact of clearing and settlement on the pattern of land use. The impact of land use changes like these depends on the degree to which they fragment existing habitat, perhaps into pieces that are too small to maintain certain species such as major predators.

Luc Anselin, pioneer in spatial analysis

Figure 15.12 Luc Anselin

Luc Anselin (Figure 15.12) is Professor at the University of Illinois, Urbana-Champaign, where he directs a spatial analysis research laboratory and holds appointments in the Departments of Agricultural and Consumer Economics, Economics, Geography, and Urban and Regional Planning. He was originally trained as an economist at the Free University of Brussels (VUB) in Belgium and obtained a Ph.D. in Regional Science from Cornell University. Most of his career has been devoted to the development, implementation, and application of statistical and econometric methods that explicitly take into account spatial effects (spatial dependence and spatial heterogeneity) that may be present in the data. His much-cited *Spatial Econometrics* (1988) was the first comprehensive treatment that outlined the methodology to test for spatial effects in regression models and how to estimate such models. Many of these methods were incorporated into his SpaceStat software package in the early 1990s. Anselin gained an interest in local statistics, particularly how they could be incorporated into software tools for exploratory spatial data analysis (ESDA). He developed the concepts of a Moran scatterplot (Figure 15.10) and LISA test (local indicators of spatial association; Figure 14.8) and implemented them in several software tools. The latest incarnation of these tools is the popular GeoDa package (Figure 14.8). 'I first realized the power of GIS and spatial analysis as a faculty member in the Department of Geography at the University of California, Santa Barbara. I was trained as an econometrician, and even though my research dealt with spatial aspects, I had not fully realized the potential of the interface between GIS and the analytical techniques I had developed. For example, I had carried out all the illustrative regressions and tests for the *Spatial Econometrics* book without ever looking at a map. When GIS tools became available, it was just mind-boggling how much more efficiently analysis could be carried out. Patterns could be readily distinguished that would take hours and hours of tabulations and regression runs in the old days. Especially now, with the geovisualization tools of desktop GIS and ESDA software such as GeoDa, there is a potential for tremendous efficiency gains during the inductive stage of an empirical analysis, in particular when theory is of little guidance or when the object of study is at the interface of multiple disciplines. It is also nothing short of revolutionary to see how in recent years, spatial analysis and the use of GIS methods have become increasingly accepted in mainstream social science.'

An obvious measure of fragmentation is simply the number of patches. If the representation is vector this will be the number of polygons, whereas if it is raster it will be computed by comparing neighboring cells to see if they have the same class. But as argued in Section 6.2.1, landscapes rarely possess well-defined natural units, so it is important that methods for measuring fragmentation take into account the fundamental uncertainty associated with concepts such as *patch*. An area can be subdivided into a given number of patches in a vast number of distinct ways and so measures in addition to the simple count of patches are needed. One simple measure is the average shape of patches, calculated using the shape measures discussed in Section 14.3.2. A somewhat more informative indicator would be a measure of average shape for each class, since we might expect shape to vary by class – the urban class might be more fragmented than the forest class.

The size distribution of patches might also be a useful indicator. Average patch size is uninformative, since it is simply the total area divided by the number of patches, and is independent of their shapes, or of the relative abundances of large and small patches. But a histogram of patch size would show the relative abundances and can be a useful basis for inferences.

The size distribution of patches is a useful indicator of fragmentation.

Summary statistics of the common boundaries of patches are also useful. The concept of fractals was introduced in Section 4.8 as a way of summarizing the relationship between apparent length and level of geographic detail, in the form of the fractional dimension D. Lines that are smooth tend to have fractional dimensions close to one, and more contorted lines have higher values. So fractional dimension has become accepted as

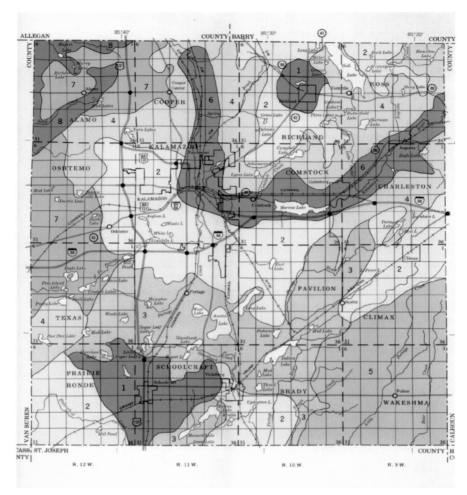

Figure 15.13 A soil map of Kalamazoo County, Michigan, USA. Soil maps such as this one show areas of uniform soil class separated by sharp boundaries (Courtesy: US Department of Agriculture)

a useful descriptive summary of the complexity of geographic objects, and tools for its evaluation have been built into many widely adopted packages. FRAGSTATS is an example – it was developed by the US Department of Agriculture for the purposes of measuring fragmentation using many alternative methods.

15.3 Optimization

The concept of using spatial analysis for design was introduced earlier in Section 15.2.1. The basic idea is to analyze patterns not for the purpose of discovering anomalies or testing hypotheses about process, as in previous sections, but with the objective of creating improved designs. These objectives might include the minimization of travel distance, or the costs of construction of some new development, or the maximization of someone's profit. The three principles of retailing are often

said to be *location, location,* and *location,* and over the years many GIS applications have been directed at problems that involve in one way or another the search for optimum designs. Several of the methods described in Section 15.2.1, including centers, were shown to have useful design-oriented properties. This section includes a discussion of a wider selection of these so-called *normative* methods, or methods developed for application to the solution of practical problems of design.

Normative methods apply well-defined objectives to the design of systems.

Design methods are often implemented as components of systems built to support decision making – so-called *spatial-decision support systems*, or SDSS. Complex decisions are often contentious, with many *stakeholders* interested in the outcome and arguing for one position or another. SDSS are specially adapted GIS that can be used during the decision-making process, to provide instant feedback on the implications of various proposals and the evaluation of 'what-if' scenarios. SDSS typically have

(A)

(B)

(C)

Figure 15.14 Three images of part of the state of Rondonia in Brazil, for 1975, 1986, and 1992, showing the expansion of settlement. Fragmentation statistics can be used to summarize the shapes of areas, their relative sizes, and relative positions (Courtesy: US Geological Survey)

special user interfaces that present only those functions relevant to the application.

The methods discussed in this section fall into several categories. The next section discusses methods for the optimum location of points and extends the method introduced earlier for the MAT. The second section discusses routing on a network and its manifestation in the *traveling-salesman problem* (TSP). The final section examines the selection of optimum paths across continuous space, for locating such facilities as power lines and highways, and for routing military vehicles such as tanks. The methods also divide between those that

are designed to locate points and routes on networks and those designed to locate points and routes in continuous space without respect to the existence of roads or other transportation links.

15.3.1 Point location

The MAT problem is an instance of location in continuous space and finds the location that minimizes total distance with respect to a number of points. The analogous problem on a network would involve finding that location on the network that minimizes total distance to a number of points, also located on the network, using routes that are constrained to the network. Figure 15.15 shows the contrast between continuous and network views and Chapter 7 discusses data models for networks.

A very useful theorem first proved by Louis Hakimi reduces the complexity of many location problems on networks. Figure 15.15B shows a typical basis for network location. The links of the network come together in *nodes*. The weighted points are also located on the network, and also form nodes. For example, the task might be to find the location that minimizes total distance to a distribution of customers, with all customers aggregated into these weighted points. In this case the weights would be counts of customers. The Hakimi theorem proves that for this problem of minimizing distance the only locations that have to be considered are the nodes – it is impossible for the optimum location to be anywhere else. It is easy to see why this should be so. Think of a trial point located in the middle of a link, away from any node. Think of moving it slightly in one direction along the link. This moves it towards some weights, and away from others, but every unit of movement results in the same increase or decrease in total weighted distance. In other words, the total distance traveled to the location is a linear function of the location along the link. Since the function is linear it cannot have a minimum mid-link, so the minimum must occur at a node.

Optimum location problems can be solved in either discrete or continuous space, depending largely on scale.

In Section 15.2 the MAT in one dimension was shown to be the location such that half of the weight is to the left and half to the right – in other words, the measure of central tendency for ordinal data discussed in Section 15.2 and known as the *median*. The MAT problem on a network is consequently known as the *1-median* problem. The *p-median* problem seeks optimum locations for any number *p* of central facilities such that the sum of the distances between each weight and the *nearest* facility is minimized. A typical practical application of this problem is in the location of central public facilities, such as libraries, schools, or agency offices, when the objective is to locate for maximum total accessibility.

Many problems of this nature have been defined for different applications, and implemented in GIS. While the median problems seek to minimize total distance, the *coverage* problems seek to minimize the *furthest* distance

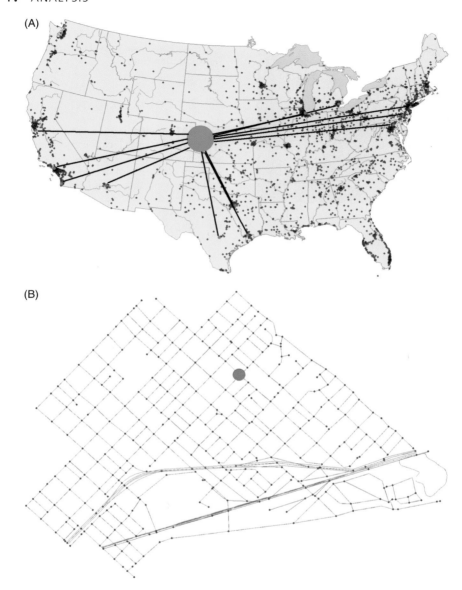

Figure 15.15 Search for the best locations for a central facility to serve dispersed customers. In (A) the problem is solved in continuous space, with straight-line travel, for a warehouse to serve the twelve largest US cities. In continuous space there is an infinite number of possible locations for the site. In (B) a similar problem is solved at the scale of a city neighborhood on a network, where Hakimi's theorem states that only junctions (nodes) in the network and places where there is weight need to be considered, making the problem much simpler, but where travel must follow the street network

traveled, on the grounds that dealing with the worst case of accessibility is often more attractive than dealing with average accessibility. For example, it may make more sense to a city fire department to locate so that a response is possible to *every* property in less than five minutes, than to worry about minimizing the *average* response time. Coverage problems find applications in the location of emergency facilities, such as fire stations (Figure 15.16), where it is desirable that every possible emergency be covered within a fixed number of minutes of response time, or when the objective is to minimize the worst-case response time, to the furthest possible point.

All of these problems are referred to as *location-allocation* problems, because they involve two types of decisions – where to *locate*, and how to *allocate* demand for service to the central facilities. A typical location-allocation problem might involve the selection of sites for supermarkets. In some cases, the allocation of demand to sites is controlled by the designer, as it is in the case of school districts when students have no choice of schools. In other cases, allocation is a matter of choice, and good designs depend on the ability to predict how consumers will choose among the available options. Models that make such predictions are known as *spatial*

Figure 15.16 GIS can be used to find locations for fire stations that result in better response times to emergencies

interaction models, and their use is an important area of GIS application in market research.

> **Location-allocation involves two types of decisions – where to locate, and how to allocate demand for service.**

15.3.2 Routing problems

Point-location problems are concerned with the design of fixed locations. Another area of optimization is in routing and scheduling, or decisions about the optimum tracks followed by vehicles. A commonly encountered example is in the routing of delivery vehicles. In these examples there is a base location, a depot that serves as the origin and final destination of delivery vehicles, and a series of stops that need to be made. There may be restrictions on the times at which stops must be made. For example, a vehicle delivering home appliances may be required to visit certain houses at certain times, when the residents are home. Vehicle routing and scheduling solutions are used by parcel delivery companies, school buses, on-demand public transport vehicles, and many other applications (see Boxes 15.4 and 15.5).

Underlying all routing problems is the concept of the *shortest path*, the path through the network between a defined origin and destination that minimizes distance, or some other measure based on distance, such as travel time. Attributes associated with the network's links, such as length, travel speed, restrictions on travel direction, and level of congestion are often taken into account. Many people are now familiar with the routine solution of the shortest path problem by websites like MapQuest.com (Section 1.5.3), which solve many millions of such problems per day for travelers, and by similar technology used by in-vehicle navigation systems. They use standard algorithms developed decades ago, long before the advent of GIS. The path that is strictly shortest is often not suitable, because it involves too many turns or uses too many narrow streets, and algorithms will often be programmed to find longer routes that use faster highways, particularly freeways. Routes in Los Angeles, USA, for example, can often be caricatured as: 1) shortest route from origin to nearest freeway; 2) follow freeway network; 3) shortest route from nearest freeway to destination, even though this route may be far from the shortest.

The simplest routing problem with multiple destinations is the so-called *traveling-salesman problem* or TSP. In this problem, there are a number of places that must be visited in a tour from the depot and the distances between pairs of places are known. The problem is to select the best tour out of all possible orderings, in order to minimize the total distance traveled. In other words, the optimum is to be selected out of the available tours. If there are n places to be visited, including the depot, then there are $(n-1)!$ possible tours (the symbol ! indicates the product of the integers from 1 up to and including the number, known as the number's *factorial*), but since it is irrelevant whether any given tour is conducted in a forwards or backwards direction the effective number of options is $(n-1)!/2$. Unfortunately this number grows very rapidly with the number of places to be visited, as Table 15.1 shows.

> **A GIS can be very effective at solving routing problems because it is able to examine vast numbers of possible solutions quickly.**

The TSP is an instance of a problem that becomes quickly unsolvable for large n. Instead, designers adopt procedures known as *heuristics*, which are algorithms designed to work quickly, and to come close to providing

Table 15.1 The number of possible tours in a traveling-salesman problem

Number of places to visit	Number of possible tours
3	1
4	3
5	12
6	60
7	360
8	2520
9	20160
10	181440

Routing service technicians for Schindler Elevator.

Schindler Elevator Corporation is a US leader in elevator maintenance. It uses a GIS to plan efficient routes for preventative-maintenance technicians. Service work for each technician is scheduled in daily routes, with emphasis on geographically compact areas of operation to allow quick response in case of emergency calls (people stuck in elevators need a quick response). Each building is visited periodically so that the technician can inspect, grease, clean, and perform periodic maintenance tasks. Additionally, visits are constrained by customers who, for example, do not want elevators out of service during the busiest times.

Schindler's GIS, named Planning Assistant for Superintendent Scheduling (PASS), went into operation in mid-1999 (Figure 15.17). It creates optimum routes, based on multiple criteria, for a number of technicians assigned to one of

Schindler's 140 offices in the USA. As well as assigning visits for multiple technicians, based on maintenance periods and customer preference, it is able to consider many other factors such as technician skill sets and permissible work hours (including overtime), legally binding break times, visit duration, and contract value. Much of this information is stored in Schindler's central SAP/R3 Enterprise Resource Planning (ERP) system, which must be queried to obtain the data before each routing operation.

The system has had a major impact on Schindler's business, saving them several millions of dollars annually (for an initial outlay of around $1 million). It has also allowed them to improve overall operational efficiency by restructuring offices based on routing requirements.

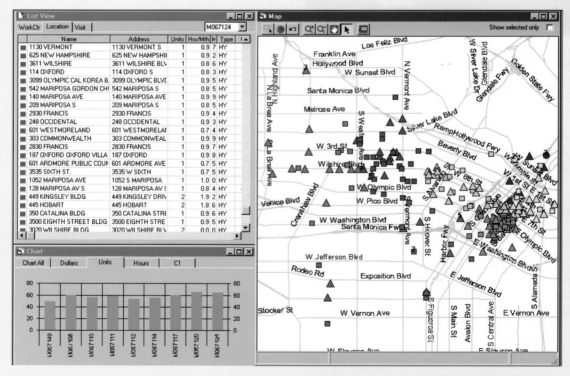

Figure 15.17 A daily routing and scheduling solution for Schindler Elevator. Each symbol on the map identifies a service stop, at the addresses listed in the table on the left

Sears Smart Toolbox Mapping software

Sears, Roebuck and Company uses GIS navigational software in 12 000 Sears Home Services vehicles. The mapping application (based on ESRI ArcObjects technology) allows technicians to review daily scheduled routes, modify routes, and receive both audible and visible turn-by-turn directions while driving to a service location. The application runs on a dockable ruggedized touch-screen laptop (Figure 15.18) and communicates to a roof-mounted Mobile Base Station device that provides GPS information as well as satellite and terrestrial wireless data communication capabilities. The real-time GPS location information allows the application to track the vehicle along the optimized route and to reroute the vehicle dynamically when off route. The wireless connection allows the navigation application to download and upload service orders automatically throughout the technician's work day.

The use of GIS in the vehicles provides a tremendous business benefit to Sears. Prior to using this system a technician would need to look up a street name on a paper map and visually determine the best route approximately 8–10 times per day. The GIS eliminates this task from a technician's daily routine and reduces the chance of getting lost on the way to the stop, resulting in significant time and cost savings for Sears. In addition to this, by calculating the most efficient route, the system helps to minimize transportation distances and times, thus realizing additional savings. Along with reduced costs, the increased efficiency has improved customer service, as technicians are now more likely to meet their scheduled customer time windows. However, if a technician is likely to miss a time window the client-side application has the ability to upload a stop and send it to a server where it may be reassigned to another technician. Conversely, if a technician has extra time in his or her day, and a high-priority service call comes in, a new stop may be wirelessly downloaded to the vehicle and added as a stop.

Figure 15.18 Screen shot from the in-vehicle Smart Toolbox Mapping software installed on 12 000 Sears service vehicles, showing the optimum route to the next four stops to be made by this vehicle in Chicago

the best answer, while not guaranteeing that the best answer will be found. One not very good heuristic for the TSP is to proceed always to the closest unvisited destination, and finally to return to the start. Many spatial optimization problems, including location-allocation and routing problems, are solved today by the use of sophisticated heuristics, and a heuristic approach was used to derive the solution shown in Figure 15.17.

There are many ways of generalizing the TSP to match practical circumstances. Often there is more than one vehicle and in these situations the division of stops between vehicles is an important decision variable – which driver should cover which stops? In other situations it is not essential that every stop be visited, and instead the problem is one of maximizing the rewards associated with visiting stops, while at the same time minimizing the total distance traveled. The is known as the *orienteering* problem, and it matches the situation faced by the driver of a mobile snack bar who must decide which of a number of building sites to visit during the course of a day. ESRI's ArcLogistics and Caliper Corp's TransCAD (**www.caliper.com**) are examples of packages built on GIS platforms to solve many of these everyday problems using heuristics.

15.3.3 Optimum paths

The last example in this section concerns the need to find paths across continuous space, for linear facilities like highways, power lines, and pipelines. Locational decisions are often highly contentious, and there have been many instances of strong local opposition to the location of high-voltage transmission lines (which are believed by many people to cause cancer, although the scientific evidence on this issue is mixed). Optimum paths across continuous space are also needed by the military, in routing tanks and other vehicles, and there has been substantial investment by the military in appropriate technology. They are used by shipping companies to minimize travel time and fuel costs in relation to currents. Finally, continuous-space routing problems are routinely solved by the airlines, in finding the best tracks for aircraft across major oceans that minimize time and fuel costs, by taking local winds into account. Aircraft flying the North Atlantic typically follow a more southerly route from west to east, and a more northerly route from east to west (see Figure 14.11).

A GIS can find the optimum path for a power line or highway, or for an aircraft flying over an ocean against prevailing winds.

Practical GIS solutions to this problem are normally solved on a raster. Each cell is assigned a *friction* value, equal to the cost or time associated with moving across the cell in the horizontal or vertical directions (compare Figure 14.18). The user selects a *move set*, or a set of allowable moves between cells, and the solution is expressed in the form of a series of such moves. The simplest possible move set is the *rook's case*, named for the allowed moves of the rook (castle) in the game of

chess. But in practice most software uses the *queen's case*, which allows moves in eight directions from each cell rather than four. Figure 15.19 shows the two cases. When a diagonal move is made the associated cost or time must be multiplied by a factor of 1.414 (i.e., $\sqrt{2}$) to account for its extra length.

Friction values can be obtained from a number of sources. Land-cover data might be used to differentiate between forest and open space – a power line might be given a higher cost in forest because of the environmental impact (felling of forest), while its cost in open space might be allocated based on visual impact. Elaborate methods have been devised for estimating costs based on many GIS layers and appropriate weighting factors. Real financial costs might be included, such as the costs of construction or of land acquisition, but most of the costs are likely to be based on intangibles, and consequently hard to measure effectively.

Economic costs of a project may be relatively easy to measure compared to impacts on the environment or on communities.

Given a cost layer, together with a defined origin and destination and a move set, the GIS is then asked to find the least-cost path. If the number of cells in the raster is very large the problem can be computationally time-consuming, so heuristics are often employed. The simplest is to solve the problem hierarchically, first on a coarsely aggregated raster to obtain an approximate corridor, and then on a detailed raster restricted to the corridor.

There is an interesting paradox with this problem that is often overlooked in GIS applications, but is nevertheless significant and provides a useful example of the difference between interval and ratio measurements (Box 3.3). Figure 15.20 shows a typical solution to a routing problem. Since the task is to minimize the sum of costs, it is clear that the costs or friction values in each cell need to be measured on interval or ratio scales. But is interval measurement sufficient, or is ratio measurement required – in other words, does there need to be an absolute zero to the scale? Many studies of optimum routing have been done using scales that clearly did not have an absolute zero. For example, people along the route

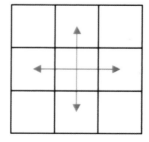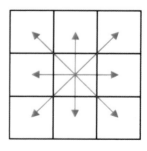

Figure 15.19 The rook's-case (left) and queen's-case (right) move sets, defining the possible moves from one cell to another in solving the problem of optimum routing across a friction surface

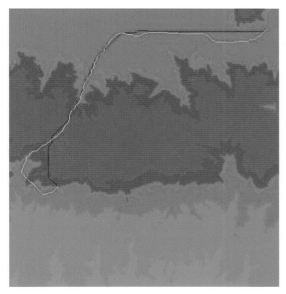

Figure 15.20 Solution of a least-cost path problem. The white line represents the optimum solution, or path of least total cost, across a friction surface represented as a raster, and using the queen's-case move set (eight possible moves). Brown denotes the highest friction. The blue line is the solution when cells are aggregated in blocks of 3 by 3, and the problem is solved on this coarser raster (Reproduced by permission of Ashton Shortridge)

might have been surveyed and asked to rate impact on a scale from 1 to 5. It is questionable whether such as scale has interval properties (are 1 and 2 as different on the scale as 2 and 3, or 4 and 5?), let alone ratio properties (there is no absolute zero and no sense that a score of 4 is twice as much of anything as a score of 2).

But why is this relevant to the problem of route location? Consider the path shown in Figure 15.20. It passes through 456 cells and includes 261 rook's-case moves and 195 diagonal moves. If the entire surface were raised by adding 1.0 to every score, the total cost of the optimum path would rise by 536.7, allowing for the greater length of the diagonal moves. But the cost of a straight path, which requires only 60 rook's-case and 216 diagonal moves, would rise by only 365.4. In other words, changing the zero point of the scale can change the optimum – and in principle the cost surface needs to be measured on a ratio scale. Interval and ordinal scales are not sufficient, despite the fact that they are frequently used.

15.4 Hypothesis testing

This last section reviews a major area of statistics – the testing of hypotheses and the drawing of inferences – and its relationship to GIS and spatial analysis. Much work in statistics is *inferential* – it uses information obtained from samples to make general conclusions about a larger

population, on the assumption that the sample came from that population. The concept of inference was introduced in Section 4.4, as a way of reasoning about the properties of a larger group from the properties of a sample. At that point several problems associated with inference from geographic data were raised, and this section revisits and elaborates on that topic and discusses the particularly thorny issue of hypothesis testing.

For example, suppose we were to take a random and independent sample of 1000 people, and ask them how they might vote in the next election. By *random and independent*, we mean that every person of voting age in the general population has an equal chance of being chosen and the choice of one person does not make the choice of any others – parents, neighbors – more or less likely. Suppose that 45% of the sample said they would support George W. Bush. Statistical theory then allows us to give 45% as the best estimate of the proportion who would vote for Bush among the general population, and also allows us to state a *margin of error*, or an estimate of how much the true proportion among the population will differ from the proportion among the sample. A suitable expression of margin of error is given by the 95% confidence limits, or the range within which the true value is expected to lie 19 times out of 20. In other words, if we took 20 different samples, all of size 1000, there would be a scatter of proportions, and 19 out of 20 of them would lie within these 95% confidence limits. In this case, a simple analysis using the *binomial distribution* shows that the 95% confidence limits are 3% – in other words, the true proportion lies between 42% and 48% 19 times out of 20.

This example illustrates the *confidence limits* approach to inference, in which the effects of sampling are expressed in the form of uncertainty about the properties of the population. An alternative, very commonly used in scientific reasoning, is the *hypothesis-testing* approach. In this case, our objective is to test some general statement about the population – for example, that 50% will support Bush in the next election (and 50% will support the other candidate – in other words, there is no real preference in the electorate). We take a sample and then ask whether the evidence from the sample supports the general statement. Because there is uncertainty associated with any sample, unless it includes the entire population, the answer is never absolutely certain. In this example and using our confidence limits approach, we know that if 55% were found to support Bush in the sample, and if the margin of error was 3%, it is highly unlikely that the true proportion in the population is as low as 50%. Alternatively, we could state the 50% proportion in the population as a *null hypothesis* (we use the term *null* to reflect the absence of something, in this case a clear choice), and determine how frequently a sample of 1000 from such a population would yield a proportion as high as 55%. The answer again is very small; in fact the probability is 0.0008. But it is not zero, and its value represents the chance of making an error of inference – of rejecting the hypothesis when in fact it is true.

Methods of inference reason from information about a sample to more general information about a larger population.

These two concepts – confidence limits and inferential tests – are the basis for statistical testing and form the core of introductory statistics texts. There is no point in reproducing those introductions here, and the reader is simply referred to them for discussions of the standard tests – the F, t, χ^2, etc. The focus here is on the problems associated with using these approaches with geographic data, in a GIS context. Several problems were already discussed in Section 4.7 as violations of the assumptions of inference from regression. The next section reviews the inferential tests associated with one popular descriptive statistic for spatial data, the Moran index of spatial dependence that was discussed in Section 4.6. The following section discusses the general issues and points to ways of resolving them.

15.4.1 Hypothesis tests on geographic data

Although inferential tests are standard practice in much of science, they are very problematic for geographic data. The reasons have to do with fundamental properties of geographic data, many of which were introduced in Chapter 4, and others have been encountered at various stages in this book.

First, many inferential tests propose the existence of a population, from which the sample has been obtained by some well-defined process. We saw in Section 4.4 how difficult it is to think of a geographic dataset as a sample of the datasets that might have been. It is equally difficult to think of a dataset as a sample of some larger area of the Earth's surface, for two major reasons.

First, the samples in standard statistical inference are obtained independently (Section 4.4). But a geographic dataset is often *all there is* in a given area – it *is* the population. Perhaps we could regard a dataset as a sample of a larger area. But in this case the sample would not have been obtained randomly – instead, it would have been obtained by systematically selecting all cases within the area of interest. Moreover, the samples would not have been independent. Because of spatial dependence, which we have understood to be a pervasive property of geographic data, it is very likely that there will be similarities between neighboring observations.

A GIS project often analyzes all the data in a given area, rather than a sample.

Figure 15.21 shows a typical instance of sampling topographic elevation. The value obtained at B could have been estimated from the values obtained at the neighboring points A and C – and this ability to estimate is of course the entire basis of spatial interpolation. We cannot have it both ways – if we believe in spatial interpolation, we cannot at the same time believe in independence of geographic samples, despite the fact that this is a basic assumption of statistical tests.

Finally, the issue of spatial *heterogeneity* also gets in the way of inferential testing. The Earth's surface is highly variable and there is no such thing as an average

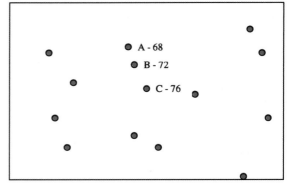

Figure 15.21 Spatial dependence in observations. The value at B, which was measured at 72, could easily have been interpolated from the values at A and C – any spatial interpolation technique would return a value close to 72 at this point

place on it. The characteristics observed on one map sheet are likely substantially different from those on other map sheets, even when the map sheets are neighbors. So the census tracts of a city are certainly not acceptable as a random and independent sample of all census tracts, even the census tracts of an entire nation. They are not independent, and they are not random. Consequently it is very risky to try to infer properties of *all* census tracts from the properties of all of the tracts in any *one* area. The concept of sampling, which is the basis for statistical inference, does not transfer easily to the spatial context.

The Earth's surface is very heterogeneous, making it difficult to take samples that are truly representative of any large region.

Before using inferential tests on geographic data, therefore, it is advisable to ask two fundamental questions:

- Can I conceive of a larger *population* about which I want to make inferences?
- Are my data acceptable as a *random* and *independent* sample of that population?

If the answer to either of these questions is *no*, then inferential tests are not appropriate.

Given these arguments, what options are available? One strategy that is sometimes used is to discard data until the proposition of independence becomes acceptable – until the remaining data points are so far apart that they can be regarded as essentially independent. But no scientist is happy throwing away good data.

Another approach is to abandon inference entirely. In this case the results obtained from the data are descriptive of the study area and no attempt is made to generalize. This approach, which uses local statistics to observe the *differences* in the results of analysis over space, represents an interesting compromise between the nomothetic and idiographic positions outlined in Section 1.3. Generalization is very tempting, but the heterogeneous nature of the Earth's surface makes it very difficult. If generalization is required, then it can be accomplished by appropriate experimental design – by replicating the study in a sufficient number of distinct areas to warrant confidence in a generalization.

Another, more successful, approach exploits the special nature of spatial analysis and its concern with detecting pattern. Consider the example in Figure 15.10, where the Moran index was computed at +0.4011, an indication that high values tend to be surrounded by high values, and low values by low values – a positive spatial autocorrelation. It is reasonable to ask whether such a value of the Moran index could have arisen by chance, since even a random arrangement of a limited number of values typically will not give the theoretical value of zero corresponding to no spatial dependence (actually the theoretical value is very slightly negative). In this case, 51 values are involved, arranged over the 51 features on the map (the 50 states plus the District of Columbia; Hawaii and Alaska are not shown). If the values were arranged randomly, how far would the resulting values of the Moran index differ from zero? Would they differ by as much as 0.4011? Intuition is not good at providing answers.

In such cases a simple test can be run by simulating random arrangements. The software used to prepare this illustration includes the ability to make such simulations, and Figure 15.22 shows the results of simulating 999 random rearrangements. It is clear that the actual value is well outside the range of what is possible for random arrangements, leading to the conclusion that the apparent spatial dependence is real.

How does this *randomization* test fit within the normal framework of statistics? The null hypothesis being evaluated is that the distribution of values over the 51 features is random, each feature receiving a value that is independent of neighboring values. The population is the set of all possible arrangements, of which the data represent a sample of one. The test then involves comparing the test statistic – the Moran index – for the

actual pattern against the distribution of values produced by the null hypothesis. So the test is a perfect example of standard hypothesis testing, adapted to the special nature of spatial data and the common objective of discovering pattern.

Randomization tests are uniquely adapted to testing hypotheses about spatial pattern.

Finally, a large amount of research has been devoted to devising versions of inferential tests that cope effectively with spatial dependence and spatial heterogeneity. Software that implements these tests is now widely available and interested readers are urged to consult the appropriate sources. GeoDa is an excellent comprehensive software environment for such tests (see **www.csiss.org** under Spatial Tools), and many of these methods are available as extensions of standard GIS packages.

15.5 Conclusion

This chapter has covered the conceptual basis of many of the more sophisticated techniques of spatial analysis that are available in GIS. The last section in particular raised some fundamental issues associated with applying methods and theories that were developed for non-spatial data to the spatial case. Spatial analysis is clearly not a simple and straightforward extension of non-spatial analysis, but instead raises many distinct problems, as well as some exciting opportunities. The two chapters on spatial analysis have only scratched the surface of this large and rapidly expanding field.

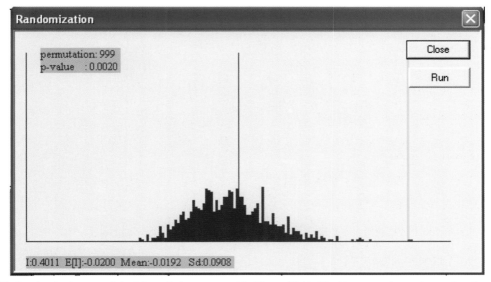

Figure 15.22 Randomization test of the Moran index computed in Figure 15.10. The histogram shows the results of computing the index for 999 rearrangements of the 51 values on the map (Hawaii and Alaska are not shown). The yellow line on the right shows the actual value, which is clearly very unlikely to occur in a random arrangement, reinforcing the conclusion that there is positive spatial dependence in the data

Questions for further study

1. Parks and other conservation areas have geometric shapes that can be measured by comparing park perimeter length to park area, using the methods reviewed in this chapter. Discuss the implications of shape for park management, in the context of a) wildlife ecology and b) neighborhood security.

2. What exactly are *multicriteria* methods? Examine one or more of the methods in the Eastman chapter referenced below, summarizing the issues associated with a) measuring variables to support multiple criteria, b) mixing variables that have been measured on different scales (e.g., dollars and distances), c) finding solutions to problems involving multiple criteria.

3. Besides being the basis for useful measures, fractals also provide interesting ways of simulating geographic phenomena and patterns. Browse the Web for sites that offer fractal simulation software, or investigate one of many commercially available packages. What other uses of fractals in GIS can you imagine?

4. Every point on the Earth's surface has an antipodal point – the point that would be reached by drilling an imaginary hole straight through the Earth's center. Britain, for example, is approximately antipodal to New Zealand. If one third of the Earth's surface is land, you might expect that one third of all of the land area would be antipodal to points that are also on land, but a quick look at an atlas will show that the proportion is actually far less than that. In fact, the only substantial areas of land that have antipodal land are in South America (and their antipodal points in China). How is spatial dependence relevant here, and why does it suggest that the Earth is not so surprising after all?

Further reading

Cliff A.D. and Ord J.K. 1973 *Spatial Autocorrelation*. London: Pion.

Eastman J.R. 1999 'Multicriteria methods.' In Longley P.A., Goodchild M.F., Maguire D.J., and Rhind D.W. (eds) *Geographical Information Systems: Principles, Techniques, Management and Applications (abridged edition (2005))*. Hoboken, NJ: Wiley.

Fotheringham A.S., Brunsdon C., and Charlton M. 2000 *Quantitative Geography: Perspectives on Spatial Data Analysis*. London: Sage.

Fotheringham A.S. and O'Kelly M.E. 1989 *Spatial Interaction Models: Formulations and Applications*. Dordrecht: Kluwer.

Ghosh A. and Rushton G. (eds) 1987 *Spatial Analysis and Location-Allocation Models*. New York: Van Nostrand Reinhold.

Mardia K.V. and Jupp P.E. 2000 *Directional Statistics*. New York: Wiley.

16 *Spatial modeling with GIS*

Models are used in many different ways, from simulations of how the world works, to evaluations of planning scenarios, to the creation of indicators of suitability or vulnerability. In all of these cases the GIS is used to carry out a series of transformations or analyses, either at one point in time or at a number of intervals. This chapter begins with the necessary definitions and presents a taxonomy of models, with examples. It addresses the difference between analysis, the subject of Chapters 14 and 15, and this chapter's subject of modeling. The alternative software environments for modeling are reviewed, along with capabilities for cataloging and sharing models, which are developing rapidly. The chapter ends with a look into the future of modeling and associated GIS developments.

Geographic Information Systems and Science, 2nd edition   Paul Longley, Michael Goodchild, David Maguire, and David Rhind.
© 2005 John Wiley & Sons, Ltd.   ISBNs: 0-470-87000-1 (HB); 0-470-87001-X (PB)

Learning Objectives

At the end of this chapter you will:

- Know what modeling means in the context of GIS;

- Be familiar with the important types of models and their applications;

- Be familiar with the software environments in which modeling takes place;

- Understand the needs of modeling, and how these are being addressed by current trends in GIS software.

16.1 Introduction

This chapter identifies many of the distinct types of models supported by GIS and gives examples of their applications. After *system* and *object*, *model* is probably one of the most overworked technical terms in the English language, with many distinct meanings in the context of GIS and even of this book. So first it is important to address the meaning of the term as it is used in this chapter.

> *Model* **is one of the most overworked terms in the English language.**

A clear distinction needs to be made between the *data models* discussed in Chapters 3 and 8 and the spatial models that are the subject of this chapter. A data model is a template for data, a framework into which specific details of relevant aspects of the Earth's surface can be fitted. For example, the raster data model forces all knowledge to be expressed as properties of the cells of a regular grid laid on the Earth. A data model is, in essence, a statement about form or about how the world *looks*, limiting the options open to the data model's user to those allowed by its template. Models in this chapter are expressions of how the world is believed to *work*, in other words they are expressions of process (see Section 1.3 on how these both relate to the science of problem solving). They may include dynamic simulation models of natural processes such as erosion, tectonic uplift, the migration of elephants, or the movement of ocean currents. They may include models of social processes, such as residential segregation or the movements of cars on a congested highway. They may include processes designed by humans to search for optimum alternatives, for example, in finding locations for a new retail store. Finally, they may include simple calculations of indicators

or predictors, such as happens when layers of geographic information are combined into measures of groundwater vulnerability or social deprivation.

The common element in all of these examples is the manipulation of geographic information in multiple stages. In some cases these stages will perform a simple transformation or analysis of inputs to create an output, and in other cases the stages will loop to simulate the development of the modeled system through time, in a series of iterations – only when all the loops or iterations are complete will there be a final output. There will be intermediate outputs along the way, and it is often desirable to save some or all of these in case the model needs to be re-run or parts of the model need to be changed.

All of the models discussed in this chapter are *digital* or *computational* models, meaning that the operations occur in a computer and are expressed in a language of 0s and 1s. In Chapter 3, representation was seen as a matter of expressing geographic form in 0s and 1s – this chapter looks for ways of expressing geographic *process* in 0s and 1s. The term *geocomputation* is often used to describe the application of computational models to geographic problems; it is also the name of a successful conference series.

All of the models discussed in this chapter are also *spatial* models. There are two key requirements of such a model:

1. There is variation across the space being manipulated by the model (an essential requirement of all GIS applications, of course); and

2. The results of modeling change when the locations of objects change – location *matters* (this is also a key requirement of spatial analysis as defined in Section 14.1).

Models do not have to be digital, and it is worth spending a few moments considering the other type, known as *analog*. The term was defined briefly in Section 3.7 as describing a representation, such as a paper map, that is a scaled physical replica of reality. The Varignon Frame discussed in Box 15.1 is an analog model of a design process, the search for the point of minimum aggregate travel. Analog models can be very efficient, and they are widely used to test engineering construction projects and proposed airplanes. They have two major disadvantages relative to digital models, however: they can be expensive to construct and operate, and unlike digital models they are difficult to copy, store, or share. Box 16.1 describes the analog experiments conducted in World War II in support of a famous bombing mission.

> **An analog model is a scaled physical representation of some aspect of reality.**

The level of detail of any analog model is measured by its *representative fraction* (Section 3.7, Box 4.2), the ratio of distance on the model to distance in the real world. Like digital data, computational models do not have a well-defined representative fraction and instead the level of detail is measured as *spatial resolution*, defined as the

Modeling the effects of bombing on a dam

In World War II, the British conceived a scheme for attacking three dams upstream of the strategically important Ruhr valley, the center of the German iron and steel industry. Dr Barnes Wallis envisioned a roughly spherical bomb that would bounce on the water upstream of the dam, settling next to the dam wall before exploding. Tests were conducted on scaled models (Figure 16.1) to prove the concept. This approach assumes that all aspects of the real problem scale equally, so that conclusions reached using the scale model are also true in reality – and because this assumption may be shaky, it is advisable to test the conclusions of the model later with full-scale prototypes. The successful attacks by the Royal Air Force's 617

Squadron were immortalized in a 1951 book (*The Dam Busters* by Paul Brickhill) and in a 1954 movie (*The Dam Busters* starring Richard Todd as Wing Commander Guy Gibson).

With modern computer power, much of the experimental work could be done quickly and cheaply using computational models instead of scaled analog models. It would be necessary first to build a digital representation of the system (the dam, the lake, and the bomb), and then to build a digital model of the processes that occur when a bomb is dropped at a shallow angle onto a lake surface, bounces on the surface, and then explodes underwater in close proximity to a concrete dam wall.

About Us   Plan a Visit   What's On   News   Collections   Exhibitions   Education   Shop

RAF✈MUSEUM

MILESTONES OF FLIGHT
A TIMELINE OF AVIATION HISTORY

ROYAL AIR FORCE MUSEUM
American Foundation

HENDON

You are here: RAF Museum Home / Hendon / Exhibitions / Dam Busters /

Model Dam Experiments

One of the models built across the Bricket Wood Stream in early 1941.
B689

While researching the means to attack Nazi industry, Barnes Wallis was assisted by the Building Research Laboratory.

A team led by Dr Norman Davey built, then blew up, several large model dams to determine the best possible point of detonation to cause maximum destruction.

Unfortunately these tests weren't to everyone's convenience as an Air Ministry press release relates 'allotment holders...were bewildered and annoyed... when a mysterious and sudden onrush of water swept down...and inundated their plots'.

Corporate Events
Research
Photographic Sales
Commercial Photography
Society of Friends
Policy & Performance
Job Vacancies
Support Us
Newsletter
DONATIONS

Figure 16.1 An analog model built in 1941 to validate the 'bouncing bomb' ideas of Dr Barnes Wallis in preparation for the successful Royal Air Force assault on German dams in 1943 (Reproduced by permission of Royal Air Force Museum)

shortest distance over which change is recorded. *Temporal resolution* is also important, being defined as the shortest time over which change is recorded and in the case of many dynamic models corresponding to the time interval between iterations.

Spatial and temporal resolution are critical factors in models. They define what is left out of the model, in the form of variation that occurs over distances or times that are less than the appropriate resolution. They also therefore define one major source of uncertainty in the model's outcomes. Uncertainty in this context can be defined best through a comparison between the model's outcomes and the outcomes of the real processes that the model seeks to emulate. Any model leaves its user uncertain to some degree about what the real world will do – a measure of uncertainty attempts to give that degree of uncertainty an explicit magnitude. Uncertainty has been discussed in the context of geographic data in Chapter 6; its meaning and treatment are discussed in the context of spatial modeling later in this chapter.

Any model leaves its user uncertain to some degree about what the real world will do.

Spatial and temporal resolution also determine the cost of acquiring the data because in general it is more costly to collect a high-resolution representation than a low-resolution one. More observations have to be made and more effort is consumed in making them. They also determine the cost of running the model, since execution time expands as more data have to be processed and more iterations have to be made. One benchmark is of critical importance for many dynamic models: they must run faster than the processes they seek to simulate if the results are to be useful for planning. One expects this to be true of computational models, but in practice the amount of computing can be so large that the model simulation slows to unacceptable speed.

Spatial resolution is a major factor in the cost both of acquiring data for modeling, and of actually running the model.

16.1.1 Why model?

Models are built for a number of reasons. First, a model might be built to support a design process, in which the user wishes to find a solution to a spatial problem that optimizes some objective. This concept was discussed in Section 15.3 in the context of locating both point-like facilities such as retail stores or fire stations, and line-like facilities such as transmission corridors or highways. Often the design process will involve multiple criteria, an issue discussed in Section 16.4.

Second, a model might be built to allow the user to experiment on a replica of the world, rather than on the real thing. This is a particularly useful approach when the costs of experimenting with the real thing are prohibitive, or when unacceptable impacts would

result, or when results can be obtained much faster with a model. Medical students now routinely learn anatomy and the basics of surgery by working with digital representations of the human body, rather than with expensive and hard-to-get cadavers. Humanity is currently conducting an unprecedented experiment on the global atmosphere by pumping vast amounts of CO_2 into it. How much better it would have been if we could have run the experiment on a digital replica, and understood the consequences of CO_2 emissions well in advance.

Experiments embody the notion of *what-if scenarios*, or policy alternatives that can be plugged into a model in order to evaluate their outcomes. The ability to examine such options quickly and effectively is one of the major reasons for modeling (Figure 16.2).

Models allow planners to experiment with 'what-if' scenarios.

Finally, models allow the user to examine dynamic outcomes, by viewing the modeled system as it evolves and responds to inputs. As discussed in the next section, such dynamic visualizations are far more compelling and convincing when shown to stakeholders and the public than descriptions of outcomes or statistical summaries. Scenarios evaluated with dynamic models are thus a very effective way of motivating and supporting debates over policies and decisions.

16.1.2 To analyze or to model?

Section 2.3.4.2 discussed the work of Tom Cova in analyzing the difficulty of evacuating neighborhoods in Santa Barbara, California, USA, in response to an emergency caused, for example, by a rapidly moving wildfire. The analysis provided useful results that could be employed by planners in developing traffic control strategies, or in persuading residents to use fewer cars when they evacuate their homes. But the results are presented in the form of static maps that must be explained to residents, who might not otherwise understand the role of the street pattern and population density in making evacuation difficult.

By contrast, Box 16.2 describes the dynamic simulations developed by Richard Church. These produce outcomes in the form of dramatic animations that are instantly meaningful to residents and planners. The researcher is able to examine what-if scenarios by varying a range of parameters, including the number of vehicles per household in the evacuation, the distribution of traffic control personnel, the use of private and normally gated exit roads, and many other factors. Simulations like these can galvanize a community into action far more effectively than static analysis.

Models can be used for dynamic simulation, providing decision makers with dramatic visualizations of alternative futures.

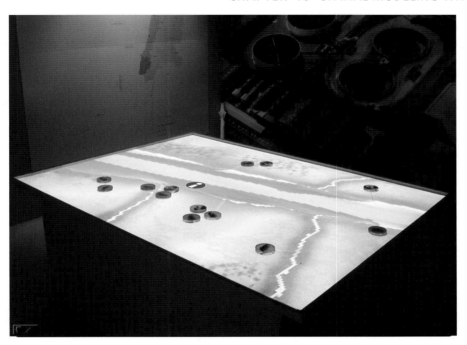

Figure 16.2 A 'live' table constructed by Antonio Câmara and his group (see Box 11.5) as a tool for examining alternative planning scenarios. The image on the table is projected from a computer and users are able to 'move' objects that represent sources of pollution around on the image by interacting directly with the display. Each new position results in the calculation of a hydrologic model of the impacts of the sources, and the display of the resulting pattern of impacts on the table (Reproduced by permission of Antonio Camara/Ydreams)

Biographical Box **16.2**

Richard Church and traffic simulation

Richard Church (Figure 16.3) received his Bachelor of Science degree from Lewis and Clark College. After finishing two majors, one in Math and one in Chemistry, he continued his graduate education in Geography and Environmental Engineering at The Johns Hopkins University where he specialized in water resource systems engineering. Currently, Rick is Professor of Geography at the University of California, Santa Barbara. He continues to be involved in environmental modeling and has specialized in forest systems modeling and ecological reserve design. He also teaches and conducts research in the areas of logistics, transportation, urban planning, and location modeling.

Rick is often motivated by real-world problems, whether it is operating a reservoir, designing an ambulance posting plan for a city, or finding the best corridor for a right-of-way.

One of his recent projects involves simulating the emergency evacuation of a neighborhood in advance of a rapidly moving fire, such as the one in the Oakland Hills in 1991 that destroyed 2900 structures and resulted in 25

Figure 16.3 Richard Church, Professor of Geography, University of California, Santa Barbara

deaths. Figure 16.4 shows the Mission Canyon area of Santa Barbara, a neighborhood that exhibits many of the same characteristics as the Oakland Hills.

Aware of the potential threat, members of the Mission Canyon neighborhood association asked if there was a way in which they could better understand the difficulty they would face in an evacuation. To address this problem Church utilized a micro-scale traffic simulation program called PARAMICS to model a number of evacuation scenarios. Figure 16.5 shows the simulated pattern some minutes into an evacuation, with cars bumper-to-bumper crowding the available exits, and unable to move should the fire advance over them. Through this work, it was evident that the biggest gain in safety was to encourage residents to use one vehicle per household in making a quick exit. Furthermore, traffic control was found to be very important

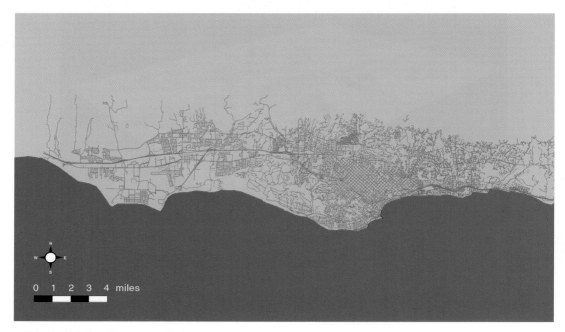

Figure 16.4 The Mission Canyon neighborhood in Santa Barbara, California, USA (highlighted in red), a critical area for evacuation in a wildfire emergency. Note the contorted network of streets in this topographically rugged area

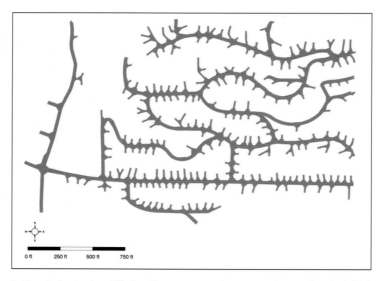

Figure 16.5 Simulation of driver behavior in a Mission Canyon evacuation some minutes after the initial evacuation order. The red dots denote the individual vehicles whose behavior is modeled in this simulation

in decreasing overall clearing time. The County Government has since agreed to pre-position traffic control officers during 'red flag' alerts (times of extreme fire danger) so that the time to respond with traffic control measures is kept as low as possible. The work reported here was sponsored by a grant from the California Department of Transportation (Caltrans).

The simulations using PARAMICS raise an important issue for GIS. The conventional representation of a street in a GIS is as a polyline, a series of points connected by straight lines. However it would be impossible to simulate driver behavior on such a representation because it would be necessary to slow to negotiate the sharp bends at each vertex. Instead PARAMICS uses a representation consisting of a series of circular arcs with no change of direction where arcs join. Integrating PARAMICS with GIS therefore requires transformation software to convert polylines to acceptable representations (Figure 16.6).

Figure 16.6 The representation problem for driver simulation. The green line shows a typical polyline representation found in a GIS. But a simulated driver would have to slow almost to a halt to negotiate the sharp corners. Traffic simulation packages such as PARAMICS require a representation of circular arcs (the black line) with no change of direction where the arcs join (at the points indicated by red lines)

In summary, analysis as described in Chapters 14 and 15 is characterized by:

- A static approach, at one point in time;
- The search for patterns or anomalies, leading to new ideas and hypotheses; and
- Manipulation of data to reveal what would otherwise be invisible.

By contrast, modeling is characterized by:

- Multiple stages, perhaps representing different points in time;
- Implementing ideas and hypotheses; and
- Experimenting with policy options and scenarios.

16.2 Types of model

16.2.1 Static models and indicators

A static model represents a single point in time and typically combines multiple inputs into a single output.

There are no time steps and no loops in a static model, but the results are often of great value as predictors or indicators. For example, the Universal Soil Loss Equation (USLE, first discussed in Section 1.3) falls into this category. It predicts soil loss at a point, based on five input variables, by evaluating the equation:

$$A = R \times K \times LS \times C \times P$$

where A denotes the predicted erosion rate, R is the Rainfall and Runoff Factor, K is the Soil Erodibility Factor, LS is the Slope Length Gradient Factor, C is the Crop/Vegetation and Management Factor, and P is the Support Practice Factor. Full definitions of each of these variables and their measurement or estimation can be found in descriptions of the USLE (see for example **www.co.dane.wi.us/landconservation/uslepg.htm**).

A static model represents a system at a single point in time.

The USLE passes the first test of a spatial model in that many if not all of its inputs will vary spatially when applied to a given area. But it does not pass the second test, since moving the points at which A is evaluated will not affect the results. Why, then, use a GIS to evaluate the USLE? There are three good reasons:

1) because some of the inputs, particularly *LS*, require a GIS for their calculation from readily available data, such as digital elevation models; 2) because the inputs and output are best expressed, visualized, and used in map form, rather than as tables of point observations; and 3) because the inputs and outputs of the USLE are often integrated with other types of data, for further analysis that may require a GIS. Nevertheless, it is possible to evaluate the USLE in a simple spreadsheet application such as Excel, and the website cited in the previous paragraph includes a downloadable Excel macro for this purpose.

Models that combine a variety of inputs to produce an output are widely used, particularly in environmental modeling. The DRASTIC model calculates an index of groundwater vulnerability from input layers by applying appropriate weights to the inputs (Figure 16.7). Box 16.3 describes another application, the calculation of a groundwater protection model, from several inputs in a karst environment (an area underlain by potentially soluble limestone, and therefore having substantial and rapid groundwater flow through cave passages). It uses ESRI's ModelBuilder software which is described later in Section 16.3.1.

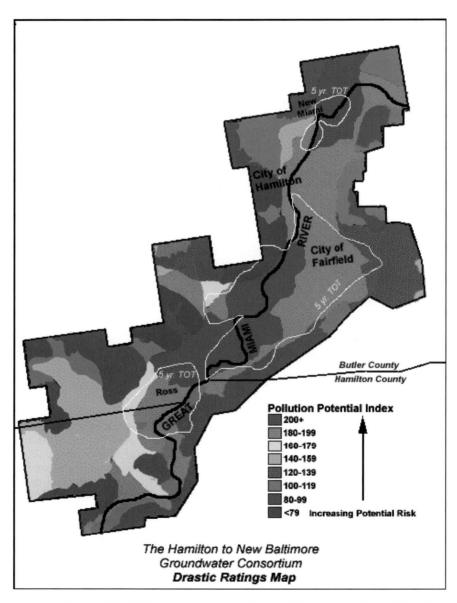

Figure 16.7 The results of using the DRASTIC groundwater vulnerability model in an area of Ohio, USA. The model combines GIS layers representing factors important in determining groundwater vulnerability, and displays the results as a map of vulnerability ratings (Reproduced by permission of Hamilton to New Baltimore Grand Water Consortium Pollution Potential (DRASTIC) map reproduced by permission of Ohio Department of Natural Resources)

16.2.2 Individual and aggregate models

The simulation model described in Box 16.2 works at the individual level, by attempting to forecast the behavior of each driver and vehicle in the study area. By contrast, it would clearly be impossible to model the behavior of every molecule in the Mammoth Cave watershed, and instead any modeling of groundwater movement must be done at an aggregate level, by predicting the movement of water as a continuous fluid. In general, models of physical systems are forced to adopt aggregate approaches because of the enormous number of individual objects involved, whereas it is much more feasible to model individuals in human systems, or in studies of

---

Applications Box **16.3**

Building a groundwater protection model in a Karst environment

Rhonda Pfaff (ESRI staff) and Alan Glennon (a doctoral candidate at the University of California, Santa Barbara) describe a simple but elegant application of modeling to the determination of groundwater vulnerability in Kentucky's Mammoth Cave watershed. Mammoth Cave is protected as a National Park, containing extensive and unique environments, but is subject to potentially damaging runoff from areas in the watershed outside park boundaries and therefore not subject to the same levels of environmental protection. Figure 16.8 shows a graphic rendering of the model in ESRI's ModelBuilder. Each operation is shown as a rectangle and each dataset as an ellipse. Reading from top left, the model first clips the slope layer to the watershed, then selects slopes greater than or equal to 5 degrees.

A landuse layer is analyzed to select fields used for growing crops, and these are then combined with the steep_slopes layer to identify crop fields on steep slopes. A dataset of streams is buffered to 300 m, and finally this is combined to form a layer identifying all areas that are crop fields, on steep slopes, within 300 m of a stream. Such areas are particularly likely to experience soil erosion and to generate runoff contaminated with agricultural chemicals, which will then impact the downstream cave environment with its endangered populations of sightless fish. Figure 16.9 shows the resulting map.

A detailed description of this application is available at **www.esri.com/news/arcuser/0704/files/modelbuilder.pdf**, and the datasets are available at **www.esri.com/news/arcuser/0704/summer2004.html**.

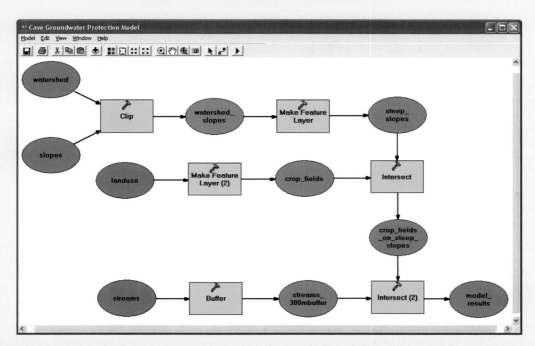

Figure 16.8 Graphic representation of the groundwater protection model developed by Rhonda Pfaff and Alan Glennon for analysis of groundwater vulnerability in the Mammoth Cave watershed, Kentucky, USA

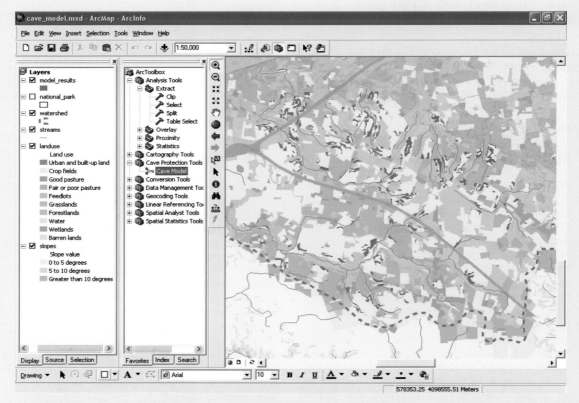

Figure 16.9 Results of the groundwater protection model. Highlighted areas are farmed for crops, on relatively steep slopes and within 300 m of streams. Such areas are particularly likely to generate runoff contaminated by agricultural chemicals and soil erosion and to impact adversely the cave environment into which the area drains

animal behavior. Even when modeling the movement of water as a continuous fluid it is still necessary to break the continuum into discrete pieces, as it is in the representation of continuous fields (see Figure 3.11). Some models adopt a raster approach and are commonly called *cellular* models (see Section 16.2.3). Other models break the world into irregular pieces or polygons, as in the case of the groundwater protection model described in Box 16.3.

Aggregate models are used when it is impossible to model the behavior of every individual element in a system.

Models of individuals are often termed *agent-based* models (ABM) or *autonomous agent* models (Box 16.4). With the massive computing power now available on the desktop, and with techniques of object-oriented programming, it is comparatively easy to build and execute ABMs, even when the number of agents is itself massive. Such models have been used to analyze and simulate many types of animal behavior, as well as the behavior of pedestrians in streets (Box 16.4), shoppers in stores, and drivers on congested roads (Box 16.2). They have also been used to model the behavior of

decision makers, not through their movements in space but through the decisions they make regarding spatial features. For example, ABMs have been constructed to model the decisions made over land use in rural areas, by formulating the rules that govern individual decisions by landowners in response to varying market conditions, policies, and regulations. The impact of these factors on the fragmentation of the landscape, with its implications for wildlife habitat, are of key interest in such models.

16.2.3 Cellular models

Cellular models, which were introduced in Section 2.3.5.1, represent the surface of the Earth as a raster. Each cell in the fixed raster has a number of possible states, which change through time as a result of the application of transition rules. Typically the rules are defined over each cell's neighborhood, and determine the outcome of each stage in the simulation based on the cell's state, the states of its neighbors, and the values of cell attributes. The study of cellular models was first popularized by the work of John Conway, Professor of Mathematics at Princeton University, who studied the properties of a model he called the Game of Life (Box 16.5).

Technical Box **16.4**

Agent-based models of movement in crowded spaces

Working at the University College London Centre for Advanced Spatial Analysis, geographer and planner Michael Batty uses GIS to simulate the disasters and emergencies that can occur when large-scale events generate congestion and panic in crowds (Figure 16.10). The events involve large concentrations of people in small spaces that can arise because of accidents, terrorist attacks, or simply by the build-up of congestion through the convergence of large numbers of people into spaces with too little capacity. He has investigated scenarios for a number of major events, such as the movement of very large numbers of people to Mecca to celebrate the Hajj (a holy event in the Muslim calendar; Figure 16.10) and the Notting Hill Carnival (Europe's largest street festival, held annually in West Central London). His work uses agent-based models that take GIScience to a finer level of granularity and incorporate temporal processes as well as spatial structure. Fundamental to agent-based modeling of such situations is the need to understand how individuals in crowds interact with each other and with the geometry of the local environment.

Batty and his colleagues have been concerned with modeling the behavior of individual human agents under a range of crowding scenarios. This style of model is called agent-based because it depends on simulating the aggregate effect of the behaviors of individual objects or persons in an inductive way (bottom-up: Section 4.9). The modeling is based on averages – our objective is modeling an aggregate outcome (such as the onset of panic) although the particular behavior of each agent is a combination of responses which are based on how routine behavior is modified by the presence of other agents and by the particular environment in which the agent finds himself or herself. It is possible to simulate a diverse range of outcomes, each based on different styles of interaction between individual agents and different geometrical configurations of constraints.

Figure 16.11 shows one of Batty's simulations of the interactions between two crowds. Here, a parade is moving around a street intersection – the central portion of the movement (walkers in white) is the parade and the walkers around this in gray/red are the watchers. This model can be used to simulate the build-up of pressure through random motion which then generates a breakthrough of the watchers into the parade, an event that often leads to disasters of the kind experienced in festivals, rock concerts, and football matches as well as ritual situations like the Hajj.

Figure 16.10 Massive crowds congregate in Mecca during the annual Hajj. On February 1, 2004 244 pilgrims were trampled when panic stampeded the crowd. This was a unique but unfortunately not a freak occurrence: 50 pilgrims were killed in 2002, 35 in 2001, 107 in 1998, and 1425 were killed in a pedestrian tunnel in 1990

▶

(A) (B)

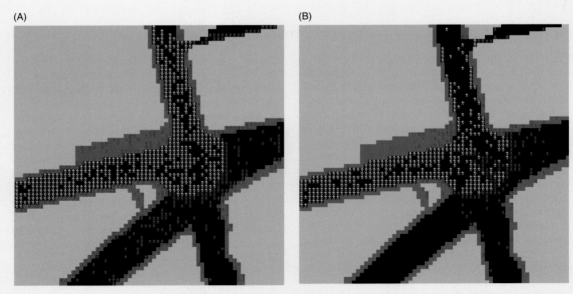

Figure 16.11 Simulation of the movement of individuals during a parade. Parade walkers are in white, watchers in red. The watchers (A) build up pressure on restraining barriers and crowd control personnel, and (B) break through into the parade (Reproduced by permission of Michael Batty)

Cellular models represent the surface of the Earth as a raster, each cell having a number of states that are changed at each iteration by the execution of rules.

Several interesting applications of cellular methods have been identified, and particularly notable are the efforts to apply them to urban growth simulation. The likelihood of a parcel of land developing depends on many factors, including its slope, access to transportation routes, status in zoning or conservation plans, but above all its proximity to other development. These models express this last factor as a simple modification of the rules of the Game of Life – the more developed the state of neighboring cells, the more likely a cell is to make the transition from undeveloped to developed. Figure 16.13 shows an illustration from one such model, that developed by Keith Clarke and his co-workers at the University of California, Santa Barbara, to predict growth patterns in the Santa Barbara area through 2040. The model iterates on an annual basis, and the effects of neighboring states in promoting infilling are clearly evident. The inputs for the model – the transportation network, the topography, and other factors – are readily

The Game of Life

The game is played on a raster. Each cell has two states, *live* and *dead*, and there are no additional cell attributes (no variables to differentiate the space), as there would be in GIS applications. Each cell has eight neighbors (the Queen's case, Figure 15.19). There are three transition rules at each time-step in the game:

1. A dead cell with exactly three live neighbors becomes a live cell;

2. A live cell with two or three live neighbors stays alive; and

3. In all other cases a cell dies or remains dead.

With these simple rules it is possible to produce an amazing array of patterns and movements, and some surprisingly simple and elegant patterns emerge out of the chaos (in the field of agent-based modeling these unexpected and simple patterns are known as *emergent properties*). The page **www.math.com/students/wonders/life/life.html** includes extensive details on the game, examples of particularly interesting patterns, and executable Java code. Figure 16.12 shows three stages in an execution of the Game of Life.

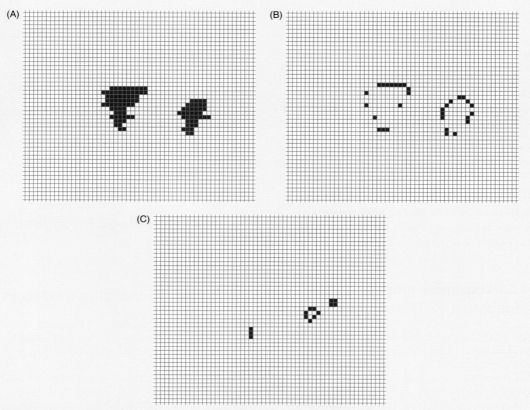

Figure 16.12 Three stages in an execution of the Game of Life: (A) the starting configuration, (B) the pattern after one time-step, and (C) the pattern after 14 time-steps. At this point all features in the pattern remain stable

Figure 16.13 Simulation of future urban growth patterns in Santa Barbara, California, USA. (Upper) growth limited by current urban growth boundary; (Lower) growth limited only by existing parks (Courtesy: Keith Clarke)

prepared in GIS, and the GIS is used to output the results of the simulation.

One of the most important issues in such modeling is calibration and validation – how do we know the rules are right, and how do we confirm the accuracy of the results? Clarke's group has calibrated the model by *brute force*, by running the model forwards from some point in the past under different sets of rules and comparing the results to the actual development history from then to the present. This method is extremely time-consuming since the model has to be run under vast numbers of combinations of rules, but it provides at least some degree of confidence in the results. The issue of accuracy is addressed in more detail, and with reference to modeling in general, later in this chapter.

A model must be calibrated against real data to determine appropriate values for its parameters and to ensure that its rules are valid.

16.2.4 Cartographic modeling and map algebra

In essence, modeling consists of combining many stages of transformation and manipulation into a single whole, for a single purpose. In the example in Box 16.3 the number of stages was quite small – only six – whereas in the Clarke urban growth model several stages are executed many times as the model iterates through its annual steps.

The individual stages can consist of a vast number of options, encompassing all of the basic transformations of which GIS is capable. In Section 14.1.2 it was argued that GIS manipulations could be organized into six categories, depending on the broad conceptual objectives of each manipulation. But the six are certainly not the only way of classifying and organizing the numerous methods of spatial analysis into a simple scheme. Perhaps the most successful is the one developed by Dana Tomlin and known as *cartographic modeling* or *map algebra*. It classifies all GIS transformations of rasters into four basic classes, and is used in several raster-centric GISs as the basis for their analysis languages:

- *local* operations examine rasters cell by cell, comparing the value in each cell in one layer with the values in the same cell in other layers;

- *focal* operations compare the value in each cell with the values in its neighboring cells – most often eight neighbors;

- *global* operations produce results that are true of the entire layer, such as its mean value;

- *zonal* operations compute results for blocks of contiguous cells that share the same value, such as the calculation of shape for contiguous areas of the same land use, and attach their results to all of the cells in each contiguous block.

With this simple schema, it is possible to express any model, such as Clarke's urban growth simulation, as a *script* in a well-defined language. The only constraint is that the model inputs and outputs be in raster form.

Map algebra provides a simple language in which to express a model as a script.

A more elaborate map algebra has been devised and implemented in the PCRaster package, developed for spatial modeling at the University of Utrecht (**pcraster.geog.uu.nl**). In this language, a symbol refers to an entire map layer, so the command $A = B + C$ takes the values in each cell of layers B and C, adds them, and stores the result as layer A.

16.3 Technology for modeling

16.3.1 Operationalizing models in GIS

Many of the ideas needed to implement models at a practical level have already been introduced. In this section they are organized more coherently in a review of the technical basis for modeling.

Models can be defined as sequences of operations, and we have already seen how such sequences can be expressed either as graphic flowcharts or as scripts. One of the oldest graphic platforms for modeling is Stella, now very widely distributed and popular as a means of conceptualizing and implementing models. Although it now has good links to GIS, its approach is conceptually non-spatial, so it will not be described in detail here. Instead, the focus will be on similar environments that are directly integrated with GIS. The first of these may have been ERDAS IMAGINE software (**www.gis.leica-geosystems.com/Products/Imagine/**), which allows the user to build complex modeling sequences from primitive operations, with a focus on the manipulations needed in image processing and remote sensing. ESRI introduced a graphic interface to modeling in ArcView 3.x Spatial Modeler, and enhanced it in ArcGIS 9.0 ModelBuilder (see Figure 16.8).

Any model can be expressed as a script, or visually as a flowchart.

Typically in these interfaces, datasets are represented as ellipses, operations as rectangles, and the sequence of the model as arrows. The user is able to modify and control the operation sequence by interacting directly with the graphic display.

Fully equivalent, but less visual and therefore more demanding on the user, are *scripts*, which express models as sequences of commands, allowing the user to execute an entire sequence by simply invoking a script. Initial efforts to introduce scripting to GIS were cumbersome, requiring the user to learn a product-specific language such as ESRI's AML (Arc Macro Language) or Avenue.

Today, it is common for GIS model scripts to be written in such industry-standard languages as Visual Basic for Applications (VBA), Perl, JScript, or Python. Scripts can be used to execute GIS operations, request input from the user, display results, or invoke other software. With interoperability standards such as Microsoft's .Net (Box 7.3), it is possible to call operations from any compliant package from the same script. For example, a script might combine GIS operations with calls to Excel for simple tabular functions, or even calls to another compliant GIS (Section 7.6).

> **With standards like Microsoft's .Net it is possible for a script to call operations from several distinct software packages.**

16.3.2 Model coupling

The previous section described the implementation of models as direct extensions of an underlying GIS, through either graphic model-building or scripts. This approach makes two assumptions: first, that all of the operations needed by the model are available in the GIS (or in another package that can be called by the model); and second, that the GIS provides sufficient performance to handle the execution of the model. In practice, a GIS will often fail to provide adequate performance, especially with very large datasets and large numbers of iterations, because it has not been designed as a modeling engine. Instead, the user is forced to resort to specialized code, written in a source language such as C. Clarke's model, for example, is programmed in C, and the GIS is used only for preparing input and visualizing output. Other models may be spatial only in the sense of having geographically differentiated inputs and outputs, as discussed in the case of the USLE in Section 16.2.1, making the use of a GIS optional rather than essential.

In reality, therefore, much spatial modeling is done by *coupling* GIS with other software. A model is said to be *loosely coupled* to a GIS when it is run as a separate piece of software and data are exchanged to and from the GIS in the form of files. Since many GIS formats are proprietary, it is common for the exchange files to be in export format, and to rely on translators at either end of the transfer. In extreme cases it may be necessary to write a special program to convert the files during transfer, when no common format is available. Clarke's is an example of a model that is loosely coupled to GIS. A model is said to be *closely coupled* to a GIS when both the model and the GIS read and write the same files, obviating any need for translation. When the model is executed as a GIS script or through a graphic GIS interface it is said to be *embedded*.

16.3.3 Cataloging and sharing models

In the digital computer everything – the program, the data, the metadata – must be expressed in the same language of 0s and 1s. The value of each bit or byte (Box 3.1) depends on its meaning, and clearly some bits are more valuable than others. A bit in Microsoft's operating system Windows XP is clearly extremely valuable, even though there are hundreds of millions of them. On the other hand, a bit in a remote sensing image that happens to have been taken on a cloudy day has almost no value at all, except perhaps to someone studying clouds. In general, one would expect bits of programs to be more valuable than bits of data, especially when those programs are useable by large numbers of people.

This line of argument suggests that the GIS scripts, models, and other representations of process are potentially valuable to many users and well worth sharing. But almost all of the investment society has made in the sharing of digital information has been devoted to data and text. In Section 11.2 we saw how geolibraries, data warehouses, and metadata standards have been devised and implemented to facilitate the sharing of geographic data, and parallel efforts by libraries, publishers, and builders of search engines have made it easy nowadays to share text via the Web. But very little effort to date has gone into making it possible to share *process objects*, digital representations of the process of GIS use.

> **A process object, such as a script, captures the process of GIS use in digital form.**

Notable exceptions include ESRI's ArcScripts, a library of GIS scripts contributed by users and maintained by the vendor as a service to its customers. The library is easily searched for scripts to perform specific types of analysis and modeling. At the time of writing, it included 3286 scripts, ranging from one to couple ArcView 3.x to the StormWater Management Model (SWMM) written in Avenue, to a set of basic tools for crime analysis and mapping. ArcGIS 9.0 includes the ability to save, catalog, and share scripts as GIS functions.

The term *GIService* is often used to describe a GIS function being offered by a server for use by any user connected to the Internet (see Sections 1.5.3 and 7.2). In essence, GIServices offer an alternative to locally installed GIS, allowing the user to send a request to a remote site instead of to his or her GIS. GIServices are particularly effective in the following circumstances:

- When it would be too complex, costly, or time-consuming to install and run the service locally;

- When the service relies on up-to-date data that would be too difficult or costly for the user to constantly update;

- When the person offering the service wishes to maintain control over his or her intellectual property. This would be important, for example, when the function is based on a substantial amount of knowledge and expertise.

To date, several standard GIS functions are available as GIServices, satisfying one or more of the above requirements. The Alexandria Digital Library offers a *gazetteer service* (**www.alexandria.ucsb.edu**), allowing a user to send a placename and to receive in return one or more coordinates of places that match the placename. Several companies offer *geocoding services*, returning coordinates in response to a street address (try **www.geocode.com**). The Geography Network (**www.geographynetwork.com**) provides a searchable catalog to GIServices.

16.4 Multicriteria methods

The model developed by Pfaff and Glennon (depicted in Figure 16.8) rates vulnerability to runoff based on three factors – cropland, slope, and distance from stream – but treats all three as simple binary measures, and produces a simple binary result (land is vulnerable if slope is greater than 5%, land use is cropping, and distance from stream is less than 300 m). In reality, of course, 5% is not a precise cutoff between non-vulnerable and vulnerable slopes, and neither is 300 m a precise cutoff between vulnerable distances and non-vulnerable distances. The issues surrounding rules such as these, and their fuzzy alternatives, were discussed at length in Section 6.2.

A more general and powerful conceptual framework for models like this would be constructed as follows. There are a number of factors that influence vulnerability, denoted by x_1 through x_n. The impact of each factor on vulnerability is determined by a transformation of the factor $f(x)$ – for example, the factor *distance* would be transformed so that its impact *decreases* with increasing distance (as in Section 4.5), whereas the impact of *slope* would be *increasing*. Then the combined impact of all of the factors is obtained by weighting and adding them, each factor i having a weight w_i:

$$I = \sum_{i=1}^{n} w_i f(x_i)$$

In this framework both the functions f and the weights w need to be determined. In the example, the f for slope was resolved by a simple step function, but it seems more likely that the function should be continuously decreasing, as shown in Figure 16.14. The hill-shaped function also shown in the figure would be appropriate in cases where the impact of a factor declines in both directions from some peak value (for example, smoke from a tall smokestack has its greatest impact at some distance downwind).

Many decisions depend on identifying relevant factors and adding their appropriately weighted values.

This approach provides a good conceptual framework both for the indicator models typified by Box 16.3 and for many models of design processes. In both cases, it is possible that multiple views might exist about appropriate functions and weights, particularly when modeling a decision over an important development with impact on the environment. Different *stakeholders* in the design process can be anticipated to have different views

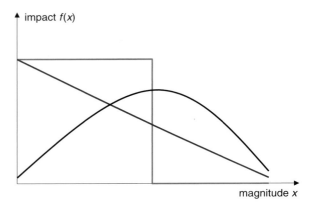

Figure 16.14 Three possible impact functions: (red) the step function used to assess slope in Figure 16.8; (blue) a decreasing linear function; and (black) a function showing impact rising to a maximum and then decreasing

about what is important, how that importance should be measured, and how the various important factors should be combined. Such processes are termed *multi-criteria decision making* or MCDM, and are commonly encountered whenever decisions are controversial. Some stakeholders may feel that environmental factors deserve high weight, others that cost factors are the most important, and still others that impact on the social life of communities is all-important.

An important maxim of MCDM is that it is better for stakeholders to argue in principle about the merits of different factors and how their impacts should be measured, than to argue in practice about alternative decisions. For example, arguing about whether slope is a more important factor than distance, and about how each should be measured, is better than arguing about the relative merits of Solution A, which might color half of John Smith's field red, over Solution B, which might color all of David Taylor's field red. Ideally, all of the controversy should be over once the factors, functions, and weights are decided, and the solution they produce should be acceptable to all, since all accepted the inputs. Would it were so!

Each stakeholder in a decision may have his or her own assessment of the importance of each relevant factor.

Although many GIS have implemented various approaches to MCDM, Clark University's IDRISI offers the most extensive functionality as well as detailed tutorials and examples (**www.clarklabs.org**; see Box 16.6). A common framework is the Analytical Hierarchy Process (AHP) devised by Thomas Saaty, which focuses on capturing each stakeholder's view of the appropriate weights to give to each impact factor. First, the impact of each factor is expressed as a function, choosing from options such as those shown in Figure 16.14. Then each stakeholder is asked to compare each pair of factors (with n factors there are $n(n-1)/2$ pairs), and to assess their relative importance in ratio form. A matrix is

Table 16.1 An example of the weights assigned to three factors by one stakeholder. For example, the entry '7' in Row 1 Column 2 (and the 1/7 in Row 2 Column 1) indicates that the stakeholder felt that Factor 1 (slope) is seven times as important as Factor 2 (land use)

	Slope	Land use	Distance from stream
Slope		7	2
Land use	1/7		1/3
Distance from stream	1/2	3	

created for each stakeholder, as in the example shown in Table 16.1. The matrices are then combined and analyzed, and a single set of weights extracted that represent a consensus view (Figure 16.15). These weights would then be inserted as parameters in the spatial model to produce a final result. The mathematical details of the method can be found in books or tutorials on the AHP.

16.5 Accuracy and validity: testing the model

Models are complex structures and their outputs are often forecasts of the future. How, then, can a model be tested, and how does one know whether its results can be trusted? Unfortunately, there is an inherent tendency to trust the results of computer models because they appear in numerical form (and numbers carry innate authority) and because they come from computers (which also appear authoritative).

Results from computers can seem to carry innate authority.

Normally, scientists test their results against reality, but in the case of forecasts reality is not yet available, and by the time future events have played out there is likely little interest in testing. So modelers must resort to other methods to verify and validate their predictions.

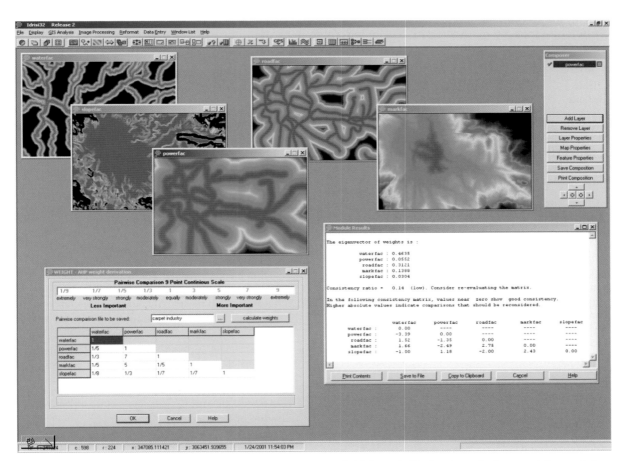

Figure 16.15 Screen shot of an AHP application using IDRISI (**www.clarklabs.org**). The five layers in the upper left part of the screen represent five factors important to the decision. In the lower left, the image shows the table of relative weights compiled by one stakeholder. All of the weights matrices are combined and analyzed to obtain the consensus weights shown in the lower right, together with measures to evaluate consistency among the stakeholders (Reproduced by permission of J. Ronald Eastman)

Biographical Box **16.6**

Ron Eastman, developer of the IDRISI GIS

Figure 16.16 Ron Eastman, developer of IDRISI

J. Ronald Eastman (Figure 16.16) is Director and lead software engineer of Clark Labs, a non-profit center within the George Perkins Marsh Institute of Clark University that produces the IDRISI GIS and image processing system, non-profit software with over 35 000 registered users worldwide. He is also Professor of Geography and former Director of the Graduate School of Geography at Clark University. He gained his Ph.D. in Geography from Boston University in 1982. His research is focused most strongly on the development of analytical procedures for the investigation and modeling of land-cover change and effective resource allocation in environmentally related decision making. He also has extensive experience in the deployment of GIS and remote sensing in the developing world, and is a Senior Special Fellow of the United Nations Institute for Training and Research. He writes: 'With a degree in perceptual psychology and a strong love of travel and the diversity of landscapes and cultures, I became interested in the power of the visual system in facilitating the analysis of geographic phenomena through the medium of maps. My graduate degrees were thus in geography/cartography. However, I had the fortune of witnessing and participating in the birth of GIS and digital image processing. It was immediately clear to me that the analytical potential of this new medium far outstripped what could be achieved by visual analysis alone, particularly in the study of our rapidly changing environment. At the same time, I was also concerned that many in the developing world had only limited access to computing resources and GIS/image processing technology – thus my motivation in developing a system focused on the analytical needs of development and global change.'

Models can often be tested by comparison with past history, by running the model not into the future, but forwards in time from some previous point. But these are often the data used to *calibrate* the model, to determine its parameters and rules, so the same data are not available for testing. Instead, many modelers resort to *cross-validation*, a process in which a subset of data is used for calibration, and the remainder for validating results. Cross-validation can be done by separating the data into two time periods, or into two areas, using one for calibration and one for validation. Both are potentially dangerous if the process being modeled is also changing through time, or across space (it is *non-stationary* in a statistical sense), but forecasting is dangerous in these circumstances as well.

Models of real-world processes can be validated by experiment, by proving that each component in the model correctly reflects reality. For example, the Game of Life would be an accurate model if some real process actually behaved according to its rules; and Clarke's urban growth model could be tested in the same way, by examining the rules that account for each past land use transition. In reality, it is unlikely that real-world processes will be found to behave as simply as model rules, and it is also unlikely that the model will capture every real-world process impacting the system.

If models are no better than approximations to reality, then are they of any value? Certainly human society is so complex that no model will ever fit perfectly.

As Ernest Rutherford, the experimental physicist and Nobel Laureate is said to have once remarked, probably in frustration with social-scientist colleagues, 'The only result that can possibly be obtained in the social sciences is, some do, and some don't'. Neither will a model of a physical system ever perfectly replicate reality. Instead, the outputs of models must always be taken advisedly, bearing in mind several important arguments:

- A model may reflect behavior under ideal circumstances and therefore provide a norm against which to compare reality. For example, many economic models assume a perfectly informed, rational decision maker. Although humans rarely behave this way, it is still useful to know what would happen if they did, as a basis for comparison.

- A model should not be measured by how closely its results match reality but by how much it reduces uncertainty about the future. If a model can narrow the options, then it is useful. It follows that any forecast should also be accompanied by a realistic measure of uncertainty.

- A model is a mechanism for assembling knowledge from a range of sources and presenting conclusions based on that knowledge in readily used form. It is often not so much a way of discovering how the world works, as a way of presenting existing knowledge in a form helpful to decision makers.

■ Just as the British politician Denis Healey once remarked that capitalism was the worst possible economic system 'apart from all of the others', so modeling often offers the only robust, transparent analytical framework that is likely to garner any respect amongst decision makers with competing objectives and interests.

Any model forecast should be accompanied by a realistic measure of uncertainty.

Several forms of uncertainty are associated with models, and it is important to distinguish between them. First, models are subject to the uncertainty present in their inputs. Uncertainty *propagation* was discussed in Section 6.4.2 – here, it refers to the impacts of uncertainty in the inputs of a model on uncertainty in the outputs. In some cases propagation may be such that an error or uncertainty in an input produces a proportionate error in an output. In other cases, a very small error in an input may produce a massive change in output; and in other cases outputs can be relatively insensitive to errors in inputs. It is important to know which of these cases holds in a given instance, and the normal approach is through repeated numerical simulation (often called Monte Carlo simulation because it mimics the random processes of that famous gambling casino), adding random distortions to inputs and observing their impacts on outputs. In some limited instances it may even be possible to determine the effects of propagation mathematically.

Second, models are subject to uncertainty over their parameters. A model builder will do the best possible job of calibrating the model, but inevitably there will be uncertainty about the correct values, and no model will fit the data available for calibration perfectly. It is important, therefore, that model users conduct some form of *sensitivity analysis*, examining each parameter in turn to see how much influence it has on the results. This can be done by raising and lowering each parameter value, for example by +10% and −10%, re-running the model and comparing the results. If changing a parameter by 10% produces a less-than-10% change in results the model can be said to be relatively insensitive to that parameter. This allows the modeler to focus attention on those parameters that produce the greatest impact on results, making every effort to ensure that their values are correct.

Sensitivity analysis tests a model's response to changes in its parameters and assumptions.

Third, and most importantly, models are subject to uncertainty because of the *labeling* of their results. Consider the model of Box 16.3, which computes an indicator of the need for groundwater protection. In truth, the areas identified by the model have three characteristics – slope greater than 5%, used for crops, and less than 300 m from a stream. Described this way, there is little uncertainty about the results, though there will be some uncertainty due to inaccuracies in the data. But once the areas selected are described as *vulnerable*, and in need of management to ensure that groundwater is *protected*, a substantial leap of logic has occurred. Whether that leap is valid or not will depend on the reputation of the modeler as an expert in groundwater hydrology and biological conservation; on the reputation of the organization sponsoring the work; and on the background science that led to the choice of parameters (5% and 300 m). In essence, this third type of uncertainty, which arises whenever labels are attached to results that may or may not correctly reflect their meaning, is related to how results are described, in other words to their metadata, rather than to any innate characteristic of the data themselves.

16.6 Conclusion

Modeling was defined at the outset of this chapter as a process involving multiple stages, often in emulation of some real physical process. Modeling is often dynamic, and current interest in modeling is stretching the capabilities of GIS software, most of which was designed for the comparatively leisurely process of analysis, rather than the intensive and rapid iterations of a dynamic model. In many ways, then, modeling represents the cutting edge of GIS, and the next few years are likely to see explosive growth both in interest in modeling on the part of GIS users, and software development on the part of GIS vendors. The results are certain to be interesting.

Questions for further study

1. Write down the set of rules that you would use to implement a simple version of the Clarke urban growth model, using the following layers: transportation (cell contains road or not), protected (cell is not available for development), already developed, and slope.

2. Review the steps you would take and the arguments you would use to justify the validity of a GIS-based model.

3. Select a field of environmental or social science, such as species habitat prediction or residential segregation. Search the Web and your library for published models of processes in this field and summarize their conceptual structure, technical implementation, and application history.

4. Compare the relative advantages of the different
   types of dynamic models discussed in this chapter
   when applied to a selected area of environmental or
   social science.

Further reading

Goodchild M.F., Parks B.O., and Steyaert, L.T. (eds)
1993 *Environmental Modeling with GIS*. New York:
Oxford University Press.

Heuvelink G.B.M. 1998 *Error Propagation in Environmental Modelling with GIS*. London: Taylor and
Francis.

Saaty T.L. 1980 *The Analytical Hierarchy Process: Planning, Priority Setting, Resource Allocation*. New York:
McGraw-Hill.

Tomlin C.D. 1990 *Geographic Information Systems and
Cartographic Modeling*. Englewood Cliffs, NJ: Prentice Hall.

Management and Policy

17 Managing GIS

Much of the material in earlier parts of this book assumes that you have a GIS and can use it effectively to meet your organization's goals. This chapter addresses the essential 'front end': the procurement and operational management of such systems. Later chapters deal with how GIS contributes to the business.

This chapter describes how to choose, implement, and manage operational GIS. It involves four key stages: the analysis of needs, the formal specification, the evaluation of alternatives, and the implementation of the chosen system. In particular, implementing GIS requires consideration of issues such as planning, support, communication, resource management, and funding. Successful on-going management of an operational GIS has five key dimensions: support for customers, operations, data management, application development, and project management.

Geographic Information Systems and Science, 2nd edition   Paul Longley, Michael Goodchild, David Maguire, and David Rhind.
© 2005 John Wiley & Sons, Ltd.   ISBNs: 0-470-87000-1 (HB); 0-470-87001-X (PB)

Learning Objectives

After reading this chapter, you will understand:

- How to go about choosing a GIS to meet your needs;

- Key GIS implementation issues;

- How to manage an operational GIS effectively with limited resources and ambitious goals;

- Why GIS projects fail – some pitfalls to avoid and some useful tips about how to succeed;

- The roles of staff members in a GIS project;

- Where to go for more detailed advice.

17.1 The big picture

This chapter is concerned with the practical aspects of managing an operational GIS. It is embedded deliberately

in the part of the book focused on high-level management concepts: success comes from combining strategy and implementation. It is the role of management in GIS projects to ensure that operations are carried out effectively and efficiently, and that a healthy, sustainable GIS can be maintained – one which meets the organization's strategic objectives.

Obtaining and running a GIS seems at first sight to be a routine and apparently 'mechanical' process. It is certainly not 'rocket science'. But neither is it simple. The consequences of failure can be catastrophic, both for the organization and for careers. Success involves constant sharing of experience and knowledge with other people, keeping good records, and making numerous judgments where the answer is not pre-ordained.

Clearly we cannot deal with all the relevant aspects of managing GIS in one chapter. This, then, is a summary and a pointer to more detailed information for those who need it. Perhaps the best 'whole book' general overview of the process of arriving at the running of a successful GIS has been produced by Roger Tomlinson, based on 40 years of experience in building and consulting on GIS for organizations across the world (see Box 17.1). He sees the initial stages of the process as having ten basic steps, encapsulated in Box 17.2.

GIS as we now recognize it began during the spring of 1962.

However, before actually starting on the process of acquiring and implementing a GIS, ask the fundamental question: *do I really need a GIS?* There are many applications where the answer is obvious – as shown in

Biographical Box **17.1**

Roger Tomlinson, GIS pioneer

GIS had many roots, but if one person can be said to be its father, it is Roger Tomlinson (see Figure 17.1: see also Section 1.4.1 and Table 1.4). Roger pioneered the use of GIS worldwide to collect, manage, and manipulate geographic data, changing the face of geography as a discipline.

GIS, as we now recognize it, could be said to have begun during the spring of 1962. While on a plane bound from Ottawa to Toronto, Tomlinson met Lee Pratt, then recently named head of the Canada Land Inventory (CLI). Tomlinson was chief of the computer mapping division at Spartan Air Services, Ottawa. The two men discussed a vast mapping project CLI was about to undertake: a multilayer land-use/planning map of Canada's inhabited and productive land – around 1 million square miles. Tomlinson told Pratt that he used computers for mapping projects and some of his ideas might work for CLI. Pratt went home and did the arithmetic, then called Tomlinson. He said: 'We better talk about this, because we've tested out how to do it manually and it's far too expensive', Tomlinson recalls. Shortly afterwards, he was employed by the Canadian government, heading its GIS development program, where he was instrumental in developing the Canada Geographic Information System or CGIS. Many other developments worldwide drew inspiration from this beginning.

Roger Tomlinson was born in Cambridge, England, in 1933. He flew planes in the Royal Air Force during the early 1950s and twice led expeditions to the Norwegian Ice Cap (1956 and 1957). He holds BAs from Nottingham University, England, and Acadia University, Nova Scotia, Canada; an MA from McGill University, Quebec, Canada; and a Ph.D. from University College, London, England. Tomlinson adopted Canada as his home country in 1957.

Dr Tomlinson established Tomlinson Associates Ltd, a firm of consulting geographers, in 1977. As a geographic consultant, Tomlinson has advised an extraordinary list of clients including the World Bank, United Nations Food and Agriculture Organization, the US departments of Commerce and Agriculture, US Geological Survey, US Forest Service, US Bureau of the Census, the Canadian Forest Service, and numerous US state and Canadian provincial and municipal government agencies – as well as many other bodies worldwide. He has been a keynote speaker at innumerable GIS conferences across the globe.

Dr Tomlinson's contributions have been recognized by a glittering array of high honors including the 2003 Gold Medal of the Royal Canadian Geographical Society, the Murchison Award of the Royal Geographical Society and – most prestigious of all – the Order of Canada.

(A)   (B)

Figure 17.1 Dr Roger Tomlinson, GIS pioneer, in 2003 and as he was in 1957 leading an expedition to the Norwegian Ice Cap

Technical Box **17.2**

The Tomlinson methodology for getting a GIS that meets your needs

Stage	Action	Commentary (after Tomlinson 2003)
1	Consider the strategic purpose	Strategic purpose is the guiding light. The system that gets implemented must be aligned with the purpose of the organization as a whole.
2	Plan for the planning	Since the GIS planning process will take time and resources, you will need to get an approval and commitment at the front end from senior managers in the organization.
3	Conduct a technology seminar	Think of the technology seminar as a sort of 'town-hall meeting' between the GIS planning team, the various staff, and other stakeholders in the organization.
4	Describe the information products	Know what you want to get out of it.
5	Define the system scope	Scoping the system means defining the actual data, hardware, software, and timing.

▶

Stage	Action	Commentary (after Tomlinson 2003)
6	Create a data design	The data landscape has changed dramatically with the advent of the Internet and the proliferation of commercial datasets. Developing a systematic procedure for safely navigating this landscape is critical.
7	Choose a logical data model	The new generation of object-oriented data models is ushering in a host of new GIS capabilities and should be considered for all new implementations. Yet the relational model is still prevalent and the savvy GIS user will be conversant with both (Chapter 8).
8	Determine system requirements	Getting the system requirements right at the outset [and providing the capacity for their evolution] is critical to successful GIS planning.
9	Benefit-cost, migration, and risk analysis	The most critical aspect of doing a realistic benefit-cost analysis is the commitment to include all the costs that will be involved. Too often managers gloss over the real costs, only to regret this later.
10	Make an implementation plan	The implementation plan should illuminate the road to GIS success.

Box 17.3. This response is especially common where other organizations in the same business have shown demonstrable benefits from operating one – and hence where competition decrees the need to mimic or surpass them (see Section 2.3.1). Important aspects of the business case which should be created formally for any GIS acquisition or major enhancement are described in Chapter 18. But at the most strategic level, there are usually two 'demand side' reasons and three general 'supply side' reasons to implement a GIS:

Managing land information in Korea through GIS

In South Korea, local government authorities administer the public land through the assessment of land prices, management of land transactions, land use planning and management, and civil services. In many cases, more than one department of a local authority produces and manages the same or similar land and property information; this has led to discrepancies in the information. With the large number of public land administration responsibilities and the control of each given to the local authorities, many problems arose; Korea is a complex and rapidly changing society (see Figure 17.2A). This led to the decision to develop a GIS-based method for sharing the information produced or required for administering land in the public and private sectors (Figure 17.2B).

The Korean Land Management Information System (LMIS) was established in 1998. The purpose of this GIS is to provide land information, increase the productivity of the public land administration, and support the operation of the land planning policies of the Korean Ministry of Construction and Transportation (MOCT). The LMIS database includes much spatial data such as topographic, cadastral, and land use district maps.

(A)

Figure 17.2 (A) Seoul by night. (B) Land information map for part of Seoul (Reproduced by permission of Corbis)

Hyunrai Kim, vice director of the Land Management Division of Seoul Metropolitan City, summarized the advantages this provides: 'By means of the Internet-based Land Information Service System, citizens can get land information easily at home. They don't have to visit the office, which may be located far from their homes.' The system has also resulted in time and cost savings. With the development of the Korean Announced Land Price Management System, it is also possible to compute land prices directly and produce maps of the variations in land price.

Initially, the focus was mainly on the administrative aspects of data management and system development; however, attention then turned to the expansion and development of a decision support system using various data analyses. It is intended that the Land Legal Information Service System will also be able to inform land users of regulations on land use. In essence, LMIS is becoming a crucial element of e-government (see Section 20.3.5). This case study highlights the role that GIS can play beyond the obvious one of information management, analysis, and dissemination. It highlights the value of GIS in enabling organizational integration and the reality of generating benefits through improved staff productivity.

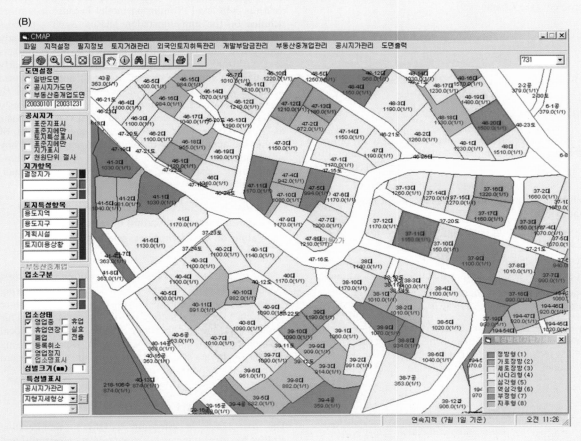

Figure 17.2 (*continued*)

■ *Cost reduction.* GIS can replace, in part or completely, many existing operations, e.g., drafting maps, locating and maintaining customers, and managing land acquisition and disposal more efficiently.

■ *Cost avoidance.* Examples of GIS use are in locating facilities away from areas at high risk from natural hazards (e.g., tornadoes, floods, or land slides) and by minimizing delivery routes (e.g., letters, refrigerators, and beer).

■ *Increased revenue.* Examples include finding and attracting new customers, making maps and data for sale and using GIS as a tool to support consultants (e.g., facility siting, natural resource conservation, and real estate management).

■ *Getting wholly new products.* The GIS may enable you to produce products which simply could not have been created previously or which would have been impossibly costly or time-consuming to produce (e.g., satellite imagery of an area draped on a three-dimensional landscape or escape routes from hazards which take into account likely congestion; see Section 2.3.4.2 and Box 16.2).

■ *Getting non-tangible (or intangible) benefits.* These benefits are difficult to measure, but they can be very important nonetheless. Examples include making better decisions, providing better service to customers and clients (which can lead to improved public image), the use of consistent information across an organization, the production of reproducible and defensible results, and the ability to document and share the processes and methodology used to solve a problem (see Chapter 19).

17.2 The process of developing a sustainable GIS

GIS projects are similar to many other large IT projects in that they can be broken down into four major life-cycle phases (Figure 17.3). For our simplified purposes, these are:

■ business planning (strategic analysis and requirements gathering);

■ system acquisition (choosing and purchasing a system);

■ system implementation (assembling all the various components and creating a functional solution); and

■ operation and maintenance (keeping a system running).

These phases are iterative. Over a decade or more, several iterations may occur, often using different generations of GIS technology and methodologies. Variations on this model include prototyping and rapid application development – but space does not permit much discussion of them here.

> **GIS projects comprise four major lifecycle phases: business planning; system acquisition; system implementation; and operation and maintenance.**

17.2.1 Choosing a GIS

Clarke has proposed a general model of how to specify, evaluate, and choose a GIS, variations of which have been used by organizations over the past 20 years or so. The model we prefer (Figure 17.4) is based on 14 steps grouped into four stages: analysis of requirements; specification of requirements; evaluation of alternatives; and implementation of system; we use it here rather than the 'top level' Tomlinson model because it is more detailed on the implementation phases. Such a process is both time-consuming and expensive. It is really only appropriate for large GIS implementations (contracts over $100 000 in initial value), where it is particularly important to have investment and risk appraisals (see Section 19.4.2 onwards). We describe the model here so that those involved with smaller systems can judiciously select those elements relevant to them. On the basis of painful experience, however, we urge the use of formalized approaches to evaluating the need for and any subsequent acquisition of a system. It is amazing how small projects, carried out quickly because they are small, evolve into big and costly ones!

> **Choosing a GIS involves four stages: analysis of requirements; specification of requirements; evaluation of alternatives; and implementation of system.**

For organizations undertaking acquisition for the first time, huge benefits can be accrued through partnering with other organizations (Chapter 20) that are more advanced, especially if they are in the same field. This is often possible in the public sector, e.g., where local governments have similar tasks to meet. But a surprising number of private sector organizations are also prepared to share their experiences and documents.

Stage 1: Analysis of requirements

The first stage in choosing a GIS is an iterative process for identifying and refining user requirements, and for determining the business case (see Chapter 19) for acquiring a GIS. The deliverable for each step is a report that should be discussed with users and management. It is important to keep records of the discussions and share them with those involved so there can be no argument at a later stage about what was agreed! The results of each report help determine successive stages.

Step 1: Definition of objectives

This is often a major decision for any organization. The rational process of choosing a GIS begins with and spins

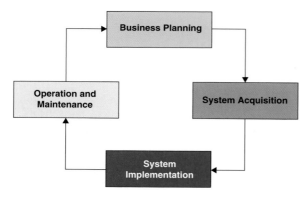

Figure 17.3 GIS project lifecycle stages

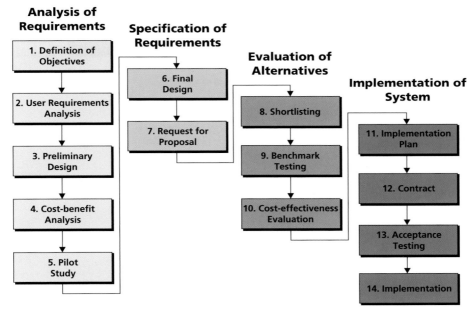

Figure 17.4 General model of the GIS acquisition process

out of the development of the organization's strategic plan (Section 19.4.1) – and an outline decision that GIS can play a role in the implementation of this plan. Strategic and tactical objectives must be stated in a form understandable to managers. The outcome from Step 1 is a document that managers and users can endorse as a plan to proceed with the acquisition, i.e., the relevant managers believe there is sufficient promise to proceed to the next step and commit the initial funding required.

Step 2: User requirements analysis

The analysis will determine how the GIS is designed and evaluated. Analysis should focus on what information is presently being used, who is using it, and how the source is being collected, stored, and maintained. This is a map of existing processes (which may possibly be improved as well as being replicated by the GIS). The necessary information can be obtained through interviews, documentation, reviews, and workshops. The report for this phase should be in the form of workflows, lists of information sources, and current operation costs. The clear definition of likely or possible change (e.g., future applications – see Figure 17.5), new information products (e.g., maps and reports – Figure 17.6) or different utilization of functions and new data requirements is essential to successful GIS implementation.

Step 3: Preliminary design

This stage of the design is based on results from Step 2. The results will be used for subsequent cost-benefit analysis (Step 4 below) and will enable specification of the pilot study. The four key tasks are: develop preliminary database specifications; create preliminary functional specifications; design preliminary system models; and survey the market for potential systems. Database specifications involve estimating the amount and type of data

APPLICATION

    Display zoning map information for a user-defined area.

FUNCTIONS USED IN THE APPLICATION:

    Review and prepare zoning changes.

DESCRIPTION OF APPLICATION:

    This application uses zoning and related parcel-based data from the database to display existing information related to zoning for a specific area that is defined by the user. The application must be available interactively at a workstation when the user invokes a request and identifies the subject land parcel. The application will define a search area based upon the search distance defined and input by the user, and will display all required data for the area within the specified distance from the outer boundary of the subject parcel.

DATA INPUTS:

User defined: Parcel identifier
                 Search distance

Database: Zoning boundaries
                 Zoning dimensions
                 Zoning codes
                 Parcel boundaries
                 Parcel dimensions
                 Parcel numbers
                 Street names
                 Addresses

PRODUCTS OUTPUT:

1.   Zoning map screen display with subject parcel highlighted, search area boundaries, search distance, all zoning data, parcel data, street names and addresses.

2.   Hard copy map of the above.

Figure 17.5 Sample application definition form

(see Chapter 9). Many consultants maintain checklists and vendors frequently publish descriptions of their systems on their websites. The choice of system model involves decisions about raster and vector data models and system

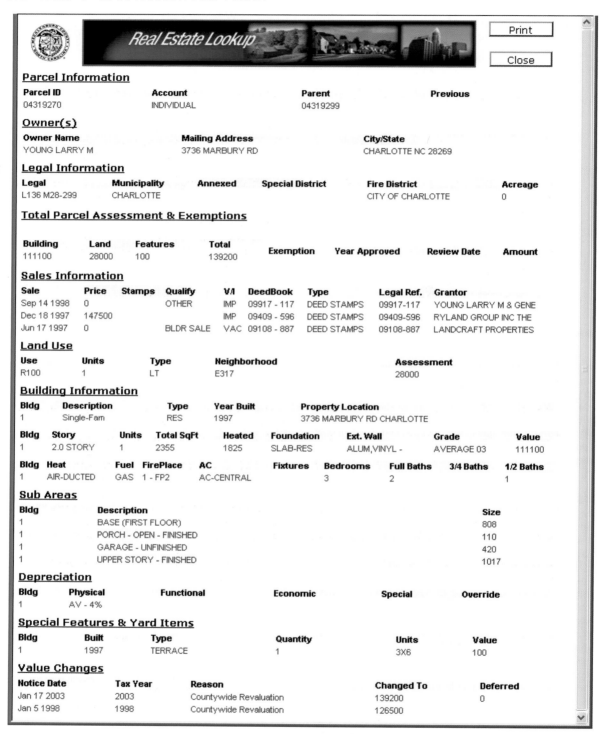

Figure 17.6 A Web-displayed report produced by a local government GIS (*Source*: Mecklenburg County, North Carolina GIS; see **maps2.co.mecklenburg.nc.us/website/realestate/viewer.htm**) (Reproduced by permission of Mecklenburg County, NCIST)

type (see Chapters 7 and 8). Finally, a market survey should be undertaken to assess the capabilities of commercial off-the-shelf (COTS) systems. This might involve a formal Request For Information (RFI) to a wide range of vendors. A balance needs to be struck in writing this between creating a document so open that the vendor has problems identifying what needs are paramount and one so prescriptive and closed that no flexibility or innovation is possible.

Whether to buy or to build a GIS used to be a major decision (Chapter 7). This occurred especially at 'green field' sites – where no GIS technology has hitherto been used – and at sites where a GIS has already been implemented but was in need of modernization. But the situation is now quite different: use of general-purpose COTS solutions are the norm. These have ongoing programs of enhancement and maintenance and can normally be used for multiple projects. Typically they are better documented and more people in the job market have experience of them. As a consequence, risk arising from loss of key personnel is reduced.

There has been a major move in GIS away from building proprietary GIS toward buying COTS solutions.

Step 4: Cost-benefit analysis

Purchase and implementation of a GIS is a non-trivial exercise, expensive in both money and staff resources (typically management time). It is quite common for organizations to undertake a cost-benefit (also called benefit-cost) analysis to justify the effort and expense, and to compare it against the alternative of continuing with the current data, processes, and products – the *status quo*. Box 17.4 summarizes the process.

Cost-benefit cases are normally presented as a spreadsheet, along with a report that summarizes the main findings and suggests whether the project should be continued or halted. Senior managers then need to assess the merits of this project in comparison with any others competing for their resources (see Section 19.4.2).

Technical Box 17.4

Cost-benefit analysis

Cost-benefit analysis begins with a determination of GIS implementation costs (hardware, software, data, and people) and the projected benefits (improved service, reduced costs, greater use of data, etc.). A monetary value is then assigned to both costs and benefits. When these are summed, the assessed benefits should exceed the projected costs over the full lifecycle if the project is to proceed. This sounds quite simple and straightforward. In practice, however, cost-benefit analysis is more of an art than a science because there are many complicating factors and assumptions to be made.

For example, it is difficult to assign a monetary value to many activities, and in many traditional public organizations fixed costs (such as staff time, centralized administration, and computing facilities) are still not taken properly into account. In addition, a positive cost-benefit analysis (benefits exceed costs) may not ensure that the project proceeds – that will also depend on whether the organization's cash flow and management time resources are adequate, whether the risks are acceptable (see Section 19.4.2), and whether another project competing for the same resources has a better cost-benefit out-turn

Table 17.1 Simple examples of GIS costs and benefits (after Obermeyer 2005, with additions)

Category	Costs	Benefits
Economic (tangible)	• Hardware • Software • Training • New staff or skills • Additional space • Data purchase or collection	• Reduced costs (e.g., of staff) • Greater throughput • Increased revenues • New market services or products
Institutional (intangible)	• Interpersonal shifts • Layoffs of low-skilled staff, partially replaced by fewer if individually more expensive skilled staff	• Improved client relationships • Better decisions • Improved morale • Better information flow • Better culture of 'achievers'

▶

The potential costs and benefits of implementing a GIS fall into two categories (Table 17.1): economic (tangible) and institutional (intangible). Unfortunately, experience suggests that the more easily measured economic benefits tend ultimately to be less important in GIS projects than the much more ill-defined institutional benefits. This is true especially of systems that perform decision-support rather than operational tasks (e.g., city planning versus postal delivery routing). But it is an important discipline to identify both in as formal a way as possible: the process helps in decision making and will inevitably be the subject of internal audit at a later stage.

Although cost-benefit analysis is the best-known method of assessing the value of a GIS, a number of alternatives have also been proposed. 'Cost-effectiveness analysis' compares the cost of providing the same service using alternative means. The 'payback period' is derived by dividing the cost of implementing a system by the estimated annual net benefits of using it. The 'value-added approach' emphasizes the new things that a GIS allows an organization to perform. In all assessments, account should be taken of the different value of money at different times.

Surprisingly, there are few readily accessible examples of GIS cost-benefit studies. Many users do not even bother to carry out such a prior evaluation, especially now that GIS is becoming a mainstream science and technology. This is short-sighted because adoption of GIS has long-term consequences – such as staff training and development needs – which go beyond simple initial financial expenditure on equipment and software. Many organizations also gain competitive advantage from GIS and do not want to publish their analyses. Moreover, cost-benefit studies are often carried out by consultants who do not want to make their approach public for fear that they will no longer be able to charge for it.

Step 5: Pilot study

A pilot study is a mini-version of the full GIS implementation that aims to test several facets of the project. The primary objective is to test a possible or likely system design before finalizing the system specification and committing significant resources. Secondary objectives are to develop the understanding and confidence of users and sponsors, to test samples of data if a data capture project is part of the implementation, and to provide a test bed for application development.

A pilot is a mini-version of a full GIS implementation designed to test as many aspects of the final system as possible.

It is normal to use existing hardware or to lease hardware similar to that which is expected to be used in the full implementation. A reasonable cross-section of all the main types of data, applications, and product deliverables should be used during the pilot. But the temptation must be resisted to try to build the whole system at this stage, irrespective of how easy 'the techies' may claim it to be! Users should be prepared to discard *everything* after the pilot if the selected technology or application style does not live up to expectations.

The outcome of a pilot study is a document containing an evaluation of the technology and approach adopted, an assessment of the cost-benefit case, and details of the project risks and impacts. Risk analysis is an important activity, even at this early stage (see Section 19.4.1.2). Assessing what can go wrong can help avoid potentially expensive disasters in the future. The risk analysis should focus on the actual acquisition processes as well as on implementation and operation.

Stage 2: Specification of requirements

The second stage is concerned with developing a formal specification that can be used in the structured process of soliciting and evaluating proposals for the system.

Step 6: Final design

This creates the final design specifications for inclusion in a Request For Proposals (RFP: also called an invitation to tender or ITT) to vendors. Key activities include finalizing the database, defining the functional and performance specifications, and creating a list of possible constraints. From these, requirements are classified as mandatory, desirable, or optional. The deliverable is the final design document. This document should provide a clear description of essential requirements – without being so prescriptive that innovation is stifled, costs escalate, or insufficient vendors feel able to respond.

Step 7: Request for proposals

The RFP document combines the final design document with the contractual requirements of the organization. These will vary from organization to organization but are likely to include legal details of copyright of the design and documentation, intellectual property ownership, payment schedules, procurement timetable, and other draft terms and conditions. Once the RFP is released to vendors by official advertisement and/or personal letter, a minimum period of several weeks is required for vendors to evaluate and respond. For complex systems, it is usual to hold an open meeting to discuss technical and business issues.

Stage 3: Evaluation of alternatives

Step 8: Short-listing

In situations where several vendors are expected to reply, it is customary to have a short-listing process. Submitted proposals must first be evaluated, normally using a weighted scoring system, and the list of potential suppliers narrowed down to between two and four. Good practice is that the scoring must be done by several individuals acting independently and the results – and the differences between the evaluations – compared. This whole process allows both the prospective purchaser and supplier organizations to allocate their resources in a focused way. Short-listed vendors are then invited to attend a benchmark-setting meeting.

Step 9: Benchmarking

The primary purpose of a benchmark is to evaluate the proposal, people, and technology of each selected vendor. Each one is expected to create a prototype of the final system that will be used to perform representative tests. The results of these tests are scored by the prospective purchaser. Scores are also assigned for the original vendor proposal and the vendor presentations about their company. Together, these scores form the basis of the final system selection. Unfortunately, benchmarks are often conducted in a rather secretive and confrontational way, with vendors expected to guess the relative priorities (and the weighting of the scores) of the prospective purchaser. Whilst it is essential to follow a fair and transparent process, maintain a good audit trail and remain completely impartial, a more open co-operative approach usually produces a better evaluation of vendors and their proposals. If vendors know which functions have the greatest value to customers they can tune their systems appropriately.

Step 10: Cost-effectiveness evaluation

Next, surviving proposals are evaluated for their cost-effectiveness. This is again more complex than it might seem. For example, GIS software systems vary quite widely in the type of hardware they use, some need additional database management system (DBMS) licenses, customization costs will vary, and maintenance will often be calculated in different ways. The goal of this stage is to normalize all the proposals to a common format for comparative purposes. The weighting used for different parts must be chosen carefully since this can have a significant impact on the final selection. Good practice involves debate within the user community – for they should have a strong say – on the weighting to be used and some sensitivity testing to check whether very different answers would have been obtained if the weights were slightly different. The deliverable from this stage is a ranking of vendors' offerings.

Stage 4: Implementation of system

The final stage is planning the implementation, contracting with the selected vendor, testing the delivered system, and actual use of the GIS.

Step 11: Implementation plan

A structured, appropriately paced, implementation plan is an essential ingredient of a successful GIS implementation. The plan commences with identification of priorities, definition of an implementation schedule, and creation of a resource budget and management plan. Typical activities that need to be included in the schedule are installation and acceptance testing, staff training, data collection, and customization. Implementation should be coordinated with both users and suppliers.

Step 12: Contract

An award is subject to final contractual negotiation to agree general and specific terms and conditions, what elements of the vendor proposal will be delivered, when they will be delivered and at what price. General conditions include contract period, payment schedule, responsibilities of the parties, insurance, warranty, indemnity, arbitration, and provision of penalties and contract termination arrangements.

Step 13: Acceptance testing

This is to ensure that the delivered GIS matches the specification agreed in the contract. Part of the payment should be withheld until this step is successfully completed. Activities include installation plus tests of functionality, performance, and reliability. It is seldom the case that a system passes all tests first time and so provision should be made to repeat aspects of the testing.

Step 14: Implementation

This is the final step at the end of what can be a long road (see also Section 17.2.1.1 below). The entire GIS acquisition period can stretch over many months or even longer. Activities include training users and support staff, data collection, system maintenance, and performance monitoring. Customers may also need to be 'educated' as well! Once the system is successfully in operation, it may be appropriate to publicize its success for enhancement of the brand image or political purposes.

17.2.1.1 Discussion of the classical acquisition model

The general model outlined above has been widely employed as the primary mechanism for large GIS procurements in public organizations. It is rare, however, that 'one size fits all' and, although it has many advantages, it also has some significant shortcomings:

- The process is expensive and time-consuming for both suppliers and vendors. A supplier can spend as much as 20% of the contract value on winning the business and a purchaser can spend a similar amount in staff time, external consultancy fees, and equipment rental. This ultimately makes systems more expensive – though competition does drive down cost.

- Because it takes a long time and because GIS is a fast-developing field, proposals can become technologically obsolete within several months.

- The short-listing process requires multiple vendors, which can end up lowering the minimum technical selection threshold in order to ensure enough bidders are available.

- In practice, the evaluation process often focuses undue attention on price rather than the long-term organizational and technical merits of the different solutions.

- This type of procurement can be highly adversarial. As a result, it can lay the foundations for an uncomfortable implementation partnership (see Chapter 20) and often does not lead to full development of the best solution. Every implementation is a trade-off between functionality, time, price, and risk. A full and frank discussion between purchaser and vendor on this subject can generate major long-term benefits.

- Many organizations have little idea about what they *really* need. Furthermore, it is very difficult to specify precisely in any contract exactly what a system must perform. As users learn more, their aspirations also rise – resulting in 'feature creep' (the addition of more capabilities) often without any acceptance of an increase in budget. On the other hand, some vendors adopt the strategy of taking a minimalist view of the capabilities of the system featured in their proposal and make all modifications during implementation and maintenance through chargeable change orders. All this makes the entire system acquisition costs far higher than was originally anticipated; the personal consequences for the budget holders concerned can be unfortunate.

- Increasingly, most organizations already have some GIS; the classical model works best in a 'green field site' situation.

As a result of these problems, this type of acquisition model is not used in small or even some larger procurements, especially where the facilities can be augmented rather than totally replaced. A less complex and formal selection method is *prototyping*. Here a vendor or pair of vendors is selected early on using a smaller version of the evaluation process outlined above. The vendor(s) is/are then funded to build a prototype in collaboration with the user organization. This fosters a close partnership to exploit the technical capabilities of systems and developers and helps to maintain system flexibility in the light of changing requirements and technology. This approach works best for those procurements – sometimes even some large ones – where there is some uncertainty about the most appropriate technical solution and where the organizations involved are mature, able to control the process, and not subject to draconian procurement rules.

Prototyping is a useful alternative to classical, linear system acquisition exercises. It is especially useful for smaller procurements where the best approach and outcome are more uncertain.

17.2.2 Implementing a GIS

The goal of this section is to provide a checklist of important management issues when implementing a GIS.

17.2.2.1 Plan effectively

Good planning is essential through the full lifecycle of all GIS projects. Both strategic and operational, or tactical, planning are important to the success of a project. Strategic planning involves reviewing overall organizational goals and setting specific GIS objectives. Operational planning is more concerned with the day-to-day management of resources. There are several general project management productivity tools available that can be used in GIS projects. Figure 17.7 shows one diagramming tool called a Gantt chart. Several other implementation techniques and tools are discussed in Box 17.5.

17.2.2.2 Obtain support

If a GIS project is to prosper, it is essential to garner support from *all* key stakeholders. This often requires establishing executive (director-level) leadership support; developing a public-relations strategy by, for example, exhibiting key information products or distributing free maps; holding an open house to explain the work on the GIS team; and participating in GIS seminars and workshops, locally and sometimes nationally.

17.2.2.3 Communicate with users

Involving users from the very earliest stages of a project will lead to a better system design and will help with user acceptance. Seminars, newsletters, and frequent updates about the status of the project are good ways to educate and involve users. Setting expectations about capabilities, throughput, and turn-around at reasonable levels is crucial to avoid any later misunderstandings with users and managers.

17.2.2.4 Anticipate and avoid obstacles

These may involve staffing, hardware, software, databases, organization/procedures, timeframe, and funding. Be prepared!

17.2.2.5 Avoid false economies

Money saved by not paying staff a reasonable (market value) wage or by insufficient training is often manifested in reduced staff efficiencies. Furthermore, poorly paid or trained staff often leave through frustration. You cannot prevent this by contractual means (see Section 19.1.2) so must do so through paying market rates and/or building a team culture where staff enjoy working for the organization.

Cutting back on hardware and software costs by, for example, obtaining less powerful systems or cancelling maintenance, may save money in the short-term but

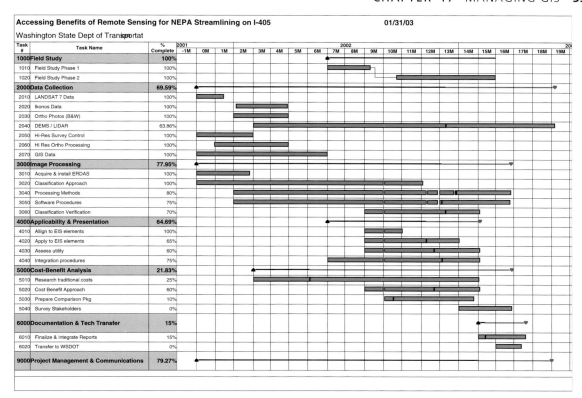

Figure 17.7 Gantt chart of a basic GIS project. This chart shows task resource requirements over time, with task dependencies. This is a relatively simple chart, with a small number of tasks (*Source*: **www.wsdot.wa.gov/environment/envinfo/docs/ RSPrj_GanttJan2003.pdf**) (Reproduced by permission of Washington State Department of Transportation)

will likely cause serious problems in the future when workloads increase and the systems get older. Failing to account for depreciation and replacement costs, i.e., by failing to amortize the GIS investment, will store up trouble ahead. The amortization period will vary greatly – hardware may be depreciated to zero value after, say, four years whilst buildings may be amortized over 30 years.

17.2.2.6 Ensure database quality and security

Investing in the database quality is essential at all stages from creation onwards. Catastrophic results may ensue if any of the up-dates or (especially) the database itself is lost in a system crash or corrupted by hacking, etc. This requires not only good precautions but also contingency (disaster recovery and business continuity) plans and periodic serious trials of them.

17.2.2.7 Accommodate GIS within the organization

Building a system to replicate ancient and inefficient ones is not a good idea; nor is it wise to go to the other extreme and expect the whole organization's ways of working to be changed to fit better with what the GIS can do! Too much change at any one time can destroy organizations

just as much as too little change can ossify them. In general, the GIS must be managed in a way that fits with the organizational aspirations and culture if it is to be a success (see Section 18.2.3). All this is especially a problem because GIS projects often blaze the trail in terms of introducing new technology, interdepartmental resource sharing, and generating new sources of income.

17.2.2.8 Avoid unreasonable timeframes and expectations

Inexperienced managers often underestimate the time it takes to implement GIS. Good tools, risk analysis, and time allocated for contingencies are important methods of mitigating potential problems. The best guide to how long a project takes is experience in other similar projects – though the differences between the organizations, staffing, tasks, etc., need to be taken into account.

17.2.2.9 Funding

Securing ongoing, stable funding is a major task of a GIS manager. Substantial GIS projects will require core funding from one or more of the stakeholders. None of these will commit to the project without a business case and risk analysis (Section 19.4). Additional funding for special projects, and from information and service sales, is likely to be less certain. It is characteristic of

Technical Box **17.5**

Implementation techniques

There are many tools and techniques available for GIS managers to use in implementation projects. These are summarized in Table 17.2.

Table 17.2 GIS implementation tools and techniques (after Heywood et al 2002, with additions)

Technique	Purpose
SWOT analysis	A management technique used to establish Strengths, Weaknesses, Opportunities, and Threats in a GIS implementation. The output is a list and narrative.
Rich picture analysis	Major participants are asked to create a schematic/picture showing their understanding of a problem using agreed conventions. These are then discussed as part of a consensus-forming process.
Demonstration systems	Many vendors and GIS project teams create prototype demonstrations to stimulate interest and educate users/funding agencies.
Interviews and data audits	These aim to define problems and determine current data holdings. The output is a report and recommendations.
Organization charts, system flow charts, and decision trees	These are all examples of flow charts that show the movement of information, the systems used, and how decisions are currently reached.
Data flow diagrams and dictionaries	These are charts that track the flow of information and computerized lists of data in an organization.
Project management tools	Gantt charts (see Figure 17.7) and PERT (Program, Evaluation, and Review Techniques) are tools for managing time and resources.
Object model diagrams	These show objects to be modeled in a GIS and the relationships between them (see for example Figure 8.16).

many GIS projects that the operational budget will change significantly over time as the system matures. The three main components are staff, goods and services, and capital investments. A commonly experienced distribution of costs between these three elements is shown in Table 17.3.

Table 17.3 Percentage distribution of GIS operational budget elements over three time periods (after Sugarbaker 2005)

Budget item	Year 1–2	Year 3–6	Year 6–12
Staff and benefits	30	46	51
Goods and services	26	30	27
Equipment and software	44	24	22
Total	100	100	100

17.2.2.10 Prevent meltdown

Avoiding the cessation of GIS activities is the ultimate responsibility of the GIS manager. According to Tomlinson, some of the main reasons for the failure of GIS projects are:

- lack of executive-level commitment;
- inadequate oversight of key participants;
- inexperienced managers;
- unsupportive organizational structure;
- political pressures, especially where these change rapidly;
- inability to demonstrate benefits;
- unrealistic deadlines;
- poor planning; and
- lack of core funding.

17.2.3 Managing a sustainable, operational GIS

Larry Sugarbaker has characterized the many operational management issues throughout the lifecycle of a GIS project as: customer support; effective operations; data management; and application development and support. Success in any one – or even all – of these areas does not guarantee project success, but they certainly help to produce a healthy project. Each is now considered in turn.

> **Success in operational management of GIS requires customer support, effective operations, data management, and application development and support.**

17.2.3.1 Customer support

In progressive organizations *all* users of a system and its products are referred to as customers. A critical function of an operational GIS is a customer support service. This could be a physical desk with support staff or, increasingly, it is a networked electronic mail and telephone service. Since this is likely to be the main interaction with GIS support staff, it is essential that the support service creates a good impression and delivers the type of service users need. The unit will typically perform key tasks including technical support and problem logging plus meeting requests for data, maps, training, and other products. Performing these tasks will require both GIS analyst-level and administrative skills. It is imperative that *all* customer interaction is logged and that procedures are put into place to handle requests and complaints in an organized and structured fashion. This is both to provide an effective service and also to correct systemic problems

Customer support is not always seen as the most glamorous of GIS activities. However, a GIS manager who recognizes the importance of this function and delivers an efficient and effective service will be rewarded with happy customers. Happy customers remain customers. Effective staff management includes finding staff with the right interests and aspirations, rotating GIS analysts through posts and setting the right (high) level of expectation in the performance of all staff. Managers can learn much by taking a turn in the hot seat of a customer support role!

17.2.3.2 Operations support

Operations support includes system administration, maintenance, security, backups, technology acquisitions, and many other support functions. In small projects, everyone is charged with some aspects of system administration and operations support. But as projects grow beyond five or more staff, it is worthwhile to designate someone specifically to fulfill what becomes a core, even crucial, role. As projects become larger, this grows into a full-time function. System administration is a highly technical and mission-critical task requiring a dedicated, properly trained and paid person.

Perhaps more than in any other role, clear written descriptions are required for this function to ensure that a high level of service is maintained. For example, large, expensive databases will require a well-organized security and backup plan to ensure that they are *never* lost or corrupted. Part of this plan should be a disaster recovery strategy. What would happen, for example, if there were a fire in the building housing the database server or some other major problem (e.g., see Box 18.1)?

17.2.3.3 Data management support

The concept that geographic data are an important part of an organization's critical infrastructure is becoming more widely accepted. Large, multi-user geographic databases use database management system (DBMS) software to allocate resources, control access, and ensure long-term usability. DBMS can be sophisticated and complicated, requiring skilled administrators for this critical function.

A database administrator (DBA) is responsible for ensuring that all data meet all of the standards of accuracy, integrity, and compatibility required by the organization. A DBA will also typically be tasked with planning future data resource requirements – derived from continuing interaction with current and potential customers – and the technology necessary to store and manage them. Similar comments to those outlined above for System Administrators also apply to this position.

17.2.3.4 Application development and support

Although a considerable amount of application development is usual at the onset of a project, it is also likely that there will be an on-going requirement for this type of work. Sources of application development work include improvements/enhancements to existing applications, as well as new users and new project areas starting to adopt GIS.

Software development tools and methodologies are constantly in a state of flux and GIS managers must invest appropriately in training and new software tools. The choice of which language to use for GIS application development is often a difficult one. Consistent with the general movement away from proprietary GIS languages, wherever possible GIS managers should try to use mainstream, open languages that appear to have a long lifetime. Ideally, application developers should be assigned full-time to a project and should become permanent members of the GIS group to ensure continuity (but often this does not occur).

17.3 Sustaining a GIS – the people and their competences

Throughout this chapter we have sometimes highlighted and sometimes hinted at the key role of staff as assets in all organizations; their education is discussed in Section 19.2. If they do not function well – individually and as a team – nothing of merit will be achieved.

17.3.1 GIS staff and the teams involved

Several different staff will carry out the operational functions of a GIS. The exact number of staff and their precise roles will vary from project to project. The same staff member may carry out several roles (e.g., it is quite common for administration and application development to be performed by a GIS technical person), and several staff members may be required for the same task (e.g., there may be many digitizing technicians and application developers). Figure 17.8 shows a generalized view of the main staff roles in medium to large GIS projects.

All significant GIS projects will be overseen by a management board built up of a senior sponsor (usually a Director or Vice-President), members of the user community, and the GIS manager. It is also useful to have one or more independent members to offer disinterested advice. Although this group may seem intimidating and restrictive to some, used in the right way it can be a superb source of funding, advice, support, and encouragement.

Typically, day-to-day GIS work involves three key groups of people: the GIS team itself; the GIS users; and external consultants. The GIS team comprises the dedicated GIS staff at the heart of the project; the GIS manager is the team leader. This individual needs to be skilled in project and staff management (see Section 17.3.2) and have sufficient understanding of GIS technology and the organization's business to handle the liaisons involved. Larger projects will have specialist staff experienced in project management, system administration, and application development.

GIS users are the customers of the system. There are two main types of user (other than the leaders of organizations who may rely upon GIS indirectly to provide information on which they base key decisions). These are professional users and clerical staff/technicians. Professional users include engineers, planners, scientists, conservationists, social workers, and technologists who utilize output from GIS for their professional work. Such users are typically well-educated in their specific field, but may lack advanced computer skills and knowledge of the GIS. They are usually able to learn how to use the

system themselves and can tolerate changes to the service. Clerical and technical users are frequently employed as part of the wider GIS project initiative to perform tasks like data collection, map creation, routing, and service call response. Typically, the members of this group have limited training and skills for solving *ad hoc* problems. They need robust, reliable support. They may also include staff and stakeholders in other departments or projects that assist the GIS project on either a full- or part-time basis, e.g., system administrators, clerical assistants, or software engineers provided from a common resource pool or managers of other databases or systems with which the GIS must interface. Finally, many GIS projects utilize the services of external consultants. They could be strategic advisors, project managers, or technical consultants able to supplement the available staffing. Although these may appear expensive at first sight, they are often well-trained and highly focused. They can be a valuable addition to a project, especially if internal knowledge and/or resources are limited and for benchmarking against approaches elsewhere. But the in-house team must not rely too heavily on consultants lest, when they go, all key knowledge and high-level experience goes with them.

> **The key groups involved in GIS are: the management board; the GIS team (headed by a GIS manager); the users; external consultants; and various customers.**

17.3.2 Project managers

A GIS project will almost certainly have several subprojects or project stages and hence require a structured approach to project management. The GIS manager may take on this role personally, although in large projects it is customary to have one or more specialist project managers. The role of the project manager is to establish

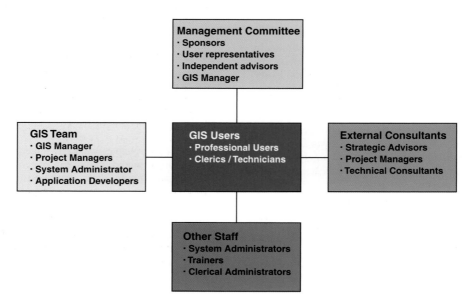

Figure 17.8 The GIS staff roles in a medium to large GIS project

user requirements, to participate in system design, and to ensure that projects are completed on time, within budget, and according to an agreed quality plan. Good project managers are rare creatures and must be nurtured for the good of the organization. One of their characteristics is that, once one project is completed, they like to move on to another so retaining them is only possible in an enterprising environment. Transferring their expertise and knowledge into the heads and files of others is a priority before they leave a project.

17.3.3 Coping with uncertainty

As we have seen earlier, geographic information varies hugely in its characteristics, and rarely is the available information ideal for the task in hand. Staff need a clear understanding of the concepts and implications of uncertainty (Chapter 6) and the related concepts of accuracy, error and sensitivity. Understanding of business risk arising from GIS use and how GIS can help reduce organizational risk is also essential (Chapter 19). This section focuses on the practical aspects relevant to managers of operational GIS – and hence on the skills, attitudes, and other competences they need to bring to their work.

Organizations must determine how much uncertainty they can tolerate before information is deemed useless. This can be difficult because it is application-specific. An error of 10 m in the location of a building is irrelevant for business geodemographic analysis but it could be critical for a water utility maintenance application that requires digging holes to locate underground pipes. Some errors in GIS can be reduced but sometimes at a considerable cost. It is common experience that trying to remove the last 10% of error typically costs 90% of the overall sum. As we concluded in Section 6.5, uncertainty in GIS-based representations is almost always something that we have to live with to a greater or lesser extent. The key issue here is identifying the amount of uncertainty that can be tolerated for a given application, and what can be done at least partially to eliminate it or ameliorate its consequences. Some of this, at least, can only be done by judgement informed by past experience.

A typology of the errors commonly encountered in GIS is also in Chapter 6. Some practical examples of errors in operational GIS are:

- Referential errors in the identity of objects, e.g., a street address could be wrong, resulting in incorrect property identification during an electricity network trace.
- Topological errors, e.g., a highway network could have missing segments or unconnected links, resulting in erroneous routing of service or delivery vehicles.
- Relative positioning errors, e.g., a gas station incorrectly located on the wrong side of a dual

carriageway road could have major implications for transportation models.

- Absolute errors in the location of objects in the real world, e.g., tests for whether factories are within a smoke control zone or floodplain could provide erroneous results if the locations are incorrect. This could lead to litigation.
- Attribute errors, e.g., incorrectly entering land use codes would give errors in agricultural production returns to government agencies.

Managing error requires use of quality assurance techniques to identify them and assess their magnitude. A key task is determining the error tolerance that is acceptable for each data layer, information product, and application. It follows that both data creators and users must make analyses of possible errors and their likely effects, based on a form of cost-benefit analysis. And sitting in the midst of all this is the GIS manager who must know enough about uncertainty to ask the right questions; if the system provides what subsequently turns out to be nonsense, he or she is likely to be the first to be blamed! In no sense is the good and long-lasting GIS manager simply someone who ensures that the 'wheels go round'. Box 17.6 presents one highly skilled GIS manager.

17.4 Conclusions

Mostly, any management function in GIS (and indeed elsewhere) is about motivating, organizing or steering, enhancing skills, and monitoring the work of other people.

Managing a GIS project *is* different to using GIS in decision making. Normally, the first requires good GIS expertise and first class project management skills. In contrast, those involved at different levels of the organization's management chain need some awareness of GIS, its capabilities, and its limitations – scientific and practical – alongside their substantial leadership skills. But our experience is that the division is not clear-cut. GIS project managers cannot succeed unless they understand the objectives of the organization, the business drivers, and the culture in which they operate (Chapter 18), plus something of how to value, exploit, and protect their assets (Chapter 19). Equally, decision makers can only make good decisions if they understand more of the scientific and technological background than they may wish to do: running or relying on a good GIS service involves much more than the networking of a few PCs running one piece of software.

So, good management of GIS requires excellent people, technical and business skills, and the capacity to ensure mutual respect and team working between the users and the experts.

Dawn Robbins, GIS manager

Dawn Robbins is the Geographic Information Officer (GIO) for Ventura County, California, USA. She has a long history in GIS that includes experience as a consultant. Ms. Robbins holds geography degrees from Dartmouth College (B.A.) and Penn State (M.S.). As GIO, she oversees county-wide geographic data coordination and provision of GIS services such as base map maintenance, system administration, database management, application design and development, and training. Her GIS department has ten staff and an annual budget of over $1m.

Dawn says:

Figure 17.9 Dawn Robbins, GIS Manager Ventura County

> As a GIS manager, my job is to focus on the business viability and day-to-day operations of a GIS Division. My daily duties include project management, review of technical issues, marketing and sales, and staff supervision. I try to get the best use of geographic data throughout the organization (that is, throughout the entire County). Sometimes the solution to a problem does not involve GIS (at least in the traditional sense), but involves a new application that happens to include maps.
>
> For example, two of the questions I recently faced as the GIO were 'Why are the Assessor and the Public Works departments performing seemingly redundant map creation and update procedures – one in GIS and one in CAD?'; and 'how can the County provide access to all land-related information (GIS, CAD, graphics systems, legacy databases, etc.) in a consistent, easy to access manner?' These questions require a 'big-picture' approach and a certain amount of political knowledge and sales ability to find a solution. Challenges that took a great deal of energy over my first year in the job included improving relationships with the other organizations in the region; firming up a budget for the GIS Division; and staff evolution.
>
> A GIO must have the basic technical knowledge of GIS, database, and IT processes. But this knowledge must be coupled with vision, political awareness, and a strong sales ability. A GIO mirrors the eclectic nature of GIS itself, as the individual must be able to work in a variety of locations (agencies, cities, etc.), and at various administrative levels. This is difficult – I don't believe I'm there yet.

Questions for further study

1. How can we assess the potential value of a proposed GIS?

2. Prepare a new sample GIS application definition form using Figure 17.5 as a guide.

3. List ten tasks critical to a GIS project that a GIS manager must perform and the roles of the main members of a GIS project team.

4. Why might a GIS project fail? Draw upon information from various chapters in this book.

Further reading

Harmon J.E. and Anderson S.J. 2003 *The Design and Implementation of Geographic Information Systems*. Hoboken, NJ: Wiley (e-book).

Heywood I., Cornelius S. and Carver S. 2002 *An Introduction to Geographical Information Systems* (2nd edn). Harlow: Prentice-Hall.

Obermeyer N.J. 2005. 'Measuring the benefits and costs of GIS'. In Longley P.A., Goodchild M.F., Maguire D.J., and Rhind D.W. (eds) *Geographical Information Systems: Principles, Techniques, Management and Applications (abridged edition)*. Hoboken, NJ: Wiley: 601–610.

Sugarbaker L.J. 2005. 'Managing an operational GIS'. In Longley P.A., Goodchild M.F., Maguire D.J., and Rhind D.W (eds) *Geographical Information Systems: Principles, Techniques, Management and Applications (abridged edition)*. Hoboken, NJ: Wiley: 611–620.

Tomlinson R. 2003 *Thinking about GIS: Geographic Information System Planning for Managers*. Redlands, CA: ESRI Press.

Thomas C. and Ospina M. 2004 *Measuring Up: The Business Case for GIS*. Redlands, CA: ESRI Press.

18 GIS and management, the Knowledge Economy, and information

It is not enough knowing how to obtain and run a GIS (Chapter 17). Potentially, GIS can now underpin virtually every decision made within and between organizations. Thus sustainable success only comes if the system as used makes a material contribution to the success of the entire enterprise; and any enterprise will only succeed if it is led and managed effectively. Therefore, effective leadership and management are crucial elements for successful GIS in government, business, 'not for profit' organizations, and academia.

The Knowledge Economy (KE – which includes GIS), and management that operates within it, is affected by the growth of networking (such as the Internet) and other technological changes. But many business principles remain unchanged, e.g., understanding the business drivers and how GIS can contribute to those that, locally, are the most important. Technology alone is only one contributor to success. To make a success of GIS, managers need to understand the Knowledge Economy, the role of innovation, what motivates people, and the unique characteristics of geographic information.

Geographic Information Systems and Science, 2nd edition   Paul Longley, Michael Goodchild, David Maguire, and David Rhind.
© 2005 John Wiley & Sons, Ltd.   ISBNs: 0-470-87000-1 (HB); 0-470-87001-X (PB)

Learning Objectives

After reading this chapter, you will understand:

- Government, academia, 'not for profits', and commerce as 'businesses';

- Management and GIS;

- Business drivers;

- GIS and the growth of the Knowledge Economy and Knowledge Industries;

- The characteristics of geographic information as a commodity;

- The impact of the Internet and the Web upon GI and management;

- GIS as a growth business in its own right.

18.1 Are we all in 'managed businesses' now?

Most readers of this book live in a capitalist society. Our tangible wealth and physical well-being – and hence freedom to do many things – come increasingly from the exploitation of knowledge or skills. Much of this exploitation is underpinned by scientific or technological innovation. Increasingly, we see the breakdown of previously discrete sectors: commercial considerations impinge more widely and there is convergence between the activities and operations of commerce and industry, government, the not-for-profit sector and academia. Increasingly everyone is concerned with explicit goals (many of which share similar traits) and with 'knowledge management'. Moreover, in the GIS world at least, we also see significant overlap in functions – for instance, both government and the private sector are producers of geographic information (GI). The not-for-profit sector increasingly acts as an agent for or supplement to government and usually operates in a very business-like fashion. The result is that previous distinctions are becoming blurred and movement of staff from one sector to another is becoming common.

For these reasons, we use 'business' as a single term to describe the corporate activities in all four sectors. Accordingly, we think of the GIS world as being driven by organizational or individual objectives, using raw material (data or information: see Chapter 1), tools (geographic information systems) and human capital (skills, knowledge, attitudes, and approaches) to create new products or services (e.g., other information or knowledge) which help us to meet the objectives.

This commonality between the different sectors should not be exaggerated. For example, in publicly owned commercial organizations in the USA, pleasing the stockholders every quarter is a commonplace requirement. The lack of profitability and resulting low price of shares in Smallworld – a pioneer – in GIS object orientation – led to its purchase by General Electric in fall 2000. Takeovers and mergers are also much more common in the commercial than the government sector – as in the purchase of the major US data firm GDT by the multinational Tele Atlas in 2004. But it is still realistic to regard all good organizations as business-like.

Geography is increasingly recognized as a central concern for all businesses. This book contains many examples of why this is so. In the real world, however, rarely is geography the *only* issue with which managers have to grapple. And rarely do managers have the luxury of being solely in charge of anything or have all the information needed to make truly excellent decisions. Box 18.1 gives a typical example of how geographic and non-geographic issues are intertwined and how many

---

The geography of coping with catastrophe

On the evening of 21 May 2001, The City University in London suffered a disaster. One section of an 1894 building (Figure 18.1A), an icon for generations of students, was engulfed in flames (Figure 18.1B). The London Fire Service was soon there and pouring water on the inferno. As the night rolled on, a helicopter parked overhead sent images of the hottest areas to the firefighters; the desperate battle to contain the flames ebbed and flowed. Soon after midnight, the immediate crisis was over although the fire service did not leave until hours later.

The university was lucky. All 250 people in the building were evacuated safely. There was no wind that evening: had there been any, many more rooms and around a thousand exam scripts stored nearby would have been lost. The fire station was nearby, ensuring the fire tenders arrived within a few minutes of the alarm being

(A)

(B)

(C)

Figure 18.1 The City University fire of 2001: an example of geographically-related management decisions taken by multiple players (mostly) in concert. (A) College Building before the fire; (B) College Building in flames; (C) the headlines in the next day's London newspaper, replicated by global TV coverage (courtesy Evening Standard)

raised. And there were no other major fires in central London that night so much assistance was on hand.

In the immediate aftermath of the fire – the next morning – both normal operations (including exams) and planning for recovery had to be managed. The geographically-related decisions that were made – some macro-level and some micro-level – by various players and processes included the following:

■ The evacuation of students and staff by security staff through safe corridors, necessitating good public records (and staff awareness) of the intricate internal geography of the building.

■ The speedy arrival of the fire tenders using address/street network-based routing systems.

■ The closing of local streets and re-routing of traffic by the police.

■ The mobilization of other suitably located fire tenders on a risk-assessment basis from across London to help fight the fire.

■ The telemetering of thermal imaging of 'hot spots' from helicopter to fire fighters.

▶

- The mapping of the areas damaged by fire, flood, or smoke the next morning to quantify the scale of the problem and act as the basis for immediate management action. This included surveys for newly revealed hazards such as exposed asbestos.

- The immediate appropriation of suitable surviving space for examinations and the setting up of posters to guide students to the new venues in good time.

- The search by commercial firms for suitable replacement office space within an acceptable distance, using their GIS and personal knowledge.

- The evaluation of temporary accommodation offered by other universities to test the feasibility of its use given geographical proximity, etc.

- The decisions on which areas of the world to concentrate advertising, given the huge publicity the fire received in the media (including CNN, broadcast worldwide). Their message was that the university had been destroyed (Figure 18.1C). This encouraged overseas students to believe they should switch to another university – which would have led in turn to huge loss of income.

- The progressive (and beneficial) restructuring of space within the university over the next three years, triggered by the fire and carried out partly on space-planning software.

separate groups make decisions, often in parallel, which impact on others.

18.2 Management is central to the successful use of GIS

18.2.1 Some basics

For many people, management implies a dull, routine job ensuring processes are followed and production targets are met. Today's manager, however, is required to anticipate future opportunities and possible disasters and take action appropriately to change the local world. He or she has to keep up-to-date with changes in organizational aims and mores – and help to shape them. Persuading colleagues and staff to give of their best and ensuring targets are achieved is essential. Such pro-active management is often more properly titled 'leadership'; and it takes place at all levels in organizations – from the highest (Box 18.2) to almost the lowest.

Management matters; luck and idiocy on the part of one's competitors usually only plays a small part in what happens. Though necessary, excellent science and technology – however good – are not sufficient conditions for success. Microsoft did not succeed simply because it produced good software. It succeeded because the organization had great management and smart people. Its leaders had a vision of what they wanted to achieve, ideas on how to do it, and superb marketing skills. They also had the ability to reprioritize and reformulate plans and activities as often as necessary. All these management abilities make the difference between success and, at best, mediocrity. It follows that business awareness and flexibility of the staff and flexibility of the GIS tools used are crucial.

There are some things which the manager can rely upon. These include:

- *People cause more problems than technology.*
- *Things change increasingly rapidly* – both the technology and the user expectations.
- *Complexity is normal but some of it is self-imposed* – and all organizations tend to believe they are unique and complicated.
- *Uncertainty is always with us.*
- *Inter-dependence is inescapable.* Creation and implementation of a GIS often have far wider ramifications than originally imagined and hence impacts on many more people, at least indirectly.
- *Customers for a system often have very fixed ideas of what they need* yet these often change by the time a system appears.
- *Detailed specifications of needs are usually seen as necessary* to hold people (contractors, managers, etc.) to account. But these then complicate (and make expensive) changes to a project.
- *There are considerable differences in national or even regional cultures.* These can have significant impacts in running GIS projects in other countries (see Box 18.3).

Management is not 'rocket science' – almost anyone can be competent at it with good training, enough practice, an understanding of 'the big picture', and a modicum of understanding of people. But it is not a trivial pastime – far from it. Few people are good at it all the time. Getting good results normally requires unremitting effort, intelligence, and an ability to welcome criticism and adapt behavior appropriately (see Box 18.3).

Management is not 'rocket science' but neither is it easy. Everyone is a manager at some stage.

Suwit Khunkitti: Deputy Prime Minister of Thailand

By background a chemist trained at the University of Kentucky, His Excellency Suwit Khunkitti is a strong advocate of GIS. This is derived in part from huge experience in the government of his country. A member of the Thai Parliament since 1983, he has at various times been Minister for Agriculture and Cooperatives, for Justice, for Natural Resources and Environment, and for Science, Technology and Environment. In addition, he chairs the National Research Council of Thailand, the Information and Communication Technology Committee, the National Water Resources Committee, the National Environment Board – and the National Geographic Information System Committee. Amongst many other roles, he has also been Chancellor of the Law Society of Thailand.
Kunkitti says:

Figure 18.2 His Excellency Suwit Khunkitti, pictured at a UN Conference in Geneva in 2004 on 'Hazardous Chemicals and Pesticides in International Trade'

As a national executive, I envision the development of information technology as fundamental for providing solutions to national challenges and for our ability to ensure the sustainable growth of the nation. Such development requires databases which are correct and must always be updated. It is vital that government agencies apply contemporary information technology to their day-to-day operations. GIS in particular is an information technology that is essential for *all* aspects of national administration. I have laid the strong foundations for the national GIS development through establishing the national GIS committee and ordering the NSDI study. These actions will allow all organizations in Thailand to develop efficient GIS systems of their own. Already GIS has been used widely in Thailand, including applications to solve problems related to the management of natural resources and the environment. GIS have been used for city planning, public utility services, management of agriculture, flood control, to support the drought protection plan – and much else.

18.2.2 Some GIS-specific challenges for managers

These arise from specific characteristics of GI or of the nature of the industry; these are discussed later. But, as an illustration of the specific challenges, consider:

■ There are many different ways of representing the world (Chapter 3) in GI form. How this is done

influences what can be achieved and the quality of the results obtained.

■ GI is fuzzy in that each and every element of it has associated imprecisions (Chapter 6); mixing data of different accuracies can lead to big problems.

■ Our techniques for describing 'data quality' are still primitive. Thus assessing its 'fitness for purpose' is often not simple, especially if the metadata (Section 11.2) are inadequate.

Cultural and political consequences of geographical errors

An illustration of the consequences of getting things wrong was given publicly by the senior geopolitical strategist at Microsoft during the four-yearly International Geographical Congress in 2004. This candor is refreshing: all organizations get some things wrong but few admit it so others can learn from their mistakes.

When Microsoft released Windows 95, the time zone map covering India did not include

Kashmir, the disputed border area between India and Pakistan. Though only a few pixels in size, this contravened Indian law. It led to the enforced withdrawal of all 200 000 copies of that system and the supply of a replacement, costing the firm millions of dollars. The solution adopted was to degrade the resolution of the map so that individual countries would not be identified.

▶

The map of Turkey in Encarta originally contained the word 'Kurdistan', denoting the cultural homeland of the Kurds. The Turkish government, which has been in armed conflict with Kurdish separatists, arrested Microsoft representatives and interrogated them. Eventually, the name was removed.

Source: as reported by Michael McCarthy in *The Independent* of 19 August 2004

Another embarrassing geographic factual error was the loss of Wales from a map of Europe on the cover of the *Eurostat Statistical Compendium* for 2004 (Figure 18.3). That Europe's statistical agency missed this off caused a political furor. The moral is clearly that a 'heavy-duty' quality control mechanism and highly aware management are essential in

Figure 18.3 A bureaucratic blunder left Wales off a map of Europe on the cover of the Eurostat Statistical Compendium for 2004 (Reproduced by permission of European Communities)

encoding geographic data, especially those with political overtones.

- In some cases, combining data gives rise to an 'ecological fallacy' which can produce false correlations between the variables (see Section 6.4.3).
- There remains a general lack of awareness of the value of GIS which means that GIS managers need to be evangelical.

18.2.3 Attitudes towards management

The idea that GIS can be implemented and run simply by following 'cookbook' instructions and that success will inevitably follow is quite wrong. It is contrary to all management experience in projects based on the implementation of high technology. It has been argued there are three different managerial approaches relevant to IT in general and GIS in particular. These can be described as technological determinism, managerial rationalism, and social interactionism. We consider these to be part of a spectrum and consider only the end points, the first and third approaches.

Successful GIS implementation and use require more than just technical solutions.

Technological determinism is a Utopian approach which stresses the inherent technical merits of an innovation. The approach can be recognized in those GIS projects which are defined in terms of equipment and software. The project is usually sold to potential users on the basis that 'There Is No Alternative (TINA); this will do everything you need and has lots of intangible benefits because of the capacity of the technology'. Relatively little emphasis is placed on the human and organizational aspects compared to fine-tuning the software and meeting the detailed technical specification.

Social interactionism involves a formal recognition of uncertainty on how organizations really work and a belief that knowledge (see Section 1.2) and culture within the institution have big impacts on the success of IT projects. In this view, decision making is an interactive, iterative, and often fast-changing process between individuals and groups – both within the organization and without – which are sometimes in conflict and sometimes in collaboration. Success comes from placing stress on the organizational and user acceptance and use of the technology, rather than on the intrinsic merits of the system.

Given the nature of contemporary business life, normally only some version of the social interactionism approach works well (though in war-time other approaches may operate successfully). In practice, of course, most organizations contain people of both persuasions so managers have to ensure they function well together (see Box 18.4).

18.2.4 Business drivers

Overall, we can think of the imperatives that drive *all* businesses (falling within our wide definition) as being:

- Saving money or avoiding new costs (e.g., due to new regulations).
- Saving time (which *may* also involve saving money but sometimes may not, e.g., in emergencies).
- Increasing efficiency, productivity, and/or accuracy, and aiding budgeting.
- Creating new assets, e.g., intellectual property rights, enhanced brands, or trust through new investment.
- Generating additional revenues or other returns from the identification, creation, and marketing of new products and services by exploitation of the assets.

Hugh Neffendorf: a 'can do' manager and expert consultant

Hugh's consultancy firm, Katalysis, is so named because of its focus on facilitation, bringing appropriate sets of skills to projects, and getting things done. He has had wide experience in the private, public, and academic sectors and has acted as a consultant on GI for many British public sector organizations and for a number of commercial organizations.

Transplanted in his youth from Texas to Northern England, his involvement in geographic information stretches back over 30 years. It began mainly in the field of transportation planning where he headed a large division of a major multi-national consultancy firm. He then applied his experience to the wider GI arena, leading many strategic and technical projects, and taking responsibility for several software and data product developments, including those creating population census software, map data files, and ADDRESS-POINT™, the standard British data source for address geography.

Figure 18.4 Hugh Neffendorf: manager and consultant

Hugh's involvement in projects is unusually varied, including areas as diverse as strategy, business management, marketing, product development, information technology, mathematical modeling, and statistics – and dealing with clients. Nowadays, his most usual role is as the leader of technical teams and as facilitator of working groups. Many of his projects relate to official statistics: he is facilitator of the UK Office for National Statistics Geography Advisory Group, an external national 'think tank' that helps to guide statistical geography policy.

Hugh leads the UK Association for Geographic Information's (AGI) Special Interest Group on Address Geography. Fitting with his view (and that of the authors of this book) that work should be fun as well as effective, he also helps to organize the annual AGI Golf Day!

- Identifying risk and reducing it to acceptable levels.
- Supporting better-informed and more effective decision making in the business.
- Ensuring effective communication with key stakeholders.

Every organization now has to listen and respond to customers, clients, or fellow stakeholders. Every organization has to listen to citizens whose power sometimes can be mobilized successfully against even the largest corporations. Every organization has to plan strategically and deliver more for less input, meeting (sometimes public) targets. Everyone is expected to be innovative and deliver successful new products or services much more frequently than in the past. Everyone has to act and be seen to be acting within the laws, regulatory frameworks, and some conventions. Finally, everyone has to be concerned with risk minimization, knowledge management, and protection of the organization's reputation and assets.

All organizations must be responsive to the needs of customers and other stakeholders and must demonstrably be effective. This is achieved through good management, rather than luck or individual genius.

These drivers have different importance in each sector. In addition, the ambitions of and drivers for *individuals* within the organization cannot be ignored, especially in enterprises dependent on clever, highly marketable people (such as some universities and GIS firms). These factors are incorporated and expanded in Table 18.1.

Understand motives and you understand why many things happen.

The relationship between the different sectors also shifts over time. Until the early 1980s, much of the available GIS software was produced by government or by individuals. Only with the arrival of significant commercial enterprises – and hence real competition – has GIS become a global reality, with the number of users at least ten thousand times greater today than it was in the 1970s. Relatively little general-purpose and widely used GIS software is now produced outside the commercial sector, one exception being the IDRISI software from Clark University (see Box 16.6). Also, in part, commercial assemblers and sellers of geographic information have taken over a role traditionally associated with governments.

All organizations and their employees are subject to similar needs and incentives, though what is most important varies locally and over time.

The context in which you manage is rarely constant for long so we now describe the changing world of knowledge creation and exploitation – a world in which GIS is prospering.

Table 18.1 Some business drivers and typical responses. Note that some of the responses are common to different drivers

Sector	Selected business drivers	Possible response	GIS example
Private	Create bottom-line profit and return part of it to shareholders. Build tangible and intangible assets of firm. Build brand awareness.	Get first-mover advantage; create or buy best possible products, hire best (and most aggressive?) staff; take over competitors or promising startups to obtain new assets; invest as much as needed in good time; ensure effective marketing and 'awareness raising' by any means possible; reduce internal cost base.	Purchase and exploitation by Autodesk of MapGuide software and subsequent development – one of earliest Web mapping tools. Engagement of ESRI in collaboration with educational sector since c.1980 – leading to 80% + penetration of that market and most students then becoming ESRI software-literate. Purchase of GDT Technologies by Tele-Atlas to obtain comprehensive, consolidated USA database and to remove major competitor.
	Control risk.	Set up risk management procedures, arrange partnerships of different skills with other firms, establish secret cartel with competitors (illegal)/gain de facto monopoly. Establish tracking of technology and of the legal and political environment. Avoid damage to the organization's 'brand image' – a key business asset.	Typically, GIS firms will partner with other information and communication technology organizations (often as the junior partner) to build, install or operate major information technology (IT) systems. Many GIS software suppliers have partners who develop core software, build value-added software to sit 'on top', act as system integrators, resellers, or consultants. Data creation and service companies often establish partnerships with like bodies in different countries to create pan-national seamless coverage. Avoid partnerships with 'fly by night operations'. Know what is coming via network of industry, government, and academic contacts.
	Get more from existing assets.	'Sweat assets', e.g., find new markets which can be met from existing data resources, re-organized if necessary.	Target marketing: use data on existing customers to identify like-minded consumers and then target them using geodemographic information systems (see Section 2.3.3).
	Create new business.	Anticipate future trends and developments – and secure them.	Go to GIS conferences and monitor developments, e.g., via competitors' staff advertisements. Buy start-ups with good ideas (e.g., ESRI and the Spatial Database Engine or SDE). Network with others in GI industry and adjacent ones and in academia to anticipate new opportunities.
Government	Seek to meet the policies and promises of elected representatives or justify actions to politicians so as to get funding from taxes.	Identify why and to what extent proposed actions will impact on policy priorities of government and lobby as necessary for tax appropriations. Obtain political champions for proposed actions: ensure they become heroes if these succeed.	Attempts to create national GIS/GI strategies and NSDIs (see Chapter 20). Impact of President Clinton's Executive Order 12906. Force the pace of progress on interoperability to meet the needs of Homeland Security (see Chapter 20).
	Provide good Value for Money (VfM) to taxpayer.	Demonstrate effectiveness (meeting specification, on time and within budget) and efficiency (via benchmarking against other organizations).	Constant reviews of VfM of British government bodies (including Ordnance Survey, Geological Survey, etc.) over 15 years, including comparison with private sector providers of services. National Performance Reviews of government (including GIS users) in USA and the impact of the 2002 e-government initiative.

Table 18.1 (*continued*)

Sector	Selected business drivers	Possible response	GIS example
	Respond to citizens' needs for information.	Identify these; set up/encourage delivery infrastructure; set laws which make some information availability mandatory.	Setting up of US National Geospatial Clearinghouse under Executive Order 12906 plus equivalent developments elsewhere (Chapter 20). Subsequent development of hundreds of GI portals worldwide.
	Act equitably and with propriety at all times.	Ensure all citizens and organizations (clients, customers, suppliers) are treated identically and that government processes are transparent, publicized, and followed strictly.	Treat all requests for information equally (unless a pricing policy permits dealing with high-value transactions as a priority). Put all suitable material on Web but also ensure material is available in other forms for citizens without access to the Internet.
Academic	Establish high research reputation.	Carry out, publish, and disseminate results of new research, advancing field. Win big research grants from prestigious sources (peer-reviewed academic or science sources normally give more prestige than those from industry or government). Run conferences attended by top-class participants, form informal 'club'. Win prizes.	Focus research on things that business does not do or cannot talk about for fear of losing competitive advantage, i.e., fundamental research (often technical), 'soft, human-focused' research or policy-relevant research. Set up research centers, get funding to attract graduate students and other researchers. Publish – sometimes with overseas colleagues – in top journals in the GIS-related field (Box 1.5), including some (but not too many) in journals outside of field. Talk to people in different disciplines to get new ideas. Attend many conferences; accept keynote speech invitations.
	Establish reputation for teaching excellence and relevance. Build international links to anticipate globalization of learning.	Provide top-class educational experience for students (increasingly important). Alumni are the best form of advertising in the long term.	Recognize that GIS principles and technology are near-identical worldwide so: • Lead/build consortia to create curriculum materials, deliver identical courses worldwide (e.g., UNIGIS model – see Box 19.4) and/or develop state-of-the-art courses which can be run on a residential basis. • Differentiate courses from others in ways suited to the local market needs. Pro-actively seek overseas students to build international base.

18.3 The Knowledge Economy, knowledge management, and GIS

Different people mean different things by the term 'Knowledge Economy'. Central to it, however, is the use of new knowledge to run things more effectively (e.g., local democracy) and/or to create still-newer knowledge and exploit it. The nature of the Knowledge Society is dynamic – knowledge is constantly being added to so keeping up to date is crucial.

18.3.1 Innovation matters

Continuous innovation and creativity are fundamental to the health of the Knowledge Economy. The British Science Minister Lord Sainsbury summarized their importance by saying:

We define innovation as 'the successful exploitation of new ideas'. Often it involves new technologies or technological applications. . . For consumers, innovation means higher quality and better value goods, more efficient services (both private and public) and higher standards of living. . .The innovative company or organisation delivers higher profits for its owners and investors. For employees, innovation means new and more interesting work, better skills and higher wages. Equally, an absence of innovation can lead to business stagnation and a loss of jobs.

For the economy as a whole, innovation is the key to higher productivity and greater prosperity for all. Innovation will also be essential for meeting the environmental challenges of the future – including moving to a low carbon economy and reducing waste. We need to find new ways to break the link between economic growth and resource depletion and environmental degradation.

Source: DTI (2003) *Competing in the Global Economy*, London, Department of Trade and Industry.

Worldwide, there is a gathering pace in commercial developments through such innovation of new products, processes, and services, and in the rapidity of their dissemination and acceptance on a national or worldwide basis (see Figure 18.5). Inventive entrepreneurs in Hong Kong take for granted the truth of the phrase 'If it works, it's obsolete'. The life cycle of new products, services, and even of knowledge itself is becoming ever shorter under competition which is ever more global. This and many other business developments have been facilitated greatly by scientific and technological developments, especially in the information and communication technologies (ICT); but these are not the only factor, and technology has not always done what was anticipated (Box 18.5).

This escalation of innovation is most rapid in the 'Knowledge Industries' where substantial national and business revenues may be at stake through licensing of new processes or products (Figure 18.6). It has long been

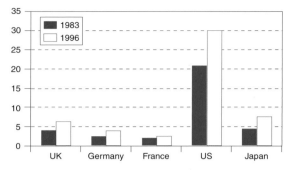

Figure 18.6 Knowledge Economy revenues from overseas earnings on royalties and license fees (US$bn). (*Source*: 'UK Competitiveness White Paper 1998'. © UK Government, Department of Trade and Industry) (Crown copyright material is reproduced with the permission of the Controller of HMSO and the Queen's Printer for Scotland)

recognized that these industries will be the key to future national and business success. As an example, the World Bank's *1998 World Development Report* said:

For countries in the vanguard of the world economy, the balance between knowledge and resources has shifted so far towards the former that knowledge has become perhaps the most important factor determining the standard of living. . . Today's most technologically advanced economies are truly knowledge-based.

GIS is involved inescapably in this innovative gallop and, irrespective of the definitions used, is part of the Knowledge Industries. All the major software vendors have to operate on a continental or global scale: there are enough commonalities of user need – at least at the tool box level – to create huge economies of scale by operating as part of a global business. Each year GIS vendors announce new functionality or new applications for their software; and multi-national mergers and acquisitions in the GI business are no longer unusual.

Those active in GIS are working in the knowledge industries and contributing to the Knowledge Economy; continuing innovation in GIS is strong and crucial to our success.

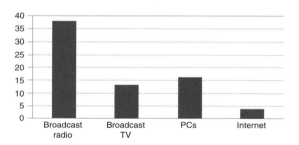

Figure 18.5 Increased speed of up-take of innovations, shown by the years needed for an innovation to be accepted by 50 million Americans (*Source*: US Department of Commerce)

Applications Box **18.5**

Globalization, technology, and unintended consequences

Globalization is not only a factor of technology: it has been driven as much by worldwide trade liberalization and the removal of restrictions on cross-border capital movements. Moreover, technology effects are not always predictable or benign: some have had unintended consequences. Deaths from hurricanes in the USA have dropped from six thousand a year in 1900 to a few dozen thanks in large measure to radar and other tools to track and predict storms (see Figure 2.16, Box 3.5). Yet insurance payouts have soared – $7.3 billion after Hurricane Andrew – in part because people felt safer in building in hurricane-prone areas.

18.3.2 Putting a value on knowledge and information

Since knowledge – however it is defined – is becoming so crucial, it is no surprise that investors increasingly recognize the growing importance of knowledge assets in the way they value enterprises. Box 18.6 demonstrates one approach to categorizing knowledge in commercial enterprises. However, intangible assets occur at least as frequently in government-based organizations, although valuing the equity in such cases is even more complex. What should be counted as intangible and what is tangible are sometimes of great importance in GIS – as shown by the debate over how to account for the National Geographic Database in the accounts of Ordnance Survey, Britain's national mapping agency. The Auditor General of the UK's National Audit Office made the following comment in his report to Parliament on the OS accounts for 2003/04:

> **Ordnance Survey's turnover of £116 million derives principally from the exploitation of data held in Ordnance Survey's geospatial databases the creation of which has been funded from public monies over many years. . . the Agency has not capitalised the costs of setting up and maintaining the data held in the geospatial databases in its Balance Sheet. In the Agency's view, the data [are] an intangible fixed asset that does not meet the conditions for capitalisation set by Financial Reporting Standard 10. In my opinion, the data held in the geospatial databases [are] a tangible fixed asset that should be capitalised in accordance with Financial Reporting Standard 15. Having taken expert advice about the valuation of the data held, in my view the value to the business is not less than £50 million. Had the data been capitalised at that value, the effect would have been to increase tangible fixed assets included in the Balance Sheet at 31 March 2004 from £36 million to £86 million.**

Also, it is obvious that intangible assets may be negative (e.g., where brand image is destroyed by loss of public acceptance of the quality or safety of data or software products or services provided by the commercial firm or government body). Intangible assets are present in extreme form in academia or small businesses where the poaching of one key member of staff or a team may destroy the reputation of the organization in that field overnight.

The most direct way to put a value on information is to seek to sell or lease it. This is done by many commercial firms and some governments (see Section 19.3.2).

18.4 Information, the currency of the Knowledge Economy

Information, then, is a key ingredient of the Knowledge Economy and Knowledge Industries and is the 'rocket fuel' for GIS. In this section, we discuss several different aspects of the role and characteristics of information in general and geographic information (GI) in particular. We

---

Applications Box **18.6**

Valuing knowledge

In fast-growing sectors like biotechnology and computer software (including some parts of GIS), a large part of the value of the company resides in the knowledge embodied in its patents and in its staff and, in turn, how the outside world sees these being used and developed to capitalize on new opportunities. A classic example is the huge growth in the difference between Sun Microsystem's stock market value (what people would pay for its stocks) and its book value (based on normal audit of its physical assets and liabilities) in 1995–96 because of the announcement of Java; this represented a major increase in its intangible assets. One way of categorizing the different types of intangible assets is shown below:

Visible equity (book value)	Intangible assets (stock price premium)		
	External structure	**Internal structure**	**Individual competence**
Tangible assets minus visible debt	Brands, customer and supplier relations (e.g., 'trust')	The organization: governance, management, legal structure, manual systems, attitudes, R&D, software	Education, experience, skills

start with how information underpins decision making. Along with projected efficiency improvements and risk moderation (see Section 18.2), this is the basis on which many GIS have been procured in the past.

18.4.1 Information and knowledge – the basis for good decision making?

Until now, we (and many others) have assumed that having more knowledge and information gives more power and the capability to make better decisions. As authors, we hold to the view that better information often *does* permit better decisions to be made. But the reality is usually much more complicated, especially in GIS – in part this is because it is rarely the *only* important input to decision makers (Section 18.1). Many decisions require the balancing of conflicting objectives and judgments about the trade offs, as known in the light of available knowledge at that moment in time. Thus the role of 'raw information' in shaping the running of the world can easily be exaggerated: managers are not information processors, with the 'right decisions' simply falling out of an analytical process. Five commonplace yet erroneous assumptions about management and information are:

■ *More information leads to better decisions.* In reality, most managers increasingly suffer from information overload, especially of irrelevant information. Mapping and geovisualization facilities (Chapters 12 and 13) provided by GIS can help to reduce the problem greatly but rarely cure it completely.

■ *Managers need the information they want.* Managerial 'wish lists' often do not match to what is actually needed to run the business.

■ *Managers are able to model the decision they wish to make.* In the real world it is unusual for a clear model to exist which identifies the relevant variables and their inter-relationships.

■ *Managers do not need to understand the information system.* In reality, understanding the strengths and weaknesses of a GIS providing evidence is key to informed management decisions – GIS are not simply 'black boxes' producing 'objective answers'. GIS managers use judgment, operate with certain ethical standards, etc. (see Section 19.2.2)

■ *Good information systems lead to enhanced communication.* In practice, the internal structures of the organization may vitiate this potential advantage. Too much information may also clog up responses.

Chapter 2 introduced the application-driven nature of much GIS. But from an institutional perspective, successful use of GIS requires an understanding of the aims, objectives, and ways of operating of the organization and focusing on what is most essential. Given the business drivers set out in Section 18.2.4, the potential benefits of GIS for managers of big projects, or even for corporate bosses, are in creating or clarifying knowledge through:

■ Providing 'factual information' about the location of resources, human or natural (*a facilitation process*).

■ Computing derived 'facts' (see Box 18.7), e.g., fastest routes through a network, population within five miles of a toxic waste dump, or on what has changed since

---

Applications Box 18.7

What are geographic 'facts'?

GIS is seen by many as creating geographic 'facts'. There is, however, a great academic argument about what constitutes 'facts' and to what extent these are universally true or are context-dependent (see Section 1.7). Therefore, managers need to be aware of what can be said to be 'true' in reporting results from GIS to their bosses.

Geographically-referenced statistics are often cited as facts, even if they are based upon measurements of variable quality or where the very concept of a single definitive answer is suspect. For example, the CIA *World Factbook* (www.cia.gov/cia/publications/factbook/geos/sf.html#Geo) claims that the South African coastline is 2798 km in length (see Box 4.5 on the measurement of length and the fractal nature of many geographic phenomena).

It is sometimes argued that the outline of a building, the bounds of a land parcel, or a line of constant elevation on a map are facts because any other person or sensor attempting to represent these facts would do it in much the same way. This completely neglects the cartographer's art and, in particular, generalization and different possible geographical representations of the same phenomena as stored in different GIS (see Section 3.8 and Chapter 8).

Fortunately, some sources of 'facts' are more circumspect: the GeoExplorer website (www.geoexplorer.co.uk/sections/geofacts/geofacts.htm), for instance, actually includes the injunction *'remember to question the accuracy of the GeoFact before you use it'*!

last year (*a facilitation process, producing new information*).

■ The selection, compression and visualization of information to facilitate decision making (*information filtration*).

■ The search for regularity and possible causality: looking for patterns and correlates of geographic distributions, e.g., does spending on certain goods coincide with particular ethnic groups or distance between shops and home? Chapter 2 gives more examples (*information filtration, leading to evidence*), and Section 4.7 examines methods.

■ The creation of added value by linkage of information from different sources. In general, such linkage extends the range of applications possible – with two information sets for a given region, we have one combination; with 20 sets, we have over a million combinations in total (*information fermentation, leading to more information, evidence, and possibly knowledge*: see Section 1.2).

■ Predicting future events which are geographically distributed (*evidence or knowledge-based operations*). This requires both initial 'state variables' and process models (see Chapter 16).

So, under what circumstances is GI valuable or even vital for decision making? We answer this by citing lots of examples throughout the book. But a better way involves understanding of the characteristics of information, how it is valued, and the sorts of processes which now work well through its use – especially those that rely upon use of the Internet which is now the main mechanism for sharing GI.

18.4.2 The characteristics of information in general

We can think of the very same information being used for two different purposes and thus being valued very differently by the same people in different circumstances. The information may be used:

■ *For consumption.* Individuals decide how much value they assign to it based upon their valuation of pleasure, time saving, or some other metric, and their awareness of the uses and the potential benefits of the information.

■ *As a factor of production*, where the information is used as part of a good or service. The end-user of that good or service will make his or her decision on the uses to which it can be put and hence its value to him or her. Value will be affected by its availability to others (it will tend to be more valuable if it is not available to others, such as competitor organizations) and the ease of substitutability. Actually measuring these values is rather difficult, so discussion thus far has typically been at a very high conceptual level (but see Box 17.4 on cost/benefit analysis).

Information, from a management perspective, has a number of unusual characteristics as a commodity. In particular, it does not wear out through use (though it may diminish in value as time passes; for example, population census data from ten years ago are less reliable for current policy-making than equivalent data from last year).

It is usually argued that information is in general a 'public good'. A 'pure' public good has very specific characteristics:

■ The marginal cost of providing an additional unit is zero. Thus, in effect, copying a small amount of GI adds nothing to the total cost of production.

■ Use by one individual does not reduce availability to others ('non-rivalry'). This characteristic is summarized in the famous Thomas Jefferson quotation that: 'He who receives an idea from me, receives instruction himself without lessening mine; as he who lights his taper at mine, receives light without darkening me'.

■ Individuals cannot be excluded from using the good or service ('non-excludability').

A pure public good is a special form of externality. Such externalities arise where production or consumption of a good (e.g., information) by one 'agent' imposes costs on or delivers benefits to other producers or consumers. In this case, provision of a pure public good for one group of users will benefit other potential users because it is not possible to exclude them. Pollution of water or air and by ambient noise form classic examples of external costs (Figure 18.7); refuse collection, education, and public health are classic examples of external benefits.

In practice, information is an optional public good, in that – unlike defense – it is possible to opt to take it or not; you may not feel the need to avail yourself of US Geological Survey data, for example! Moreover, the accessibility and cost of the systems to permit use of information influence whether, in practice, it is a public good in any practical sense. Since maybe only 0.03% of the world's population has ready access to GIS tools, scarcely can it be argued that geographic information in digital form is yet a universal public good in the traditional sense. Finally – to be pedantic – it may also be best to define information as a quasi-public good since it may be non-rival (see above) but its consumption can in certain circumstances be excluded and controlled. The business cases and vast investments of a number of major commercial GI purveyors such as Space Imaging Inc. and DigitalGlobe (Table 1.4) are based on this proposition; if everything they produced could be copied for free and re-distributed at will by anyone, their business would be untenable. Thus the pecuniary value of information may well depend on restricting its availability whilst its social value may be enhanced by precisely the opposite approach.

Traditionally, geographic information has been seen as a type of public good but its characteristics differ in different countries and in different sectors of the economy.

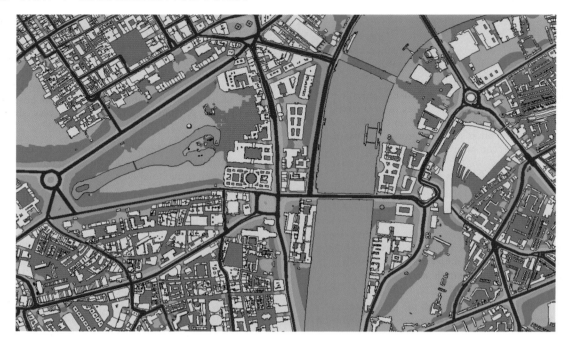

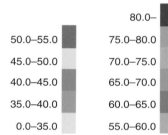

Figure 18.7 The geography of a negative externality: patterns of noise pollution in decibels east from Buckingham Palace (extreme left) in London (*Source*: Department for the Environment, Food and Rural Affairs. Crown copyright reserved)

What is certainly true, however, is that information in digital form – however expensive it is to collect or create in the first instance and even to update – can be copied and distributed via the Internet at near-zero marginal cost. Its widespread availability has many positive (and some negative) externalities which arise far beyond the original creator. In regard to information, the types of externalities can be summarized as:

- Ensuring consistency in the collection of information creates *producer externalities*. Inconsistency can raise the costs of creating and using data and limit the range of applications.

- Providing users access to the same information produces *network externalities*. Network benefits exist for any user of software or data being widely used by others (e.g., Microsoft software). It is, for example, important that all the emergency services use the same geographic framework data.

- Promoting efficiency of decision making generates *consumer externalities*. One example is where access to consistent information allows pressure groups to be

more effective in influencing government policy and in monitoring activities in regard to pollution.

Often, a set of information is also an 'experience good' which consumers find hard to value unless they have used it before. Finally, there is no morality about information itself: often, its source (especially GI) can be readily disguised – for example, by changing the representation of roads from single lines to cased ones, changing their color, removing some sinuosity, or computing tables of distance between places. Provenance of data is an important characteristic both because some data producers are trusted more than others and because information theft can undermine businesses.

18.4.3 The characteristics of GI

GI has the standard characteristics of information described above but also some others that mark it out. For instance, central to the perceived value of GIS is that it can be used to integrate information for the same region derived from many different sources. Such data linkage has at least three potential benefits:

■ it permits infilling of missing data, thereby cutting the cost of data collection in some cases;

■ it facilitates verifying the quality of individual datasets through a check on their consistency;

■ it provides added value, almost for nothing. The number of permutations rises very rapidly as the number of input datasets rises (see Section 18.4.1 and Box 18.8). Hence, when datasets are linked together, many more applications can be tackled and new products and services created than when they are held separately.

Although this presents a colossal advantage, it brings associated problems. Generally, we have to take data 'as they come' from other people. Their classification or level of accuracy may not be what you would like in an ideal world (see Section 6.2.3). There is also a risk in that the results may be provably wrong – and lead to humiliation or being sued (being proved wrong is much easier than being proved right in GIS, just as in most scientific endeavors: see Chapter 6 on uncertainty). Taking information from multiple sources can lead to later conflict over the ownership of products if prior agreements have not been forged.

User attitudes to GI are often that it is, or should be, a free good (provided free or at very low cost and without restriction on its use) as well as a public good. This perception arises from two sources – it has been historically provided free or at low cost by governments to taxpayers and some detailed geographic information has the characteristics of a natural monopoly. The latter effectively means that the first person with the infrastructure wins because normally others can never justify duplication of facilities. Obviously, such monopolies are antipathetic to competition, which demands multiple choices. The obvious parallel for such GI is with telephone lines before the advent of cellular telephones. For instance, assume that only one organization has a complete set of information for every house in the country, that it is unwilling to share it freely, and that the majority of that organization's costs are sunk ones (i.e., they have been spent and cannot be recovered). In such circumstances – not uncommon even now in some parts of the world with government-held information – it is unlikely that the collection and sale of duplicate information can be made competitive.

Some geographic information has the characteristics of a natural monopoly.

Perhaps surprisingly, some competition may not be in the national interest so far as geographic framework information is concerned: it can give rise to negative externalities whilst a monopoly can produce network externalities (Section 18.4.2). Consider, for instance, the situation where information drawn from different sources is brought together and that information was originally compiled on different frameworks or base maps. Even where spatial coincidence occurs in the real world, this would not occur in the GIS because of differences in generalization, resolution, etc., between the two frameworks (see Chapter 5). Putting a value on such externalities is difficult, especially since the costs are visited on those at the end of supply chains and are often only discovered long after analysis of the data has led to mistaken conclusions. This provides great opportunities for lawyers! But it does demonstrate that sometimes there can be a public interest in a natural monopoly in GI in regard to everyone using the same framework or base map – how effective that monopoly is would depend on how well the situation was regulated.

Another factor in making GI special is the relative lack of transience of much traditional geographic information, which is analogous to 'reference books'. Typically, the ground features in topographic maps only change slowly except in very small areas. Satellite imagery of the same area may change much more because of crop change, imperfect conflation of the images sensed at different times, and the application of any classifier used – but any particular business may very well be completely uninterested in this change for its own particular purposes and therefore one purchase per decade may be adequate.

Everyone sharing one 'geographic framework' avoids many problems.

In Chapter 10 we stressed the added value from data linkage; but success in this requires that the results are meaningful and hence the data have been 'fit for purpose'. Measuring the quality of many GI sets and hence of their 'fitness for purpose' is not easy. Moreover, we do not have a good idea of the likely consequences of using 'best available (but probably not good enough)' data for there have been relatively few legal tests as yet of liability involving GI in the courts.

Finally, the role of governments in GI and GIS is, and has been, important. In most parts of the world, the bulk of geographic information has until the last few years been created by governments and these are still major data custodians and providers. Many existing commercial products (e.g., road center-line files) also originated within government, typically with value added by the private sector. Governments also introduce market distortions – varying by country – due to subsidies and legal constraints. Most of this is somewhat different from other Knowledge Industry situations. For example, in the case of pharmaceuticals, government is not directly involved in production; multiple drugs produced by different multi-nationals sometimes compete and government's main role is to regulate safety and ensure that citizens are not being held to ransom by extortionate charging. Quite apart from their own involvement in data creation, GI enterprise is more difficult for governments to regulate than, say, electricity, water, gas, or even telecoms. The same GI may be manifested in many different forms by re-sampling or generalization. There is no single constant unit of GI quantity (c.f. water). Thus the capacity for gross profiteering, creation of monopolies, or even fraud is greater than in some other industries, especially with an immature market.

In summary, GI fits somewhat awkwardly within standard economic classifications. Its market (including social) value – and hence use – are difficult to predict because of its source and other characteristics. This should be seen as an opportunity as much as a problem!

GI is a tradeable commodity and a strategic resource. Its supply and use is governed by public policy, economics, contractual, financial, and other considerations but regulating the market is difficult.

18.4.4 The Internet, World Wide Web, and GI

The Internet has been hailed as a distribution channel, a communications tool, a marketplace, and an information system – all in one. It is said to have altered almost everything managers do, from finding suppliers to co-ordinating projects to collecting and managing customer and client data. Its use is global and expanding rapidly (see Figure 18.8).

Some people argue that the change is more apparent than real, at least so far as commerce and the economics of information are concerned. They claim that there is no need for a new set of principles to guide business strategy and public policy in the Internet age. Have you read, they ask rhetorically, the standard literature on differential pricing, bundling, signaling, licensing, lock in, or network economics – have you studied the history of the telephone system or the battles between IBM and Microsoft with the US Justice Department? On this argument, technologies change but economic laws do not.

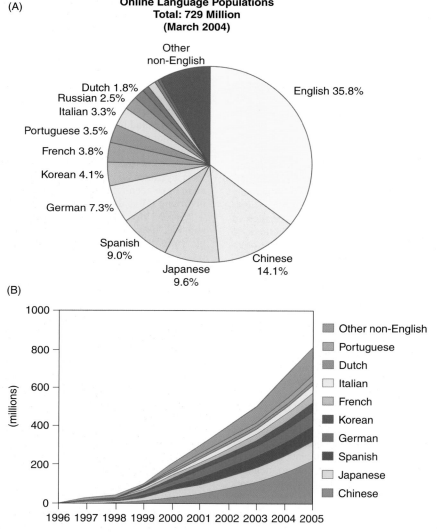

Figure 18.8 Use of the Internet, according to the GlobalReach consultancy in March 2004. (A) the use made by different language groups in 2004; (B) growth of use amongst non-English-speaking groups (*Source*: GlobalReach March 2004 **global-reach.biz/globstats/evol.html**) (Reproduced by permission of GlobalReach)

Some Internet hype

Extravagant claims for the Internet have been made by MIT's Media Lab staff, e.g., Nicholas Negroponte said: '[thanks to the Internet, the children of the future] are not going to know what nationalism is'; his colleague Michael Dertouzos has claimed that digital communications will bring 'computer-aided peace…which may help to stave off future flare ups of ethnic hatred and national breakups'. All of this is reminiscent of earlier quotes: a telecoms enthusiast said: 'it is impossible that old prejudices and hostilities should longer exist, while such an instrument has been created for the exchange of thought between all the nations of the Earth'. He spoke in 1858 and referred to the advent of the first transatlantic telegraph cable.

Certainly there have been some extravagant claims made for the Internet (see Box 18.8). There remains a substantial 'digital divide' between those who have access to it and those who do not. Whilst there are circumstances in which sales and distribution via the Internet really have changed an industry, this particularly applies to goods in finite supply (e.g., airline seats) and those that are highly 'perishable' (airline seats are valueless after take-off). Much GI is not very perishable.

So far as GIS/GI users and managers are concerned, however, there are a number of very real or apparent advantages of the Internet. These are:

- The ease of setting up and using 'information location tools' (portals and clearinghouses or catalogs for digital libraries: see Chapters 11 and 20).

- The possibility to preview or 'taste' simplified versions of the information to check its suitability (essential in what is for many an 'experience good' – see Section 18.4.2).

- The capacity for mass customization, i.e., meeting the needs of the individual through tailored use of automated tools and standard data – hence segmenting the market to a huge degree and permitting differential pricing (see Figure 18.9).

(A)

(B)

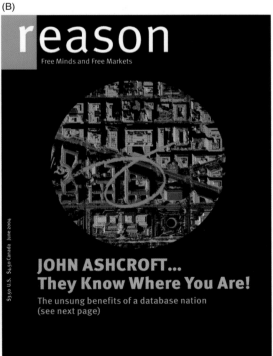

Figure 18.9 Customized covers of the June 2004 issue of *Reason* magazine, with an air photo of the home of each one of the 40 000 individual postal subscribers shown on the copy they received. This was achieved by merging their subscription address through a geo-coded address matching process with suitably geo-referenced aerial photography and generating customized, clipped digital images for the digital printing process. It heralds 'hyper-individualized' publications but also highlights concerns about privacy (see Section 19.1.4). *Source*: **reason.com/june-2004/samples.shtml**

- Its ability to transfer data (generally voluminous in the GI world) at very low cost.

- The ability to transfer costs to the users from the producers. Instead of paying for staff to deal with orders, package and dispatch material, etc., use of the Internet requires that users spend their own time on explicit selection from a menu of options and invest in whatever communication costs are involved.

- The reduction in search time for those who do not live near to good libraries and the immediacy of obtaining the results.

- The seamless way in which information generally may be obtained.

- If charging is implemented, it can be done through simple, low-cost operations like the use of credit cards.

- Initial supply of the information may be tracked and conditions of use enforced on the initial user through his or her inescapable agreement to terms and conditions of use before the information is down-loaded (as for software).

- The astonishing and rapidly growing familiarity with, and use by many people of, the Internet produces network externalities (see Chapter 11, Section 18.4.2, and Figure 18.8).

The Internet facilitates improved marketing, supply, and formatting of GI.

In the GI case, the greatest of all benefits come from customized mass production – where sub-sets of the data are selected to meet particular needs – and from the 'added value' potentially obtained by linking datasets together. Figure 18.9 shows where these two capabilities have been brought together. But the benefit is not because of the Internet *per se* but rather because of the digital form in which that information is held, together with its geographic identifiers. The Internet thus is a 'simplifier' of problems rather than a necessary condition for extracting added value. More importantly, in a world where fast answers are expected, the Internet and World Wide Web may make the delays so short that for the first time they are tolerable for managers and users.

18.5 GIS as a business and as a business stimulant

Thus far, we have treated GIS as a factor in achieving business success in commercial, government, not-for-profit, or academic organizations. But just how big is the GIS business itself – indeed, how can we measure it?

18.5.1 The supply side

First, let us consider the spending on the purchase of software, hardware needed to run it, data, and human resources (people). Truly comparable 'supply side' figures are hard to come by since published accounts of different firms often group activities in different ways reflecting their particular organizational structure. Not surprisingly, firms also portray their figures in the best possible light. The structure of the GIS industry mutates constantly as software vendors provide new services (such as data, publishing or education and training) to strengthen their hold on customers, often through partnership with other players (sometimes also as competitors – see Chapter 20) so a definition of GIS can never be definitive. However, Figure 18.10 shows how global GIS revenues are said to have increased, according to a consultancy specializing in measuring market performance. The figures should be treated with some caution but they show that software sales alone have more than doubled in five years.

An alternative approach is to consider one company and break down its revenues and other indicators of success such as licenses. ESRI – perhaps the largest GIS company worldwide – has no less than 125 000 organizations worldwide with signed license agreements; it has more than a million 'seats' for its different products. It is estimated that this translates to about half a million active users at any one time. ESRI gross incomes have risen between 10% and 20% annually over the last decade, with the financial year 2003 gross revenues being $500 million; if the revenues of the partly owned ESRI 'franchises' in other countries are added in, the total revenue comfortably exceeds $700 million. Another indicator of growth has been the numbers attending

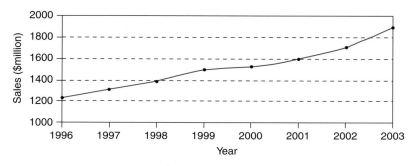

Figure 18.10 The growth in size of the global GIS market, according to Daratech Inc. (© Daratech)

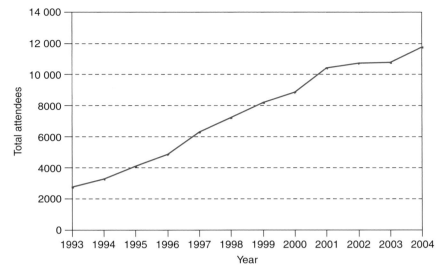

Figure 18.11 Growth in the numbers of attendees at the annual ESRI User Conference, which reflects growth in the GIS community worldwide

the annual ESRI conference: Figure 18.11 shows how this has grown from 23 attendees in 1981 to about 12 000 in 2004 – in many ways an allegory for GIS as a whole. Also the company estimates that their activities leverage at least 15 times the revenues they get in terms of staff, hardware, training, and other expenditures on the part of the users. All of this suggests that ESRI alone generates GI-related expenditures of over $7 billion annually.

If the world population had grown at the same rate as attendance at ESRI conferences since 1981, the global population would now be 2000 billion people, instead of 6.4 billion!

Figure 7.10 shows the proportions of the total global GIS software market achieved by various different players. If we assume each has the same leverage factor as ESRI, the total expenditure on GIS and related activities worldwide cannot be much less than $15 to 20 billion. Depending on the width of the definition of GIS, it could be much higher still (e.g., if the finances of both commercial and military satellites are included). Given the number of ESRI licenses, those of other commercial firms, use of 'free software' provided by governments and other parties plus the illegal, unlicensed copying and use of commercial software which still occurs, together with the contributions of the education sector and other factors, the likely number of active GIS professional users must be well over two million people worldwide. At least double that number of individuals will have had some direct experience of GIS and perhaps an order of magnitude more people (i.e., well over 10 million) will have heard about it (e.g., through such events as GIS Day – see Box 20.1). Yet more will, sometimes unknowingly, have used elementary GIS capabilities in passive Web services such as local mapping.

18.5.2 The demand side

Other ways of quantifying activities are to consider the expenditures of public bodies, all of which – other than security agencies – generally are reported. In the European Union, for example, the civilian national mapping agencies alone spend between $1.5 and $2 billion in total whilst land registration and other map and GI-using public sector bodies spend several times this amount. Worldwide, this figure can be multiplied by perhaps at least ten. The military expenditure in this area is very considerable: the US National Geospatial-Intelligence Agency (NGA) has a publicly stated budget of about $1 billion. How much of all this is on GIS and GI is partly a matter of definition. But, from all this 'user demand side' information, we can get some approximate confirmation of the software supply side estimates in the previous paragraphs. Cary and Associates stated that in 2003 geotechnology spending reported to the Federal Procurement Data System by US federal agencies exceeded $6 billion for the first time. This represents an increase of nearly eight percent over 2002.

The amount of other economic benefits underpinned by this GIS activity is not clear; few studies have been done and disentangling the interactions is difficult. One such study by consultant economists, however, has suggested that the $200 million annual expenditure on keeping up-to-date the national mapping of Britain supports and distributes products about one thousand times that sum in the Gross Value Added to the economy. But in no sense is the connection absolute: other sources of information – or guesswork – would be used if the national Ordnance Survey mapping were not available. Nevertheless, even if wrong (either way) by a factor of three this does demonstrate the scale of the GI/GIS business when considered on a wide definition.

18.6 Discussion

In this chapter, we have seen that GIS is now nothing less than an increasingly useful part of management's weaponry across almost all sectors of human activity. GIS is used in commerce to obtain a comparative advantage for one group over others; it is used as a means of evaluating policy and practice options in both the public and private sectors. Also, it is used as a highly effective information sharing device where that is deemed desirable.

To be useful to an individual manager, such generalizations need to be refined to fit the business environments faced by different organizations. Manifestly, these differ on many detailed counts: by virtue of the level of operations (e.g., international, national, or local); sector (private/public and government of different levels/not-for-profit' bodies); by the level of competition (formal and informal, as occurs in many governments); by business culture (e.g., willingness to collaborate with other bodies); and other factors. Some environments are much more conducive to success than others: for instance, the constraints of working with poorly skilled staff, on out-of-date equipment, and to impossible deadlines are non-trivial. As always, it is the interaction of the various elements of the business environment and the incentive structures and operating constraints that pose the greatest challenges for management in general and GIS in particular.

Questions for further study

1. Write down – without looking it up on your Intranet or in paper documentation – the five or six key strategic aims of your organization. Are the most important drivers in the organization in accord with these aims?

2. What are the special characteristics of geographic information as a commodity?

3. What potential benefits does GIS bring to businesses of different types (as defined in Section 18.1)?

4. Imagine that tomorrow you are to take over responsibility for five demoralized and under-funded staff in a GIS group in academia, business, or government. What would be your first actions? How do you decide whether someone is doing a good job and what do you do if they are not?

Further reading

Brown J.S. and Duguid P. 2000 *The Social Life of Information*. Cambridge, MA: Harvard Business School Press.

Shapiro C. and Varian H.R. 1999 *Information Rules: A Strategic Guide to the Network Economy*. Cambridge, MA: Harvard Business School Press.

Thomas C. and Ospina M. 2004 *Measuring Up: The Business Case for GIS*. Redlands, CA: ESRI Press.

19 *Exploiting GIS assets and navigating constraints*

Recent years have seen a big increase in the value of 'g-business' in commerce, government, not-for-profit organizations and academia. This has been built on use of both tangible and intangible business assets such as legal ownership or rights to physical and intellectual property, the caliber and skills of staff, the customer or client base, brand image and organizational reputation, and organizational compliance with accepted standards. Sometimes however the same factors can act as constraints. This chapter considers the importance of three major factors both as assets and constraints – the law relating to GIS, staff competences and the availability and accessibility of 'core' datasets. GIS is inevitably disruptive to most organizations in the early stages of its adoption even if it is crucial to their long term success. And events (such as 9/11) or new technologies (such as the Web) can reshape our worlds remarkably rapidly. From all this, we demonstrate the need for an organization-wide GIS strategy and implementation plan and the need to identify and reduce risk to acceptable levels.

Geographic Information Systems and Science, 2nd edition   Paul Longley, Michael Goodchild, David Maguire, and David Rhind.
© 2005 John Wiley & Sons, Ltd.   ISBNs: 0-470-87000-1 (HB); 0-470-87001-X (PB)

Learning Objectives

After reading this chapter, you will understand:

■ How the law impacts on GIS assets, plans, and operations;

■ The need for particular staff skills and attitudes plus the reasons why professional certification is advocated by some experts;

■ How availability, accessibility, and quality of core GI act as business constraints;

■ The issues about making public sector GI freely available to all users;

■ How a GIS strategy helps us to exploit assets, cope with constraints, and reduce risk.

Figure 19.1 The scales of justice: the essence of the law. Actions of individuals or organizations are assessed against the laws of the land and judgments handed down (Reproduced by permission of Richard Bailey)

19.1 GIS and the law

The law touches everything (Figure 19.1). During a career in GIS, we may have to deal with several manifestations of the law. These could include copyright and other intellectual property rights (IPR), data protection laws, public access issues enabled, for example, through Freedom of Information Acts (FOIA), and legal liability issues. But since laws of various sorts have several roles – to regulate and incentivize the behavior of citizens and to help resolve disputes and protect the individual citizen (Figure 19.2) – almost all aspects of the operations of organizations and individuals are steered or constrained by them. One complication in areas such as GIS is that the law is always doomed to trail behind the development of new technology; laws only get enacted after (sometimes long after) a technology appears.

There is also a geography of the law. The legal framework varies from country to country. For example, the Swedish tradition of open records on land ownership and many personal records dates back to the 17th century, but much less open frameworks exist even in other countries of Europe. The creation, maintenance, and dissemination of 'official' (government-produced) geographic information (see Section 19.3) is strongly influenced by national laws and practice. Box 19.1

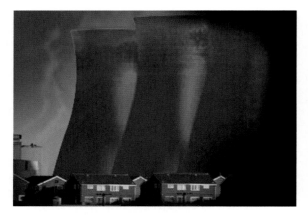

Figure 19.2 Corporations – or governments – can often act in ways which infringe the rights of individuals, such as building highly intrusive structures adjacent to homes. The law can give protection against certain acts of this kind (Courtesy: Landmark Information Group)

summarizes the situation and the way in which commerce has to adjust to the legal framework of each country in which it operates – though supra-national bodies like the World Trade Organization and the European Commission increasingly force greater homogeneity in certain laws and regulations.

Given this heterogeneity and on-going change, none of what follows should be taken as other than a general discussion of the issues. It is vital that everyone with management responsibilities working in GIS reads more detailed treatments of GIS and the law (see the guide to further reading at the end of this chapter) *and* seeks local, professional legal advice at an early stage wherever the law might be involved in a problem.

The law touches everything. Be prepared.

19.1.1 Protecting information assets

Innovation is key to GIS success (Section 18.3.1) Through it, individuals or organizations may create something better than has been done before. It is a normal reaction to consider how to achieve the greatest benefit from that investment of time and other resources. One response is to make it generally and freely available – either because that is 'good science', because wider availability will benefit the owners more, or because laws (e.g., those under which the US Federal Government operates) dictate that should happen. In many commercial or artistic operations, however, protection of one's own creative works (e.g., through licensing of the intellectual property rights) is the very basis of their operations. Governments have long recognized the need for such protection of investment – witness the Berne Convention of 1886, subsequently modified on several occasions. This treaty has been signed by most countries.

At the other end of the spectrum, there is clearly a societal interest in ensuring that innovations diffuse widely, rather than society being 'held to ransom' indefinitely by an unscrupulous monopolist. The law attempts to balance this right to exploit results of creativity and of investment against the need to foster society-wide gains. Most typically this is done by limiting the time of protection enjoyed by an innovation, such as for patents for new drugs or copyright on published material. A variety of means of legal protection exist – through legislation on copyright, designs, patents, trademarks, and trade secrets. Because of its nature, most of the protected material in GIS has been safeguarded by copyright legislation. The GIS world is becoming increasingly litigious: lawyers will continue to benefit.

Copyright is a universally agreed form of protecting investment – but copyright owners can opt not to enforce it.

Patenting is also a process impinging increasingly on the IT industry generally and GI in particular. Traditionally given for a new manufactured good, some patents have been awarded for software but a recent development has been their award for processes. There is a geography to this – what can be patented in the USA sometimes cannot be so protected in Britain or Australia. The reality, however, is that patenting is much more widely applicable than is commonly supposed and it is possible for anything which is useful, novel, and non-obvious – including business methods. In mid-2000, Etak – a pioneer in car guidance systems – was awarded four patents relevant to GIS. These included a method for storing map data in a database using space-filling curves and a system for path-finding computation. An example of a GIS-related patent is shown in Box 19.2. Firms owning patented processes commonly seek royalties from others who use arguably identical approaches. In such cases, the choices are either to give in – avoiding legal costs but paying for something which may have been developed independently – or fight the case in court arguing independent development and prior use. The latter course ties up senior management, incurs legal costs, and may end in failure and awards of punitive damages. A number of enterprises have judged it better to license rather than challenge claimed infringement of patents.

Applications Box **19.1**

Globalization of commerce and the law

Commerce is global. Law, for the most part, is not...never has that truth caused greater uncertainty than now, as global business clashes with local and national law in the borderless new world of electronic commerce... all the uncertainty is good for lawyers but not for the growth of e-commerce...The threshold question is: whose laws apply? ... the question of jurisdiction remains unresolved, both internationally and within the US: what consumer laws, contract law, privacy laws and other laws apply to e-commerce applications? Where does a transaction take place? How will conflicts in law be determined?

Financial Times, 23 December 1999

Applications Box **19.2**

Interactive automated mapping patent

Patent Number: 5,214,757

Name: Interactive Automated Mapping

Issuing office: United States Patent Office

Date of Patent: May 25,1993

Inventors: Thad Mauney, Aglaia C. F. Kong, Douglas B. Richardson, all of Yellowstone County, Montana

Abstract: An automated, fully transportable mapping system utilizes position information gathered from a Global Positioning Satellite (GPS) capture program to create new maps or annotate existing maps contained in a Geographic Information System (GIS) database in real time. In addition, the present invention displays position information captured by GPS in real time, enabling users to track the path on which they are traveling. Attributes related to the position information may also be entered in real time, and are stored in a file for subsequent inclusion in a GIS database.

The rest of the patent document describes the specific inventions claimed, the technical field, background art (context), and detailed description of the preferred embodiment.

Society has ordained that there must be a balance between the right to benefit from one's innovations and protection against long-term monopolies.

19.1.2 Some big information ownership questions

In this section, we show how some frequently asked questions about GIS assets and the law can be answered.

Can geographic data, information, evidence, and knowledge be regarded as property?

The answer to this is normally yes, at least under certain legal systems. The situation is most obviously true in regard to data or information – under either copyright or database protection (see below). Who owns the data is sometimes difficult to define unequivocally, notably in aggregated personal data. An example is bills at a supermarket which have been paid by credit card and subsequently geocoded to generate aggregate purchasing characteristics for people in a zipcode (or postcode) area.

Can geographic information always be legally protected?

There is little argument about copyright where it is manifestly based on great originality and creativity, e.g., in relation to a painting. Rarely, however, is the situation as clear-cut, especially in the GIS world where some information is widely held to be in the public domain, some is regarded as 'facts' (see Section 18.4.1), and some underpins either the commercial viability or the public rationale of government organizations.

In the USA, the Supreme Court's ruling in the famous *Feist* case of 1991 has been widely taken to mean that factual information gathered by 'the sweat of the brow' – as opposed to original, creative activities – is not protectable by copyright law. Names and addresses are regarded widely as 'geographic facts'. Despite this, several jurisdictions in the USA and elsewhere have found ways to protect compilations of facts, provided that these demonstrate creativity and originality. The real argument is about which compilations are sufficiently original to merit copyright protection. A number of post-*Feist* instances have occurred where maps have been recognized by courts not as facts but as creative works, involving originality in selection and arrangement of information and the reconciliation of conflicting alternatives. Thus uncertainty still exists in US law about what GI can be protected.

In Europe, different arrangements pertain – both copyright and database protection exists. For copyright protection to apply to a database, it must have originality in the selection or arrangement of the contents and for a database right to apply, the database must be the result of substantial investment. It is, of course, entirely possible that a database will satisfy both these requirements so that both copyright and database rights may apply.

A database right is in many ways very similar to copyright so that, for example, there is no need for registration – it is an automatic right like copyright and commences as soon as the material that can be protected exists in a recorded form. Such a right can apply to both paper and electronic databases. It lasts for 15 years from making but, if published during this time, then the term is 15 years from publication. Protection may be extended if the database is refreshed substantially within the period. As stressed earlier, geographic differences in the law have potential geographic consequences for commercial GI activity. Protection equivalent to the database right will not necessarily exist in the rest of the world, although all members of the World Trade Organization (WTO) do have an obligation to provide copyright protection for some databases.

Can information collected directly by machine, such as a satellite sensor, be legally protected?

Given the need for originality, it might be assumed that automated sensing should not be protectable by copyright because it contains no originality or creativity. Clearly this interpretation of copyright law is not the view taken by the major players, such as Space Imaging Inc., OrbImage, and DigitalGlobe (see Section 20.5.1.2 and Table 1.4) since they include copyright claims for their material and demand contractual acceptance of this before selling products to firms or individuals. It helps that there are at present only a few suppliers of satellite imagery. The cost of getting into the business is high; thus any re-publication or re-sale of an image could be tracked back to source relatively easily. In addition, there is some evidence that some of the key customers (notably the military) are interested primarily in near-real-time results. In any event, imagery up to 15 years old would be protected under the EC's Copyright Directive, at least within the European Union area itself.

How can tacit geographic and process knowledge – such as that held in the heads of employees and gained by experience – be legally protected?

'Know how' is a foot-loose element. Suppose that the lead designer of a new software system is seduced away by a large software vendor. Though no code is actually transported, the essence of the software and the lessons learned in creating the first version would facilitate greatly the creation of a competitor. This situation has happened in GIS/GI organizations from the earliest times. The only formal way to protect tacit knowledge is to write some appropriate obligations into the contract of all members of staff – and be prepared to sue the individual if he or she abrogates that agreement. Human rights legislation may complicate winning such action. Keeping your staff busy and happy is a better solution!

> **You can rely upon someone copying your work. If you don't want that to happen without your permission, fingerprint your data and be ready to sue.**

How can you prove theft of your data or information?

It is now quite common to take other parties to court for alleged theft of your information. One of the authors previously led an organization which successfully litigated against the Automobile Association in the UK for illegal copying, amending, and reproducing (in millions of copies) portions of Ordnance Survey (OS) maps. This was despite the terms, conditions of use, and charges for such mapping being published and well-known: the OS received over $30 million in back payment.

The legal process in such circumstances is an adversarial one. Therefore, it is crucial to have good evidence to substantiate one's case and to have good advice on the laws as they exist and as they have been previously interpreted by courts. How then do you prove that some GI is really yours and that any differences which actually exist do not simply demonstrate different provenances – especially since digital technology makes it easy to disguise the 'look and feel' of your GI, and reproduce products to very different specifications, perhaps generalized?

The solution is both proactive and reactive. In the first instance, the data may be watermarked in both obvious and non-visible ways. Though watermarks can be removed by duplicitous individuals, they serve as a warning. A series of groups of small numbers of colored pixels scattered apparently randomly throughout a raster dataset ('salting') can be effective. Proof acceptable to courts requires a good audit trail within one's own organization to demonstrate that these particular pixels were established by management action for that purpose. In addition to watermarking, finger-printing can also be used: at least one major US commercial mapping organization is said to add occasional fictitious roads to its road maps for this purpose. In areas subject to temporal change (like tides or vegetation change), it is very unlikely that any two surveys or uses of different aerial photographs could ever have produced the same result in detail. Such evidence should, in the right hands, be enough to demonstrate a good case though rather different techniques may be called for in regard to imagery-based datasets of slow-changing areas.

Who owns information derived by adding new material to source information produced by another party?

Suppose you find that you have obtained a dataset that is useful but which, by the addition of an extra element, becomes immensely more valuable or useful. This is common: the 'First Law of GIS Management' says that you get something for nothing by adding geographic information together. The number of data combinations and potential uses rise very rapidly as you add a few more 'data layers' – some of which may have been derived from pre-existing maps (Section 18.4.1 and Figure 19.3).

> **The 'First Law of GIS Management' says you get something for nothing by bringing together GI from different sources and using it in combination.**

So who owns and can exploit the results in derivative works? The answer is both you and the originator of the first dataset. Without his or her input, yours would not have been possible. This is true even if the original data are only implicit – if, for instance, you trace off only a few key features and dump all the rest of the original. Moreover, you may be in deep trouble if you combine your data with the original data and omit to ask the originator for permission to do so. Where 'moral rights'

Figure 19.3 The concept of adding layers of geographic information (e.g., that stored in different maps) to get added value (Courtesy: Landmark Information Group)

exist, e.g., in European countries, the author has the right to integrity (i.e., to object to any distortion or modification of his or her work which might be prejudicial to his or her honor). He or she also has the right to attribution as the author, to decide when a work will be published, and to withdraw a work from publication under defined circumstances.

19.1.3 Legal liability and GIS

Liability is a creation of the law to support a range of important social goals, such as avoidance of injurious behavior, encouraging the fulfillment of obligations established by contracts, and the distribution of losses to those responsible for them. It is a huge and complicated issue. Cho identifies Alaska as the only US state that has specific provisions for limiting liability for GIS. In many cases liability in data, products, and services related to GIS will be determined by resort to contract law and warranty issues. This assumes that gross negligence has not

occurred, when the situation can be far worse and individuals as well as corporations can be liable for damages. Other liability burdens may also arise under legislation relating to specific substantive topics such as intellectual property rights, privacy rights, anti-trust laws (or non-competition principles in a European context), and open records laws.

Clearly all this is a serious issue for all GIS practitioners. Consider, for instance, the following realistic scenario. You create some software and assemble some data from numerous sources in your spare time and for your own interest. A friend admires the result and asks for a copy: you are glad to oblige, knowing he is doing some important work on behalf of the local community. Decisions are made using these tools which subsequently are shown to be founded on results that are in error because of poor programming and inadequate, error-prone data. You get ostracized and sued by the community, by your friend – and by data suppliers who are annoyed that you have ripped off their data and, by implication, also brought them into disrepute.

Minimizing losses for users of geographic software and data products and reducing liability exposure for creators and distributors of such products is achieved primarily through performing competent work and keeping all parties informed of their obligations. If you have a contract to provide software, data, services, or consultancy to a client, that contract should make the limits of your responsibility clear. Disclaimers rarely count for much.

19.1.4 GIS and privacy

Societies capable of constant and pervasive surveillance are being built rapidly around us, sometimes with our co-operation, more often without our knowledge. GIS is part of all this activity. For example, the use of postal addresses of individuals provides some form of geographic identification for many purposes. Adding value (Section 18.4.1) via the bringing together of geographic information and personal information can lead to unverifiable conjectures about the characteristics of human individuals or households which are then subsequently used in marketing and elsewhere. In some eyes, this is unacceptable and immoral. Typically, civil rights campaigners favor an 'opt in' approach. In general terms, industry favors 'opt out' provisions – the data subject must say he or she wants to be excluded from having their records used by firms.

Some of the reasons for this move are that technically it is now possible; but others arise from security considerations. *The Economist* of 23 January 2003 summarized the situation as follows:

. . .since the terrorist attacks of September 11th 2001. . .governments everywhere, not just in the United States, have rushed to expand their powers of surveillance (see Box 20.6). Even the European Union, which has long been at loggerheads with America about data protection, performed a sharp U-turn, relaxing legal restraints on the collection

and use of information by governments. In June 2002 the EU adopted a new directive to allow member states to require firms to retain data on everyone using mobile phones, landlines, e-mails, chatrooms, the Internet or any other electronic device. In America, the USA Patriot Act, passed in the wake of September 11th, made it much easier for the authorities to obtain court permission to monitor Internet traffic and obtain wiretaps. The technology to do this is being built into the Internet's infrastructure.

There is no escape from the uncomfortable reality: GI and GIS are in the front line so far as both security and sales and marketing are concerned and their effective use can erode privacy. The only practical (though modest) amelioration is through the use of agreed codes of professional conduct and, where appropriate, laws (where the situation continues to mutate).

19.2 GIS people and their skills

In many businesses, especially those forming part of the Knowledge Economy (Section 18.3), staff are usually seen as the prime business asset and the factor differentiating one organization from another. The supply of GIS experts and GIS-literate people is also a crucial factor in determining the rapidity of up-take of the existing facilities and their successful use.

Actually, we know very little about the skills and other competences and characteristics of the millions of people working in the GIS industry worldwide at present. But anecdotal evidence suggests that most see themselves and their careers as being technology-focused. Of the 134 jobs advertised on the GIS Jobs Clearinghouse in a 2 month period in summer/fall 2004, some 40% were for GIS analysts, 25% were for technicians, 22% for managers, and 10% for programmers. It seems that the historical origins of GIS have led to many GIS-trained employees being highly skilled technically but being somewhat ill at ease and unfamiliar with business issues – or even with the user community outside of that one in which they were first employed. This situation calls for improvement; appropriate education, training, and personal development is essential.

Most GIS industry people think of themselves as technical experts. An over-strong focus in this area is not in the long-term interests of GIS as a whole.

19.2.1 Existing GIS education and training

Despite very different cultures and ways of operating worldwide, there are at least three reasons why we would expect global similarities in GIS curricula and educational approach. The first is the pervasive influence of the software vendors whose systems have harmonized some GIS concepts and terminology. The second is the effect of the GIS Core Curriculum created by a team led from the University of California Santa Barbara in 1990. This curriculum was made available, cheaply, worldwide and has been used in over 1000 universities. Finally, there have been a number of subsequent attempts to create international consortia delivering the GIS education material in different countries, most notably the UNIGIS consortium, with nodes in 14 countries (see Box 19.4). Web-based material is enabling many universities to deliver courses to individuals irrespective of where they are physically located. The different types of courses operating are shown in Table 19.1.

Typically, however, much learning in GIS is achieved in substantial measure by 'doing', rather than by copying and reciting 'facts' given in lectures. That 'doing' is best guided or supported by a tutor or mentor. Such tutoring can be provided in a rote-learning situation by a software wizard. More complex educational issues, where judgment is concerned, require a human being to act in this capacity.

19.2.2 Professional accreditation in GIS

From the public's point of view, there is an attraction in having properly certified or accredited learning or training – especially where safety-critical activities are involved, for example, in engineering. Globalization is leading to globally acceptable certification in some fields. At present – and despite early moves (often by land surveyors) – there is no known professional certification of GIS/GI courses analogous to those in other professions. But there are areas of GIS where professional skills are being exercised, where its exploitation is not simply a matter of pressing buttons, and where the results may impact strongly on people's lives. In such circumstances, the whole question of judgment, quality of work, and liability (see Section 19.1.3) comes into play. Two approaches exist – to do it through existing professional or statutory bodies (as for accountants, surveyors, or those in the insurance community) or to set up a scheme administered centrally by individuals or bodies experienced in GIS and GI.

Within the USA, the Urban and Regional Information Systems Association (URISA) launched a certification program in 2004 whose purpose is to provide those professionals who work in the field of geographic information systems with a formal process that will:

- allow them to be recognized by their colleagues and peers for having demonstrated exemplary professional practice and integrity in the field;
- establish and maintain high standards of both professional practice and ethical conduct;
- encourage aspiring GIS professionals to work towards certification for the purpose of professional development and advancement;

Table 19.1 A simple summary of GIS/GI-related training and education courses

	Main delivery mode	
	Face-to-face courses	'Distance learning'
Software training courses	Commonplace – all vendors provide this themselves or via licensed partners. Can be very remunerative.	Becoming ever more important, e.g., ESRI Virtual Campus
Software development and customization	Varies greatly. Fundamental training is provided by university computer science courses. Vendors may provide specific courses, e.g., on their macro language.	As above
School-level education	GIS was once part of the UK government's Core Curriculum for schools. Elsewhere the school curriculum is typically defined locally, e.g., see Box 19.3.	
Undergraduate-level education	GIS is taught in many hundreds of undergraduate programs around the world but in many instances as modules in degree courses in geography and non-geography courses (e.g., forestry or environmental sciences). Only a few universities run undergraduate courses entirely in GIS.	The UNIGIS consortium (see Box 19.4) has set up a distance-learning course with a common 'pick and mix' curriculum now used by universities around the world. This benefits from some face-to-face teaching.
Postgraduate-level university education	Numerous Masters programs exist	Some of these programs can be taken in a distance mode but relatively few have been designed from the outset for that.
Short courses for professionals (i.e., continuing professional development or CPD) – the systematic maintenance, improvement, and broadening of knowledge and skill, and the development of personal qualities necessary for the execution of professional and technical duties throughout one's working life.	This is a matter of rapidly growing importance. Taking CPD is often mandatory to remain a member of a professional body.	An increasing number of these courses are given by distance-based means.

GIS in schools: a Texas success story

From being a failing school a few years ago to winning one of the top 20 Blue Ribbon schools awards in summer 2004 – against entries from 1200 schools across the USA – is a stunning achievement. This demonstrates leadership of a high order. But GIS helped as well. The Bishop Dunne Catholic K–12 School, Dallas, won an INTEL Scholastic twenty-first century schools of distinction award for their innovative use of GIS technology (**www.bdhs.org/gis/**) – along with a slew of other awards (Figure 19.4A).

The GeoTech Research Lab at Bishop Dunne Catholic School is a unique place where students become the researchers, active community stakeholders, and the creators of geographic and environmental solutions. For example, the Bishop Dunne students partnered with the Dallas Police Department for several projects. In the early stages of the partnership, students produced maps for neighborhood crime watch organizations. The maps illustrated where crimes were and what types of crimes were occurring on a monthly basis. Later projects included producing maps to help determine where the Dallas Police Department should deploy its Robbery Task Force.

Fire Ants are a growing environmental problem in the North Texas area. Students of Bishop Dunne Catholic School's GIS program studied fire ant colonization in the areas surrounding the school, beginning with field investigations on campus, and expanding later to nearby Kiest Park. Figure 19.4B shows some

of the results. In this case, the students' community partners were City of Dallas Parks & Recreation, the Department of Agriculture – at Texas A&M University, and the USDA–Dallas County Extension Agents.

The use of GIS to achieve results of real value for the community, to enhance student skills, and as a means of inter-working with various organizations is an example that with value could be cloned worldwide.

(A)

(B)

Figure 19.4 (A) Some of the students (and teachers) at Bishop Dunne Catholic School, Texas, who have won a host of national awards for their GIS work with local communities. They are seen here receiving a major award at the 2004 ESRI User Conference. (B) The clustering of fire ants around the school, as based upon student surveys and aggregated using a GIS

■ encourage established GIS professionals to continue to hone their professional skills and ethical performance even as GIS technology changes.

The URISA aim is to provide a formal system to evaluate the competency of GIS professionals. This is a noble aspiration. In practice, it was felt essential to: be voluntary and open to all; be flexible; use existing GIS educational bodies; be collaborative; and include a code of ethics. As implemented through the GIS Certification Institute (**www.gisci.org/**), it is not achieved through any test but by self-certification based on points calculated from educational attainment, professional experience, and contributions to the profession. Benchmarks in each of these three areas are, respectively, a Bachelor's degree with some GIS courses (or equivalent), four years in GIS application or data development (or equivalent), and annual membership and modest participation in a GIS professional association (or equivalent). A committee adjudicates upon each application for certification based on the fit of the application to the benchmarks and each successful applicant must sign a code of ethics.

Such a process can be scorned easily: it is based largely on self-certification, measures inputs (e.g., hours in classes) not outcomes, and the process for coping with unprofessional behavior is not yet clear. Moreover, it implies the acceptability of some 'authorized' course contents rather than other material; this has led in other

professional certification programs to ossification of what is acceptable and much-reduced innovation. All that said, this is the first major, national attempt to put GIS certification in place.

> **Certification of GIS practitioners could have the benefit that it may enhance professional standards – but it may also stunt innovation.**

19.2.3 Fixing the shortcomings of current GIS education and learning

The 'survey' of GIS staff characteristics quoted earlier plus other evidence suggests that a technical fixation remains current in GIS. This creates a self-perpetuating role for GIS as a pursuit for those 'with clever fingers'. It is no surprise, therefore, that relatively few GIS people have yet made it to the highest levels in major organizations. If we wish to have GIS skills and approaches more deeply involved in high-level decision making, we need to modify the nature of our education and learning.

The reality – and one both espoused and noted throughout this book – is that GIS is now a much 'broader church' than simply a collection of technical expertise. As a result, it seems unwise to have a substantial GIS/GI

Biographical Box **19.4**

Josef Strobl and UNIGIS

Josef Strobl (Figure 19.5) is an associate professor of Geography at the University of Salzburg in Austria and Director of the Centre for GeoInformatics (**www.zgis.at**). His interests are focused very much internationally, working with colleagues in many other countries.

But our particular interest here is in his role as co-founder, and his leadership of, the UNIGIS Consortium (**www.unigis.net**). UNIGIS is a worldwide network, now of 15 universities, co-operating since 1990 in the design and delivery of distance learning in GIS and GIScience. UNIGIS host universities currently operate in Austria, Canada, the Czech Republic, Ecuador, Hungary, India, the Netherlands, Poland, Portugal, Russia, South Africa, Spain, the UK, and the USA. The network offers Postgraduate Certificate, Diploma, and Masters courses. The courses are supported by an international network of qualified academics and other professionals and use industry-standard software to give appropriate practical experience. UNIGIS courses are modular and flexible study programs. The content varies to meet the needs of local students. Optional courses provide the opportunity to tailor the program to meet individual student's needs, and residential workshops are offered to support communication and networking. Local centers provide administrative support and interaction with experienced tutors. Tutoring and assessment is by electronic communication. Dedicated websites provide virtual e–Learning environments and support at all stages of the learning process.

Figure 19.5 Josef Strobl, GIS educator

UNIGIS is a unique entity in GIS and has enrolled about 6200 students since it began. Josef and his colleagues have shared generously their experience in running the network, including the development of online teaching material for use in diverse educational and national cultures.

course which does not include at least some treatment of the following:

- entrepreneurial skill development and leadership;
- the principles of geographic science;
- understanding of and familiarity with GIS technologies;
- understanding organizations;
- finance, investment criteria, and risk management;
- human resources policies and practice;
- legal constraints to local operations;
- cultural differences between disciplines;
- awareness of international differences in culture, legal practice, and policy priorities;
- formal management training, including staff development and presentational and analytic skills.

Clearly not all courses and learning needs to include all of this material. Some will be particularly relevant to those engaged (as all professionals should be) in continuing professional development short courses. The introduction of locally related legal, cultural, and application-related elements – as well as buttressing the global technical, business, and management issues – will be to the benefit of GIS practitioners, the discipline, and business (used in our wide sense) alike.

19.3 Availability of 'core' geographic information

If GI is needed but not available, someone will be faced with a substantial cost in creating or buying it. Without such 'rocket fuel', vendors have much more difficulty in selling software. And politicians who require speedy guidance on what to do after natural disasters or quick answers in regard to a policy issue simply will not be able to use GIS. It follows, therefore, that data existence, availability, and ease and cost of access have major impacts on the entire GIS enterprise and on its future potential. Central to all this is the availability of the 'core GI' – the geographic framework on which all other information is assembled (see Boxes 19.6 and 19.7). Typically, geographic information is collected by many different agencies, usually within the boundaries of national territories; and, historically, most of these agencies have collected data primarily to suit their own local purposes. Thus, if the benefits of added value are to be obtained from data integration within a GIS (see Section 18.4.1 and Box 19.5), there must be a mechanism (or spatial 'key') to register the different sets of information together in their correct

A value-adding GI organization

Landmark Information Group is a leading supplier of digital mapping, property, and environmental risk information. The organization has forged data supply agreements and partnerships with a number of government bodies. Its business is founded upon two principles. The first is that all property-related investment decisions are affected by potential environmental risks and liabilities. The second is that combining and consolidating environmental data from a wide variety of sources not only improves the availability of data and makes them more convenient to access, it also creates new types of information services which can provide new solutions to old problems. In other words, it creates added

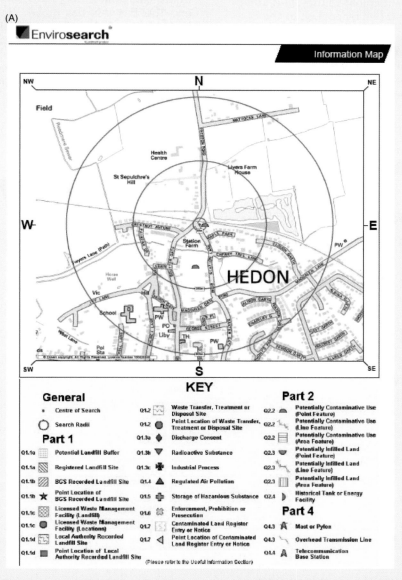

Figure 19.6 (A) Hazards in an area within specified distances of a property which is being purchased. The results are based on bringing together environmental data from many sources to give added value. (B) Flood hazards in the same area as shown in (A). The results are again derived by bringing together environmental data from many sources (Courtesy: Landmark Information Group – see **www.landmarkinfo.co.uk**)

value. On this basis, a wide range of nationally complete digital GI has been assembled from different sources and geo-referenced so it fits together with the Ordnance Survey framework data (see Box 19.6). These include:

- One million historical Ordnance Survey maps dating back to the 1850s, together with a record including address and coordinates for every address in the country.
- A century-long set of Trade Directory Entries and Business Census data indicating sites with potentially contaminative uses, plus other such sites identified by analysis of the OS maps.
- Areas at risk of coastal or riverine flooding.
- Natural hazard areas (e.g., those liable to coal mining subsidence, radon-affected areas, and groundwater vulnerability areas).

- Planning consents and site information (e.g., discharge consents into streams and rivers, hazardous substance consents, registered landfill sites).
- Air Pollution Control records, i.e., authorizations.

From all this and other information integrated within their GIS, Landmark offers a variety of services tailored to different markets – to environmental professionals, real estate professionals, property lawyers, local governments, utility companies, and individual house owners. The business has grown from its origins in 1995 to a $80 m turnover business in 2004. A sample of two parts of the standard report for house owners on the potential environmental hazards surrounding their home is shown in Figure 19.6A and B; the full report is produced for $50.

Figure 19.6 (*continued*)

relative positions. The geographic framework provides this ubiquitous data linkage key through the use of co-ordinates or (sometimes) via implicit geographic coding (e.g., zip or postcodes).

> **Geographic framework information underpins all use of GIS.**

19.3.1 The nature of the framework

What constitutes the framework varies between different countries. In the USA, the Federal Geographic Data Committee has defined the information content of the US framework as consisting of:

- geodetic control (control points, datums, etc.);
- elevation data;
- hydrography;
- public land cadastral information;
- digital orthoimagery (see Section 9.3.2.4);
- transportation;
- the geography of governmental/administrative units.

Similar definitions – including placenames – would in practice be found in many countries. Most typically, the framework is manifested in user terms through topo-graphic mapping – the 'topographic template'. Normally, such mapping will contain all this information but partly in implicit form (e.g., geodetic control). It has long been produced by national mapping agencies; as arms of government, these have provided continuity of provision over decades or even centuries in many countries.

> **Generally, the framework's quality is publicly defined and well known and its governmental source is likely to stay in operation to update and guarantee it.**

Important differences do exist between national frame-works. These are tailored to different major types of application, were created in response to different legal and governmental environments, and have persisted in a his-toric form for different periods. Typically, only one or two public-sector organizations are responsible for providing the overall national framework (as in Britain, described in Box 19.6; and also France, the Nordic countries, India, and Japan). But, in countries with federal governments, the normal situation is that the national government produces relatively low-resolution materials (see Box 19.7) and local and more detailed versions of the framework are produced by other (e.g., state or local) governments (as in Australia, Germany, and the USA). Detailed parts of the framework may also be unavailable in some countries for reasons of state security or lack of investment. Moreover, at present, much framework information is still only available in paper map or coordinate list form in some countries.

Primarily, to date, the private sector has been involved in marketing some of this framework data or as contractors in its collection. However, as the availability and use of digital imagery from commercial satellites becomes more common, the situation is changing quite rapidly. The world of national mapping agencies is also changing rapidly: the new US approach to creating The National Map is extraordinarily different to the traditional one, involving sewing together material from many different sources and the use of volunteers to update it using GPS (see Box 19.7).

19.3.2 Making the framework available

The quality, currency, ready accessibility, and price of a framework are also important constraints for users. Commercial imperatives in selling framework data are readily understood. But a number of governments are also

---

Applications Box 19.6

The British national framework

Britain's physical and governance characteristics and its history ensured that its national mapping differs significantly from countries having federal government structures, such as the USA: Britain is much smaller (one-fortieth of the USA in area) but the level of geographic resolution provided is over 100 times greater than the basic map information made available by USGS and is continuously updated. Almost all of the geographic framework is crafted by one central government body, Ordnance Survey (OS). Unlike US Federal Government bodies, this charges for its products and services (see Section 19.3.2); numerous partnerships exist with commercial organizations which are

licensed to add value and distribute OS data (e.g., see Figure 19.6).

The core product from the national topo-graphic database is MasterMap which has been produced by re-engineering the origi-nal OS large-scale (highly detailed) database, completed – as a world first – in 1995, into an object-oriented form. It comprises detailed topo-graphic, cartographic, administrative bound-ary, postal address, and network features positioned on the National Grid plus an Imagery Layer. Every OS MasterMap feature has a unique identifier called a TOpographic IDentifier (TOID®), which is used to refer to the feature. Key features are complete

▶

up-to-date coverage, the seamless data (i.e., no map tiles), orthorectified aerial imagery, topographic area features, and a topologically structured transport network.

All 440 million features are codified on a new classification basis with its unique identifier, a version number, and a defined life cycle linked to the 'real world object'. Data are available by theme as well as by geographic area; the customer can be supplied with 'change only updates'. From MasterMap data, other products can be spun off at lower levels of resolution. Figure 19.7 shows an example of some of the data layers.

(A) (B)

(C) (D)

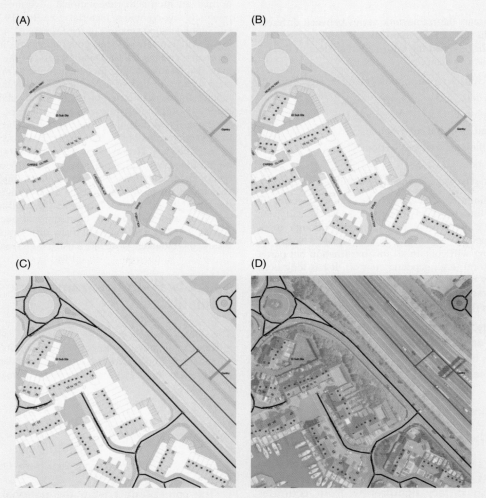

Figure 19.7 Sample of part of the Ordnance Survey MasterMap, showing various layers built up from the topography (A), with addresses added (B), with integrated transport links added (C), and with the orthophoto layer (D) (Courtesy: Ordnance Survey; Crown copyright reserved)

seeking to recover some of the heavy up-front costs of information (especially GI) collection through charging for its use.

Some of this has been brought about through governments worldwide reviewing their roles, responsibilities, and taxation policies and reforming their public services as a consequence. In some cases this has led to privatizations of government functions or at least out-sourcing. The imperative for some reviews has been financial, with a search for greater efficiency and wiping out of duplication (e.g., the US e-government initiative). In others, the imperative was ideological – that the state should do less and the private sector should do more.

The reform of government has led to two approaches in regard to GI. The first has been the outsourcing of production: thus large parts of the imagery created by many national mapping organizations (e.g., US Geological Survey) are now produced under contract by the private

The US national framework

Essentially, the US situation has two parts – the traditional approach and a new one. The traditional manifestation is what is available out of the US National Mapping Program (NMP). The NMP incorporates the documented needs of 40 Federal agencies and the 50 States, which are solicited and analyzed as part of a continuous requirements assessment process. Products available include:

- Topographic quadrangle maps, derived originally from digitizing of 1:24 000-scale paper maps and available in vector and raster forms. On average, the contents of these maps were last substantially up-dated around 25 years ago. Less-detailed national coverage is also produced.

- digital elevation models, orthophotoquads, satellite images derived from a variety of sources, and the Geographic Names Information System (GNIS). The latter forms the official national record of domestic geographic names information.

Given its difficulty in addressing all the myriad needs for up-to-date framework data and the increasing role of other players in providing such data across the nation, USGS announced in 2001 the concept of The National Map (TNM). This plan involves a novel partnership approach drawing data from multiple (other federal agencies, states, counties, academic, and private sector) sources to create, make consistent, and distribute much-more-current framework data. A highly innovative USGS initiative was to seek to estimate the creation costs and the application benefits of building a US national map in this fashion; though some of the estimates made are (necessarily) heroic and the projected costs seem modest, the simulation study was ground-breaking and suggested a break-even time for a successful TNM of 14 years and a net benefit of between \$1 and \$3bn over a 30 year lifetime. At the time of writing, the ambitious TNM vision was fairly clear but the reality is embryonic (see **nationalmap.usgs.gov/**). The relationship between this and various other initiatives is described in Chapter 20.

sector. Another approach has been simultaneously to force economies within government and get the user to pay for the bulk of GI. In some European countries, for instance, land records and the cadaster are paid for totally through user fees; since many people do not own land or property (around 30% in the UK), it is deemed unfair for them to contribute to the costs via general taxation. Also, it is worth stressing that this charging is nothing new in principle: historically, almost all government organizations have made *some* charge for data, including Federal Government bodies in the USA. The only issue is for what elements users should be charged, e.g., the whole product or only the costs of documentation and distribution. The use of the Internet has dramatically changed this equation (see Section 18.4.4).

The advantages and disadvantages of the 'charging for everything' and 'free access' approaches are summarized in Box 19.8, though these are subject to much argument. In Britain, the on-going costs of running the national mapping organization are met totally from revenues raised from users. Perhaps as a consequence, the main elements of the database are guaranteed never to be more than six months out-of-date. Figure 19.8 shows how the proportion of funding arising from the users rose over many years. That said, the policy also has some downsides, as suggested in Box 19.8, and policy shifts have happened in various countries – sometimes several times (see Box 19.9).

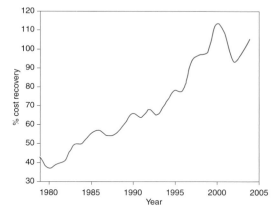

Figure 19.8 Cost recovery by Ordnance Survey which operates a 'user pays' model for digital geographic information and paper maps. Prior to 1999, Ordnance Survey was set a percentage cost recovery target by government that rose steadily towards 100% from a start of 25% in the mid 1970s. Since then, Ordnance Survey has been tasked with a target of achieving a surplus on operating activities. In the graph, post-1999 cost recovery has been calculated as trading revenue as a percentage of operating costs. Exceptional costs (provision for substantial early retirement programs), exceptional revenues (settlement with the Automobile Association's new owners – see Section 19.1.2), and net interest movements have been excluded from the calculation. The dip below 100% since 1999 was the result of a planned investment program

Applications Box **19.8**

The advantages and disadvantages of the 'user pays' and 'free access' philosophies (*and some contrary views*)

Arguments for cost recovery from users:

- Charging which reflects the cost of collecting, checking, and packaging data actually measures 'real need', reduces market distortions, and forces organizations to establish their real priorities (*but not all organizations have equivalent purchasing power, e.g., utility companies can pay more than charities; differential pricing invites legal challenge*).

- Users exert more pressure where they are paying for data and, as a consequence, data quality is usually higher and the products are more 'fit for purpose'.

- Charging minimizes the problem of subsidy of users at the expense of taxpayers – about a 30 000 to 1 ratio globally (*but some users are acting on behalf of the whole populace, such as local governments*).

- Empirical evidence shows that governments are more prepared to part-fund data collection where users are prepared to contribute meaningful parts of the cost. Hence full data coverage and update is achieved more rapidly: the average age of the traditional US basic scale mapping is over 30 times that in the UK (*but once the principle is conceded, government usually seeks to raise the proportion recovered inexorably*).

- It minimizes frivolous or trivial requests, each consuming part of fixed resources. (*This rationale is now less relevant given the user-driven access to the Internet in many countries*).

- It enables government to reduce taxes.

  Arguments for dissemination of data at zero or copying cost:

- The data are already paid for hence any new charge is a second charge on the taxpayer for the same goods (*but taxpayers are not the same as users – see above*).

- The cost of collecting revenue may be large in relation to the total gains (*not true in the private sector – why should it be so in the public sector?*).

- Maximum value to the citizenry comes from widespread use of the data through intangible benefits or through taxes paid by private sector added-value organizations (*this contention is neither provable or disprovable because of the many factors involved*).

- The citizen should have unfettered access to any information held by his/her government, other than that which is sensitive for security or environmental protection reasons (e.g., the location of nests of endangered species).

19.4 Navigating the constraints

We have seen that various legal, staff skills, information availability, and other constraints can complicate the full exploitation of a GIS severely. The introduction of a GIS will introduce risk to an organization – though in the longer term it may well reduce it. Each constraint or problem can be tackled individually but a much better approach is to have a GIS strategy and carry out regular reviews of risk. By being proactive, many constraints can be avoided or minimized.

19.4.1 The role of the GIS strategy

Strategies set direction, based upon the organization's aims, its resources, and current position. It will be rare for any organization to have a wholly free-standing GIS strategy: normally, it will have to fit within and support the organization's overall strategy and business plan. Moreover, it will in turn also need to fit and be supported by other strategies with which it will be associated – such as the finance, human resources, information systems, and services and estates strategies, and possibly others such as external relations/publicity and international ones. Sometimes the GIS strategy may simply be embedded, even buried, within the information technology, systems, or services strategy. We think that this is often a mistake because of the cross-cutting nature of GIS and (especially) GI across many levels and (potentially) most of the organization's activities – part of why we stated that 'spatial is special' in Section 1.1.1.

Here we treat the GIS strategy as a free-standing entity for illustrative purposes. The particular form of that strategy will vary depending on how GIS is seen within the organization (see Chapter 7.3.1). Finally, strategies are useless without an implementation plan, senior management commitments to proceed, and milestones at which progress is monitored. Most of these issues are dealt with in more detail in Chapter 18.

Swings and roundabouts of public policy on GI

Following the policies of the British government of the day, Ordnance Survey took action to prevent 'free riding' or illegal use of the government's maps at least as early as 1817. They warned expressly anyone infringing their copyright that this would lead to legal action (see Figure 19.9). In the 1930s world of mapping, however, public policy changed to a 'charge only for ink and paper' basis; in the 1960s it changed back again to charging users a fraction of the total costs of operations. In 1999, the move to a Trading Fund status effectively led to OS having to become profitable and thus meet full costs and interest on capital (but see Section 18.3.2 on whether all the assets were capitalized).

TRIGONOMETRICAL SURVEY OF GREAT BRITAIN

It having been represented to the Master-General and Principal Officers of His Majesty's Ordnance, that certain mapsellers and others have, through inadvertence or otherwise, copied, reduced, or incorporated into other works and published, parts of the "Trigonometrical Survey of Great Britain," a work executed under the immediate orders of the said Master-General and Board, the said Master-General and Board have thought proper to direct, that public notice be given to all mapsellers and others, cautioning them against copying, reducing, or incorporating into other works and publishing, all or any part of the said "Trigonometrical Survey," or of the Ordnance maps which may have been or may be engraven therefrom.

"Every offender after this notice given, will be proceeded against according to the provisions of the Act of Parliament made for the protection of property of this kind."

By order of the Board,
R.H.Crew, Secretary

Office of Ordnance, 24th February 1817

The London Gazette, Saturday March 1, 1817, number 17225, page 498

Figure 19.9 The earliest known warning against infringing GI copyright: the warning printed by the Board of Ordnance in the *London Gazette* of 1 March 1817

19.4.1.1 Getting to the GIS strategy and the implementation plan

Organizations achieve things routinely through standard processes. An illustrative top-down process view of how to get to a GIS strategy is shown in Figure 19.10. To aid realism, we use a hypothetical example of a mid-size local government, with perhaps a thousand or so employees. Sometimes things happen in such a top-down way but often, in practice, GIS strategies are grown partly from the 'grass-roots'. And the strategy would be different in some respects in a commercial enterprise or a government one which sells data; a pricing sub-strategy would contribute to the overall GIS and organizational one (see Box 19.10 for some elements of it).

In essence, getting to a GIS (and corporate) strategy requires a statement of the current situation, agreed by all the key players as an accurate summary. This needs to cover the reason why the organization exists, its strengths, and how well it is achieving its current aims. Also, it needs to describe any major shortcomings and the causes of them (based largely on customer, client, or 'owner's' views). From this must come some vision of where the organization should go. The strategy must include summaries of the threats from others and the new opportunities available, plus the constraints that will need to be circumnavigated. In addition, a new strategy needs to be underpinned by a coherent summary of the organization's culture and values – or new ones if these need to be changed. To enable implementation of the strategy, an explicit description of the present structure and operational management and a process map of the organization are required. All of this needs to be as dispassionate as possible – which is why consultants are often called in to help.

Any strategy will fail if it does not contain a statement of all the major changes necessary and a comparison of these with the financial, staffing, and other constraints plus an assessment of their ramifications. These may involve changing the numbers, disposition, and skills of staff, changes of software, the need to acquire or create additional data/information or technology, and much else. Finally, it is crucial to get 'buy in' from senior managers and the staff concerned.

19.4.1.2 Identifying risk

Risk identification and management is fundamental to the success of any organization – and to its strategies. Risk can arise internally from new initiatives, simply from on-going operations, or because of external changes. It is important to realize that, inherently, some risks are

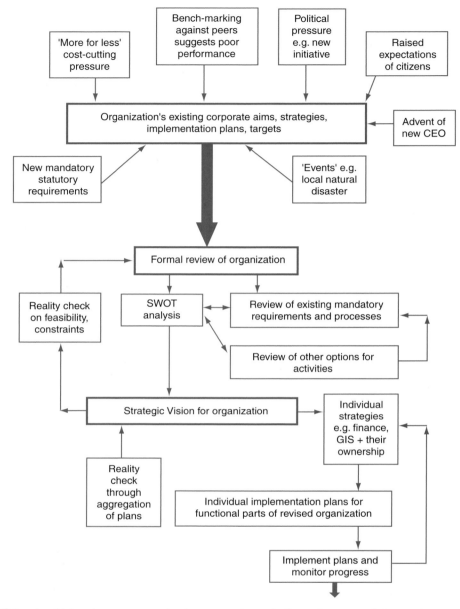

Figure 19.10 External and internal pressures on a local government organization and consequential moves to a corporate and a GIS strategy and implementation plans

unpredictable in detail though their aggregate likelihood can sometimes be insured against (e.g., via professional indemnity insurance). Failure to control risk can lead to financial disaster, to public humiliation for the organization, and the loss of jobs, especially of senior personnel. Central to the notion of risk is accountability. It is essential on a practical basis to 'lock in' your superiors in the project, obtaining their agreement to the strategy and key operational decisions.

Different types of risk need to be considered separately (Table 19.2). In general, risks are greater when GIS are being used for wholly new activities within organizations, rather than when such systems are being used to do

existing things better or cheaper, especially if they are introduced incrementally (see Section 17.1).

On the other side, GIS is now a major contributor to the estimation and control of risk in the world's largest insurers – in tasks ranging from estimates of credit-worthiness of individuals to monitoring the distribution and frequency of natural hazard events, especially major natural disasters (see Section 21.4). Indeed, technology is changing the very nature of how risk is assessed: 'pay-as-you-go' insurance, with charges varying by the speed at which you are driving and the levels of traffic and road conditions, is now possible thanks to the integration of GPS and GIS (Figure 19.11).

Building sub-strategies – an information pricing example

Getting pricing wrong is usually disastrous. In general, the information owner should set a strategy which:

- involves setting prices by perceived value, rather than cost of production;
- achieves cost leadership through economies of scale and scope; and/or
- differentiates the products, i.e., creates multiple versions of the same product differing in accuracy, appearance, and currency and creates different products by linking different datasets together and extracting 'new' information goods (see Box 19.5). Avoid letting the information become a commodity, i.e., where everyone is selling the same good. In that case, fierce price competition will decide who wins.
- personalizes the pricing as well as the product. One way to achieve this is to group prices (e.g., for all local governments) via some kind of sector-wide deal or Service Level Agreement, thereby permitting bulk discounts quite legally. Ordnance Survey

Great Britain now has one deal – the Pan-Government Agreement – with 280 departments, agencies, and other central government bodies in the UK and has parallel collective deals with all the utility companies and all the local governments in Britain.

- seizes 'first mover' advantages (i.e., gets into the market first and establishes brand image, customer base, etc.) but continues to innovate. Avoid greed if you are the market leader. For example, employ 'limit pricing', which are prices set just low enough to discourage competitors from investing to join the market since they will never make a suitable return on their employed capital. And 'play tough' – a new potential competitor should know that you are willing to lower your prices aggressively if necessary and sue for any infringement of intellectual property rights.
- is built upon understanding of the markets and their current profitability and use of promotions to measure demand.

Table 19.2 A classification of risk faced by organizations

Major risk category	Sub-category	GIS-related example
Business risk	Product/service risk	Implementation of the system over-runs financial or time budgets. Analyses, outputs, outcomes found flawed or unconvincing due to software inadequacy or staff incompetence
	Economic risk (via change of markets)	No-one wants 'line maps' when up-to-date image maps available?
	Technology risk (out-of-date c.f. competitors)	Competitors can achieve the same results (e.g., analyses, maps) faster, better, more cheaply
Event risk	Reputation risk	Erroneous or delayed results lead to collapse in confidence in organization, fanned by the media (e.g., 'State denounces GIS firm for unprofessional work.')
	Legal risks	As above, leading to legal action to recover damages and/or costs
	Disaster risks	Software incapable of doing what was promised; fire destroys all facilities including data and results to date
	Regulatory or policy shift risk	Government decrees that information policy has changed or that all materials contributing to end results have to be in public domain
	Political risk	Elected official supports competitor
Financial risk	Credit, market, liquidity, operational, price, delivery, etc., risks and fraud	Other 'normal' business risks – all must be guarded against

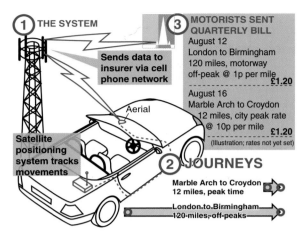

Figure 19.11 The pay-as-you-go car insurance scheme being piloted by Norwich Union in the UK – an enterprising combination of new technology and digital maps to differentiate charging for different customers with different driving habits (based on a risk assessment)

19.4.2 Managing risks in GIS projects

Managing risk within significant GIS projects requires at least two appraisals, carried out at different stages. These may overlap with – and be part of – the business case for a GIS. The appraisals are:

■ a sound investment appraisal before commitment is made to proceed;

■ a risk appraisal to identify the significant factors which may go wrong in implementation of the project.

The investment appraisal is needed because someone above you in the management chain needs to analyze your logic and test it to destruction. The opportunity costs of going ahead also need to be exposed. In many organizations, there is only a finite amount of financial resource available for new developments. This limitation may arise from an unwillingness to add to the tax burden (in the public sector) or from restrictions on bank or market support (in the private sector). Such investment appraisals also need to be done when money is not a major problem: the most precious commodity in many businesses is senior management skills and it may be necessary to use these to take forward or to rescue another project. In short, the GIS business case is a necessary but not sufficient case for proceeding – a more wide-ranging investment appraisal is needed and rarely can this be done by GIS staff alone. It needs at least to be sanctioned by the finance director or his/her equivalent.

> **Managing risk requires both an investment appraisal and a risk appraisal.**

An investment appraisal needs to cover at least:

■ The predicted costs of people, materials, hardware and software, training, and other organizational costs insofar as they can be quantified, all broken down by time.

■ The predicted costs and benefits (see Box 17.4) and a comparison with those of the *status quo*, i.e., the 'do nothing option'.

■ Sensitivity analyses of the impacts of variations on the plan on the business case. Such variations could include cost over-runs, failure to achieve the predicted benefits in the predicted time, etc.

■ Conclusions on the basis of all the above, preferably agreed individually with as many of the key players as possible before the formal consideration of the appraisal.

Anticipation of risk requires a formal process and people charged to identify and manage it. No project should start without a table of potential risks, classified by probability and anticipated magnitude of impact (which may be very different) and by steps to be taken to reduce the probability and/or minimize the impact. Moreover, the risk assessment is iterative: after every major event or even periodically, the process shown in Figure 19.12 must be repeated.

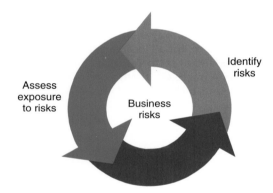

Figure 19.12 The iterative nature of risk identification, assessment, and management

19.5 Conclusions

Throughout this book we have argued that GIS can – and increasingly does – underpin many of the activities of businesses of all types, whether academic, governmental, not-for-profit, or commercial. To succeed, these businesses need to exploit their assets, including all the capabilities that GIS can provide – and avoid disaster through following a strategy, anticipating risk, and navigating past the constraints, i.e., being business-like is essential. Such plans and actions need to be reviewed frequently: the impacts of technology change, the advent of new competitors, and changing expectations on the part of the business's stakeholders ensure that tomorrow will not be the same as today.

Questions for further study

1. Who owns the geographic information you use? What rights do you have in it? How would you protect your organization's GI against misuse by competitors?

2. If you were being educated for your own job, what would be the ideal components of a GIS course? Is distance-based learning for GIS a satisfactory approach? For what types of personal development is it best suited?

3. Suppose tomorrow morning you receive a writ for damages from a commercial firm for whom you have carried out consultancy. It alleges your results – on which they acted – are now demonstrably false and you did not take due care and attention. What do you and your lawyer do first? How do you 'cover your back'?

4. How would you go about creating a new GIS strategy for your organization and what are the main risks that you have to guard against?

Further reading

Cho G. 2005 *Geographic Information Science: Mastering the Legal Issues*. John Wiley & Sons Ltd., Chichester, UK.

Photogrammetric Engineering and Remote Sensing. Special Issue on The National Map. 2003 (69).

Shapiro C. and Varian H.R. 1999 *Information Rules*. Cambridge, MA: Harvard Business School Press.

20 *GIS partnerships*

'Business' (as we define it) may operate within commerce, government, the not-for-profit sector, and in academia – or between them. Many aspects of it are often highly competitive, even (though it might be denied) in government. But competition is not always the right way: frequently collaboration pays. This chapter examines how collaboration has occurred in GIS and GI, the extent to which it has worked, and the factors that sometimes make it fail. It covers collaborations at the local, national, and the global levels – and seeks to separate the hype from the reality. The US National Spatial Data Infrastructure – one of the truly 'big ideas' in the GIS world – is reviewed and lessons are drawn. Finally, the impact of overwhelming external events as business drivers is considered, based on the events of 11 September 2001 and its aftermath. These have raised crucial issues about the balance between collaboration and coercion in GIS – and beyond.

Geographic Information Systems and Science, 2nd edition   Paul Longley, Michael Goodchild, David Maguire, and David Rhind.
© 2005 John Wiley & Sons, Ltd.   ISBNs: 0-470-87000-1 (HB); 0-470-87001-X (PB)

Learning Objectives

After reading this chapter, you should know about the value of GIS partnerships through familiarity with:

■ The characteristics, benefits, and some of the potential pitfalls of collaboration;

■ National spatial data infrastructures;

■ The need for globally consistent GI;

■ The importance of political support for GIS;

■ The role of extreme events as business drivers.

and driven by short-term events. Figure 20.1 shows a locust infestation which demanded collaboration in data collection and pest eradication between FAO and various African countries in August 2004, with financial support by other members of the United Nations (**www.fao.org**). In this chapter, we look at some major enterprises based on collaboration – including one that failed and another that faded away – to attempt to draw some general conclusions; and we consider how global terrorism seems to be changing some aspects of GIS partnerships.

> **Partnerships can add vital staff skills, technology, marketing know-how, and brand image to your capabilities, leading to fresh insights on the real user needs, new products, and good publicity. It can also lead to cost- and risk-sharing.**

20.1 Introduction

Chapters 18 and 19 have emphasized the competitive nature of our world. But there are many occasions in GIS on which we find it best to collaborate or partner with other institutions or people. The need for partnerships applies at the personal, institutional, local area, national, and global levels – so there is great scope for different approaches. The form of the partnerships can range from the highly formal, based on contract, to the informal where participation is entirely voluntary. A good example of the latter is GIS Day (see Box 20.1). Some partnerships result from events which span national boundaries (like Chernobyl), and some are both multi-national and created

20.2 Collaborations at the local level

There are around 80 000 different public-sector governing bodies in the USA (e.g., counties, school boards) – perhaps the most complex structure anywhere in the world, with different jurisdictions overlapping in geography. The size of these organizations and the resources available to them ensure that, for some administrations, investment in GIS is a heavy (or sometimes impossible) burden: Figure 20.3 illustrates the town hall in the county seat in one small county, which is open by appointment. The scope for collaboration or partnerships at this level – typically involving operational matters or sharing of knowledge – is considerable.

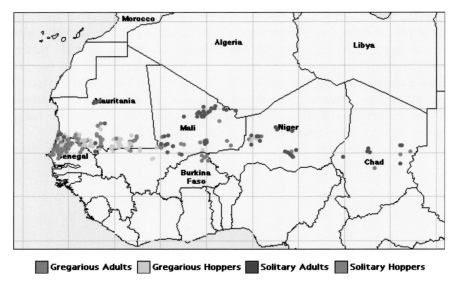

Gregarious Adults　Gregarious Hoppers　Solitary Adults　Solitary Hoppers

Figure 20.1 Locust 'hot spots' in West African countries in August 2004 (*Source*: FAO)

GIS Day: a global but informal partnership

GIS Day is an annual event which began in 1999. It is part of the US National Geographic Society's Geography Awareness Week, designed to promote geographic literacy in schools, communities, and organizations. The aim of GIS Day is for GIS users and vendors to open their doors to schools, businesses, and the general public to showcase real-world applications of the technology. News of the event is spread by use of the Internet and by advertising. Any organization can host such an event. Potential visitors can search for the nearest site to themselves. No less than 2400 organizations hosted GIS Day events in more than 76 different countries in 2003 (see Figure 20.2(A) and 20.2(B)). It was estimated that over two million children and adults were educated on GIS technology through geography on that day.

(A)

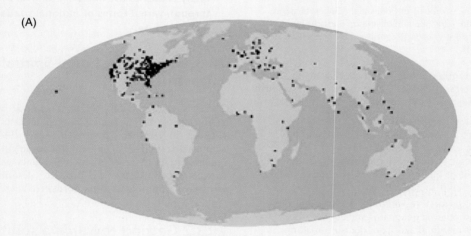

Worldwide GIS Day events

(B)

Figure 20.2 (A) The distribution of worldwide GIS Day events in 2003. (B) A GIS Day event underway at Kyrgyz State University of Construction, Transportation and Architecture (KSUCTA), Bishkek, Kyrgyzstan

Figure 20.3 The town hall in Franconia, Minnesota where meetings with the part-time administration are held by appointment when it is open (Reproduced by permission of Francis Harvey)

Nowhere is this more significant than in Public Participation GIS (PPGIS). These have been variously defined (see Section 13.4), including:

■ being an interdisciplinary research, community development and environmental stewardship tool grounded in value and ethical frameworks that promote social justice, ecological sustainability, improvement of quality of life, redistributive justice, nurturing of civil society, etc.;

■ endeavoring to involve youth, elders, women, First Nations, and other segments of society that are traditionally marginalized from decision-making processes.

Box 20.2 provides an example of such community-based GIS projects. Central to successful work in this area is good communication: once data and analysis are credible (not necessarily highly accurate or complex), having a sound strategy to introduce that information to targeted audiences is essential. The power of high-quality, well-designed maps to persuade decision makers is a crucial asset.

20.3 Working together at the national level

In some cases, national associations of local bodies have been set up to share knowledge and information and even to act as a co-ordinating mechanism. The 400 or so local governments in the UK, for instance, formed a Geographic Information House to trade in information and have forged a partnership with the Land Registry and other central government partners to create a National Land Information Service (**www.nlis.org.uk/docs/index.htm**).

Beyond such associations of like bodies, we now see truly national partnership initiatives in a number of countries. Most are based around some form of national spatial

data infrastructure (or NSDI) – one of the 'big ideas' in GIS. It is believed that all of the countries in Table 20.1 have or are implementing partnerships through forms of NSDI. Web addresses of these NSDIs are given in the Global Spatial Data Infrastructure (GSDI) website (**www.gsdi.org/SDILinks.asp**). The nature of these initiatives, their title, who is involved (e.g., see Box 20.3), and much else differs from country to country but there are also a number of striking similarities – reflecting global convergence in the nature of some of the biggest problems faced in many countries.

We use the US example as a case study since this was probably the earliest coherent scheme, it has influenced many others, and it was initially underpinned by the leader of the nation state.

Many countries are implementing GI partnerships through varied forms of national spatial data infrastructure.

20.3.1 The US National Spatial Data Infrastructure

The US 'Strategy for NSDI' document was initially published by the Federal Geographic Data Committee (FGDC) in 1994 and revised in April 1997. Box 20.4 summarizes FGDC's view of the problem NSDI was intended to fix. As ever, a few individuals were important to what actually happened: Box 20.5 describes how one key player contributed to the formative stages of NSDI.

20.3.2 Getting towards a solution

The origins of the NSDI in the early 1990s involved the Mapping Science Committee (MSC) of the United States National Research Council and the US Geological Survey. From the outset, NSDI was seen as a comprehensive and co-ordinated environment for the production, management, dissemination, and use of geospatial data, involving the totality of the relevant policies, technology, institutions, data, and individuals. A 1994 MSC report urged specifically the use of partnerships in creating the NSDI.

Table 20.1 Countries with NSDIs in August 2004 according to the Global Spatial Data Infrastructure website (**www.gsdi. org/SDILinks.asp**)

Australia	France	Cambodia	South Africa
Canada	Germany	Malaysia	Spain
Chile	Hungary	Nicaragua	Sweden
China	India	Norway	Switzerland
Colombia	Indonesia	Philippines	The Netherlands
Cuba	Ireland	Poland	United Kingdom
Czech Republic	Iceland	Portugal	United States
Denmark	Israel	Russia	Uruguay
Dominican Republic	Italy	Salvador	Venezuela
Finland	Japan	Slovenia	

Applications Box **20.2**

A successful PPGIS

Larry Orman summarized his experience in PPGIS at the 2003 conference on that topic, together with his work in GreenInfo Network (**www.greeninfo.org/**). Figure 20.4(A) is an example of a GreenInfo map which helped conservation advocates to show what could happen if suburban sprawl were to continue and forced answers to the question: 'should

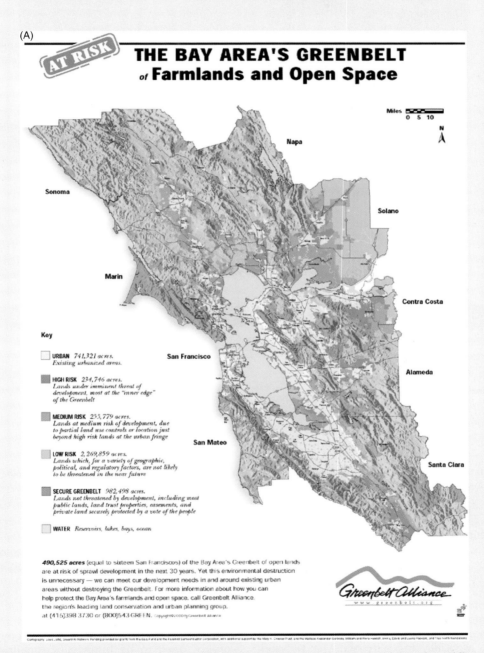

Figure 20.4 (A) Map showing threats from advancing urbanization in the Bay Area produced through a public participation GIS project carried out by the GreenInfo Network, the source of the map. (B) A simulation of wind turbines on a Scottish hillside to enable the local population to visualize a development proposal (*Source*: Macaulay Institute, Aberdeen)

(B)

Figure 20.4 (*continued*)

Support for Different Types of Projects and/or Organizations

Increasing the opportunities for leveraging

Grants to nonprofits
working on specific
individual conservation
issues/projects

Support projects by
one conservation
organization that
can benefit
multiple organizations

Grants to intermediaries
who meet the needs of
many organizations

Supporting knowledge
building and
networking activities

Support for
development of data
standards or other
sector wide impact projects

Figure 20.5 The differential leverage of GIS public participation projects. Greater long-term benefit tends to come from working towards the right end of the scale (Reproduced by permission of GreenInfo Network)

the future be this way?'. Highly compelling visual design helped the project to obtain good coverage in the news media.

For many public interest groups, GIS software and data are still too complicated, expensive, and frustrating. To circumvent this problem, GreenInfo Network was set up to be a service center for non-governmental organizations (NGOs), especially in California. The Network reached a critical mass of staff and began producing highly influential work approximately

three years later. Financial stability is crucial for such organizations: getting grants or gifts for hardware, etc., is relatively easy initially but these soon dry up so project funding must be obtained. But 'one off' projects often have only a very local effect: Orman argues for choosing to do projects which leverage the effort in various ways (Figure 20.5). In short, community groups need to be business-like in how they operate if they are to continue to prosper (see Section 18.1)

Typically, GreenInfo Network works in the areas of: conservation for land trusts, public agencies, advocacy organizations, and various foundations; for a range of advocacy groups in environmental matters; and for research organizations, service providers, and others in social well-being matters. It helps to link up client groups with projects and information from other clients, building broader education about geographic literacy and engagement. The organization's staff of eight GIS specialists usually works with 30 to 40 groups at any one time and collaborated with over 300 in the period 1999–2004. Most projects are in the $1000–10 000 range, with a few in the $25 000–75 000 range.

At the time of writing, there were two other independent non-profit GIS service centers besides GreenInfo (New York – **www.nonprofit-maps.org**, and Seattle – **www.commenspace.org**), both founded in the late 1990s. There are, of course, also many thousands of community groups working in the USA and elsewhere globally. Therefore, use of GIS tools to visualize proposed developments and communicate these to the public (e.g., see Figure 20.4(B)) is becoming common, helping to sustain local democracy. But, typically, such success is only achieved through successful partnering and investment in people.

Biographical Box **20.3**

Ian Masser: European SDI pioneer

Ian Masser's early work was as a computer-literate urban planner with a human geography background. He became involved in matters relating to GI policy as the national coordinator of the UK Regional Research Laboratory (RRL) initiative in the late 1980s. This coincided with the British Government's publication of the pioneering 'Chorley report' on *Handling Geographic Information* in 1987. This report was accepted immediately by the UK government and was an important milestone in the development of thinking about GI policy.

In 1993 he became co-director of the European Science Foundation's GISDATA scientific program. This was funded by 15 National Science Councils and also had strong links with the US National Center for Geographic Information and Analysis. An important spin-off from the program was the establishment of the Association of Geographic Information Laboratories in Europe (AGILE) in 1998. He was involved in the creation of an independent European Umbrella Organisation for Geographic Information (EUROGI) in 1993, the debates regarding a

Figure 20.6 Ian Masser

European Geographic Information Infrastructure between 1995 and 1999, and the first GSDI Conference in Bonn in 1996. He published a comparative evaluation of NSDIs in Australia, Britain, the Netherlands, and the US in 1998. Based in the Netherlands from 1998, Ian undertook missions to places as far afield as China, Colombia, India, Peru, and Vietnam.

EUROGI (see above) has 22 member organizations which collectively represent more than 6500 organizations in total; it has been actively involved in various European Commission-funded SDI-type projects. Ian Masser was its President from 1999 to 2003. He became President of the Global Spatial Data Infrastructure Association in 2002 with a remit to make it operational rather than simply a 'nice idea'.

A lever for realizing the vision was the Clinton Administration's wish to 'reinvent' the federal government in early 1993. Staff of the US Federal Geographic Data Committee (FGDC) outlined the concept of the NSDI as a means to foster better intergovernmental relations, to empower state and local governments in the development of geospatial datasets, and to improve the performance of the federal government. In September 1993, the NSDI was listed as one of the National Performance Review (NPR) initiatives to reinvent that government.

Applications Box **20.4**

FGDC's statement of the problem

In the United States, geographic data collection is a multibillion-dollar business. In many cases, however, data are duplicated. For a given piece of geography, such as a state or a watershed, there may be many organizations and individuals collecting the same data. Networked telecommunications technologies, in theory, permit data to be shared, but sharing data is difficult. Data created for one application may not be easily translated into another application. The problems are not just technical – institutions are not accustomed to working together. The best data may be collected on the local level, but they are unavailable to state and federal government planners. State governments and federal agencies may not be willing to share data with one another or with local governments. If sharing data among organizations were easier, millions could be saved annually, and governments and businesses could become more efficient and effective.

Public access to data is also a concern. Many government agencies have public access mandates. Private companies and some state and local governments see public access as a way to generate a revenue stream or to recover the costs of data collection. While geographic data have been successfully provided to the public through the Internet, current approaches suffer from invisibility. In an ocean of unrelated and poorly organized digital flotsam, the occasional site offering valuable geographic data to the public cannot easily be found.

Once found, digital data may be incomplete or incompatible, but the user may not know this because many datasets are poorly documented. The lack of metadata or information on the 'who, what, when, where, why, and how' of databases inhibits one's ability to find and use data, and consequently, makes data sharing among organizations harder. If finding and sharing geographic data were easier and more widespread, the economic benefits to the nation could be enormous.

Biographical Box **20.5**

Nancy Tosta: NSDI pioneer

Nancy Tosta is a Vice President at Ross & Associates Environmental Consulting, Ltd in Seattle, Washington, USA, where she seeks to assist federal government agencies with their environmental, public health, and information challenges. From 1992–1994, however, she was centrally involved at the birth of the US National Spatial Data Infrastructure (NSDI). As the first Staff Director to the Federal Geographic Data Committee (FGDC), housed within the US Department of the Interior, she was a major source of thought and energy behind the initial conceptualization and implementation of the US NSDI.

She was the lead drafter of the April 1994 Presidential Executive Order 12906 that established the components of the NSDI as a combination of partnerships, standards, framework data, and data clearinghouses. She oversaw the FGDC development and approval of the first geospatial metadata standard. She was a firm advocate of the 'bottom-up, patchwork quilt' approach to spatial data infrastructures, believing strongly in

Figure 20.7 Nancy Tosta

the need for spatial data efforts to be rooted in serving community needs – be they in cities, counties, watersheds, parks, states, regional, or national jurisdictions. She recognized that the value of the NSDI was in helping to promote a culture of data sharing and that the NSDI was not a tangible 'thing' that would ever be constructed or finished, but more a continually evolving 'state of mind'.

Nancy has since given talks worldwide about the value of sharing geospatial information and conceptualizing a common and coordinated approach for doing so – continuing to give impetus to the NSDI movement.

In April 1994, Executive Order 12906: 'Co-ordinating Geographic Data Acquisition and Access: The National Spatial Data Infrastructure' was signed by President Clinton. This directed that federal agencies, coordinated by FGDC, carry out specified tasks to implement the NSDI; it created an environment within which new partnerships were not only encouraged, but required. Also, it raised the national and international political visibility of geographic information collection, management, and use. Given the nature of the US constitution and the many parties involved from very different sectors, no one body was ever seen as being 'in charge'.

20.3.3 What has been the reality of the NSDI?

The Presidential Executive Order identified three primary areas to promote development of the NSDI: (i) the development of standards; (ii) improvement of access to, and sharing of, data by developing a National Geospatial Data Clearinghouse; and (iii) the development of the National Digital Geospatial Data Framework.

The Federal Geographic Data Committee, the stipulated coordinating mechanism, has operated as a bureaucracy through a series of subcommittees based on different themes of geospatial data (e.g., soils, transportation,

cadastral), each chaired by a different federal agency. Figure 20.8 illustrates the original sub-committees and who in the federal government led each one.

Many of these groups developed standards for data collection and content, classifications, data presentation, and data management to facilitate data sharing. For example, a Standards Working Group developed the metadata standard, which was adopted formally by the FGDC in mid-1994 after an extensive public review process. It has since been adapted in the light of international and other national developments (see Section 11.2). The NSDI Executive Order mandated that all federal agencies must use all FGDC-adopted standards.

The second activity area was intended to facilitate access to data, minimize duplication, and assist partnerships for data production where common needs exist. This has been done both by advertising the availability of existing data and institutional willingness to create new data through the National Geospatial Data Clearinghouse. Agencies producing geographic information describe its existence with metadata and serve those metadata on the Internet, to be accessed by commonly used commercial Internet search and query tools (see Section 11.2). As a partial result of all this, nearly all federal agencies, as well as most states and numerous local jurisdictions, have become active users of the Internet for disseminating geographic information. This model does not assume

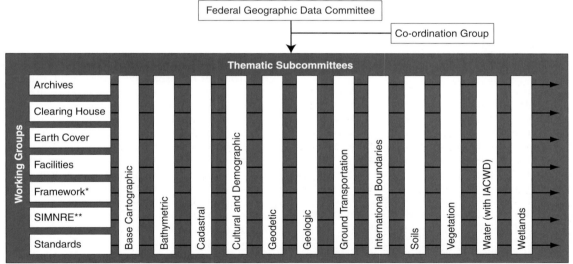

*Includes representatives of state and local government

**Sample Inventory and Monitoring of Natural Resources and the Environment

FGDC Subcommittees and Lead Departments

Base Cartographic - Interior	Geologic - Interior	Water - co-sponsored with the Interagency Advisory Committee on Water Data
Bathymetric - Commerce	Ground Transportation - Transportation	
Cadastral - Interior	International Boundaries - State	Wetlands - Interior
Cultural and Demographic - Commerce	Soils - Agriculture	
Geodetic - Commerce	Vegetation - Agriculture	

Figure 20.8 The original structure of the Federal Geographic Data Committee sub-committees and lead responsibilities for them (Reproduced by permission of the FGDC)

necessarily that GI will be distributed for free. Obtaining some of the datasets requires the payment of a fee, while others are free.

The third NSDI activity area is the conceptualization and development of a US digital geospatial framework dataset. This has proved somewhat more difficult than the other two activities (see discussion of The US National Map in Box 19.7).

The NSDI is not a concrete 'thing' but is more of a vision, a state of mind, a campaign, and an enabler for better use of scarce resources.

20.3.4 The NSDI stakeholders

The stakeholders are now defined to include public sector bodies, within federal, state, and local government, and the private sector. In general terms, the US NSDI is steered by representatives of the following organizations: the National States Geographic Information Council, the National Association of Counties, the Open Geospatial Consortium, the University Consortium for Geographic Information Science, the National League of Cities, the FGDC, cooperating state councils, the International City/County Management Association (ICMA), and the Intertribal GIS Council. Engagement with the private sector was slow to occur. This is now channeled through the Open Geospatial Consortium, itself a partnership (see Section 20.5.2)

20.3.5 Has the US NSDI been a success?

Since the creation of America's NSDI, many more organizations – from different levels of government and occasionally from the private sector – have formed consortia in their geographic area to build and maintain digital geospatial datasets. Examples include various cities in the US where regional efforts have developed among major cities and surrounding jurisdictions (e.g., Dallas, Texas), between city and county governments (e.g., San Diego, California), and between state and federal agencies (e.g., in Utah). The characteristics of these partnerships vary depending on the level of technology development within the partner jurisdictions, on institutional relations, on the funding, and on the type of problems being addressed.

Inevitably, many problems have arisen in such an ambitious project involving huge numbers of people and organizations. Incentivizing different organizations to work together and produce benefits for all those organizations incurring costs is not a simple matter – especially when there is no agreed measure of success or even progress and 'no-one is in charge'. As indicated above, initially the inputs of the private sector were modest except in delivering metadata software – which they needed to do in any case to meet the needs of their government customers. The NSDI concept of 'bottom up' aggregation of data from many (perhaps thousands of)

sources to form quality national datasets is intrinsically complex, even daunting. Also it is surprising that NSDI was not more concerned with education and training in its early stages – these have been key influences on the development and use of GIS and GI (see Section 1.5.5).

Typically, government initiatives wane after their authors have moved on. NSDI's profile and prestige diminished somewhat around 2000 when it was no longer new. Internal dissension within the federal government about who led on what, some shortcomings in agreements between federal, state, and local government and with the private sector on how to progress, budget constraints, a host of other new initiatives demanding staff and resources, all contributed to a slowdown of progress of the NSDI in its heartland. A new 'second phase' was triggered, however, by the US e-government initiative launched in 2002, an initiative designed to improve the value for money of government. Driven by the Office of Management and Budget, it led directly to the Geospatial One-Stop geoportal (see Box 11.2). As part of this, the NSDI has been re-examined in Congress by the Committee on Government Reform (Box 20.6). The greatly increased focus on Homeland Security after 9/11 is also forcing a re-think of some aspects of NSDI (see Section 20.6).

The impetus for greater government efficiency and effectiveness and the advent of Homeland Security have re-ignited the cause of NSDI.

Confusion about what all the different initiatives were and how they related to each other began to become common in 2003/04. A paper published by senior USGS and FGDC staff in April 2004 sought to define the relationships: according to this, the US National Map program (Box 19.7) enables the provision of content via integration of data from different partners, the FGDC (Section 20.3.2) provides tested standards, and Geospatial OneStop (see Box 11.2) provides the access tool to a range of federal and some state GI. The FGDC also produced a thoughtful paper in mid-2004 entitled 'Towards a National Geospatial Strategy and Implementation Plan'. This argued for 'partnerships with purpose' between commerce, academia, and all levels of government, led by FGDC; the standardization of geographic framework themes; and a need to communicate better the message about NSDI through a business case, a strategic communications plan, and training programs. All these were accompanied by proposed targets to be achieved by different dates in the period up to 2007.

Highly partisan views exist on what NSDI has achieved and what should be done next. This is not surprising given its ambitious scope, its nature, and the lack of simple measures of success.

Shortly afterwards, however, the USGS Director reorganized all of the components described above to be within a new National Geospatial Programs Office within his own organization, under the direction of the former Geographic Information Officer (now an Associate Director for Geospatial Information and Chief Information

The Putnam NSDI hearings in Congress

At the time of writing, Congressman Putnam's Sub-Committee on technology, information policy, intergovernmental relations, and the Census (which neatly covers many of the concerns of this book) had held two hearings with witnesses. The first (in June 2003) is summarized at **reform.house.gov/TIPRC/Hearings/Event-Single.aspx?EventID=510**. The second, a year later, was entitled 'Geospatial Information: Are we headed in the right direction or are we lost?' and is summarized at **reform.house.gov/TIPRC/Hearings/Event-Single.aspx?EventID=1150**.

Essentially, the first hearing was setting the scene. In it, a senior official of the Office for Management and Budget (OMB) conjectured that up to half of all spending on geospatial information across all levels of government was duplicative – a contention later disputed. Amongst the conflicting views expressed at the second meeting (by representatives from the bodies in parentheses) were:

■ A complete and up-to-date strategic plan is missing, federal agencies are not complying with General Accounting Office (GAO) direction on coordination and OMB's oversight methods to prevent duplication have proved ineffective (GAO).

■ Considerable progress has been made but more needs to be done (OMB).

■ The Geospatial OneStop (GOS) portal in mid-2004 held information from 155 federal and state sources and received visits from 6600 different individuals per month. It will continue to be developed (USGS).

■ The National Geospatial-Intelligence Agency of the Department of Defense works to coordinate Defense involvement in GOS and spends about $0.5 m per annum directly on it. Because of its statutes and regulation, NGA cannot provide geospatial intelligence support to state, county, tribal, and other political administrations (NGA).

■ Data sharing between local governments, states, and federal government has to meet the most exacting requirements – for local detail – and hence is more expensive than data created for federal agencies alone but is more up-to-date... A single model does not fit all states in regard to NSDI... The US needs a single federal agency, with cross-cutting

authority, that can direct and speak for all federal agencies on geospatial development and coordination. Legislation constrains collaboration, e.g., the Federal Advisory Committee Act prevents federal agencies from putting local, state, or tribal organizations on their advisory bodies (National States Geographic Information Council).

■ Federal government geospatial programs would benefit from the private sector being a full partner in developing plans and policies... A well-funded business plan is required for the NSDI (Spatial Technologies Industry Association).

■ We are not proceeding with a good road map. Numerous studies detail the lack of coordination of federal mapping and geospatial activities and government's duplication with the private sector (Management Association for Private Surveyors).

■ We are basically on the right track but some mid-course corrections are needed... policy makers have overlooked the importance of the Open Geospatial Consortium (OGC)'s interoperability standards effort ... the FGDC and other federal agencies need to continue to participate in the OGC to ensure that unfinished standards reflect the needs of the public and the requirements of the government agencies entrusted to serve the public interest (OGC).

■ The FGDC has done a good job, especially in metadata standards... [but] voluntary partnerships are not working and the federal government must find new carrots and sticks to realize the potential of the NSDI (Mapping Science Committee).

It is obvious that partisan, and strongly held, views were expressed. In so far as this is representative of wider views amongst the stakeholders, little consistency of view exists – other than agreement on the fact that an NSDI is (actually or potentially) a good thing and the federal government has an important role (though there is no clear consensus on what the latter should be). It is obvious that the NSDI is controlled by statutory, political, legal, commercial/financial, and other factors as well as technical ones.

Officer). The clear aim was to reduce any confusion and foster inter-working, at least within key components of the most central federal government agency (USGS) so far as GI is concerned.

20.3.6 Conclusions on the progress of the US NSDI

It is no surprise that NSDI matters have not gone smoothly. Anything so broadly defined and forming such an intangible asset (Box 18.6), whose creation involves huge numbers of different organizations and individuals – each with their own views, values, and objectives – and having 'no-one in charge' was always going to be difficult to achieve. At a high level, there are few who dispute the merits of achieving data sharing, reduction of duplication, and risk minimization through the better use of good-quality GI. But how to make it happen for real is a different matter.

> **Something as diffuse as NSDI will never be seen as a success by everyone, but it has been a catalyst for many positive developments.**

Notwithstanding all these complications, in its short lifetime NSDI has generated high levels of interest in the USA and beyond. Some real successes have been achieved, as indicated above. Perhaps its greatest success, however, has been as a catalyst, acting as a policy focus, publicizing the importance of geographic information, and focusing attention on the benefits of collaboration. This is especially important in a country as large and governmentally complex as the USA.

20.4 Multi-national collaborations

There are many multi-national partnership arrangements in GIS, especially trans-border and bi-lateral ones. Here we concentrate on two such examples – the GIS and GI developments in the European Union and in Southeast Asia. The similarities and the differences between them and with the USA are obvious.

20.4.1 The European dimension

After the accession of 10 new member countries in 2004 – taking the total to 25 countries – the European Union (EU) contains some 456 million people. The European Commission (EC), the public service of the EU, has been considering a variety of GIS and GI issues since the 1980s. Progress has been slow, complicated by the decentralized, multi-cultural, and multi-lingual nature of decision making amongst the countries involved and the relatively low priority accorded to GI matters in EC work. A serious attempt to define a GI policy framework

for Europe through the GI2000 initiative failed. There have been some achievements in GIS research terms via the coordination and funding provided by the EC's Joint Research Centre in Ispra, Italy, including work by EUROGI and AGILE (see Box 20.3). More productive thus far has been the incorporation of GI within broader programs. Examples of this include the eEurope and eContent initiatives to foster European industry and the use of new technologies and content by citizens. The EC's ideas on public sector information, which highlighted the need for much easier access to information by citizens, may have long-running consequences since much information of this type is geographic. Once initiatives to take forward policies are agreed and financed by the European Union, they then acquire considerable momentum: an EC Directive approved by the European Parliament and the Council of Ministers must be implemented in national law within a defined period (normally two years).

A major initiative of the EC is the INSPIRE program (**www.ec-gis.org/inspire/**). This is a Directive proposed formally by the European Commission to establish a legal framework for the establishment and operation of an Infrastructure for Spatial Information in Europe. Based on much work by the Directorate of the Environment, Eurostat (Europe's statistical agency), and the Joint Research Centre at Ispra, this is designed to meet both community and local needs. The value of what is proposed is anticipated to be greatest to policy-makers such as legislators and public authorities but citizens, academics, and business are also expected to benefit if INSPIRE is adopted. The legitimation for INSPIRE is similar to that of the US NSDI (see Section 20.3.1) – even if the process for taking it forward is very different. The proposal was produced after making impact and risk assessments, having a consultation period and a public hearing, and taking account of the EU principles of subsidiarity (i.e., that nothing should be done at Community level which can be done more efficiently at national or lower level).

The proposed directive covers spatial datasets held by public authorities or their agents and spatial data services which employ these datasets. It mandates the use of a Community geoportal (Section 11.2) by all member states to supply their data to all other member states plus national and Europe-wide coordination structures and the creation of metadata. Procedures are laid down for establishing harmonized data specifications, so as to facilitate safe added-value operations, including unique identifiers for spatial objects across the continent. Most specifically, member states of the EU will be mandated (after acceptance of the Directive) to provide certain classes of services. These are discovery services permitting search for spatial datasets and related services, viewing services to display and navigate data, download services, data transformation services, and the capacity for invoking spatial data services. The first two of these must be provided free of charge though public access to the others may be curtailed for a number of reasons (e.g., national security and statistical or commercial confidentiality). One public authority in each country is responsible for liaising with the EC on the directive.

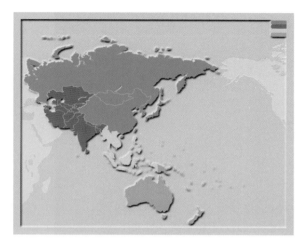

Figure 20.9 The member countries of the Permanent Committee on GIS Infrastructure for Asia and the Pacific (*Source*: PCGIAP)

INSPIRE was created following many political, scientific, and administrative battles over previous attempts to create an NSDI for an enterprise almost with twice the size of population of the USA and one in which Presidential Executive Orders (see Section 20.3.2) do not exist. The progress of the directive through the approval processes will not be easy but could make a material difference to how European GIS and GI operate. In microcosm, the attempts to build US and European NSDIs illustrate how differently the USA and the European Union operate and the importance of political leadership.

20.4.2 The Permanent Committee on GIS Infrastructure for Asia and the Pacific (PCGIAP)

This is the largest multi-national GIS body known at present. It was set up under UN auspices in 1994 and involves 55 countries (Figure 20.9). Its aims include maximizing the economic, social, and environmental benefits of geographic information in accordance with Agenda 21 (Section 20.5.1.3) by providing a forum for nations from Asia and the Pacific to cooperate in the development of a regional geographic information infrastructure and share experiences. 'GIS infrastructure' is taken as including the institutional framework (which defines the policy, legislative, and administrative arrangement for building, maintaining, accessing, and applying standards and fundamental datasets), technical standards, fundamental datasets, and a framework which enables users to access them. For the members of PCGIAP, key applications of GIS are national or regional land administration, land rights and tenure, resource management and conservation, and economic development. Considerable GIS activity is underway across the member nations, including the development of numerous NSDIs, and this information is shared through numerous meetings and newsletters (**www.gsi.go.jp/PCGIAP/**).

Continent-wide NSDIs are much more difficult to launch in Europe than in the USA because of the power structures.

20.5 Nationalism, globalization, politics, and GIS

We have shown above how the USA and groupings of other nation states have attempted to enhance the quality, utility, accessibility, and awareness of geographic information – and its use. This has been carried out through institutional structures and 'umbrella bodies', holding together disparate partners. But much GI has hitherto been created to suit national needs (at best). Its wider use is constrained by its historical legacy and the need for continuity through time for comparative purposes.

Other than that collected via satellite remote sensing, truly global coverage GI is currently little more (and sometimes less) than the sum of the national parts and is not always readily available. Yet the need for globally or regionally available, easily accessible, and consistent GI – especially the 'geographic framework' variety (see Section 19.3) – is increasingly evident. Even though few people currently have management responsibilities which are global, the present shortcomings in global GI have severe consequences for many actual and potential users who operate beyond national frameworks. The United Nations Regional Director of Development for the Pacific argued forcefully for the need for greater partnering and coordination of data collection and supply on a regional and global basis. He emphasized that development of such datasets is labor-intensive and costly and, although these datasets could support a wide variety of applications, generally no single use can justify the full cost of development.

Other than for imagery, global GI is currently little more than the sum of the highly varied national parts and detailed information is not readily available.

20.5.1 Global databases

There are two obvious approaches to creating global databases: to use existing mapping or to create something wholly new from imagery or other sources (though typically the products are often later used in combination). The latter has developed hugely in recent years through imagery produced by NASA, ESA, and similar bodies (e.g., see Figure 20.10), and commercial organizations.

20.5.1.1 Databases and maps

Global mapping has progressed rather more slowly than satellite imagery as a data source. And there have been false steps along the way: the International Map of

Figure 20.10 One section of a 'night time map' of the world produced as a composite of hundreds of images made by the Defense Meteorological Satellite Program which can detect low levels of visible/near-infrared radiance at night including lights from cities, towns, industrial sites, gas flares, and ephemeral events such as fires. Clouds have been removed. It was compiled from the October 1994–March 1995 data collected when moonlight was low (Courtesy NASA and US National Oceanic and Atmospheric Administration (NOAA))

the World was assembled over 60 years to provide a common geographic framework worldwide. Some 750 map sheets were published eventually but IMW is now simply history. The primary reasons for its ultimate failure were organizational and political. There was a lack of commitment of finance by those who agreed to participate – national mapping agencies – and conflicts in priority with national objectives. There was a lack of clearly articulated needs which the IMW was designed to meet, and few demonstrations of success from its use, and a lack of clear management responsibility for overall success. Finally, there was much duplication of work because of the technologies then available: much of the same mapping had to be created separately on both national and global map projections (see Section 5.7).

In the 1990s, in particular, a number of developments occurred in global topographic mapping – the traditional form of a geographic framework and still crucial. The US and allied military organizations (see below) created and distributed at low cost the Digital Chart of the World (DCW), subsequently known as VMap Level 0. Derived from the paper Operational Navigation Chart (ONC) 1:1 million scale map of the world created by NATO partners, this has been widely used in a variety of applications – including in low-price hand-held GPS receivers. Its popularity is despite well-known shortcomings in the quality of the original cartographic source material. This project was brought to reality because one organization led it strongly, there was a clear operational need, and financing of it could be justified on that basis. The release of DCW in 1993 led to robust discussions between a number of national mapping organizations (NMOs) and the then US Defense Mapping Agency (now part of the US National Geospatial-Intelligence Agency or NGA) about

infringements of national Intellectual Property Rights (see Section 19.1). Since then, an agreement has been forged for subsequent VMap products. This ensures that, where NMO-sourced material is used in NGA products, the providing nation has the option of restricting availability to NATO military use and making other arrangements (if desired) for commercial and public dissemination of its material.

Finally, there have been many other plans to build global databases which include mapping – all based on some form of partnership. The most advanced map-based project is that originally proposed by the Japanese NMO, initially for a new 1:1 million scale global digital map. An International Steering Committee for Global Mapping (ISCGM) has produced a full specification. The committee's initial proposals involve using a combination of VMap level 0, 30 arc second (i.e., approximately 1 km resolution) digital elevation model data and the Global Land Characteristics Database. Thus ISCGM proposed using vector and raster data already in the public domain, though not hitherto linked together, to provide the default dataset. The plan was that this would be replaced on a country-by-country basis wherever NMOs agreed to provide higher resolution data – though inevitably the final result would be somewhat inconsistent, especially at national boundaries in areas of territorial disputes. Recognizing such inherent difficulties, the ISCGM's specification ensures that – where any difference occurs – both sets of international boundaries will be stored. Encouragement to include data deriving from more detailed 1:250 000 scale mapping has now occurred. Figure 20.11 shows the countries that had committed to an involvement in this project by 2004. Although a sizeable number of these countries is not yet able or willing to supply

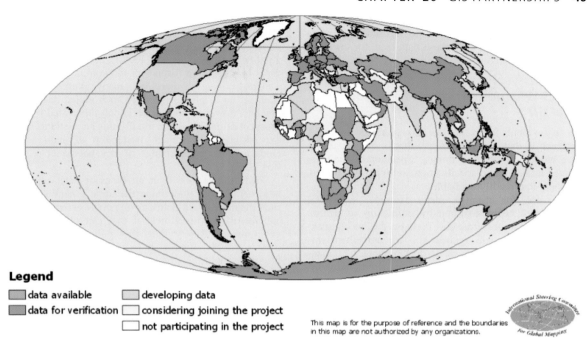

Legend

▒ data available	☐ developing data
▓ data for verification	☐ considering joining the project
	☐ not participating in the project

This map is for the purpose of reference and the boundaries in this map are not authorized by any organizations.

Figure 20.11 The countries that had committed themselves to the Global Mapping Project by August 2004

data in practice, materially this project has improved the availability of strategic-level GI for some areas of the world.

In some instances, certain aspects of maps have been abstracted and are being enhanced and up-dated by GPS and other means. The prime example is roads databases. Here the key players have almost all been private-sector bodies – even though many obtained raw data from national mapping organizations (NMOs). Navigation Technologies (NAVTEQ), a company operating in North America and Europe, has been building databases for well over a decade for the intelligent transportation industry and for in-vehicle navigation. These are built to one worldwide specification. Every road segment in the NAVTEQ database can have up to 150 attributes (pieces of information) attached to it, including information such as street names, address ranges, and turn restrictions (see Section 8.2.3.3). In addition, the database contains Points of Interest information in more than 40 categories. The information has multiple sources and several NMOs have relationships with Navigation Technologies. For example, the Institut Géographique National in France bought shares in NAVTEQ. A competitor of NAVTEQ is Tele Atlas which has coverage in Singapore, Hong Kong, and large parts of Australia as well as North America and Europe; the industrial scale of the applications of such data can be seen from the client lists. For Tele Atlas this includes Blaupunkt, BMW, Daimler Chrysler, Ericsson, ESRI, Michelin, Microsoft, and Siemens. All of these firms have business partners who may include surveying firms, national mapping organizations, value-added resellers, and much else.

Public and private sector partnerships are creating new GI which, in time, will probably be global in coverage.

20.5.1.2 Databases from satellite imaging

The variety of global databases now available from satellite-based sensing is extraordinary. One of the most relevant for many GIS users is that arising from NASA's $220 million Shuttle Radar Topography Mission launched in February 2000. The mission's aim was to obtain the most complete 'high resolution' (by the standards of what previously existed) digital topographic database of the Earth. The primary intended use of this data is for scientific research. It used C-band and X-band interferometric synthetic aperture radars to acquire topographic data between latitudes 60°N and 56°S during the 11-day mission. The resulting grid of heights is available on CD. The data released into the public domain are 3 arc seconds resolution (i.e., about 90 meters at the Equator) for the land areas of the Earth; however, 1 arc second data are available for the continental USA.

The first high-resolution civilian satellite was launched by Space Imaging Inc. in September 1999. This produced various imagery sets, the most detailed being 1 meter resolution. Space Imaging also collects and distributes Earth imagery from a range of partner organizations in Canada, Europe, India, and the USA. In essence, it is setting out to be a global geo-information, products, and services organization, working with a variety of regional partners across the world. Other competitors such as DigitalGlobe have joined Space Imaging and data resolutions as detailed as 60 cm are now available from these commercial providers for many areas of the world (see Figure 20.12), with 42 cm resolution planned by 2006/07. A variety of public and private sector channels exist for obtaining access to these data (see, for instance, **www.terraserver.com**)

20.5.1.3 Why do we not yet have good global GI?

What is evident from the last 20 years or more of experience is that everyone thinks that digital databases of the world are 'a good thing', especially if they are produced to a standard specification, sufficiently detailed, up-to-date, and readily available.

There are, for instance, many examples where GI and GIS have been identified by politicians as central to improving some situation. Perhaps the most striking example is that of Agenda 21. This agreement was designed to facilitate sustainable development. The manifesto was agreed in Rio in 1992 at the 'Earth Summit' and was supported by some 175 of the governments of the world. It is noteworthy that eight chapters of the Agenda 21 plan dealt (in part) with the need to provide geographic information. In particular, Chapter 40 aimed at decreasing the gap in availability, quality, standardization, and accessibility of data between nations. This was reinforced by the Special Session of the United Nations General Assembly on the Implementation of Agenda 21 held in June 1997. The report of this session includes specific mention of the need for global mapping, stressing the importance of public access to information and international cooperation in making it available. The subsequent UN World Summit on Sustainable Development in 2002 included many GIS activities and demonstrations to the 40 000 participants, including leading politicians. Again there were public calls for spatial data to be regarded as a public good (see Section 18.4.2), accessible to potential users in a common format for the common good.

Unsurprisingly, therefore, many partnerships have been set up to plan or produce such global databases. In practice, however, commerce is only interested if the real prospect of profit occurs; usually this necessitates military expenditure. Government leadership of such schemes tends to falter with change of key personnel (e.g., see Box 20.7).

Past failures to deliver promises or meet aspirations, however, do not undermine the large potential value of such databases; and political support is essential (if often ephemeral) for investment if that can only come from governments. Rarely does this political support come from a simple interest in technical matters or even a fascination with geography on the part of the politicians. To obtain support requires both fitting the GIS agenda to the existing political agendas and highly professional raising of awareness and lobbying.

20.5.2 Global partnerships for standards

There are already a number of well-resourced and long-term groupings which bring together countries for particular purposes and which have direct or indirect impacts on GIS – such as the International Standards Organization (ISO), the World Intellectual Property Organization

Figure 20.12 Part of a Quickbird 62 cm resolution image of Buckingham Palace and the Mall, London (© Quickbird)

Applications Box **20.7**

The fading of Digital Earth

'Digital Earth' was first proposed by US Vice-President Gore in his 1992 book *Earth in the Balance*. Re-launching the concept in 1998, he said a pressing problem was how to turn the huge amounts of geospatial data that were being collected into information that people could use. He stated that:

- there was no effective connection (interoperability) between the many datasets required to accomplish better management of the Earth's resources and minimization of the problems of natural disasters;

- there was no common vision to activate cooperation among the many elements of the technology base and the national and international community;

- there was low awareness and little accessibility by the 'common person' to the vast information resources about our planet and its people.

Gore's analysis was a damning indictment of what had been achieved to that date. His solution was a mechanism for tying together data from multiple, networked sources and making that integrated resource available visually as a 'browsable', 3-D version of the planet via an enhanced Internet. In the longer term, he envisaged having a complete and consistent digital map of the Earth at 1 meter resolution.

Following Gore's defeat in the 2000 US Presidential elections, Digital Earth rapidly faded from view as a project. However, some Chinese activity in this sphere continues, many relevant (but re-titled) developments continue in NASA, and commercial developments like *ArcGlobe* and *Earthviewer* (**www.earthviewer.com**) are practical demonstrations of some of Gore's aspirations.

(WIPO), and the World Trade Organization (WTO). These are voluntary groupings bound through signature of treaties or through the workings of a consensus process. But other groupings of bodies exist which have specific importance for GIS. Perhaps the most important of these is the Open Geospatial Consortium (OGC), formerly the Open GIS Consortium (see also Section 11.2).

OGC is a voluntary consensus standards organization operating in the GIS arena. The OGC is a not-for-profit, global industry association founded in 1994 to address geospatial information-sharing challenges. Its membership is primarily US-based, reflecting its origins and also the US dominance of the GIS marketplace, but the geographic spread of membership has continually increased. OGC's worldwide membership, which totals 260 entities, includes geospatial software vendors, government integrators, information technology platform providers, US federal agencies, agencies of other national and local governments, and universities. The OGC's President has claimed that the network of public/private partnerships embodied by the organization has accomplished for geospatial information what the US railroad companies had accomplished by 1886, when they achieved consensus on the adoption of a common rail gauge.

20.5.3 The Global Spatial Data Infrastructure

The other approach to building worldwide consistent facilities has been to seek to create a Global Spatial Data Infrastructure. GSDI is the result of a voluntary

coming together of national mapping agencies and a variety of other individuals (see Box 20.3) and bodies. The Steering Committee defined the GSDI as 'the broad policy, organizational, technical, and financial arrangements necessary to support global access to geographic information'. Initially, the organization sought to define and facilitate creation of a GSDI by learning from national experiences. An early scoping study showed that the reality of the benefits was difficult to prove, largely because it is very difficult to define the product that governments and other agencies might be asked to fund. Since then, perhaps wisely, the GSDI Association (**www.gsdi.org/**) has concentrated on

- serving as a point of contact and effective voice for those in the global community involved in developing, implementing, and advancing spatial data infrastructure concepts;

- fostering spatial data infrastructures that support sustainable social, economic, and environmental systems integrated from local to global scales; and

- promoting the informed and responsible use of geographic information and spatial technologies for the benefit of society.

All this has been done through a series of partnerships with many other bodies (such as OGC and regional GIS organizations). Certainly, the result has broadened the knowledge of GIS use and GI challenges. In addition, such GSDI publications as the GSDI 'cookbook' on how to create a SDI have helped many countries (**www.gsdi.org/docs2004/Cookbook/cookbookV2.0. pdf**).

Building a Global Spatial Data Infrastructure would be even more difficult than a national one, but discussions on GSDI have proved especially helpful to those outside the wealthier countries.

20.6 Extreme events can change everything

Much of what has been written above has arisen from a situation where it is essentially 'business as usual'. But the events of 11 September 2001, seared into the memory of millions of people across the globe (Figure 20.13), have also begun to change many aspects of GIS and GI.

The death and disruption caused by the Al Queda hijacking of four aircraft and their use as guided missiles ruined lives and caused billions of dollars of harm to the US and other Western economies. The immediate aftermath was marked by heroic efforts, not least amongst the GIS teams who recovered information with which to support the rescue and reconstruction efforts in New York (Box 1.1). But the event itself also led to an intense questioning of how we can best protect ourselves from further acts of terrorism – without turning ourselves into a police state. One manifestation of this debate was the creation of the US Department of Homeland Security (see Box 20.8) but the reactions were not restricted to the USA: countries as far apart as India, Russia, and Colombia have also taken action as they too have been

Figure 20.13 Newspapers from around the world on 12 September 2001 reporting the terrorist attacks on the World Trade Center. (*Source*: **www.september11news.com/index.html**. Copyright in the images subsists in the newspapers shown; *Herald Sun* page reproduced by permission of Herald & Weekly Times Pty Ltd; *Dagens Naeringsliv* page reproduced by permission of Dagens Naeringsliv)

The creation of the US Department of Homeland Security

The Office of Homeland Security was set up by the US President on 8 October 2001 and the Department of the same name was formed by an Act passed by the Congress some three months later. The Department of Homeland Security (DHS) has three primary missions: to prevent terrorist attacks within the United States, reduce America's vulnerability to terrorism, and minimize the damage from potential attacks and natural disasters. It was

formed by merging all or parts of 22 pre-existing agencies under new integrated management. It was also set up from the outset to have a Directorate of Information Analysis and Infrastructure Protection. The Secretary of the department has sweeping powers of access to virtually any information which he or she may consider relevant to the department's tasks, whether held by the federal government or elsewhere (see **www.dhs.gov**).

---

affected by terrorist action. All this highlights the need for flows of intelligence across institutional and national boundaries – requiring new forms of trust and partnership which must operate speedily and without failure.

Since everything happens somewhere and since targets of terrorism are picked to cause maximum effect, GI and systems to deal with it are at the forefront of governmental efforts to carry out their first duty – to protect their citizens. To aid this mission, we must first know how GIS can potentially help terrorists so we can guard against such threats. We must also know how best to use such systems to minimize the consequences of any terrorist actions. And we need to work out the best form of partnerships in these new circumstances.

20.6.1 How GIS can potentially help attackers

In principle, GIS can help a potential terrorist in at least four ways:

- finding locations of a particular type (e.g., major 'choke points' in roads, population centers, or highly critical places such as nuclear plants);
- adding data layers to enrich the understanding of that location's 'uniqueness' or 'connectedness';
- modeling of the likely effects of a disruption (in three dimensions and time, if desired) to the facility and of escape routes for the attackers;
- providing descriptive material or GPS coordinates to lead the attackers to the spot and to exit after the attack.

20.6.2 Is public access to GI adding to the problem?

The nature of Western society is that much information is made readily available: that is a fundamental tenet of democracy. The situation is perhaps most extreme in the USA where the huge bulk of federal government

information is made available without constraints on who can get it or on its further dissemination (and commercial exploitation – see Section 19.3). Table 20.2 summarizes the critical physical infrastructure of the USA as recognized in the national strategy for its protection.

Immediately after 9/11, a number of US government websites withdrew access to various datasets (see Box 20.9). Much was restored after quick reviews. The National Geospatial-Intelligence Agency, with US Geological Survey support, sponsored a major study by the Rand Corporation to assess the homeland security implications of publicly available geospatial information. This reviewed the types of data which might be helpful to terrorists in different ways – the 'demand side' – and studied a large sample of publicly available websites – the 'supply side' (see Table 20.2). Their conclusions can be summarized as follows:

- Publicly accessible geospatial information has the potential to be useful to terrorists for selecting a target and determining its location but is probably not the first choice to fulfill their needs – there are many alternative sources.

- Less than 6% of the publicly available GI from federal government agencies appeared as if it would be useful to potential attackers and none appeared of critical importance to them; fewer than 1% appeared both useful and unique.

- Much of the information needs of potential attackers can be met from non-governmental websites – such as industry and business, academia, NGOs, state and local governments, international sources, and even private individuals.

- The societal costs of restricting access to geospatial information are difficult to quantify but are large.

- The federal government should develop a comprehensive model for addressing the security of geospatial information, focusing on what level of protection should be afforded to data about a particular location. In addition, it should raise public awareness of the potential sensitivity of GI.

Table 20.2 US critical infrastructure sectors and accessibility to information about individual instances. About 85% of the US critical infrastructure is owned by the private sector

Sector	Example assets	Degree of public accessibility
Agriculture	Grain storage elevators	Medium to substantial
Food	Meat processing plants	Medium to substantial
Water	Drinking water facilities	Substantial
	Dams	Limited to substantial
Public health	Hospitals	Substantial
	National pharmaceutical stockpiles and supplies	Limited
Emergency services	Emergency operations centers	Medium to substantial
Government	Government agency headquarters	Limited to medium
	Regional offices	Medium to substantial
Defense industrial base	Military equipment manufacturing plants	Medium to substantial
Information and telecommunications	Transmission sites	Medium
	Internet backbone facilities	Medium to substantial
Energy	Nuclear power plants	Limited to medium
	Oil refineries	Limited to medium
Transportation	Bridges	Substantial
	Tunnels	Substantial
	Pipelines	Substantial
Banking and finance	Major financial exchanges	Substantial
	Financial utilities	Medium
Chemical industry and hazardous materials	Chemical processing plants	Limited to medium
	Hazmat material transportation	Medium to substantial
Postal and shipping	Mail processing centers	Medium to substantial

Source: Table D.1 in Baker J., Lachman B., Frelinger D., O'Connell K., Hou A., Tseng M., Orletsky D. and Yost C. 2004 *Mapping the Risks: Assessing the Homeland Security Implications of Publicly Available Geospatial Information.* Rand Corporation Monograph (see **www.rand.org/publications/MG/MG142**)

Applications Box **20.9**

The withdrawal of public geospatial information after 9/11

Before September 11, 2001, most federal groups viewed the Internet as a place to store their vast library of public documents and as a way to network with community leaders in a timelier manner. But in the days that followed 9/11, authorities realized that terrorists also used the Web. Government agencies suddenly scrambled to assess what they released into cyberspace, vetting it for any sign that it could be used to exploit structural or security vulnerabilities.

Beth Hayden, spokeswoman for the Nuclear Regulatory Commission, one of more than a dozen federal agencies that altered their sites. . . said the agency's entire site initially was stripped from the Internet on October 11 but was restored October 17. The Environmental Protection Agency acted within 48 hours of the 9/11 attacks, opting to remove risk-management plans that offer specific outlines

of the nation's chemical plants, including how they're handled and scenarios for possible releases. The plans had been posted on the Web so state or city officials and the public could view them in the event of an emergency.

Source: **CNN.com**, 10 September 2002

One state took its clearinghouse offline for two months while it removed 'tiles' that contained sensitive information. There were 500 sensitive sites on 170 quads and 1,600 orthophoto tiles. . . As of last November all the quads were back online. Ortho tiles are in the process of being 'blurred'. The state called upon a local university to develop an algorithm to blur just the footprints of the sensitive areas, again, attempting not to draw attention to them. All of the data was [sic] widely available elsewhere, but both states felt it politically correct to alter at least some of it.

Source: GISmonitor 1 April 2004

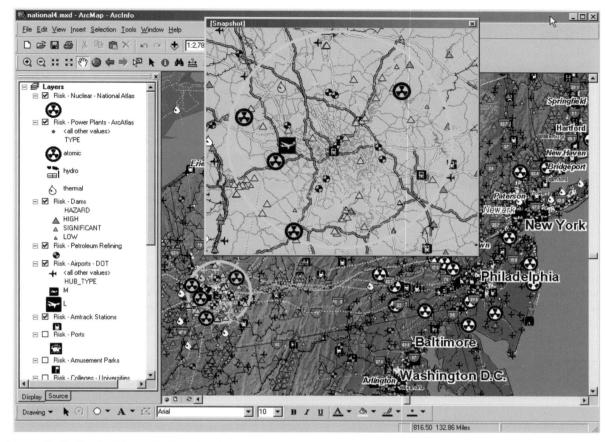

Figure 20.14 Key facilities in the North East USA, as assessed in the joint Homeland Security project of USGS and the Department of Homeland Security (Courtesy US Geological Survey)

The study's overall conclusion was that, whilst it might be useful, geographic information was unlikely to be of *critical* value to terrorists and is, in any case, so widely available that nothing much could be done to minimize the risk except in a few cases. This is comforting – but seems somewhat surprising given the final recommendation above.

20.6.3 Coping with terrorism and the GIS contribution

We can think of dealing with terrorism or some major national disasters as having five phases, each of which we review below. In each case GIS and GI can play a valuable role and partnerships between (often many) players are required.

20.6.3.1 Risk assessment

In principle, this is identical to all good business planning (see Section 19.4.1). It involves assessing the likelihood and possible impacts (on life, property, and other assets and the environment) of terrorist activity, an emergency, or disaster – and then communicating this to

the appropriate authorities. It involves identifying risks and their potential impacts (Figure 20.14), which organizations should be involved, and the necessary mitigation, response, and recovery procedures – and testing the procedures.

20.6.3.2 Preparedness

This covers those activities necessary because terrorist attack or other hazards cannot be confidently anticipated and obviated. It involves creating a set of operational plans to deal with a set of defined scenarios, including who is to be in charge (e.g., in different zones) for the different scenarios. Inventories of governmental and other resources which may be mobilized need to be created and shared on a 'need to know' basis. Early warning systems need to be installed. Training exercises will be carried out. Emergency supplies of food and medical supplies need to be stockpiled and their delivery planned. All of these involve use of geographic information and GIS (see Figures 20.15 and 20.16).

20.6.3.3 Mitigation

This involves identifying activities that will reduce or obviate the probability and/or impact of an event, e.g.,

Figure 20.15 A scene from a training exercise simulating a chemical attack on the financial district of London (Reproduced by permission of PA photos)

maintaining an inventory of the contents and locations of hazardous materials and mapping the geography and sensitivity of environmental dissemination of

contaminants (e.g., reservoirs and water distribution facilities – see Figure 20.14).

20.6.3.4 Response

These are the activities immediately following a terrorist attack, or other emergency, and are designed to provide emergency assistance for victims, stabilize the situation, reduce the probability of secondary damage, and speed up recovery activities. They include providing search and rescue, emergency shelter, medical care, and mass feeding; controlling transport links so as to prevent further injury, looting, or other problems (see Figure 20.17); shutting off contaminated utilities; and/or carrying out a damage assessment (see Figure 20.18).

Organizing evacuations may be essential (Sections 2.3.4.2 and 16.1.2) – as in areas affected by hurricanes (e.g., see Figure 2.14) – but it is also planned and periodically tested in most major cities as a prudent safeguard. In London, for instance, plans for evacuation of areas containing several hundred thousand people exist and tests of readiness against various attacks have been made; the transport system GIS plays a significant role.

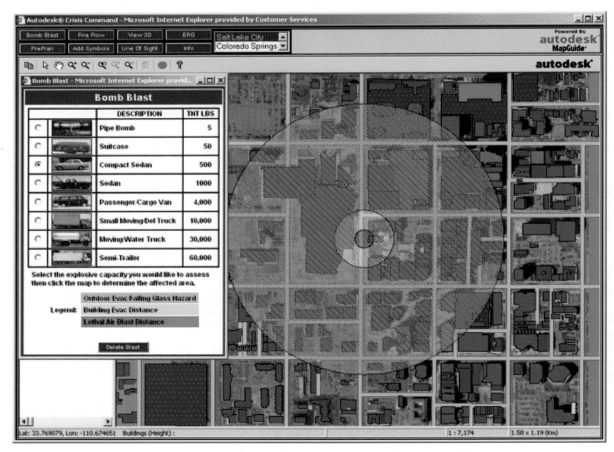

Figure 20.16 The potential impact of explosions within Colorado Springs, based on the size of an explosion (that selected is a simulation of a sedan packed with 500 pounds of TNT). The red area is the lethal air blast distance, yellow and green are the indoor and outdoor evacuation areas (Reproduced by permission of Autodesk, Inc. © 2004. All rights reserved.)

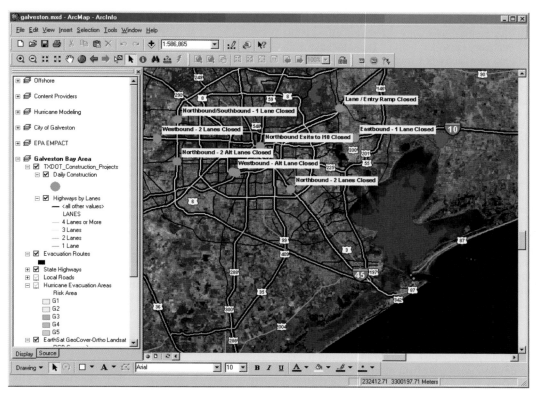

Figure 20.17 Evacuation routes and roads temporarily closed due to construction in the Galveston Bay area after an incident simulated in the joint Homeland Security project of USGS and the Department of Homeland Security (Courtesy: US Geological Survey)

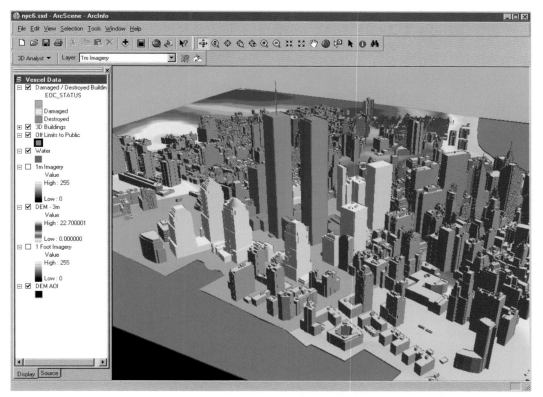

Figure 20.18 The damage caused at Ground Zero. Map created in the Homeland Security project of USGS and the Department of Homeland Security

20.6.3.5 Recovery

These are the actions necessary to return all systems to the pre-recovery state or better. They include both short- and long-term activities such as creating an agreed and shared 'status map'. 'Cleaning up', provision of temporary housing, return of power and water supplies and allocation of government assistance are essential parts of aiding recovery.

20.6.4 Does GIS help or hinder?

There is some doubt about the utility of GI and GIS to those planning terrorist attacks. But there is no doubt that informed and expert use of such assets is enormously beneficial to the populace, both in anticipating and planning for possible attacks and in recovering from these and from natural disasters. Such extreme events place great reliance on system and information inter-operability (see Section 11.2), on frequent practices, and good inter-institutional relationships and partnerships.

> **Geographic Information – and the ability to assemble, analyze, depict, and communicate it – is key to dealing with terrorism attacks or most other emergencies.**

20.7 Conclusions

Despite all the problems of partnerships due to the sometimes different agendas of different players, often there is no alternative to being part of them, e.g., where public and private sectors each provide some crucial element. It can bring vital components – staff skills, technology, marketing skills, and brand image – to add to yours and make a new product or service. It can bring fresh insights on the real user needs and can create superb support and publicity from fellow citizens. It can lead to cost- and risk-sharing.

The probability of success in partnering is raised by keeping the objectives clear, ensuring overall responsibility is agreed and that adequate resourcing exists, and keeping the numbers of partners to the minimum possible, consistent with achieving the desired aims. Remember that ultimately the interactions and trust between the people involved will determine how well the partnership succeeds. But the changes in the world post-9/11 may well ensure that our previous ways of partnership-working may well be modified in the need to safeguard our very way of life: someone has to be in overall charge and procedures enforced. In that sense, the partnerships are more like military ones than the voluntary ones common until now in the GIS community. If so, this could strongly affect NSDI and much else in GIS.

Questions for further study

1. Imagine that you are trying to get three local parties – a local government, a utility company, and a commercial supplier of GI – to work together to produce a Web-accessible GI database for free use. What do you imagine would be the problems? How would these change as you move up the spatial scale to the nation, to the region (e.g., the Americas or Europe), and to the world as a whole?

2. Do a one hour search by whatever means you have available locally for a global database in one thematic area (e.g., land use). How many sources of supply have you found? How many organizations were involved, in what sort of partnerships? What is the relationship between the datasets? What is the internal coherence and quality of what you have found?

3. Find out what preparations have been made in your area to guard against terrorism or major disasters and what role GIS is playing in them

4. How would you 'sell' the benefits of GIS and GI to a politician?

Further reading

Masser I. 2005 *Spatial Data Infrastructure: An Introduction*. Redlands, CA: ESRI Press.

NRC (2004) *Licensing geographic data and services*. National Research Council of the National Academies, Division of Earth and Life Sciences. National Academies Press, Washington DC.

Rhind D. 2005 'National and international geospatial data policies'. In Longley P.A., Goodchild M.F., Maguire D.J., and Rhind D.W. (eds.) *Geographical information Systems: Principles, Techniques, Management and Applications (abridged edition)*. 767–787, Hoboken NJ: Wiley.

21 *Epilog*

This final chapter revisits and summarizes the key ideas introduced in this book and offers some predictions about the future of GIS. We assess the implications of GIS as a technology, as a framework for applications, as an important management tool, and as a fast-developing science. Simply considering any one of these in isolation understates the crucial integrating capacity of GIS that makes it hugely valuable in the real world. We develop this theme of enduring issues and practices in a brief critique of current preoccupations with 'geospatial' systems and technologies. In focusing upon explicitly *geographic* information systems and *geographic* science we also identify some grand challenges that will likely shape the near future of the field, and assess the remit of GIS in the developing global Knowledge Economy.

Geographic Information Systems and Science, 2nd edition   Paul Longley, Michael Goodchild, David Maguire, and David Rhind.
© 2005 John Wiley & Sons, Ltd.   ISBNs: 0-470-87000-1 (HB); 0-470-87001-X (PB)

Learning Objectives

After reading this chapter you will:

- Understand some of the main themes that link GISystems, GIScience, and GIStudies, including
  - The impact of the Internet on GIS;
  - How GIS and management interact;
  - How GIS can be applied in real-world practical contexts;
  - The interdependencies between GISystems and GIScience;

- Understand the core organizing themes that link the principles, techniques, analysis, and practice of GIS;

- Be familiar with how the current usage of the term 'geospatial' relates to GIS;

- Be aware of some of the grand challenges that face GIS, and understand some of the ways in which GIS is likely to develop in the next few years.

21.1 Introduction

In *Geographic Information Systems and Science*, we have tried to define the content and context of GIS. For the many that are already experienced in the field we have introduced some different ideas and new ways of thinking about GIS as systems, science, technology, and as a management tool. For those new to GIS, we have tried to write a comprehensive primer that introduces prevailing thinking about the subject. There is compelling evidence that GIS in the 21st century is an immensely important field of endeavor and that rapid progress is taking place on several fronts. With heightened awareness of GIS comes increased uptake and wider participation, informed critical analysis, clearer scientific thinking, and greater realization that management is central to successfully achieving business and societal aims. In this epilog, we close the book with a brief review of some recurring themes, and some personal views on what the future might hold for GIS.

21.2 A consolidation of some recurring themes

21.2.1 Systems, science, and studies

In Chapter 1, we identified geographic information science (GIScience) as the set of fundamental organizing principles and methods of analysis that arise from the use of geographic information systems (GISystems). This definition is helpful, because it suggests that people using GIS must inevitably ask questions about such topics as scale, accuracy, inference, and the ways in which humans interact with computers. Answers to questions like these require systematic scientific inquiry, and access to canons of knowledge such as spatial statistics and cognitive science. Expertise at the cutting edge of GIScience is allied to an understanding of state-of-the-art technology, and we have tried to illustrate how GISystems provide efficient and effective tools to assemble, manage, and display geographic data. For almost all applications, appropriate technology and analysis are necessary but not sufficient preconditions for success in GIS. Therefore, we have emphasized the tactical and strategic importance of management in governing the way in which GIS applications are conceived, resourced, and executed. This is the way in which geographic information *systems support and drive science* in unique applications contexts (Figure 21.1).

Geographic information systems support and drive science.

There are few new principles amongst the many we have enunciated in this book – indeed, many of the issues raised by GIS have been recognized for centuries. Humans have accumulated vast stores of relevant theory and technique, and only the most precocious of GIScientists would suggest that today's GIScience is an activity that has become cut free of its diverse disciplinary roots. Core to this activity is the geographer's ability to relate unique applications to generalized understanding of the ways in which the world works (Section 1.3). Cartographic design principles, so important for effective communication using GIS (Chapter 12), originated long before the advent of either digital computers or GIS. The unique

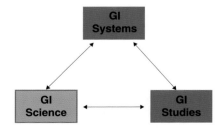

Figure 21.1 The interrelatedness of GI Systems, Science, and Studies

properties of spatial data present daunting challenges for the use of classical statistics, but the established framework of statistics nevertheless provides a starting point for understanding the nature of spatial data and the specification, estimation, and testing of geographic relationships (Chapter 14). The goal of broadening the user base of GIS requires our understanding of the established cognitive principles that guide user interaction. Taken together, this suggests a symbiotic relationship between science and systems: successful applications drive science, as noted above, but geographic information *systems also implement the storehouse of knowledge known as geographic information science.*

Geographic information science grounds successful applications in established scientific practices.

We should, of course, accept that science itself and the exploitation of technology are changing. Most of the exciting developments are now at the boundaries between what we used to think of as different branches of science – notably in the bio-sciences and social sciences.

Thus inter- and multi-disciplinary science is at the center of crucial developments (see Box 21.1). The traditional linear model of technology exploitation – where curiosity-driven research spawned new fundamental findings, which were then taken up by engineers and made into diverse products which marketers sought to sell to the public – is now seen as increasingly irrelevant. It is much closer involvement of all parties at all stages and much greater iterative experimentation that works now. The outward-looking but integrative nature of geography as an academic subject, allied to the technical and scientific foundations of GIS, ensure that people educated in both have the right sets of skills to prosper in this changing world.

While GISystems originated in the 1960s, the term *geographic information science* was first coined in the 1990s. Its remit is to make old science more relevant to new systems, and to motivate new research in old areas of science – in order to solve problems that wider changes in data supply and access have brought to the fore. From this perspective GISystems mine GIScience,

---

Biographical Box **21.1**

Dr Rita Colwell, scientist

Dr Rita Colwell (Figure 21.2) is one of the world's most distinguished scientists. She has authored or co-authored 16 books and more than 600 scientific publications, and produced the award-winning film *Invisible Seas*. She has also received 26 honorary degrees from institutions of higher education, and a geological site in Antarctica, Colwell Massif, was so named in recognition of her work in the Polar Regions. A marine microbiologist by training, she became the first woman to be Director of the US National Science Foundation (NSF) in 1998, a post she held for six years serving both Presidents Clinton and Bush. The NSF initiated its first interdisciplinary research program in the 1970s, but Dr Colwell increased the organization's focus on interdisciplinary research as a result of highly compelling evidence that this was essential in contemporary science and society.

Figure 21.2 Rita Colwell, scientist

Her own research has demonstrated this commitment to working with scientists and practitioners from many disciplines. Dr Colwell's work in Bangladesh spans a quarter of a century and has focused upon solutions to the continuing cholera outbreaks that have plagued the local people. This is a different setting in time, scale, and circumstance to the pioneering work of Dr John Snow (Box 14.1), and today's solutions rely on an interdisciplinary approach that integrates and shares information from medical, environmental, and socioeconomic resources, among others. The language that enables effective communication among the various disciplines is geography, which Dr Colwell has described as the 'ultimate, original multi-disciplinary language', through which data lead to wisdom (see Section 1.2). It is by bringing together and analyzing data from many sources that geography articulated through GIS 'contributes to science and technology and improves the human condition'.

Dr Colwell's own applications have examined the relationships between weather, sea temperature, the population of copepods (shrimp-like microscopic organisms in which cholera bacteria naturally occur), and the ways in which people in Bangladesh live and interact with their environment. These enabled her to devise simple interventions that have helped to reduce seasonal cholera outbreaks by nearly 50 percent – such as filtering water with a folded piece of cloth, typically used in making women's clothing, to reduce the number of copepods and other particulates.

and the future of GISystems depends on a constant flow of new ideas channeled through GIScience. While many of these ideas are not new, the environment of GIS is at last realizing many of the objectives of the 1960s spatial-science paradigm – in an environment that is data rich, offers excellent processing and visualization capabilities, and provides increasingly robust spatial-analysis tools. As such, GIScience extends well beyond the immediate set of issues motivated by today's GISystems, into areas such as cognitive science and information science, and tomorrow's discoveries may make possible a new generation of GISystems that bear little relationship to today's designs. The core organizing principles of today's GIScience may predate GISystems, but together they nevertheless bring huge resources of accumulated knowledge to the operational setting of GIS.

Geographic information science drives the development of systems.

21.2.2 What's in a name? GIS and geospatial

We have already noted (Sections 1.4.2 and 1.6) that the interdisciplinary lexicon includes a number of different terms to describe the activity that we call 'GIS'. The term 'geospatial' (see Section 1.1.1) was first coined in the mid-1990s to describe the data activities of a number of US federal government agencies, and has since entered common parlance, at least among spatially aware professionals. Today, it is also used by some private-sector companies and by the Open GIS Consortium (now the Open Geospatial Consortium: an organization dedicated to interoperability between GISystems – see Section 11.1). The popularity of the term appears to reside in its convenient contraction of geographic ('relating to the Earth') and spatial (relating to any coordinate system at any scale, as in medical imaging or plotting multi-dimensional statistical relationships). This is appealing to some for a number of reasons – the connotation of ownership of GIS by the academic geography community (which has been responsible for rather few technical developments in GIS) is avoided, and this loosening of ties may be taken to suggest that applications need no longer be restricted to geographic scales of measurement (usually taken as ranging from the architectural to the global). It is also a convenient label to use for instruments and technologies (such as GPS and remote sensing) that have been associated with academic disciplines other than geography, and also conveys to some a greater sense of 'hard science'. Last, and not least, it breaks commercial links with the past and allows new companies to compete more effectively.

Our own interim view is that while this terminology succeeds in conveying what is 'special about spatial' (Section 1.1), it fails to convey the achievement of bringing together the power of computer-based generalization with sensitivity to the uniqueness of the real places that are the focus of GIS applications. Much of the thrust of research in GIScience is concerned with making GIS-based representations sensitive to context, and

representing an exciting world that is vivid, unique, and real. 'Geospatial' has been used to imply commonalties of approach with GIS in describing the very small minority of applications that are neither at geographic scales of measurement nor are adequately described by other discipline-specific labels (such as computer-aided design and photogrammetry). Our own view is that the richness and diversity of GIScience may appear compromised by a term that implies that space is sterile and amorphous rather than variegated and rich in texture. We think that this may also be the view of academics working in the range of geographically enabled disciplines, generations of whom have dedicated their careers to measuring and understanding the variable characteristics of the Earth's surface. In short, we have concerns that a focus upon the 'geospatial' implies a preoccupation with technology and low-order data concepts, rather than the higher levels of the chain of understanding (evidence, knowledge, and wisdom) described in Section 1.2. Still others would contend that, in a different sense, it is graphicacy (visual interpretation of explicitly geographic phenomena) that is most immediate, exciting, and central to GIS-based analysis.

The term 'geospatial' acknowledges what is 'special about spatial' but neglects what is 'magic [and unique] about geographic'.

This point assumes more than semantic importance when we consider its implications for GI education (Section 1.5.5). Given the long heritage of the term 'GIS', its popularity in universities, and formal place in the academic curriculum, this would seem to be the preferred term for an important scientific, technical, and business field. It is also a term that is recognized increasingly in high schools, not least through outreach activities such as GIS Day. Geography does not always make the best of its opportunities to promote either GIScience or real-world relevance, but considerable intellectual capital is tied up in a discipline that enjoys a privileged status on the (high school and university) curriculum in many parts of the world, and has considerable momentum behind it in others. Promulgation of a newer term, which apparently fails to confer any unequivocal advantages, is likely to confuse the uninitiated and, worse still, present applications as devoid of real-world scientific context.

21.2.3 Geography's revenge – and GIStudies

The Internet evolved out of projects funded by the US Defense Advanced Research Projects Agency (DARPA), beginning in the late 1960s, to tie computers together with high-speed communications links. It was designed to be highly redundant, so that several possible paths could be followed between any pair of computers, ensuring that the network would survive limited attack. Thus, an email sent from one computer to another will be broken into packets by the Internet's protocols, and individual packets may follow several different routes, being reassembled on arrival at their destination. It was also designed to overcome geography, to make it possible to communicate without respect to distance. Today, few Internet protocol (IP) addresses give any indication of the user's location (Section 11.3.2). This ability of the Internet to overcome distance inspired much extravagant prose in the early years of the network's popularization, culminating in 1997 with the publication of the first edition of Frances Cairncross's book *The Death of Distance: How the Communications Revolution Will Change Our Lives*.

However, it soon became obvious that while electronic communication had major benefits in its ability to overcome distance – allowing people to send email to friends on other continents or to access foreign websites at speeds that bore no relationship to distance – there were still major reasons why location and distance remained important. Most people who work at home still need to visit the office occasionally; most people are more likely to send email to nearby addresses than distant ones; and information on nearby restaurants is usually more useful than information on restaurants in other countries. Businesses were quick to recognize this, and to find ways to determine the location of a customer's computer or cellphone, using the methods discussed in Section 11.3.2. An article titled 'The Revenge of Geography' appeared in *The Economist* in March 2003, and laid out the reasons for a resurgence of interest in distance and location in Internet applications: 'It was naive to imagine that the global reach of the Internet would make geography irrelevant. Wireline and wireless technologies have bound the virtual and physical worlds closer than ever'.

Thus far our discussion has not addressed the important need to reflect on GISystems, their impact on society, and their appropriate place within it. We noted in Section 1.7 that any technology inevitably reflects in part the nature of the society that invents and uses it, and in turn influences the way that society evolves. GIStudies is not about the technology itself, or about the science that makes that technology possible. Rather it concerns how that technology is embedded in a societal context, and how it affects such issues as the importance of distance and location to that society. Careful reflection is a vital part of the role of the academy, and GIStudies is a vital part of the academy's relationship to GISystems. As its title suggests, this book has tended to focus more on the technical and scientific aspects of GISystems and GIScience than on the reflective aspects of GIStudies, or the fast-developing field of public participation in GIS. There is an expanding literature expressing a critical perspective on GIS and its role in science and society. We welcome this as long as there remains recognition that applied science creates value, and that there is a net worth to activities that nevertheless inevitably benefit different groups in society to differing extents.

GIStudies is concerned with the ways in which systems and science interact with society.

However important it may seem, information technology is inevitably limited in its ability to represent the complexity of the real world, or to address and solve its problems. Understanding its limitations has been one of the themes of this book, notably in the chapters on representation, uncertainty, and management. Sometimes even the most skeptical of GISystems users may fail to 'see the wood for the trees', and it is important that GIStudies continue to attract people who are less immersed in the technology and can think deeply about the meaning of GIS and its role in society, and about GIS weaknesses and limitations. Yet we believe that all users need a good grounding in principles and techniques in order to clarify rather than obscure geographic issues.

21.2.4 GIS and technology

One of our main motivations for writing this book is to rebalance the preoccupation of many in the GIS field with technology. Technology is important to GIS and one of the drivers for change in a rapidly evolving field. Appropriate technology is clearly necessary, but alone it is not sufficient for success in GIS. For some, technology is a constraint on their ability to participate in GIS – adequate and appropriate hardware, software, data, and network connections are not universally accessible – and we have commented elsewhere on digital differentiation between GIS haves and have-nots. For others, GIS technology provides opportunities – most obviously the business opportunities of the GIS vendor and consultant community, but also for those interested in adapting technology to problem solving.

Technology continues to be a major GIS driver and constraint on GIS access.

In the four years since the publication of the first edition of this book, there have been significant changes in technology. Computers and networks are now approximately five times faster, there have been multiple major software releases from each of the main GIS vendors, and the problems of data availability have lessened. As we look forward, some of the key technology trends that will shape GIS in the rest of this decade include:

- GIS servers – a move from the desktop to centrally managed GIS servers possessing similar functionality will provide new ways of delivering GIS content and capabilities to a much enlarged user community. GIS servers make possible the vision of a 'GIS Appliance' – everything in a single package to 'do GIS' (hardware, software, data, and applications). Users will only need to plug their GIS Appliance into the power supply and network to be up and running.

- Services-oriented architectures – GIS enabled Web services will make it much easier to integrate GIS into the information systems strategies of still more organizations, and will drive the acceptance and adoption of GIS by the IT mainstream.

- Spatial data infrastructures (SDI) and geoportals – access to data is still a constraint on GIS, and top-down programs that organize data access via Web geoportals will alleviate these problems.

- Sensor webs – new hardware and network developments now allow low-cost unmanned sensors to be placed in many more remote locations (e.g., on flying drones, on shipping containers, in vehicles, or on people). These can track and monitor objects that move or change state, and can relay this information to central GIS logging and analysis systems. These open up many new applications for GIS, especially in the military and business logistics.

■ Modeling and analysis – our ability to perform GIScience is being enhanced by the availability of improved software for spatial analysis and modeling. Key developments in GIS-based exploratory spatial data analysis (ESDA), geostatistics, and process flow modeling are all stimulating interest in GIS in the traditional sciences, as well as improving our inferential abilities.

■ Data management – GIS now have the capability to manage some of the world's largest databases (multi-terabyte: Section 1.2 and Table 1.1), to provide multi-user read-write transactional access, and to coordinate distributed (federated) geographic databases. This will be one of the pillars of future 'massive GIS'.

■ Visualization – the ability to visualize geographic data in conventional 2-D maps, 3-D whole Earth views, and a variety of graphical and cartographic visualizations is already providing new scientific insights, as well as helping to popularize GIS.

21.2.5 GIS and management

In a more applications-specific sense, it is clear that a critical understanding of management context is important to ascertain the most efficient and effective means of deploying GIS in organizations. There are two important dimensions to GIS and management: the management of GIS and the use of GIS for management.

> **Poor management remains the Number One reason why GIS projects fail.**

GIS provides an increasingly useful part of management's weaponry in all 'businesses' – whether in commerce, government, not-for-profit organizations, or even academia. It can provide competitive advantage (Section 2.3) and is a useful means of evaluating policy and practice options in both the public and the private sectors. GIS also has an important role to play in communication and information sharing, and in public-policy formulation and implementation.

Although business leaders often do not realize it, most businesses are organized geographically and geography plays an important role in success or failure. Government services departments such as road and park maintenance, retail businesses such as banks and high-street stores, and transportation distribution companies are all critically dependent on geography to organize their business activities. For many of these, overcoming the constraints of geography is central to their strategic and operational decision making, and often gives them a competitive edge. Through GIS, businesses can save money and time, increase efficiency and productivity, improve

communication and collaboration, and generate revenue. It can be important to think of GIS as much as a source of revenue as an outflow of expenditure. Generally applicable and accessible return on investment and cost-benefit analysis templates, as well as case studies, will be of considerable value in propelling GIS use forward.

21.2.6 GIS applications

Over the past 10 years, many have begun to realize that a significant number of the world's problems are inherently geographic, both in terms of their causes and their impacts. At the same time, investigation has revealed that cross-cutting interdisciplinary science and social science are required in order to seek and implement solutions. Global warming, the hole in the ozone layer, the AIDS epidemic, desertification, homeland security, social unrest, and other massive problems can all benefit from using GIS to manage data and support investigation. GIS can help integrate data from multiple sources, record how events and processes change over space and time, support the evaluation of alternative explanatory models, and communicate results in an effective and understandable way. Some representative and successful applications include the following.

■ *Military geointelligence.* Military organizations throughout the world are changing rapidly and there is a new emphasis on gathering, storing, analyzing, and visualizing intelligence data about territory both overseas and at home. Important sources of intelligence include satellite images, surveillance from aircraft and pilotless drones, images collected from ground-based video cameras, and records of the locations of credit-card and bank transactions (Section 11.3.2). A key aspect of virtually all of this intelligence is its geographic location. Indeed, geography is commonly used to organize and index intelligence information. During the Iraq war, for example, General Tommy Franks – the commanding officer of the USA-led forces – was briefed daily using geointelligence in a GIS. During the recent civil war in the Sudan–Chad border region, humanitarian relief was coordinated using intelligence entered into mobile GIS (see Figure 13.14). An important tool for intelligence gathering is *geoparsing*, a term used to describe tools that scan massive amounts of text, searching for references to places, and using those references to link the text to maps and to other information about the referenced place. MetaCarta (**www.metacarta.com**) is an advanced geoparsing technology, developed in collaboration with US intelligence agencies, that can geoparse extensive collections of Web documents and other material.

■ *Megan's Laws.* In many jurisdictions, particularly in the US, laws have been passed requiring local authorities to provide public access to information about the residential locations of sex offenders. They are known as Megan's Laws, named for the young victim of a repeat sex offender who had been resettled in a community without the community's knowledge, and who reverted to previous behavior. Communities find GIS, and Web-based GIS maps, to be invaluable in meeting the requirements of Megan's Laws, since they give citizens ready access to information in an easily understood form. Figure 21.3 shows an example site maintained by the Sheriff's Department of Pierce County, Washington, which covers Metropolitan Tacoma.

- *West Nile Virus.* West Nile Virus was first recorded in Uganda, but has since spread into many other countries. In North America, it is spread by a species of mosquito, and infects both birds (typically crows and jays) and humans. Research has yielded detailed and accurate models of the relationships between

mosquito prevalence, bird infection, and human infection, with the result that effective early warning systems are now available. They require GIS-based mapping of bird habitat and mosquito habitat, records of dead and infected birds, and records of human infections. Figure 21.4 shows an example map of the status of West Nile virus in Pennsylvania, USA. Sean Ahearn of Hunter College of the City University of New York has developed the DYCAST system, which uses GIS functions to monitor the progress of West Nile outbreaks and to provide early warning to health agencies.

- *Agricultural policy.* The European Union Common Agricultural Policy (CAP) has played a very

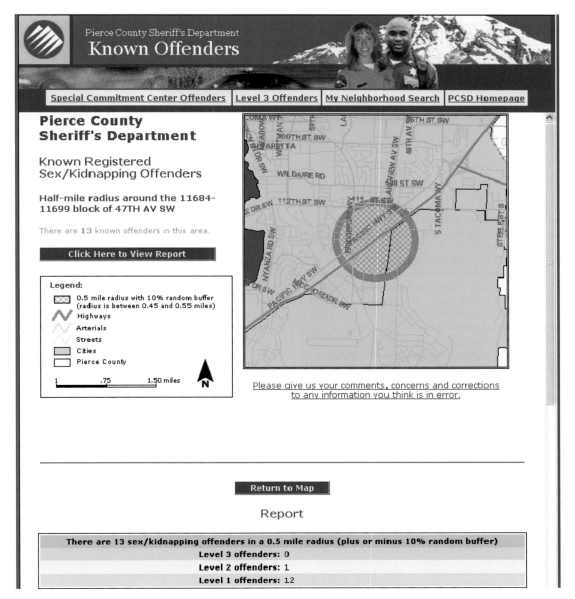

Figure 21.3 Sheriff's Department of Pierce County, Washington, Known registered sex/kidnapping offenders map (**www.co.pierce.wa.us/pc/abtus/ourorg/sheriff/sexoff/sormain.htm**)

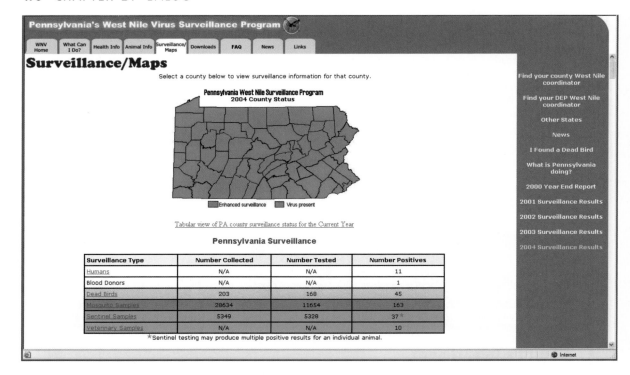

Figure 21.4 Pennsylvania West Nile Virus Surveillance Program website showing 2004 County Status Statistics

significant role in the decision-making process of farmers through enforcement of production quotas, and market subsidies. It has caused a restructuring of farm systems in countries from Ireland in the west to the new succession countries in the east. GIS has been used to spot fraudulent claims from EU farmers that declare more land than they have in order to increase the subsidies they receive. Satellite imagery combined with land parcel boundaries has been used to assess the crop area under production, and is then compared with the census returns from the farmers. This type of land resource assessment is a classic use of GIS that was employed in the first system that created a land inventory for Canada in the 1960s (Section 1.4.1). GIS automates the process of assessing and measuring, standardizes the operations, and builds a database for long-term use. In Italy alone, the system is understood to have narrowed the difference between the area declared for subsidies and the actual area from 9 to 2 per cent since 1999, saving the taxpayers millions of euros. In the enlarged EU, it is expected that over six million farmers will declare fifty million fields every year (see **www.euractiv.com**). GIS is also equally applicable to other agricultural policies – such as systems that monitor whether farmers are safeguarding their land, or the extent of land opened up for recreational activities.

■ The United Nations Office on Drugs and Crime and the Myanmar government are using GIS to monitor the poppy harvest in Myanmar, the main opium producer in Southeast Asia. Using a combination of satellite imagery and ground survey, they demonstrated that there was a 29% reduction in crop area from 2003 to 2004. Only by effective monitoring can the size of the problem and success of eradication programs be measured. Figure 21.5 shows one of the maps produced from this survey that are used to communicate the status of the program to key decision makers and stakeholders.

GIS can play a significant role in helping to resolve many types of global problems – but only if it works 'with the grain' inside other initiatives.

These examples are by no means unique – indeed they are repeated over and over again in many application areas. Each illustrates visibly how GIS is literally being used to save and change people's lives, as well as to deliver significant economic, social, environmental, and security benefits.

21.3 Ten 'grand challenges' for GIS

Clearly, GIS has come a very long way in its four decades of existence. Nevertheless, there is still much to be done

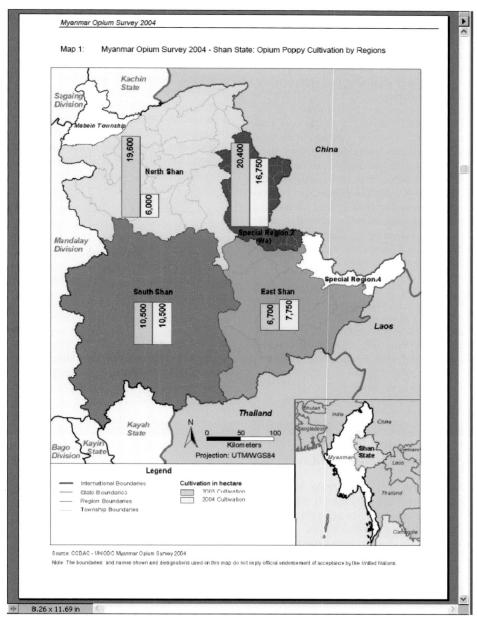

Figure 21.5 Myanmar Opium Survey 2004 – Shan State: Opium Poppy cultivation by regions (Reproduced by permission of CCDAC, Union of Myanmar)

if GIS is to realize its full potential in the vast majority of public and private organizations. Here we outline a series of 'grand challenges', in no particular order, which we believe represent the most significant and urgent action items. Each of these grand challenges can be achieved within the next five years.

21.3.1 Global data layers

Without data, GIS can make little or no contribution to the suite of problems which face us. Over the past 100

years, there have been many attempts to assemble global data layers. In Chapter 20, we described the painfully slow progress to date. The fact that higher resolution data is available for the USA as compared with much of the rest of the world reflects not a technology issue but rather a political challenge – centered on the question of whether one nation should produce data for another without the latter's agreement, the scale at which data might be created and commercially exploited, and the data standards that should be adhered to.

There is no doubt in our minds that GIS use is being held back by a lack of basic data layers, and there remains much wasteful duplication of effort and expense. Many of the outstanding problems are not technical, but (as indicated above) relate to commercial issues, coordination, agreement on standards, and politics. We feel that wider use of GIS would be facilitated if global layers were widely available for: street centerlines (at a positional accuracy of 200 m for 1:100 000 topographic mapping); surface images at a similar or better resolution (this would be around 1 petabyte of data for the land area alone); a terrain dataset; a comprehensive gazetteer (placename layer); hydrography data; data on cultural features to an agreed global specification; and geodetic control data. It is easy to say what is needed, especially from the perspective of academics without exposure to risk from getting conjectures wrong. The reality is that things happen when the incentive mechanisms are strong enough (Chapter 18). As we showed in Chapter 20, such incentives are usually commercial, related to defense and homeland security, or are political grand gestures. Altruism usually plays little part and the last driver is often ephemeral (but see Section 21.4). Fulfillment of this agenda thus requires that those making the decisions understand that there are advantages to their missions in creating data and subsequently sharing them. This will only occur through carefully planned and targeted strategies and persistent lobbying.

21.3.2 The GIS profession

Historically, GIS has grown bottom up as a loose assemblage of university academics, government employees, commercial workers, and other interested individuals. The academic discipline of geography has contributed many of the intellectual underpinnings of GIS, but other academic and non-academic societies have also contributed a great deal. GIS has now reached a critical mass that has prompted the creation of a code of ethics, and a certification scheme. These could be considered the first stages on the road to establishing a GIS profession, following much the same pattern established in the fields of, for example, geology, chemistry, land surveying, and law. Internationally, this community would embrace the places in the world where GIS has a long history, together with those that are relative newcomers to GIS as an area of activity (see Box 21.2)

For this to happen, the presently disparate GIS-related societies would need to set aside their competitive instincts towards each other and pool resources to create a charter that is endorsed by public bodies of some significance. The purpose of this professional body would be to establish standards for craftsmanship, best practices, and basic qualifications, to disseminate information, and to promote the field. Whether this can be done outside a national context is dubious: hitherto, national laws and conventions have strongly shaped professional practice in

---

Biographical Box 21.2

Dr Young-Pyo Kim, Korean Pioneer of NGIS

After a successful career in the Korea Marine Corps, Young-Pyo Kim (Figure 21.6) joined the Korea Research Institute for Human Settlements (KRIHS) in 1979. Over the next decade, he defined the concept and method of Land Assessment in Korea and focused upon a range of land planning and policy issues. In 1986, he became director of the Computing Center in KRIHS, and a champion of using GIS to devise more efficient and effective measures for land planning and policy. Implementing this in a country that, at the time, had little GIS expertise was far from straightforward, and strenuous efforts were necessary to introduce the technology and build capacity. In 1993, Dr Kim organized an official government meeting to establish policy directions for GIS development in Korea. This meeting brought the National GIS project in Korea into being, and he subsequently played a key role in making it happen.

Figure 21.6 Young-Pyo Kim, Korean Pioneer of NGIS

Today Young-Pyo Kim is director of the GIS Center in KRIHS, where he also acts as a coordinator of the Korean National GIS (NGIS) Steering Committee. His leading role in the National GIS project is fundamentally pragmatic, but he believes that GIS is much more than a toolbox. Speaking of the social and philosophical ramifications of GIS he says: 'All things in the universe are restricted by the axes of time and space, and these are recognized by humans. In Eastern classical philosophy, space, time, and humans have been called the three fundamental elements of the universe. These elements cannot be completely integrated in the real world because of the constraints of time and space. However, human construction of three-dimensional cyber-geospace can overcome the separation of time and space and create movement towards a more ubiquitous world. GIS will be a foundation technology in this project.'

different countries (for instance, teaching qualifications and professional accreditation achieved in one country are still not recognized in others). Since GIS impacts on many different disciplines, some of which already have codes of practice and conduct (e.g., engineering), we envisage that the GIS practitioner will be formally assessed upon a wide range of criteria. Establishing a formal profession of GIS is not essential within the field, but will create credibility outside of it. It will draw attention to the science, systems, and studies of GIS in ways that are not currently possible.

21.3.3 The GIS curriculum

One of the major road-blocks to growth of GIS is the paucity of suitably qualified GIS people: technicians, application analysts, and managers. Even though the tertiary education sector supply-side capability has increased several-fold in the past decade and now exceeds 150 000 per annum at over 10 000 institutions, there is considerable variability in content and quality. Moreover, we need GIS 'plugged in' more systematically to a wide range of different discipline-led courses. A global curriculum for GIS will help institutions to create courses in many disciplines, and will also lead to better core GIS courses. It will establish and implement programs reflective of the modern GIS era. In advocating this we recognize the importance of plurality of approach and diversity of outcome;

education must be sensitive to different cultural contexts. We believe that much can be common, however, and that economies of scale can be achieved without overly inhibiting further technical and intellectual development.

Thus, our vision is of a select core of spatial principles and techniques that will be used in outreach from geography to GIS-enable the widest range of relevant disciplines in the social and environmental sciences. The skills that qualify professionals to work with GIS are in many respects similar to those needed by professionals who work with any kind of visual information – in short they are the skills of spatial reasoning. Such skills are too rarely taught, but instead must often be acquired informally when students encounter spatial data and visual display, whether in physics, mathematics, or graphic arts. In our vision GIS will form a core component of a systematic approach to the development of spatial skills. This approach – of broadening the outreach of GIS to new disciplinary markets and enhancing the core courses (see Section 19.2.1) – nevertheless only satisfies part of the educational needs. Even more important is the progressive introduction of GIS into schools across the globe. There are already good examples where this has happened, not just in affluent areas of the world (see, for example, Box 21.3).

Although much material taught in schools across the world is of national importance (such as in the teaching of History), an equivalent local focus is essential to

---

Applications Box **21.3**

GIS in Indian schools

On 5 October 2004, *The Financial Express*, an Indian newspaper, described how the Indian government has empowered around 250 students of 20 schools in the Almora district in Uttaranchal to create digital maps and landscapes of their neighborhood (Figure 21.7). Equipped with hand-held computing devices supported with satellite imagery, the children have been compiling content for a GIS. The pilot program, called 'Mapping the Neighbourhood,' was initiated jointly by the Department of Science and Technology (DST), the Department of Space, and the Centre for Spatial Database Management and Solutions (CSDMS), a non-government organization. The children gathered data on resources, community demographics, livelihood information, land use, infrastructure, and much else, primarily through mapping and questionnaire surveys in villages; they produced – with help – maps and analyses of the data. The aim was to foster greater linkage between young people and their local communities and, hopefully, to solve some of the local problems of water supply and distribution.

In the words of Amitabha Pande, Joint Secretary, Department of Science and Technology, Government of India: 'As a dream, the programme has the potential of transforming the design and structure of education, making it into a creative, interactive, participatory activity which is at the same time socially relevant and useful.'

Figure 21.7 School children in India working on the 'Mapping the Neighbourhood' GIS project (Reproduced by permission of CSDMS)

encourage engagement in the community and to foster a civil society and better inter-generational harmony. As it happens, GIS and geography together have much to offer school-level education now that the PC is ubiquitous and inexpensive software is widely available. Students can gain understanding of current IT through 'doing GIS'. At the same time, they can tackle projects of manifest importance and interest to their parents. The integrative nature of both GIS and geography ensures that many different factors can be brought to bear; understanding the trade offs between multiple factors is excellent training for many problem solving needs in later life. Because of all this, we are convinced that there is a real prospect of introducing more project-led GIS in schools across the globe. A key requirement for success is suitable case studies in different national and regional contexts for teachers, and suitable education and training of the teachers by their peers.

21.3.4 Near-universal empowerment of GIS users

At a routine level, much has been achieved in developed countries through the wide use of Web-enabled technologies, but there is much to be done. At one level the problem centers upon exposure to geographic concepts through education and familiarity with technology through experience – appropriately educated users will typically use a technology first to solicit information, before likely using it to conduct transactions and eventually feeling confident to use it to participate in collaborative decision making, or so one line of thought goes. In practice, however, there is some evidence that the networks that govern access to and control of decision making fail to empower wider user communities, and there is a need to research the way that such characteristics reinforce other sources of structural disadvantage in society. These aggregate perspectives need to be complemented by extending work in public participation in GIS (PPGIS) to GIScience (PPGISc). Such work includes investigation of how user interfaces and exploratory spatial data analysis techniques might foster greater intelligibility and wider understanding. There is a need to develop software that is both easy and safe to use by the overwhelming majority users in any setting – from the street, to the classroom, to the public appeal arising out of development proposals. In the longer term, there is a need to develop our understanding of social and collective cognition, and thus develop our understanding of how GIS can accommodate the widest range of institutions and practices. Collectively, people need to be able to understand both the power of generalization and the mitigating effects of uncertainty.

21.3.5 Global population statistics

Our global knowledge about the most basic of things – how many people there are in different parts of the world and how they live, their state of health, their vulnerability to disasters – is surprisingly poor. Even in Westernized countries, the collection of demographic data by censuses is now under challenge as individuals become less ready to fill in forms. Tallying the population through linking together administrative records is technically demanding and only works well in certain cultures (e.g., in Scandinavia). In other parts of the world, knowledge of populations and (sometimes massive) migrations is often years out of date and of highly questionable accuracy; in most cases, the data published only pertain to large areas, complicating vastly the tackling of very real problems even if helpful in drawing attention to 'big picture' problems.

There have also been many attempts to estimate population numbers from aerial photographs or satellite imagery – using good local knowledge and plausible modeling methods (e.g., see Section 13.3.3). GIS is already making a significant contribution to creating global population data but the source data are highly varied in their quality. Heroic assumptions are necessarily made in the modeling, often derived from data with a spatially variable correlation with population numbers. We must not denigrate the efforts made thus far, but neither should we condone data hoarding beyond the public gaze. If a plethora of new data sources can be brought into the public domain, it should be possible to improve data infrastructures through statistical and GIS skills, honed to particular cultural contexts (e.g., through the national statistical agencies). A global partnership, led by statisticians with GIS skills, could and should benefit everyone. We think this is not just desirable but possible, that progress in the global information economy can make the underpinnings to statistical estimation more secure, and that United Nations agencies may have an important role to play in this regard.

21.3.6 Development of richer geodemographic data infrastructures

We know too little about the functioning of contemporary societies. If robust and fair rules controlling data supply and access can be developed and widely accepted, there is no reason why the tremendous potential of rich government and private-sector datasets should not be unlocked for the greater good. Uninteresting, zonally coarse, infrequently collected surrogate indicators of society are simply no longer good enough for tactical and strategic decision making in consumer-led markets. Nevertheless, huge resource expenditure (especially in the public sector) continues to be authorized using such data. Organizations with international management responsibilities also need more data that traverse national boundaries, but at fine levels of granularity. There is also a need to extend the success of business applications to the challenges of fair and equitable access to public services. We need to unlock the potential of geographically referenced data, many of which are collected at the local level, to represent living conditions, in terms of health, law enforcement, and education. Too much policy is still top down and we need to empower real communities to develop and negotiate their desires and preferences in order to establish priorities in the provision of public goods. The speed bumps on the information highway manifest an unwillingness to share information, very often because of unfounded fears of disclosure of individual circumstances. We need to lighten up in our views of data supply and access. One domain in which there is evident need concerns the application of geodemographics to the take-up of educational opportunity (Box 21.4).

21.3.7 The transition from geo-centered to ego-centered mapping

The innovation of location based services is providing new challenges to the cartography and mapping communities, driven by restrictions in the viewable areas of

---

The geodemography of geography

Many of tomorrow's GIS experts will gain their first exposure to GIS through undergraduate university programs, in geography itself or in disciplines that are at least partly spatially enabled. In the UK, the issue of access to all higher-education courses is presently a hot political topic, as increasing attention is focused upon the supplementation of educational attainment with other access criteria designed to broaden the social base of those that benefit from an education after high school.

As we have seen, 'doing GIScience' is in part a socially constructed activity. What then are the social backgrounds of its future practitioners? Working at the UK Universities Central Admissions Service (UCAS), researcher Alex Singleton has investigated some of the ways in which geography is a socially divided discipline. Figure 21.8 presents a geodemographic profile (see Section 2.3.3) created using geographic and small-area data of the profiles of parents of UK geography undergraduates admitted in 2003.

Those whose parents are within the wealthiest group (dark blue) are much more likely to be amongst those accepted onto geography courses, and other well-to-do groups (light blue and gray) are similarly over-represented. A final

group that is over-represented has parents that are rural as well as affluent (dark green). To what extent is this because aspiring geographers are motivated by the physical environment (e.g., people growing up in rural areas are more likely to become interested in Earth surface processes and landforms) or because they come from affluent social environments? A social engineer might wish to target the high-status groups to broaden the social base of geography, but this would also likely penalize candidates whose choices arise out of subject concerns with the environment, or who have been motivated by the travel opportunities that affluence brings. These are the sorts of issues that are central to understanding fairness, equity, and access, but which are usually either overlooked or crassly simplified when discussing participation in higher education. Geography is pivotal to the creation of modern geodemographic classifications, and the geography of access to different universities within nations and (increasingly) internationally is an increasingly important concern. These topics will be hotly debated in the years to come. Perhaps you have views of your own, based on the characteristics of your own contemporaries at university! Geography is key to understanding geodemography, and vice versa.

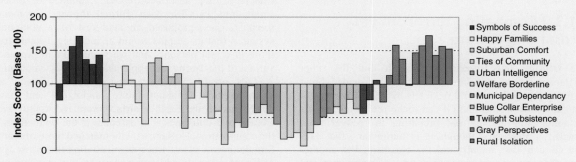

Figure 21.8 Deviations away from a base score of 100 for acceptances to study Geography in 2003, according to Type and Group of the 2003 Mosaic geodemographic system

devices. This technical constraint is associated with a change in the goal of mapping from supplying *products* to providing *services* that are *tailored* to the precise needs of users, and a shift in the application of the cartographer's skill set from developer- to user-centered visualization. Instead of the conventional *bird's eye* view oriented to North and framed by arbitrary lines of latitude and longitude, the user is likely to prefer views that are centered on his or her current position, oriented in the direction of travel (*heads-up*), and with a ground-based perspective (such views are now common in in-car navigation systems). Different classes of users will have different needs, but common characteristics will center upon the need for greater immediacy and ready intelligibility in field situations.

21.3.8 Support data models for a complete range of geographic phenomena

Throughout the book, and particularly in Chapters 3 and 8, we have encountered limitations to the ability of current GIS to represent real geographic phenomena. These range from the emphasis on polylines, which limits the ability of GIS to represent the continuous curves of rivers and roads, to a comparative dearth of methods for representing time and dynamics. In the best of all possible worlds, GIS would include data models for all types of geographic phenomena, including those that are well beyond the range of currently supported data models. It is impossible, for example, to build effective representations today of phenomena such as storms that simultaneously move, change shape, and change internal structure. Similarly, there are no methods for representing watersheds, since a watershed exists for every point on a landscape. This latter example is an interesting hybrid between fields and objects, since every point on a terrain (a field) maps to a unique area (a discrete object) – the term *object field* has been suggested for this kind of phenomenon. Many more examples exist, leading us to suggest that the development of data models for complex geographic phenomena should be given high priority by the GIScience research community.

21.3.9 Combating terrorism yet preserving culture

We treasure our freedom of speech and the other characteristics of democracy. But these are under threat in many parts of the world from terrorism. In Chapter 20, we showed how GIS and GI can make real contributions to anticipating threats and preparing to recover from attacks; it may be – though it is unproven – that GIS and GI can also actually aid the attackers. Put another way, the approach we have espoused – of being open and publishing strategies and plans, working in partnership through consensus between various (sometimes many) parties, sharing and publishing data – might actually

benefit attackers. The agents of the state charged with protecting life and limb (e.g., in homeland security, in the intelligence services, and the military) broadly take the view that more directed action to prepare for such eventualities and 'sensible safeguarding of information' is required. An incidental benefit of this concern is that huge resources are being invested in new (e.g., surveillance) technologies, especially by the military, and these are likely to find their way into civilian use at a later stage – just as high-resolution satellites became civilianized. Yet the price of being a more closed and secretive society is a heavy one and such a move may not even be possible: the Internet and Web have empowered people to such an extent that geographic (and other) information is readily available from multiple sources. As individuals, we have little ability to influence the bigger picture and events. As individuals working in GIS, however, we have an obligation to seek ways of reconciling the need to protect human life and assets whilst at the same time maintaining our way of life. Furthermore, as GIS professionals, we have an obligation to act ethically. Formal GIS accreditation (see Section 19.2.2) should be extended to encompass these dilemmas.

21.3.10 Support a wide range of types of geographic simulation

If humans are to be successful in their efforts to improve life on Earth, then an understanding of how the Earth system *works* is essential – any amount of mapping and description of how the Earth *looks* cannot substitute. How much better it would have been, for example, to have understood how the atmosphere's concentration of CO_2 and other greenhouse gases affected climate *before* we embarked on the wholesale burning of hydrocarbons, or to have understood the relationship between forest cover and climate *before* we embarked on wholesale clearing, whether it was of the Amazon basin in the late 20th century, the eastern United States in the early 19th century, or western Europe in the 17th and 18th centuries? Ideas of process are best tested in computational environments such as GIS before they are the subject of human-induced experiment, and in Chapter 16 we described several examples of process simulations in GIS. Alan Turing, the early computer scientist, once proposed an acid test of a computing system – that its outputs would be indistinguishable from those of a real person – and we would like to propose a variant of the Turing test applicable to process simulation. A GIS would pass our version of the Turing test if its predictions about future landscapes and future conditions on the Earth's surface were indistinguishable from actual outcomes – in essence, if the predictions of the computational system exactly matched those of reality. Simulations of many processes, including those discussed in Chapter 16, already come close to this standard, and we anticipate substantial progress in this direction in the next few years as tools for simulation become more and more prevalent in GIS.

The recent advances in GIS modeling and analysis discussed in Chapters 14–16 are already beginning to help us model the world better and test hypotheses about the future in a digital simulation setting. The requirement looking forward is not so much technical as scientific. We need scientific understanding to create and interpret models and scientific data to parameterize and calibrate them.

21.4 Conclusions

The journey that this book has taken us on is an excursion into the work of Geographic Information Systems and Science. We have tried to define the terms GISystems and GIScience and show how they interact closely. For us, both are critically juxtaposed and closely interwoven. One relies intimately on the other and once separated both are less useful and more prone to misinterpretation. We have also discussed the relationship to GIStudies and GIServices. The former connotes the embedding of GIS in society, the latter a mechanism to deliver GIS more effectively to a wider group of users than has heretofore been technically and commercially feasible.

It will be obvious, both from the first 20 chapters but especially from the Epilog, that we are convinced of the intrinsic merits of a geographic perspective on understanding the functioning of the Earth and the societies living on it. Repeatedly, we have demonstrated many examples where GIS has proven valuable to science

and society. The potential value is much greater still, however, as horrific natural disasters like the devastation wrought by the tsunami in South East Asia of December 2004 has demonstrated: the use of formal and informal geographic information in the search for survivors and delivering aid to them was crucial. Figure 21.9 shows the scale of the devastation and the need for international contributions and partnerships in the face of such disasters.

Only through GIS can we bring together efficiently information from many different sources, integrate it, and assess the trade offs between different actions. In this sense, GIS is the language of geography since it facilitates communication between the constituent parts of the field.

Although we are confident of a bright future for GIS, we are far from complacent; GIS is still used by only a tiny fraction of human beings. The data and models necessary for evaluating problems and possible solutions rarely exist in the right form. Furthermore, political, economic, and other factors often constrain or deny what is possible given GIS technology and human skills. We see the next decade as one in which technology will inevitably improve still further, but one in which human factors in GIS will become ever more important.

The future of GIS lies in the hands of GIS – not systems, science, studies, or services, but geographic information *students* – the next generation of software developers, application analysts, managers, researchers, and users. We hope that this book will be a contribution to their future and, through their use of GIS, to all our futures. It is our firm view that without GIS the future of the third rock from the sun will be more uncertain and less well managed.

(A)

Figure 21.9 (A) Banda Aceh, Sumatra, Indonesia on 23 June 2004, six months before the tsunami struck. (B) The same area on 28 December 2004, two days after the earthquake nearby and the consequent tsunami. High resolution satellite images such as these are of value in directing disaster relief. They were also reproduced by many newspapers around the world and helped to stimulate understanding of the scale of the problem and financial and other support for the survivors. (Images courtesy of DigitalGlobe, **www.digitalglobe.com.**)

(B)

Figure 21.9 (*continued*)

Questions for further study

1. What range of interests do your student contemporaries bring to the study of GIS? Do you think that a subject of study with diverse roots attracts students with diverse interests and motivations? Give reasons for your answer.

2. What do you think is the most significant impediment to the future success of GIS?

3. With reference to the definitions of GIS in Section 1.4, discuss the view that GIS and GIScience is a socially constructed subject.

4. Choose an application domain and give specific examples of what is?, where is?, why is?, and what if? questions. What datasets would you need to answer these questions successfully?

Further reading

Cairncross F. 2001 *The Death of Distance 2.0: How the Communications Revolution Will Change Our Lives.* Harvard: Harvard Business School Press.

Johnston R.J. 2005 'Geography and GIS'. In Longley P.A., Goodchild M.F., Maguire D.J., and Rhind D.W. (eds) *Geographical Information Systems: Principles, Techniques, Management and Applications (abridged edition).* Hoboken: Wiley: 39–47.

Nyerges T. 2001 *Geographic Information Systems for Group Decision Making.* London: Taylor and Francis.

Index

Page numbers in *italics* refer to figures; page numbers in **bold** refer to tables; ***bold, italicized*** page numbers refer to boxes.

Geographic Information Systems and Science, 2nd edition   Paul Longley, Michael Goodchild, David Maguire, and David Rhind.
© 2005 John Wiley & Sons, Ltd.   ISBNs: 0-470-87000-1 (HB); 0-470-87001-X (PB)